IET ENERGY ENGINEERING SERIES 125

Wind Energy Modeling and Simulation

Other volumes in this series:

Volume 1 **Power Circuit Breaker Theory and Design** C.H. Flurscheim (Editor)
Volume 4 **Industrial Microwave Heating** A.C. Metaxas and R.J. Meredith
Volume 7 **Insulators for High Voltages** J.S.T. Looms
Volume 8 **Variable Frequency AC Motor Drive Systems** D. Finney
Volume 10 **SF_6 Switchgear** H.M. Ryan and G.R. Jones
Volume 11 **Conduction and Induction Heating** E.J. Davies
Volume 13 **Statistical Techniques for High Voltage Engineering** W. Hauschild and W. Mosch
Volume 14 **Uninterruptible Power Supplies** J. Platts and J.D. St Aubyn (Editors)
Volume 15 **Digital Protection for Power Systems** A.T. Johns and S.K. Salman
Volume 16 **Electricity Economics and Planning** T.W. Berrie
Volume 18 **Vacuum Switchgear** A. Greenwood
Volume 19 **Electrical Safety: A guide to causes and prevention of hazards** J. Maxwell Adams
Volume 21 **Electricity Distribution Network Design, 2nd Edition** E. Lakervi and E.J. Holmes
Volume 22 **Artificial Intelligence Techniques in Power Systems** K. Warwick, A.O. Ekwue and R. Aggarwal (Editors)
Volume 24 **Power System Commissioning and Maintenance Practice** K. Harker
Volume 25 **Engineers' Handbook of Industrial Microwave Heating** R.J. Meredith
Volume 26 **Small Electric Motors** H. Moczala *et al.*
Volume 27 **AC–DC Power System Analysis** J. Arrillaga and B.C. Smith
Volume 29 **High Voltage Direct Current Transmission, 2nd Edition** J. Arrillaga
Volume 30 **Flexible AC Transmission Systems (FACTS)** Y.-H. Song (Editor)
Volume 31 **Embedded Generation** N. Jenkins *et al.*
Volume 32 **High Voltage Engineering and Testing, 2nd Edition** H.M. Ryan (Editor)
Volume 33 **Overvoltage Protection of Low-Voltage Systems, Revised Edition** P. Hasse
Volume 36 **Voltage Quality in Electrical Power Systems** J. Schlabbach *et al.*
Volume 37 **Electrical Steels for Rotating Machines** P. Beckley
Volume 38 **The Electric Car: Development and future of battery, hybrid and fuel-cell cars** M. Westbrook
Volume 39 **Power Systems Electromagnetic Transients Simulation** J. Arrillaga and N. Watson
Volume 40 **Advances in High Voltage Engineering** M. Haddad and D. Warne
Volume 41 **Electrical Operation of Electrostatic Precipitators** K. Parker
Volume 43 **Thermal Power Plant Simulation and Control** D. Flynn
Volume 44 **Economic Evaluation of Projects in the Electricity Supply Industry** H. Khatib
Volume 45 **Propulsion Systems for Hybrid Vehicles** J. Miller
Volume 46 **Distribution Switchgear** S. Stewart
Volume 47 **Protection of Electricity Distribution Networks, 2nd Edition** J. Gers and E. Holmes
Volume 48 **Wood Pole Overhead Lines** B. Wareing
Volume 49 **Electric Fuses, 3rd Edition** A. Wright and G. Newbery
Volume 50 **Wind Power Integration: Connection and system operational aspects** B. Fox *et al.*
Volume 51 **Short Circuit Currents** J. Schlabbach
Volume 52 **Nuclear Power** J. Wood
Volume 53 **Condition Assessment of High Voltage Insulation in Power System Equipment** R.E. James and Q. Su
Volume 55 **Local Energy: Distributed generation of heat and power** J. Wood
Volume 56 **Condition Monitoring of Rotating Electrical Machines** P. Tavner, L. Ran, J. Penman and H. Sedding
Volume 57 **The Control Techniques Drives and Controls Handbook, 2nd Edition** B. Drury
Volume 58 **Lightning Protection** V. Cooray (Editor)
Volume 59 **Ultracapacitor Applications** J.M. Miller
Volume 62 **Lightning Electromagnetics** V. Cooray

Volume 63 **Energy Storage for Power Systems, 2nd Edition** A. Ter-Gazarian

Volume 65 **Protection of Electricity Distribution Networks, 3rd Edition** J. Gers

Volume 66 **High Voltage Engineering Testing, 3rd Edition** H. Ryan (Editor)

Volume 67 **Multicore Simulation of Power System Transients** F.M. Uriate

Volume 68 **Distribution System Analysis and Automation** J. Gers

Volume 69 **The Lightening Flash, 2nd Edition** V. Cooray (Editor)

Volume 70 **Economic Evaluation of Projects in the Electricity Supply Industry, 3rd Edition** H. Khatib

Volume 72 **Control Circuits in Power Electronics: Practical issues in design and implementation** M. Castilla (Editor)

Volume 73 **Wide Area Monitoring, Protection and Control Systems: The enabler for smarter grids** A. Vaccaro and A. Zobaa (Editors)

Volume 74 **Power Electronic Converters and Systems: Frontiers and applications** A.M. Trzynadlowski (Editor)

Volume 75 **Power Distribution Automation** B. Das (Editor)

Volume 76 **Power System Stability: Modelling, analysis and control** B. Om P. Malik

Volume 78 **Numerical Analysis of Power System Transients and Dynamics** A. Ametani (Editor)

Volume 79 **Vehicle-to-Grid: Linking electric vehicles to the smart grid** J. Lu and J. Hossain (Editors)

Volume 81 **Cyber-Physical-Social Systems and Constructs in Electric Power Engineering** S. Suryanarayanan, R. Roche and T.M. Hansen (Editors)

Volume 82 **Periodic Control of Power Electronic Converters** F. Blaabjerg, K. Zhou, D. Wang and Y. Yang

Volume 86 **Advances in Power System Modelling, Control and Stability Analysis** F. Milano (Editor)

Volume 87 **Cogeneration: Technologies, optimisation and implementation** C.A. Frangopoulos (Editor)

Volume 88 **Smarter Energy: From smart metering to the smart grid** H. Sun, N. Hatziargyriou, H.V. Poor, L. Carpanini and M.A. Sánchez Fornié (Editors)

Volume 89 **Hydrogen Production, Separation and Purification for Energy** A. Basile, F. Dalena, J. Tong and T.N. Veziroğlu (Editors)

Volume 90 **Clean Energy Microgrids** S. Obara and J. Morel (Editors)

Volume 91 **Fuzzy Logic Control in Energy Systems with Design Applications in Matlab/Simulink®** İ.H. Altaş

Volume 92 **Power Quality in Future Electrical Power Systems** A.F. Zobaa and S.H.E.A. Aleem (Editors)

Volume 93 **Cogeneration and District Energy Systems: Modelling, analysis and optimization** M.A. Rosen and S. Koohi-Fayegh

Volume 94 **Introduction to the Smart Grid: Concepts, technologies and evolution** S.K. Salman

Volume 95 **Communication, Control and Security Challenges for the Smart Grid** S.M. Muyeen and S. Rahman (Editors)

Volume 96 **Industrial Power Systems with Distributed and Embedded Generation** R. Belu

Volume 97 **Synchronized Phasor Measurements for Smart Grids** M.J.B. Reddy and D.K. Mohanta (Editors)

Volume 98 **Large Scale Grid Integration of Renewable Energy Sources** A. Moreno-Munoz (Editor)

Volume 100 **Modeling and Dynamic Behaviour of Hydropower Plants** N. Kishor and J. Fraile-Ardanuy (Editors)

Volume 101 **Methane and Hydrogen for Energy Storage** R. Carriveau and D.S.-K. Ting

Volume 104 **Power Transformer Condition Monitoring and Diagnosis** A. Abu-Siada (Editor)

Volume 106 **Surface Passivation of Industrial Crystalline Silicon Solar Cells** J. John (Editor)

Volume 107 **Bifacial Photovoltaics: Technology, applications and economics** J. Libal and R. Kopecek (Editors)
Volume 108 **Fault Diagnosis of Induction Motors** J. Faiz, V. Ghorbanian and G. Joksimović
Volume 110 **High Voltage Power Network Construction** K. Harker
Volume 111 **Energy Storage at Different Voltage Levels: Technology, integration, and market aspects** A.F. Zobaa, P.F. Ribeiro, S.H.A. Aleem and S.N. Afifi (Editors)
Volume 112 **Wireless Power Transfer: Theory, technology and application** N. Shinohara
Volume 115 **DC Distribution Systems and Microgrids** T. Dragièević, F. Blaabjerg and P. Wheeler
Volume 117 **Structural Control and Fault Detection of Wind Turbine Systems** H.R. Karimi
Volume 119 **Thermal Power Plant Control and Instrumentation: The control of boilers and HRSGs, 2nd Edition** D. Lindsley, J. Grist and D. Parker
Volume 120 **Fault Diagnosis for Robust Inverter Power Drives** A. Ginart (Editor)
Volume 123 **Power Systems Electromagnetic Transients Simulation, 2nd Edition** N. Watson and J. Arrillaga
Volume 124 **Power Market Transformation** B. Murray
Volume 126 **Diagnosis and Fault Tolerance of Electrical Machines, Power Electronics and Drives** A.J.M. Cardoso
Volume 128 **Characterization of Wide Bandgap Power Semiconductor Devices** F. Wang, Z. Zhang and E.A. Jones
Volume 129 **Renewable Energy from the Oceans: From wave, tidal and gradient systems to offshore wind and solar** D. Coiro and T. Sant (Editors)
Volume 130 **Wind and Solar Based Energy Systems for Communities** R. Carriveau and D.S.-K. Ting (Editors)
Volume 131 **Metaheuristic Optimization in Power Engineering** J. Radosavljević
Volume 132 **Power Line Communication Systems for Smart Grids** I.R.S. Casella and A. Anpalagan
Volume 139 **Variability, Scalability and Stability of Microgrids** S.M. Muyeen, S.M. Islam and F. Blaabjerg (Editors)
Volume 155 **Energy Generation and Efficiency Technologies for Green Residential Buildings** D. Ting and R. Carriveau (Editors)
Volume 157 **Electrical Steels, 2 Volumes** A. Moses, K. Jenkins, P. Anderson and H. Stanbury
Volume 905 **Power System Protection, 4 Volumes**

Wind Energy Modeling and Simulation

Volume 1: Atmosphere and plant

Edited by

Paul Veers

The Institution of Engineering and Technology

Published by The Institution of Engineering and Technology, London, United Kingdom

The Institution of Engineering and Technology is registered as a Charity in England & Wales (no. 211014) and Scotland (no. SC038698).

First published 2019

The Institution of Engineering and Technology
Michael Faraday House
Six Hills Way, Stevenage
Herts, SG1 2AY, United Kingdom

www.theiet.org

British Library Cataloguing in Publication Data
A catalogue record for this product is available from the British Library

ISBN 978-1-78561-521-4 (hardback Volume 1)
ISBN 978-1-78561-522-1 (PDF Volume 1)
ISBN 978-1-78561-523-8 (hardback Volume 2)
ISBN 978-1-78561-524-5 (PDF Volume 2)
ISBN 978-1-78561-528-3 (hardback 2-Volume set)

Typeset in India by MPS Ltd

Disclaimer

The following chapters were authored, in part, by the National Renewable Energy Laboratory, operated by Alliance for Sustainable Energy, LLC, for the U.S. Department of Energy (DOE) under Contract No. DE-AC36-08GO28308. Funding provided by the U.S. Department of Energy Office of Energy Efficiency and Renewable Energy Wind Energy Technologies Office. The views expressed in these article do not necessarily represent the views of the DOE or the U.S. Government. The U.S. Government retains and the publisher, by accepting the article for publication, acknowledges that the U.S. Government retains a nonexclusive, paid-up, irrevocable, worldwide license to publish or reproduce the published form of these chapters, or allow others to do so, for U.S. Government purposes.

Volume 1:

Preface

Chapter 1
Looking forward: the promise and challenge of exascale computing
Michael C. Robinson and Michael A. Sprague

Chapter 2
Blade-resolved modeling with fluid–structure interaction
Ganesh Vijayakumar and James G. Brasseur

Chapter 6
Modeling and simulation of wind-farm flows
Matthew J. Churchfield and Patrick J. Moriarty

Chapter 7
Wind-plant-controller design
Bart Doekemeijer, Sjoerd Boersma, Jennifer King, Paul Fleming, and Jan-Willem van Wingerden

Chapter 9
Cost of wind energy modeling
M. Maureen Hand, Volker Berkhout, Paul Schwabe, David Weir and Ryan Wiser

Contents

Preface

People are modelers. Every human being uses mental models to capture expertise and facilitate success. Mental models allow us to think ahead, essentially running thought simulations of possible outcomes, evaluating what might happen in the future given different choices in the present. Specific models are created for each location, specialty or vocation. The mental calculations that go on are automatic, almost subconscious in some ways. The models are based on experience and are updated as these experiences expand. The models, over time and experience, become more comprehensive and complex.

People use mental models to process the endless streams of data acquired by the senses every second to organize these data, make sense of what is happening around us, and plan how we are going to respond. When I go into my kitchen to get a drink, find some food, begin a meal, or even to do some cleaning, I know what is behind closed doors, how different implements are used, and where to find the food. When a snake found its way into the middle of the kitchen floor, there was no place in this mental model for such an object. It was interpreted as a ribbon, perhaps a belt, or part of an apron. Once its motions confirmed the reality, my mental model of objects in the kitchen was modified forever—slender serpentine objects on the floor now trigger different emotional responses than they ever did before. Mental models developed by one set of experiences (data sets) need to be continuously revised with the expansion of experience and exposure to new data.

Scientific discovery is a process of model development and enhancement. Well-designed experiments are aimed at uncovering parts of the models that are weak or missing, so that the more complete representation will enable successful application. The new information is used to modify the existing model so it can grow to apply where it might not have in the past. In some cases, the added information shows that the model did not just have a gap, but the new discovery simply does not fit the paradigm. The reality may be that the model is fatally flawed and cannot contain the new information. It may have to be scrapped and a new model more consistent with the new discovery will have to be created. In this way, the model reveals that the pervious way of thinking was inadequate. The fact that the model failed to contain the new information makes it abundantly clear that the old understanding must go.

From a personal perspective, it is this model-driven understanding of physical processes, applied in engineering coursework, which drew me to the applied engineering sciences as a profession. While proficiency in many areas of study depends on the copious memorization and regurgitation of individual facts (and I am grateful for those who possess and exercise this proclivity in their fields), engineering science is a field wherein success depends much more heavily on understanding the fundamental concepts than memorizing the individual facts. Understanding how to formulate and solve the differential equation for elastic continua is much more powerful than remembering the equations for maximum deflection of a dozen different configurations of beams. The former unlocks the analysis of an infinite array of elastic structures, whereas the latter simply explains a few specific situations.

The human use of scientific knowledge is therefore one of accessing and exercising the various models we possess of the way the natural world works. Human engineering extends these models to the physics of constructed systems and devices built for individual or societal benefits. When engineers and scientists work on the advancement of technology, they not only exercise both the innate mental models that come from learning and experience, but also use mathematical models that enable them to work beyond what can be held in a single human mind. Math is the language of scientific knowledge; it is widely recognized that if you cannot describe a phenomenon with a mathematical model, you really don't understand it.

Mathematical models of the natural world go way back in history, as far back as there are human records. The ancient Egyptians and Babylonians were doing sophisticated mathematical analyses of the structures they were building. Leonardo da Vinci, in the fifteenth century, expressed the view that math is essential to the study of science. He noted, "There is no certainty in sciences where mathematics cannot be applied."[1] However, he was also convinced that nature, appearing continuous in character, is not well suited to computational solutions. He wrote, "Arithmetic is a computational science in its calculation, with true and perfect units, but it is of no avail in dealing with continuous quantity."[2] Leonardo therefore focused on geometry to describe the phenomena he discovered. It was not until Newton and Leibnitz invented the calculus that mathematical descriptions of continuous natural phenomena through differential equations brought the continuous into the realm of the discrete. However, without large-scale computing, the use of differential equations was limited to those with closed-form solutions, or at best very simple numerical approximations. The direct numerical integration of the continuous equations of physics remained intractable. Yet, the differential view of dynamics did presage the eventual use of computational systems that chop up the continuous fields and timelines of nature into large numbers of simple equations amenable to large-scale numerical solution. But the ubiquitous use of discretized computational methods needed to wait for further centuries for the development of computers that were up to the task.

Long before computers of sufficient size to solve meaningful continuum mechanics problems were created, people were dreaming of ways that discretizations could lead to enhanced solutions. Haupt *et al.*, in Chapter 3 of Volume 1 of this work, tell of a dream of Lewis Fry Richardson, atmospheric scientist and namesake of the

[1] Paris Manuscript, Notebooks/J.P. Richter, 1158,3: James McCabe, "Leonardo do Vinci's De Ludo Geometrica," Ph.D. Dissertation, UCLA, 1972, as quoted in Isaacson, *Leonardo da Vinci*, Simon and Shuster, New York, 2017, p 200.

[2] Codex Atlanticus (1478-1518) Biblioteca Ambrosiana, Milan, as quoted in Isaacson, *Leonardo da Vinci*, Simon and Shuster, New York, 2017, p 201.

Richardson number, who as far back as 1904 was suggesting how these massive computational systems might work for numerical weather prediction. He suggested a thought experiment of a massive round theater (picture the galactic senate scene in the Star Wars movies) wherein each seat was a location on the globe, and the occupant was solving a single equation. Their solution would be displayed so those in the seats surrounding them could see their answers and use them in their own computations, thus connecting the calculations step by step around the entire planet. In the realm of structural analysis, massive matrix equations describing the truss-work structure of giant Zeppelins were being solved by sheer brute force as teams of arithmeticians used mechanical adding machines and slide rules to invert the matrices. Similar to the way those who used machines to write with fixed typefaces were known as "typewriters," the massive groups using simple machines to do individual arithmetic operations in concert were known as "computers." So, in an irony of progress, while in the past, we tried to make people act like computers, we now try to make computers act as people.

The last few decades have revealed an amazing transformation in the practice of science and engineering driven by the continuous development of larger and faster computers. This progress has been sustained over so many decades that it hardly seems to be a breakthrough, but it has certainly been a revolution. These massive computational capabilities make it possible to envision the creation of models of unprecedented scope and resolution. Parallel architectures with ever faster processors make it appear that Richardson's vision has indeed become a reality, but with each seat containing the computational power that just a couple decades ago did not exist in even the fastest of single cores. Many phenomena we would like to model are described by their governing equations and defined by unique sets of boundary conditions so complex that the outcome cannot be visualized directly by the human brain. The solution is only made observable when the computational models are solved, and the results are displayed in graphical form. Computational fluid dynamics is one such area in which the use of color-coded numerical results has brought previously unimaginable complexity to light, so much so that practitioners now joke that uninformed use of the techniques, resulting in questionable results, is actually delivering "colorful fluid dynamics." When used properly, computational models are now capable of delivering a high enough resolution to reveal intricacies and phenomena that previously lay hidden within the mathematics.

The computational models and problems being solved have now progressed to the level where computational results, well validated in controlled experiments where measurements can be obtained, provide insight into domains where experimental results are unattainable. Phenomena that lay hidden where instrumentation is not capable of revealing the details of interest are revealed with a computational model that can estimate the finer details by refining the resolution and time steps as far as necessary.

And yet, we are limited. As a graduate student, I took classes in finite element methods from one of the leaders of the era: Thomas J.R. Hughes. He liked to tell students that the size and speed of computers are constant: they are always too small and always too slow and will always be that way. The nature of technological progress is to continue to present problems that push the boundaries of what is currently possible. The problem of atmospheric turbulence is a good example. We know that the atmosphere contains flow structures ranging in size from the largest synoptic scales down to the tiniest eddies of turbulence. Berg and Kelly, in their Volume 1 chapter on turbulence modeling, note that using direct solution of the Navier–Stokes equations to resolve the smallest eddies in an atmospheric flow with a domain large enough also to capture the major length scales will require discretization of at least a billion-billion (10^{18}) cells—well out of the reach of even the exascale machines that are still on the drawing board. Therefore, we need to focus models on capturing the primary scales of interest for particular investigations so that the discretization remains tractable. Not all phenomena can be included in every computational investigation. This is certainly the case now with mesoscale models separate from microscale models which are also separate from turbine models, which in turn are separate from detailed models of turbine subsystems, from blade materials to bearings. While computational models of each scale are already well developed, the interaction between the scales is still a significant challenge. These interfaces are often quite low fidelity and have embedded approximations that limit the accuracy of the total cascade from the largest to the smallest scales.

The opportunity now exists to use the increasing size and speed of computers to again expand the scope and complexity of the models by attempting to bridge scales and combine solutions of previously separate problems into a single modeling domain. It is this opportunity that is driving a revolution in modeling and simulation for wind energy technology. It creates a grand challenge in computational science, which is the topic of the first chapter of Volume 1 by Robinson and Sprague. The mathematical models that capture the physics from one scale and use that information to provide boundary conditions for the next higher resolution scale are forced to make compromises and simplifications that will inherently lose information. By combining the scales, the information is implicitly contained in the computation, thereby providing the hope of truly enhanced fidelity of the combined result.

History of modeling and simulation for design

Failure can be a great source of insight. Many first-generation wind machines experienced early and frequent failures, sometimes due to inability to withstand extreme winds, but more often just due to the repeated cyclic fatigue loads generated during normal operation. The earliest wind research efforts were focused predominantly on the aerodynamic performance of the rotor. The structure was viewed as simply a structural/mechanical design exercise. For example, when the first machine at a major research institution in the United States was built, the prototype was mounted on the roof of the main administration building. Only after the fact did anyone think to mount some strain gages on the blades to check on structural loads. The measurements showed that the metal blades were experiencing yielding of the material on every rotation, a load level that usually results in failure within a thousand load cycles. After this revelation, the machine was moved to a safer location to complete the testing.

The first successful generation of commercial wind turbines was based on combined models of the aerodynamics and the structures—aeroelastic models. My personal exposure to these models in the early 1980s was to use them to attempt to predict fatigue lifetime of the blades, many of which were failing in the field. When simulated load time histories were compared with measured load time histories, it became readily apparent that the simplified steady-wind input to the aeroelastic models, which included wind shear but no turbulence, was completely inadequate for matching the complexity revealed by the measurements.

Although these early aeroelastic models had successfully coupled aerodynamic and structural elements, they needed to couple another element—atmospheric turbulence—to be predictive of turbine loads and hence fatigue durability. It was not until computational models were constructed that combined the atmospheric turbulence and shear, the rotational dynamics of the blades moving through the wind, the aerodynamics that generate a force on the blades, and the structural dynamics that cause amplification of those loads at natural frequencies that the nature of the measurements could be reproduced. Only then was it possible to even consider the fatigue implication operating in the atmosphere. This began the long road of adding physical processes into the models of wind turbines, so that the bundled collection could accurately represent the actual environments of the turbine and be a basis for the design loads it must survive.

Content of the book

It would be relatively simple to combine elements into a larger modeling structure if the effects of one element on the next were merely cascading, from wind to aerodynamics to structure to fatigue. In reality, there is often two-way coupling between the elements that must be approximated when the models are done separately. This is especially important at the intersection between the turbine and the atmosphere where each impacts the other. When the models are combined, the coupling can be expressed explicitly, although the modeling challenge to do so is significant.

Wind energy achieves much of its cost-effectiveness from the fact that extracting energy from the air relies on the continuum properties of the fluid. The solid rotor does not need to actually contact particles of air to extract energy from them. The airfoil produces a lift force that is felt throughout the continuum and extracts energy from the moving mass of air far from the structure. Therefore, a very slender blade structure can rapidly sweep through a mass of air, creating a pressure reaction that slows the air across the entire swept area both downwind and upwind of the rotor. Even the earliest aerodynamic models recognized this two-way coupling by including the "induction effect" in the flow of the air into the rotor, correcting for the reduced velocity of the upstream air, as described by Hansen in Chapter 1 of Volume 2. However, when turbulence models were combined, the induction was often still approximated as relatively steady and uniform across the rotor, even though the inflow and the aerodynamic forces in the new combined models are far from uniform. A more accurate coupling of the inflow and the turbine requires individual blade forces to be applied to the turbulent fluid, as described by Moriarty and Churchfield in Chapter 6 of Volume 1. The blade is modeled as a lifting line, approximating the true three-dimensional flow around the blade with two-dimensional simplifications using lift and drag correction factors evolved through decades of research on both helicopters and wind turbines. The result has been a significant improvement in the understanding of the interaction of the turbine with turbulent wind fields, especially in how the wake is generated and convected downstream. These full-wind-plant models are creating new understanding of the nature of the flow through and around multiple wind turbines in a plant and initiating a flurry of activity around what to do with it.

Two-dimensional airfoil theory was used very successfully early in the development of wind turbines and remains the basis for most design tools used today, as described in Chapter 1 of Volume 2. Early researchers also coupled the elastic distortions of the structure due to the aerodynamic loads into the relative velocities and angles of attack, which have resulted in the aeroelastic models described in Chapter 2 of Volume 2. Validation studies have shown that corrections to two-dimensional airfoil approximations can be very accurate when applied within the domain of their prior validation and tuning. They also have the tendency to produce results of variable accuracy when taken out of their comfort zone and applied to different configurations. More accurate representation of the blade aerodynamics requires coupling that resolves the actual blade shape and computes the three-dimensional flow around it as it rotates through the turbulent winds. Vijayakumar and Brasseur in Chapter 2 of Volume 1 give some insight into how to implement these "blade-resolved" models in conjunction with a realistic inflow. Coupling from the flow field down to the surface of the blade removes approximations of two-dimensional airfoil theory and attempts to solve for the surface pressures everywhere on the blades. Removal of the two-dimensional airfoil simplification enables the modeling of innovative concepts not well captured by those assumptions by using a higher fidelity, three-dimensional computational simulation. The high-fidelity model has the possibility of providing more accurate results in applications where the models were not tuned by resolving the physics directly. The size and computational complexity of blade-resolved models make them capable of simulating even the most complex design alternatives, but they can't be used to define the response over the entire design envelope. Only a few inflow situations can be simulated with a high-fidelity model, whereas typical design calculations require that a turbine structure be evaluated with respect to thousands of potential inflow and control realizations. There is still a need to capture the fundamental physics of the machine, but with less computational intensity as found in the modeling and simulation capabilities focused on detailed design, as described in Volume 2.

Modeling a wind plant accurately requires not only coupling down to the millimeter-thick boundary layer of the airfoils on the blades but upward to where the flows originate in the large-scale forces that drive the atmospheric mesoscale, which covers hundreds of kilometers of area and encompasses the planetary boundary layer. Large-scale energy transfer mechanisms, thermal mixing, high- and low-level jets, and other mesoscale effects are what drive the winds at the surface and determine the nature of the turbulence as well. Haupt *et al.* provide a sweeping and comprehensive explanation of the nature of the weather modeling challenge as it has progressed throughout the last few decades and how it now relates to wind energy in Chapter 3 of Volume 1. Mirocha, in Chapter 4 of Volume 1, describes how the meso-to-microscales are bridged to bring the atmospheric flow features down to the scale of the turbines and why that is important. This pair of chapters is the foundation for the inputs to all the other chapters on the book.

Another bridge between mesoscale effects and the wind plant is captured in the models that forecast wind plant power production for use in planning and operating the electrical grid. Weather forecasting is a rapidly improving field with greater accuracy driven by the revolution in computer size and speed. But wind plant power production depends on a portion of the weather forecast, wind speed at 100–200 m above ground, which has not been the central focus of weather modeling. In addition, the plant power output depends on many factors controlling the dynamics of plant production, such as turbine power versus wind speed characteristics, wake interactions, terrain effects, and other major challenges to wind plant modeling. In Chapter 8 of Volume 1, Zack shows how these forecasts are derived and highlights some of the challenges.

Of potentially greater value than simply forecasting output is controlling output. van Wingerden *et al.*, in Chapter 7 of Volume 1, describe the relatively new field of wind-plant control. Appropriate attention to plant control can increase efficiency of the plant by better managing the way the individual turbines operate with respect to each other, managing the wind resource and turbine-operating parameters to create a plant-wide objective that is greater than the sum of the parts. Actively managing the wind by steering wakes has shown potential to increase plant output. And actively controlling the wind-power output to meet grid-driven demands, such as ramping and fault ride through, is just beginning to be considered in grid-operating strategies. With advances in both wind plant control and wind plant optimization (Ning and Dykes, Chapter 7 of Volume 2), it becomes possible to consider the features and capabilities of the plant controller in the design of the wind plant, a capability rarely exercised at this point but showing great promise.

Predicting the flow of forces downstream from the atmosphere to the turbine to inside the turbine depends on the aeroelastic model (see Hansen, Chapter 2, Volume 2), which calculates the system loads driven by the atmospheric turbulence (see Berg and Kelly, Chapter 5, Volume 1). Models of machine elastic dynamics themselves are highly complex and nonlinear. These aeroelastic models have become sufficiently accurate to where they have established the scientific underpinnings for the highly successful march of turbine technology to low cost and high reliability. Turbulence models used in design, simple as they are, have also been highly effective at approximating the atmospheric environments used to define designs that keep the primary structure safe from extreme loads with adequate fatigue durability. This is achieved by coupling the models of the machine dynamics to models of the individual components: the rotor, drivetrain, foundation, and turbine controller. Each of these subsystems are addressed in separate chapters within Volume 2.

Internal loads at critical interface locations are taken from the aeroelastic model and applied to the individual subsystems for detailed design evaluation. Within a full-system simulation, each of the subsystems is represented with only a few key degrees of freedom while suppressing all the details. This results in significant simplifications but does accurately represent the interface loads between subsystems. Designers of the subsystems can then take those interface loads and use them to estimate detailed requirements within the subsystem. It has thus been a highly effective practice for subsystem experts to focus their modeling efforts on the particular issues within their specialty. The drivetrain can be designed with a comprehensive definition of the loads on the low-speed shaft as described by Zhang *et al.* in Chapter 4 of Volume 2. Bearing location and configuration, gearbox design features, and generator selection are all driven by the rotor loads. The foundations of land-based systems are similarly designed given the loads on the tower. One place where a separation has been shown to be problematic is for offshore floating systems. There is enough interaction between the motions of the floating foundation and the rotor that the aerodynamics of the wind and hydrodynamics of the waves need to be considered together, as described by Matha *et al.* in Chapter 5 of Volume 2.

The rotor, however, is the system that generates all the atmosphere-driven loads that are then passed to other subsystems and therefore requires special attention for optimization. Bottasso and Bortolotti in Chapter 3 of Volume 2 describe the level of detail and integration between the aeroelastic and structural modeling required to conduct detailed design optimization of the rotor subsystem. Ning and Dykes continue the theme in Chapter 4 of Volume 2 with a detailed look at optimization techniques that are especially useful to enable the rotor optimization to be both tractable and linked to the greater challenge of fully integrated turbine system design.

Every design simulation must integrate the controller into the design load calculations from the very start. One cannot model the aeroelastic response of a wind turbine in the absence of its controller. Wright *et al.* describe the control system at many levels, from power only to loads attenuating control to ones that engage the complexity of floating systems wherein the inputs are from both the top and bottom (see Chapter 6 of Volume 2).

The objective of a wind power plant is to generate electricity and feed that power into the local grid, which is connected to the continent-wide electrical grid, often described as the largest machine ever built. The electrical generators and associated power electronics have their own dynamics that interact with the individual machines, the intraplant collection system, and the grid to which it is connected. Muljadi and Gevorgian in Chapter 8 of Volume 2 provide insight into how the various types of generators are modeled as well as how these dynamics feed up to the wind plant interconnection. Miller in Chapter 9 of Volume 2 looks outward from the interconnection and explains the issues around modeling the plant interconnectivity with the regional grid. These two chapters present the tremendous progress that has been made in managing the integration of very large variable generation plants into an electrical system that must balance load and generation in real time. They also provide some insight into the grid of the future, which is likely to be dominated by inputs with very little classical inertia-dominated generation. Modeling this grid system and capturing the emerging capabilities of wind plant control will be essential to meeting the challenge of high-penetration wind energy.

Ning and Dykes in Chapter 7 of Volume 2 describe the full wind plant optimization problem. Multidisciplinary analysis and optimization techniques are borrowed from aerospace beginnings and applied to the wind plant design challenge. These modeling frameworks are beginning to pull together all the disparate models that make up the subsystems of the wind plant and bring them into a unified system so that trade-offs can be assessed and significant innovative approaches to wind plant design, control, operations, and maintenance can be evaluated. It is only by understanding the principal elements of how to model each of the subsystems that these larger full-system optimization frameworks can be made to work.

System optimization modeling at the power plant level, as well as at the turbine level, requires models that reflect the capital cost, maintenance costs, and financing costs (reflecting the risks and uncertainties), as well as estimates of the financial return through energy production and sales. The chapter on cost of energy modeling by Hand *et al.* (Chapter 9 of Volume 1) brings out all the subtleties of that process. Cost models are crucial for evaluating technology innovation opportunities and are also essential elements of plant financial decision-making as well as for assessing the effects of local and national policy.

Book structure

The ability to compute solutions to the governing equations describing the physics of wind power plant operation is growing by leaps and bounds. The entire breadth of the system is now routinely modeled in the design process, and the sophistication of the models and computers that solve them continues to grow as well. It is these computational simulations that help designers understand the complexity of how the parts of the system interact and hence allow them to solve problems of system interaction in ways that could not be considered before these computational resources became available.

The breadth of the wind energy system problem has drawn the attention of major computational science organizations, declaring it to be a "grand challenge". The scales of fluid dynamics alone range from the boundary layer of wind turbine blades to the regional flows and large-scale eddies of the planetary boundary layer. The structures and mechanical parts of a turbine include flexible blades and towers up to 100 m in length as well as the tribology of the interfaces in bearings and gears at a fraction of a millimeter. Electrically, models must cover the range from the air gaps within the generator to the electrical grid that wind plants must dynamically support.

The modeling and simulation approaches used in each subsystem as well as the system-wide solution methods to optimize across subsystem boundaries are described in this book. Chapters are written by technical experts in each field to describe the current state of the art in modeling and simulation for wind plant design. Special attention is directed at the fundamental challenges and issues to be solved to extend the content beyond simply describing current practice to a level that will provide long-lasting insight into the methods that will need to be developed as the technology matures.

There are too many individual chapters to be included in a single volume. However, it is not entirely natural to divide the book because the topics are intricately interrelated. The separation of the content in to these two volumes is based on the following two main themes—Volume 1: Atmosphere and plant and Volume 2: Turbine and system.

Volume 1: Atmosphere and plant

The nature of modeling the atmosphere is inherently based on the flow of air from the mesoscale down to the surface, and then into, around, and through the many rotors of a wind plant. Models of the flow share the characteristic that the domain can be intensely discretized and solved over the entire computational domain in one numerical package. Models that focus on one area or another simplify the influence of effects outside the immediate domain, often with sub-grid phenomena that are below the primary scale of interest. Massively parallel computational facilities have opened the door to increasingly high resolution over increasingly larger domains. The promise of moving from petascale to exascale computing offers the ability to bundle more of these models together, as the first few chapters of Volume 1 describe. The opportunity is now before us to bridge traditional modeling boundaries with the promise that the improved representation of the physics will drive discovery, enhance our understanding, and enable innovation in design. The chapters included in Volume 1 therefore are those that are highly driven by high-performance computing, atmospheric science, fluid dynamics, and full-wind-power-plant representation.

Volume 2: Turbine and system

The design of the actual hardware placed in the field to capture the energy, convert it to electricity, and usefully integrate that electricity into the electric grid is dependent on the models in the second volume. To be useful in design and design optimization, these models need to be simple and fast enough to simulate the thousands of design conditions that complex atmospheric flows generate and to do this in an optimization framework. These models need to capture fundamental physics in ways that are often conceptionally different from approaches based on the intense discretization of the flow as described in Volume 1. Therefore, they are likely to draw insight from the high-fidelity models and use that insight to drive multiphysics, parametric, or phenomenological models capable of design calculations. The chapters in Volume 2 focus on detailed descriptions of individual components, system optimization capabilities for turbines, and plants and the electrical systems with which the wind plants interconnect.

The philosophy of the book

The design of a successful wind power plant involves a suite of scientific disciplines ranging far beyond what is typically included in a single product or technology.

Integrated wind energy systems will require experts in each specialty who also possess a relatively good understanding of all the other areas to infer how their choices will impact the rest of the system. Because of the vast sweep of topics and models, it is intended that each be described with sufficient detail to permit a skilled modeler from another area to understand the topic. This book describes the scope of the modeling challenges within each area of expertise, but also provides insight into the boundaries between modeling and simulation topics and how adjacent models can reflect the characteristics of neighboring modeling domains. As computational assets improve and grow, the modeling and simulation of wind energy systems will perhaps begin to encompass more of what are now separate domains into combined models with greater accuracy. It is the hope of the authors that this book will facilitate continual expansion of modeling and simulation capabilities and be an engine for innovation in wind power plants that meet the energy challenges of the future.

List of acronyms

Abbreviation	meaning
ABL	atmospheric boundary layer
AE	analog ensemble
AEP	annual energy production
AnEn	Analog Ensemble
ANN	artificial neural networks
ARPS	Advanced Regional Prediction System
ASL	atmospheric surface layer
BEM	blade element momentum
BEMT	blade element momentum theory
BP	buoyant production or destruction
BS	Brier score
BTM	behind-the-meter
CAD	computer aided design
CAPEX	capital expenditures
CART	classification and regression tree
CFD	computational fluid dynamics
CFL	Courant–Friedrich–Lewy
COAMPS	Coupled Ocean/Atmosphere Mesoscale Prediction System
CPU	central processing unit
CRPS	continuous ranked probability score
CSD	computational structural dynamics
CSI	critical success index
CT	central time
DA	data assimilation
DNS	direct numerical simulation
DOE	Department of Energy
DRM	dynamic reconstruction model
DTU	Technical University of Denmark
DWM	dynamic wake meandering
E3SM	Energy Earth Exascale Model
EC	Environment Canada
ECM	Ensemble Composite Models
ECMWF	European Centre for Medium Range Weather Forecasts

ELRAS	ERCOT Large Ramp Alert System
ENIAC	Electronic Numerical Integrator and Computer
EnKF	ensemble Kalman filter
ERCOT	Electric Reliability Council of Texas
ESA	ensemble sensitivity analysis
FFT	fast Fourier transform
FLOPS	floating-point operations per second
FLORIS	FLOw Redirection and Induction in Steady State
FSI	fluid-structure interaction
FV3	Finite-Volume Cubed-Sphere Dynamical Core
GBM	Gradient Boosted Machine
GDPS	Global Deterministic Prediction System
GFS	Global Forecast System
GMRES	generalized minimum residual
GPU	graphical processing unit
HFM	high-fidelity modelling
HPC	high-performance computing
HR	hit rate
HRRR	high resolution rapid refresh
HST	Hawaiian Standard Time
IBM	immersed boundary method
IEC	International Electrotechnical Commission
ITCZ	Inter-Tropical Convergence Zone
KF	Kalman filter
kW	kilowatt
LAM	limited area model
LBC	lateral boundary condition
LBC	lateral boundary conditions
LCOE	levelized cost of energy
LES	large-eddy-simulation
LETKF	local ensemble transform Kalman filter
LHN	latent heat nudging
LLJ	low-level jet
LMP	locational-marginal price
LSM	Land surface model
LT	local time
LVC	linear variance calibration
MAE	mean absolute error
MDAO	multidisciplinary analysis and optimization
MET	Model Evaluation Tools
MLP	Multi-Layer Perceptron
MLR	Multiple Linear Regression
MLR	model linear regression
MM	Meso Monster

MMC	mesoscale-to-microscale coupling
M–O	Monin–Obukhov
MODE	Method for Object-based Diagnostic Evaluation
MOS	model output statistics
MOST	Monin–Obukhov similarity theory
MPAS	Model for Prediction Across Scales
MR	miss rate
MW	megawatt
MYNN	Mellor–Yamada–Nakanishi–Niino
NAM	North American Mesoscale
NBA	nonlinear backscatter and anisotropy
NCAR	National Center for Atmospheric Research
NCEP	National Centers for Environmental Prediction
NLLJ	nocturnal low-level jet
NREL	National Renewable Energy Laboratory
N–S	Navier–Stokes
NUMA	non-hydrostatic unified model of the atmosphere
NWP	numerical weather prediction
O&M	operation and maintenance
OPEX	operating expenditures
PBL	planetary boundary layer
PCE	polynomial chaos expansion
PCM	pseudocanopy model
POE	probability of exceedance
PT	pressure transport
PV	photovoltaic
RAMS	Regional Atmospheric Modeling System
RANS	Reynolds-averaged Navier–Stokes
RaP	Rapid Refresh
RASS	radio acoustic sounding system
RDPS	Regional Deterministic Prediction System
RF	random forests
RMSD	root mean square difference
RMSE	root mean squared error
RWP	radar wind profiler
SCADA	supervisory control and data acquisition
SCPM	stochastic cell perturbation method
SFS	subfilter-scale
SGS	subgrid-scale
SMART	System Management of the Atmospheric Resource through Technology
SOWFA	Simulator fOr Wind Farm Applications
SP	shear production
SVM	support vector machines
SVR	support vector regression

SWiFT	Scaled Wind Farm Technology
TKE	turbulent kinetic energy
TT	turbulent transport
TWRA	Tehachapi Wind Resource Area
UCAR	University Center for Atmospheric Research
URANS	unsteady Reynolds-averaged Navier–Stokes
USNWS	United States National Weather Service
V&V	verification and validation
WACC	weighted average cost of capital
WETO	Wind Energy Technologies Office
WMLES	wall-modeled LES
WRF	Weather Research and Forecasting
WRLES	wall-resolved large-eddy simulation

Chapter 1

Looking forward: the promise and challenge of exascale computing

Michael C. Robinson[1] and Michael A. Sprague[2]

1.1 Introduction

High-performance computing (HPC) advances within wind energy sciences are being driven by the simulation requirements of maturing technologies and by the integrated-systems-analysis demands of a robust, highly competitive power-generation market. Utility-scale wind turbines have rapidly evolved from kilowatt to megawatt generation capacities with future benchmark feasibility studies addressing the challenges in obtaining a 50-MW turbine [1]. The transition in technology size has been unprecedented. Once relatively small in comparison with atmospheric flow phenomena, turbine-rotor and -tower sizes are approaching the physical dimensions of the planetary boundary layer (PBL) and the complex flow structures associated with weather-transition events. Deployment paradigms have also transitioned from small clusters of turbines remotely spaced to very large wind plant arrays composed of multimegawatt machines spanning hundreds of square kilometres both on land and offshore. These radical changes place ever-increasing demands on the simulation, design, and analysis capabilities needed by scientists and engineers to advance modern wind turbine and plant technologies for optimal performance.

Simultaneously, the cost of wind energy has significantly decreased [2] for both land-based turbines and offshore turbines deployed in shallow water. Lighter weight (per rotor size), dynamically active, and optimized systems, enabled by utilizing high-fidelity modelling (HFM) in the design and development process, have reduced turbine and system-integration costs and increased wind power plant productivity. Similar trends are driving deep-water offshore floating-technology development, for which cost-effective turbine architectures require substantial innovation given the entirely new system dynamics, deployment, and operational constraints in marine environments. Ultimately, market requirements set wind technology cost and performance objectives. HPC and HFM are the essential capabilities needed to glean a

[1]Institutional Planning, Integration and Development, National Renewable Energy Laboratory, Golden, United States
[2]National Wind Technology Center, National Renewable Energy Laboratory, Boulder, United States

better understanding of the underlying physics driving performance and to provide insight for technology advances and new architectures through predictive design and performance/cost analyses to meet market demands.

1.1.1 Future wind plant technology

The future cost of wind energy must continue to decrease below conventional power generation for wind to remain competitive. The *Wind Vision* [3] study systematically quantified the benefits derived from wind energy and provided a technology-development vision required for sustained deployment. Plant cost and performance are the primary drivers for continued impactful strategic deployment. Transitioning from a 'turbine'-centric to a 'plant'-centric design and development philosophy to reduce the integrated plant system cost of energy raises new research and development challenges. Innovative wind plant technologies and fully coupled integrated system optimization must consider site variant atmospheric resources, topography, turbine array interaction, and grid interconnection – all impacting cost and performance.

SMART (System Management of the Atmospheric Resource through Technology) wind plants that interactively adapt to the unique site-specific design and resource conditions to achieve optimized cost and performance are wind technology's future-development pathway. *Enabling the SMART Wind Power Plant of the Future through Science Based Innovation* [4] provides a scenario analysis to reduce the cost of wind energy by 50 per cent or more below 2017 levels. Achieving this goal requires a better understanding and quantification of the physical processes impacting wind plant performance as well as the adoption of new strategies to actively control and convert the wind energy resource and seamlessly integrate this energy into the grid. The underlying physics, including the complex nature of the atmospheric processes and interactions within the wind plant, determine the ultimate plant performance.

SMART wind plant technologies are predicated on the ability to predict the wind plant atmospheric inflow conditions and actively control the dynamic turbine/inflow interactions as the wind passes through the wind plant. Modelling the physical processes and interactions involved in, and developing the technologies required for, wind-resource active control to optimize cost and performance require a significant investment in HPC and HFM, focusing on integrated plant systems design and optimization.

1.1.2 Physical scales driving HFM and HPC

The potential for wind power to play a strategic role in meeting future energy demands requires a holistic vision for technology evolution considering all facets of development, deployment, and operation. The atmospheric energy resource and the advanced technology options that maximize energy extraction, mitigate operational loads for increased longevity, and reduce operation and maintenance (O&M) costs must be assessed, verified, and validated as a coupled plant system. This requires advanced HFM that numerically incorporates all of the relevant physical processes, including fluid–structure interaction (FSI), that drive momentum extraction from the wind resource.

Our ability to quantify the flow and turbine physics of interest, develop and implement new turbine technologies and plant-level control strategies, and accurately predict plant performance is a multi-scale and multi-physics computational challenge requiring HFM and HPC. Capturing the relevant flow physics includes coupling atmospheric physical phenomena ranging in scale from regional weather systems down to the turbine-blade boundary layers with all their turbulent temporal and spatial variability. There is significant uncertainty in resolving the flow physics at each scale and, to an even greater extent, transitioning the physical-flow dynamics and turbulence characteristics between scales. Ranging from the mesoscale (10^5 m) down to the blade-boundary-layer scale (10^{-5} m), each domain is dominated by a different set of dynamics and turbulence processes that must be modelled appropriately to predict both power generation and turbine loads.

Capturing the appropriate temporal scales is equally challenging. The relevant scales impacting performance range from wind-resource production at the mesoscale (minutes to hours) to the extraction of energy through turbine boundary-layer dynamics (milliseconds). Throughout the wind plant computational domain, complex turbulence is introduced through a variety of physical processes (e.g. atmospheric mixing, dissipation, convection, diffusion, complex terrain, and turbine wake development), each with unique temporal and spatial characteristics. All elements of the energy production and extraction process must be modelled as integrated and coupled continua to achieve accurate and predictive simulation results.

Fully coupled HFM simulations using petascale (today) and exascale (future) computing capabilities are needed to model the physical phenomena driving energy production in lieu of empiricism to achieve a predictive capability. When detailed atmospheric processes are appropriately captured, including the relevant turbulence sources, dramatic improvements in our understanding of the science phenomenology driving flow within and across scales will lay the foundation for disruptive innovation in wind turbine and plant design and control.

1.1.3 Turbine technology changes requiring HFM and HPC

Offshore wind-turbine technology crossing the 10-MW generation threshold highlights the rapid change in size of utility-class technology and the need to evolve new computational capabilities to support future development at this scale. Utility-scale multimegawatt architectures commonly incorporate multiple technology innovations that reduced total mass (per rotor diameter), while maintaining required stiffness margins for stability and tower clearance. Lighter-weight designs utilize advanced composite materials and structural architectures to reduce rotor mass [5]. These lighter-weight designs are dynamically more active and significantly more flexible, resulting in large, complex blade deformations (e.g. flap and bend-twist coupling). The increase in rotor scale (e.g. 220 m rotors) and associated tower heights (e.g. 150 m towers), combined with greater flexibility, increases aeroelastic coupling from inflow/rotor interactions and dynamic tower/rotor coupling.

At these physical scales, the turbine-load drivers vary substantially depending on the local atmospheric conditions and physical state of the PBL. Dependent upon the local atmospheric conditions, diverse atmospheric phenomena (e.g. Kelvin–Helmholtz waves, nocturnal jets, marine jets, boundary-layer stability) are routine. Each phenomenon has diverse and unique inflow characteristics varying radically, both temporally and spatially, yielding strong three-dimensional (3-D) variability in veer, shear, and turbulence intensity. Energy extraction is a function of the unsteady flow field over the rotor, and predicting turbine and plant performance at these scales requires more comprehensive computational-modelling capabilities and simulations at fidelities beyond the empirical model representations that have been used in the past.

2-D modelling assumptions implicit in actuator-disc and actuator-line theory (blade element momentum) ascertain rotor performance from an integration of sectional blade characteristics across the blade span. Performance is derived empirically from look-up tables that capture the sectional aerodynamic inflow interaction (lift and drag) as a function of local flow velocity and blade angle of attack. Additional modifications are made to accommodate dynamic inflow, stall, tip loss and other effects having a physical basis that are also implemented empirically through simplified models. The assumptions and modelling-implementation approaches currently being utilized likely cannot fully capture the interactions of a complex 3-D inflow interacting with an extremely soft and dynamically active rotating blade at PBL length scales. Readers are referred to [6] for more discussion on actuator-disc and -line methods.

At multimegawatt-turbine length scales, the complex 3-D inflow driving a highly aeroelastic and dynamically active rotor in a rotational frame requires a robust 3-D computational analysis capability incorporating computational fluid dynamics (CFD) and detailed FSI. The energy capture and transient loads from the inflow/rotor interaction as well as the rotor wake and downstream meandering characteristics must be resolved for all turbines in a plant to ascertain the macro plant performance. The extent to which the flow within the plant must be fully resolved will be a trade between simulation fidelity, FSI accuracy, and computational time constraints. Best estimates for future computational resources to capture relevant atmospheric structure, model detailed inflow characteristics, and fully resolve the 3-D inflow/rotor interaction and wake development require petascale computational resources for a single turbine and exascale resources for multiple (e.g. 100+) turbines deployed as a plant system [7]. Advanced numerical methods, HFM, and parallel HPC processing will be essential to achieve the simulation fidelity required to assess future SMART wind plants composed of multimegawatt turbine architectures.

Floating turbine architectures represent the last frontier for wind technology development and the greatest challenge in creating cost-competitive designs. Offshore turbines are typically much larger due to the high cost and installation challenge of the mounting superstructure. Hence, the complexities encountered modelling multimegawatt land-based turbines including aeroelastic coupling and the driving turbulent phenomena at scale are exacerbated in the offshore marine environment. An additional six degrees of freedom (for large rigid-body platform motion) must be considered in addition to the loads from hydrodynamics and air/sea interaction on tethered floating support structures. Total turbine and installation cost is ultimately a function of the total system mass. Offshore floating turbine designs will minimize the mass above the water line to reduce the additional mass required for buoyancy. Finally, the operating environment is often dominated by a stable marine boundary layer. The turbine-wake mixing and dissipation that occurs with unstable boundary layers through interactive turbulent mixing is largely absent. Wakes remain more coherent and produce stronger turbine/turbine interactions. Resolving the turbine dynamic response, loads, and system level power generation requires CFD with HFM (at petascale HPC) that dynamically couples all aerodynamic and hydrodynamic forces and the structural dynamic interactions in a six-degree-of-freedom computational domain.

1.1.4 Wind plant performance

Today's utility-scale wind power plants are grid interconnected though a substation. The total instantaneous wind plant production as a function of the total electrical load on the grid determines the level of penetration. Currently, wind plants function as a collection of individual turbines with each machine operating in isolation based on the local wind condition as measured on the nacelle hub. Energy capture is maximized based upon the measured hub inflow velocity by changing the turbine operating state (rotation speed, blade pitch, and/or yaw) to extract maximum energy. Additional control algorithms are incorporated to seek optimal settings that maximize energy extraction over a set operating period (minutes) to ensure designed state transition and actuator duty cycles are maintained. Depending on the grid capacity, wind-penetration level, and load, as well as other ancillary service requirements, the power generated is passed through the substation to the grid. Occasionally, load-control mitigation (curtailment) of the turbines within a wind plant is necessary to balance the wind resource and load-demand variability.

This turbine-centric operating paradigm is expected to significantly change with future plants incorporating SMART plant technologies. The concept of integrated plant control is based on current research [8] wherein both simulation and field experiments have demonstrated the ability to manipulate rotor wakes through active yaw control. Altering the wake-deficit properties and wake trajectories can minimize the wake-interaction impacts on downstream turbines. The ultimate objective is to coordinate and actively control multiple turbines simultaneously within the plant to optimize total-energy capture, while minimizing O&M costs driven by excessive loads. Although simple in concept, the wake development and propagation characteristics are a complex coupling of turbulent-flow interactions and processes. Varying atmospheric conditions, inflow/turbine interaction, wake development, propagation, and mixing in a multi-turbine wind plant operating environment must all be captured and simulated faster than real time (with reduced-order control-dynamics models) for integrated proactive plant control. Active plant control is complex and requires advanced HFM and HPC to both resolve the wake development and interaction physics and to develop and validate the real-time control models needed for active control implementation. The integrated power performance and turbine-load reductions are promising. The early performance improvements demonstrated open a viable pathway for future commercial industry development.

Advancing our understanding of the flow physics within wind plants is necessary to assess new turbine technologies and the viability of active control, as well as provide predictive performance modelling and analysis tools for industry that have been systematically verified and validated. Resolving the multiple physical processes at multiple scales for wind plant simulation requires petascale to exascale computational resources and innovative HFM simulation capabilities. Significant investments are also required in measurement campaigns that provide the detailed science and engineering data for verification and validation in synergy with HFM and HPC development programs. Predictive modelling of rotor-wake development and meandering processes will facilitate new strategies for active wake control by individual turbines as part of an integrated plant system-control strategy. HPC and HFM will also be essential for developing, validating, and calibrating the control codes and systems models that will make the integrated system control feasible. Advancing our current computational capabilities to glean new scientific insight and open technology development opportunities are the key to future innovation and commercialization.

1.2 Mathematical and numerical modelling pathways

When considering possible pathways to predictive wind-energy simulations using HPC, one must first consider the mathematical models for the physical systems, the numerical discretization of those models that enables solution on a computer, and the 'software stack' of tools/libraries for performing the simulation. It is the combination of models, numerics, and software that determines directly the performance on HPC. In this section, we describe the trade space in regard to choices in mathematical models, associated numerical methods, and implications for HPC performance.

As described in the Introduction, predictive computational simulations of wind turbine and plant dynamics will require models that span a multitude of spatial and temporal scales. At one end of the spatial-scale spectrum, considering that large wind plants can cover areas of 10 km by 10 km, predictive simulations will need to capture large, mesoscale weather flows, which have characteristic scales of hundreds of kilometres. At the other end of the spectrum, boundary layers at the wind turbine blade surfaces will be mm or sub-mm in thickness.

In 2015, the US Department of Energy sponsored a strategic planning meeting of experts to define a path forward for predictive wind energy simulations [9]. From that meeting, there emerged a consensus that a wind plant simulation capability that is truly predictive will have the following features:

- Blade structural dynamics models that include complicated composite structure and large, non-linear deflections that can address, e.g. extreme flap deflections and bend-twist coupling.
- Blade/nacelle/tower body-fitted fluid meshes that deform with structural deflections.
- Overset and/or sliding fluid mesh capabilities that accommodate the rotor rotation, blade pitch, and nacelle yaw.

- Fluid meshes and wall models that accommodate complex terrain.
- Hybrid Reynolds-averaged-Navier–Stokes (RANS)/large-eddy-simulation (LES) turbulence modelling, where LES captures the dynamics of wakes and RANS captures sufficiently the boundary layers at the blade surfaces.
- Coupling of mean flow and turbulence from the mesoscale via, e.g. numerical weather prediction.

Further, simulations will require extensive storage (models with $O(10^9)$ to $O(10^{11})$ grid points will be common), *in situ* data-analysis capabilities, and tools for uncertainty quantification. Although computational structural dynamics models are required for the many turbine components (e.g. shaft, tower, and blades), the most challenging structural dynamics to model, from a whole-turbine perspective, are the large, non-linear blade deflections. Those deformations can be well modelled by geometrically non-linear beam finite elements, whose sectional properties capture complicated composite-material layups and cross sections (see, e.g. [10,11]). Although predictive wind turbine and plant simulations require coupled CFD and computational structural dynamics, CFD dominates computational cost and presents the most significant barriers to predictively simulate the multiple flow scales in wind energy.

Being derived from first principles (e.g. conservation of mass and momentum), mathematical models for wind energy flows take the form of coupled, non-linear partial differential equations for field variables (e.g. velocity and pressure) defined continuously in space and time. Through numerical discretization, those continuum equations are transformed into a typically huge number (e.g. billions) of coupled non-linear and/or linear algebraic equations that are amenable to computational solution. The solution to those equations is an approximation to the true continuum solution of the foundational mathematical model, and the error in that approximation can be reduced through further refinement of the discrete model. For example, if one is using a second-order-accurate spatial discretization, error should be proportional to h^2, where h is a characteristic length of the discrete model, e.g. element size. Reducing h by half would reduce the error by factor of four (i.e. $(h/2)^2$). However, in a 3-D discrete model, the size of the model would increase by a *factor of eight* due to the doubling of elements in three directions. Further, reducing element sizes may well require a reduction in time-step sizes (due to accuracy and/or stability constraints), thereby increasing the number of time steps required. Therein lies the rub: one must balance the computational cost of refining the model (increasing model size) with a limited increase in modelling accuracy.

When choosing a fluid-modelling pathway, perhaps most fundamental is the choice between compressible- and incompressible-flow models, as the associated solver stacks are very different. In this chapter, we are focused on simulating accurately the aerodynamics of wind turbines (rather than aeroacoustics). For a given flow, the impact of compressibility on the aerodynamics is indicated by the Mach number, Ma, which is the ratio of the advective flow speed to the acoustic wave speed (roughly 340 m/s for air). Wind turbines operate at wind mean speed between 4 and 25 m/s. The low range threshold was set by a net positive energy production and the upper limit by structural aerodynamic rotor loading. The corresponding Mach number range of 0.01–0.07 is well within the incompressible flow limit. The highest Mach numbers for an operating turbine occur at the rotating-blade tip and are typically limited to a value below 0.3 to minimize aeroacoustic emissions. Again, this is below the limit requiring a compressible formulation [12]. As such, wind energy flows can be accurately modelled by either the acoustically incompressible or the fully compressible Navier–Stokes equations. Both approaches have advantages and disadvantages.

In a compressible-flow modelling approach, the model captures both acoustic waves and advective flow, which can have orders of magnitude different velocities around a wind turbine. It is well known (see, e.g. [13]) that the numerically discrete system associated with a compressible-flow model becomes 'stiff' as $Ma \rightarrow 0$ due to the large disparity in sound and advective speeds. Further, standard compressible-flow numerical schemes suffer efficiency and/or convergence issues in the low-Mach-number limit [14] and require low-Mach-number preconditioning in that flow regime [15]. In the low-Mach-number regime, because acoustic waves do not influence aerodynamics, compressible-flow solvers often rely on implicit time-update methods that enable time-step sizes at a large Courant–Friedrich–Lewy number, $c\Delta t/h \gg 1$, where h is the smallest grid dimension, Δt is the discrete time step, and c is the sound speed. In other words, compressible solvers rely on the numerical discretization to filter out unimportant acoustic waves.

In an incompressible-flow modelling approach (or, more generally, a low-Mach-number formulation), acoustic waves are mathematically filtered (from the underlying model), and the fastest scales are the advective fluid speeds. Associated numerical schemes are typically 'pressure projection' methods, for which the pressure field is calculated in a manner that maintains mass continuity (i.e. $\nabla \cdot \mathbf{u} = 0$, where $\mathbf{u}$ is the velocity field). The divergence-free constraint leads to an elliptic Poisson-type equation for the pressure field, the setup and solution to which often constitute the most expensive operations in incompressible-flow CFD.

The fluid flows around wind turbines are highly turbulent, for which a particular challenge is the non-linear cascade of turbulent energy from large eddy scales to the small dissipation scales (i.e. the Kolmogorov microscales), at which turbulent energy is converted to heat. These scales are typically separated by many orders of magnitude. Consider Figure 1.1, which shows a notional energy spectrum as a function of wave number for a turbulent flow. Although energy is being injected on the left side of the spectrum (at large spatial scales), that energy is converted to smaller scales (via the breakup of larger eddies into smaller ones) down to the Kolmogorov scale. A simulation that captures/resolves all of these motion scales and the transfer of energy from large eddies to smaller eddies is called direct numerical simulation (DNS). DNS is only feasible for sufficiently simple problems, and is wholly impractical for wind turbine or plant simulations due to the associated spatial and temporal grid-refinement requirements.

With DNS being out of the question, predictive wind simulations require additional turbulence models for the subgrid-scale unresolved dynamics. At the lower-fidelity end of the turbulence-modelling spectrum are RANS approaches, which are incapable of predicting inherently unsteady and multi-scale wind plant flows. At

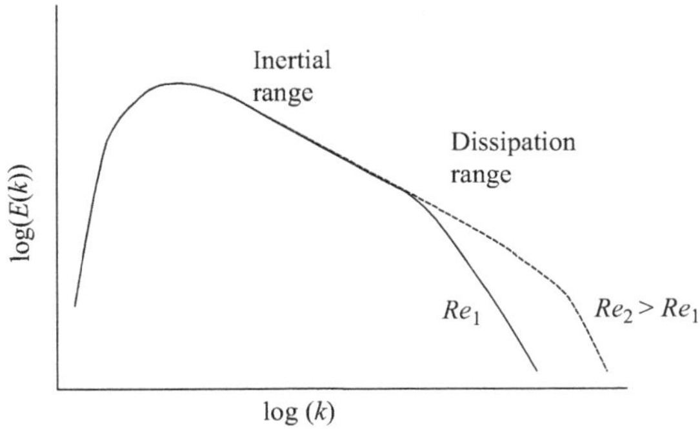

Figure 1.1 A symbolic description of a turbulent kinetic energy spectrum E(k) as a function of wave number k for two Reynolds numbers (Re_1 and Re_2). The spectrum can span many orders of magnitude, and, as Re increases, that span increases. (Image reproduced from [7])

the higher-fidelity end of the spectrum is wall-resolved LES (WRLES). In blade-resolved simulations, the small scales of the boundary layer at the blade surface make WRLES resolution requirements also impractical. The boundary-layer region necessitates either wall-modelled LES or a local RANS treatment (in the context of a larger RANS/LES or detached-eddy-simulation approach). Addressing the transition zone between RANS and LES is an active research area, as is wall modelling, especially in complex terrain. For more details on turbulence modelling for wind energy, we refer readers to [16].

Structured- or unstructured-grid approaches can be used to spatially discretize the fluid domain around a wind turbine. Each approach impacts the numerical-simulation performance in different ways. Structured grids utilize nodes on either a Cartesian grid or body-fitted curvilinear grids and have a common nodal-connectivity pattern that facilitates highly efficient computation (and parallelization), including access to geometric-multigrid solvers and preconditioners. In a Cartesian approach, complex geometries can be addressed through 'immersed-boundary'-type (IB) methods [17]. IB methods are well suited for complex moving geometries but can suffer accuracy issues at the boundaries. Hence, body-fitted, multi-block structured grids (for which structured-grid domains are combined) are appealing for sufficiently simple geometries but are problematic for a complex wind turbine system.

Alternatively, unstructured grids have an arbitrary connectivity between nodes and are well suited for resolving complex geometries. However, unstructured grids neither have the computational efficiency of structured grids nor are they amenable to geometric multigrid solvers/preconditioners but can employ less-efficient algebraic multigrid solvers/preconditioners. Readers are referred to [16] for more discussion and examples of discretization approaches for blade-resolved simulations.

The complex geometry and moving components of a land-based wind turbine (e.g. blade pitch and rotation, nacelle yaw, and proposed two-bladed hub teeter for very large machines), are particularly challenging for body-resolving meshes (both body-fitted-structured and unstructured meshes). The ultimate challenge is in modelling offshore floating turbine structures where two-phase flow (air/sea interface), tethered mooring dynamics, and six degrees of freedom for the integrated turbine and floating platform must be considered.

The grid meshing required for even simplified model conditions becomes rapidly complex. For the restricted case where yaw and blade pitch are fixed, and the tower is rigid, one could mesh the rotor in a cylinder or 'disc' of fluid that has a well-defined sliding interface (see, e.g. [18]) between the rotating fluid and the static background mesh. An example blade-resolved mesh with 176 million nodes and a sliding-mesh interface is shown in Figure 1.2, for the National Renewable Energy Laboratory (NREL) 5-MW reference turbine [19]. Also shown in Figure 1.2 is a simulation result for a coarse mesh (25 million nodes) that lacks the full physics required for predictivity.

The sliding-mesh approach faces obstacles when considering a turbine's full mobility, wherein each turbine component moves (e.g. blades pitch, nacelle yaws, tower bends, and integrated-platform/foundation moves). Overset-mesh approaches are better suited, wherein each turbine component has its own body-fitted mesh that can arbitrarily overlap with other meshes. This is appealing for both mesh creation (only the surface of the inset body needs to be resolved well) and in connecting the meshes. Although limited in application, the sliding mesh can be more efficient in which only a restricted set of nodes can interact, whereas the nodal connectivity in overset is arbitrary, and node interconnection must be continuously resolved at each time step (for moving meshes). Further, in the case of a sliding-mesh interface for the rotor with blades undergoing deformation, an arbitrary-Lagrangian–Eulerian model would typically be required to redistribute the discrete spatial grid to maintain good element quality. Alternatively, with overset meshes, the whole component-level mesh could move in an *a priori* prescribed manner with, e.g. blade deformation.

The biggest challenge of moving meshes (sliding or overset) is the computational time required to search for nodal connectivity and to initialize and reconstruct the associated discrete numerical systems (e.g. matrices and preconditioners) at every time step. In contrast, static-grid models resolve these numerical systems once for the simulation duration, and that cost is amortized over the simulation duration. The implicit-solver preconditioner selection must consider the total aggregated performance impact of each time step. It may be advantageous to use a less-optimal preconditioner if setup time cannot be offset by a sufficiently reduced solve time.

Modelling challenges are increased significantly when considering offshore wind, with the need to model the ocean environment. Without getting into details, predictive simulations for offshore wind will require capabilities for land-based systems with the additional wave/current CFD models, wave-air coupling, and support-structure models (including those for mooring lines). As described earlier, floating-platform simulations further exemplify the need for moving overset meshes, as the whole turbine will now have six-degree-of-freedom motion because of the platform motions.

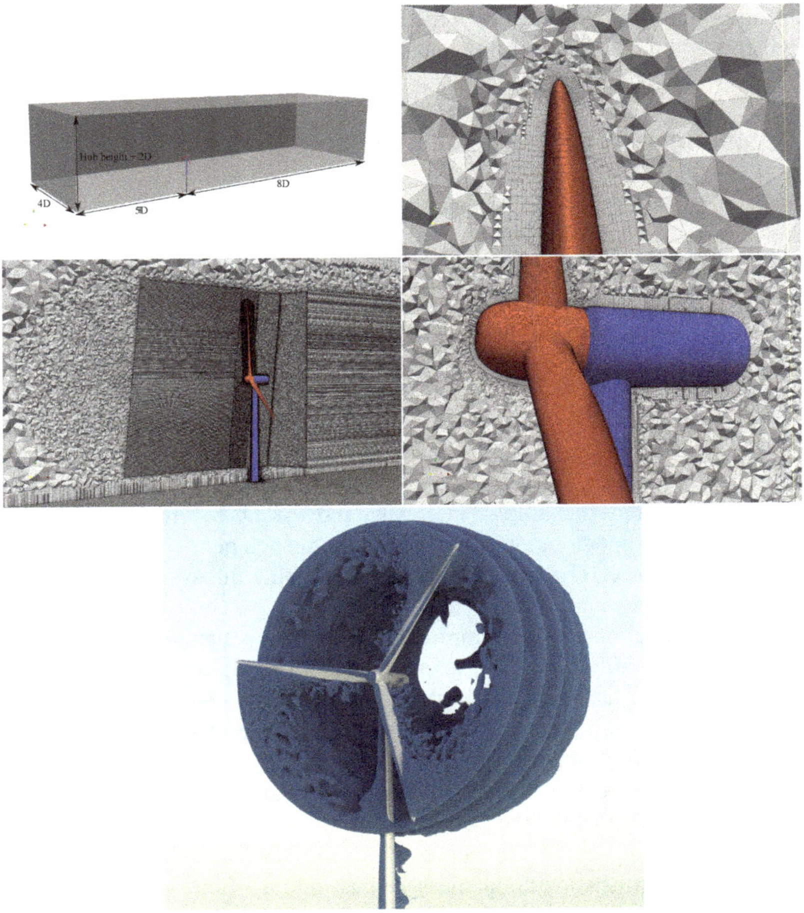

Figure 1.2 *(Top) Schematic of simulation domain and three cut-away views of a body-resolved mesh (176 million nodes) of the NREL 5-MW reference turbine, for which rotor motion is captured with a sliding-mesh interface. (Bottom) Image of the NREL 5-MW turbine simulation result (for a coarser, 25-million node mesh) under uniform inflow of 8 m/s. Shown is the wake after the turbine has completed 2.3 revolutions at 14 s of simulated time. The wake is visualized by showing the contours of velocity magnitude of 5.5 m/s. For details, see [20]*

Overall, the keys to impactful high-fidelity wind energy simulation are accuracy and short time to solution. Accuracy is achieved with appropriate mathematical and numerical-discretization models combined with enough grid resolution in space and time. Short time to solution is obtained by effectively exploiting HPC, minimizing the wall-clock time per time step by employing optimized linear- and non-linear-system solvers, and running simulations with the largest time-step size allowed by accuracy requirements and numerical-stability constraints.

1.3 Challenges at petascale and the need for exascale

As of this writing, computational scientists are living in the era of petascale computing, and the fastest supercomputers are running with maximum operating speed up to about 100 petaFLOPS, where a petaFLOPS is 10^{15} floating-point operations per second. According to the Top500 list [21], the world's fastest super computer in June 2018 was Summit at the US Department of Energy Oak Ridge National Computing Facility, which demonstrated a maximum speed of 122 petaFLOPS on the LINPACK benchmarks. The LINPACK software library is used to measure computation speed in solving large, dense linear systems. Other top-ten systems include Sunway TaihuLight (93 petaFLOPS) at the National Supercomputing Center in Wuxi, China; the AI Bridging Cloud Infrastructure (18.8 petaFLOPS) at the National Institute of Advanced Industrial Science and Technology in Japan; and Piz Daint (19.6 petaFLOPS) at the Swiss National Supercomputer Center. Although such petascale systems provide ample computing power, challenges remain in harnessing full systems, especially for predictive wind energy simulations.

For CFD simulations with an implicit time-domain solver, computation time is dominated by the underlying linear-system solvers. Consider a simulation of a single megawatt-scale wind turbine wherein the fluid domain is discretized with a body-resolved unstructured-grid finite-volume model. A well-resolved model could have a billion gridpoints, and time-step sizes would be on the order of milliseconds [22]. Solution is only feasible with a large cluster-type computer, wherein the degrees of freedom are distributed and solved across a multitude of processing elements (PEs) (i.e. central-processing-unit [CPU] cores) with the communication (data transfer) between PEs accomplished with the message passing interface (MPI). MPI is a standardized protocol for communication between PEs in parallel computing, which is implemented in several libraries including MPICH and OpenMPI.

For a single-turbine simulation, the huge, discrete numerical systems would be 'easy' to solve if the many hundreds of thousands of PEs on a modern supercomputer could be fully utilized. However, the strong-scaling limit impedes full-machine use for CFD computations. In strong scaling, the problem size is fixed (gridpoints required), and the number of PEs is increased. Ideally, doubling the number of PEs would decrease simulation wall-clock time by half. The strong-scaling limit corresponds to the minimum number of gridpoints per PE at which time per time step is minimized; going to fewer gridpoints per PE (i.e. using more of a machine) will show no further reduction in time per time step and may well increase that time. A particular limit is a function of numerical methods, libraries, and computational hardware, especially the communication fabric between nodes.

Good strong-scaling performance is only seen when the parallelizable workload for a given PE is more costly than the communication overhead and other non-parallelizable work [23]. For typical CFD solvers, good strong scaling is seen

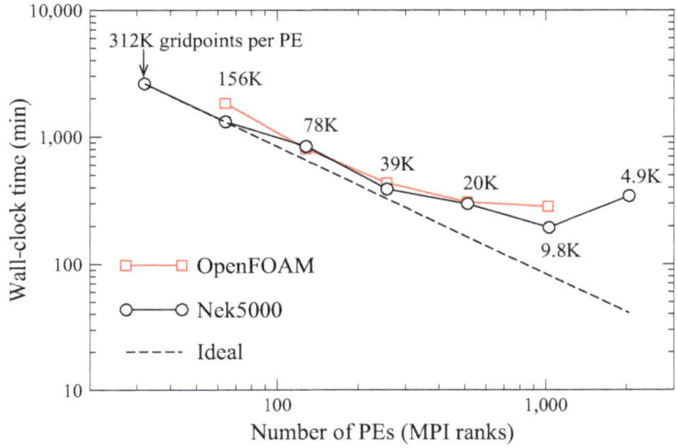

Figure 1.3 Example of strong-scaling results showing wall-clock time as a function of the number of PEs (and MPI ranks) for simulations in OpenFOAM and Nek5000 of the turbulent channel flow problem defined by [27]. Numbers above data denote the number of gridpoints (i.e. pressure DOFs) per PE. Details found in [24]

only down to about 40K gridpoints per PE. Going to fewer model gridpoints per PE (by increasing number of PEs) will show diminishing returns (i.e. parallel efficiency degrades) as the strong-scaling limit is approached. Consider, for example, the strong-scaling results of [24] shown in Figure 1.3, which shows results for OpenFOAM [25] and Nek5000 [26] models of the turbulent channel-flow problem described by Moser *et al.* [27]. OpenFOAM employs low-order finite-volume discretization, whereas Nek5000 employs high-order spectral finite elements. Each model had about 10^7 gridpoints, and timing results were taken for 100 simulated seconds. Both codes showed good strong scaling down to about 40K gridpoints per PE, with diminishing returns at fewer gridpoints per PE, and both models had a strong-scaling limit of about 10K gridpoints per core.

The strong-scaling limit is perhaps the most significant barrier to predictive wind simulations. Let us consider a notional single-turbine simulation with one-billion gridpoints on the Mira supercomputer at the Argonne Leadership Computing Facility. Mira is a 10 petaFLOPS machine with 786,432 PEs, where a PE is one core on a 16-core 1.6 GHz IBM PowerPC A2 processor. Although Mira provides ample computational power, one is challenged to access that full power in CFD. Running at an approximate strong-scaling limit of about 10K gridpoints per PE requires about 100K PEs, which is only about 13 per cent of those available on Mira. In other words, one would only be utilizing a bit more than one petaFLOPS of the ten available. If each time step (of a millisecond) takes a few seconds to solve (an optimistic estimate), simulating 10 min could well take *weeks* of wall clock time. Pushing the strong-scaling limit to fewer gridpoints per PE would directly improve time to solution, enabling researchers to explore longer durations (e.g. atmospheric-change events) and configurations.

Although the strong-scaling limit of CFD solvers presents a challenge to utilize a full petascale-class supercomputer for a single-turbine simulation, there are opportunities on petascale machines for *weak scaling* of wind energy problems. In weak scaling, one increases the size of the problem (e.g. through grid refinement, increased domain size, and/or additional physics) and commensurately increases the number of PEs. Under perfect weak scaling, for example, doubling the domain size as well as the number of PEs would show the same wall-clock time for a given time simulated. Weak scaling faces two of its own challenges. First, CFD solvers rarely exhibit perfect weak scaling. Second, increasing domain size typically increases the required simulated time as the physics must propagate over a larger domain. Third, refining the grid will typically reduce the allowable time-step size, thereby increasing the number of required time steps for a given simulated duration. However, returning to our example, single-turbine simulation on Mira, there is a clear opportunity to weakly scale the problem with the addition of more turbines and/or more resolution.

In addition to 'hero-class' simulations (i.e. capability simulations) of a single parameter/input configuration that use a large fraction (preferably all) of a machine, petascale computing provides an opportunity for *ensembles* of simulations. Given the many uncertainties in wind energy atmospheric flows and the inherent chaotic nature of turbulence, the impact of having simulation results for many conditions cannot be overemphasized. Additional runs with a suite of multi-fidelity models can expose the sensitivities of results of hero-class simulations on input parameters and modelling choices [28].

Although there remain challenges exploiting the full capability of today's petascale systems, the biggest impacts of predictive simulations on wind energy will come with next-generation exascale-class computing. The two obvious opportunity areas are hero-class simulations of a whole wind plant and tightly coupled ensembles of petascale simulations. In regard to hero-class wind plant simulations, possibilities include full-physics modelling (e.g. FSI, hybrid-RANS/LES), complex terrain, extreme resolution, and, possibly, coupling to mesoscale numerical weather prediction. Going beyond land-based wind plants, there is a myriad of new physics to model in considering offshore systems, including wave dynamics, floating-structure dynamics, and air-wave coupling. Such exascale simulations are exciting prospects that will enable new understanding of, for example, wake evolution and impingement on downwind turbines. Finally, there is the possibility of coupling high-fidelity wind plant codes with utility-grid codes for understanding the grid impacts of, e.g. atmospheric events.

1.4 The challenge of exascale computing

Across multiple disciplines, including wind energy, is the acknowledgement that exascale-computing capabilities are necessary to solve the grand-challenge problems facing scientists and engineers today, as well as addressing national security issues. The need for exascale computing is broadly accepted, with the United States, the

European Union, Japan, and China all being committed to establishing exascale-class supercomputers in the next few years. Within the United States, for example, the commitment to exascale computing is perhaps best evidenced by the number of workshops and reports directed at exascale computing, including the DOE Scientific Grand Challenges Workshops that were convened in the 2008–10 time frame [29–36], the Advanced Scientific Computing Advisory Committee report [37] on *The Opportunities and Challenges of Exascale Computing*, the 2014 *Workshop on Exascale Applied Mathematics* [38], and the 2015 *Turbulent Flow Simulation at the Exascale: Opportunities and Challenges Workshop* [7].

Exascale-class supercomputers will run the LINPACK benchmarks at 10^{18} FLOPS (an exaFLOPS) or faster. As of October 2017, Japan and China are expected to establish the first exascale systems in 2021, with the United States unveiling an 'initial' exascale system in 2021, and a 'capable' exascale system in 2023 [39] under the Exascale Computing Project. Here, 'capable' indicates a system that operates in the 20–30-MW power envelope, has fault rates of less than one per week (i.e. is resilient), and includes a software stack that supports a broad set of applications [39].

The primary challenge in establishing an exascale-class system is in minimizing system power requirements, and it is those requirements that will dictate the design of next-generation computer architectures. For example, in 2010, when there was a flurry of activity in defining the future challenges and opportunities of exascale computing, an exascale system simply scaled up from state-of-the-art technology would have required more than a gigawatt of power, roughly the output of the Hoover dam [37]. Such a power requirement is infeasible, of course, which motivates the development of new low-power, HPC architectures. Since 2010, there has been much progress in reducing power requirements in supercomputing. This trend is shown in Figure 1.4(a) and (b), which show the overall maximum performance and the efficiency, respectively, for the Top500-list fastest machines for 2010–18. Whereas the fastest supercomputer in 2010 had an efficiency of 0.25 petaFLOPS/MW, the 2018 fastest machine, Summit, has an efficiency of about 14 petaFLOPS/MW.

The increase in the efficiency of the fastest supercomputers has come with a significant change in architecture. With the approaching end of Moore's law, clock-rate improvements of CPUs have stalled. This stall is illustrated in Figure 1.4(c), which shows the clock speed of the CPUs in each of the number-one Top500-list machines in 2010–18. With the stall in CPU clock speed, increased performance is being achieved through massive concurrency, i.e. many more PEs [38]. The trend is moving away from homogeneous-architecture CPU-only compute nodes and towards heterogeneous nodes with CPUs plus accelerators, which are much more power efficient. Two prominent next-generation accelerator architectures are general-purpose graphical-processing units (GPUs) and many-core processors.

Examples of the evolving architecture are provided by the Top500 top-ranking supercomputers. Although the 2010–11 number-one systems had homogeneous nodes (all CPUs), the November 2012 Top500 number-one system, Titan (at the Oak Ridge Leadership Computing Facility), was introduced with 18,688 heterogeneous nodes, each having one 16-core AMD Opteron CPU and one NVIDIA Kepler GPU. The Kepler GPU was designed with a focus on energy efficiency in addition to

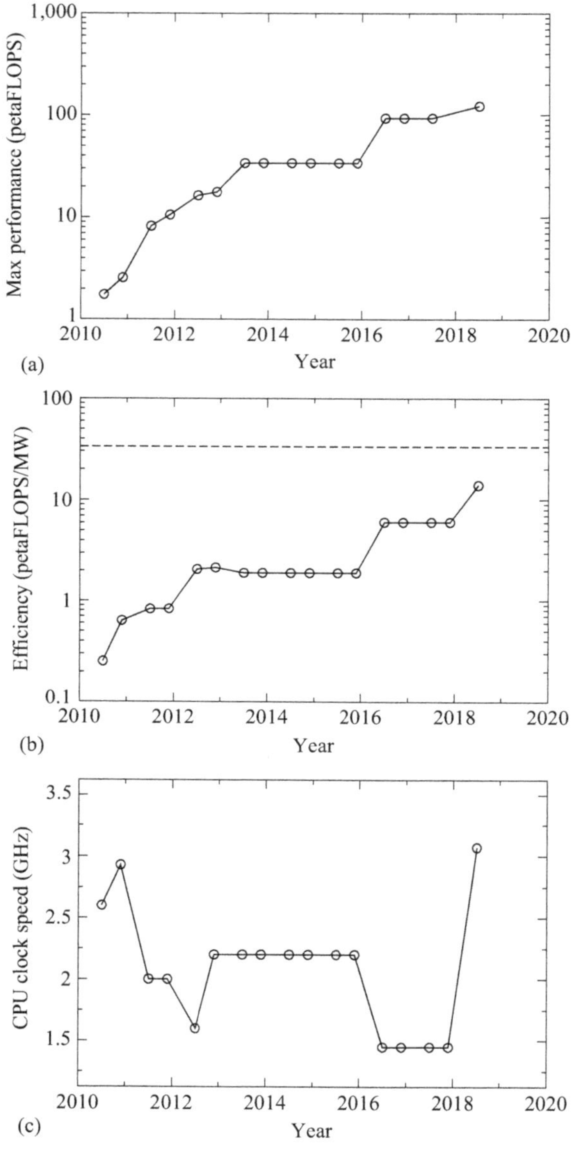

Figure 1.4 (a) Maximum computing performance, (b) efficiency, and (c) CPU clock speed for the fastest supercomputers according to the Top500 list [21] from 2010 to 2018. In (b), the horizontal dashed line indicates 33 petaFLOPS/MW, which is the minimum target for the first US capable exascale-class system [39]

performance and programmability. On the many-core-accelerator front, the June 2013 number-one system, Tianhe-2 (National Super Computer Center in Guangzhou), had 16,000 heterogeneous nodes, each with two 12-core Intel Xeon IvyBridge processors and three Xeon Phi 57-core (i.e. many-core) processors. In both of these systems, each node has one or more multicore conventional CPUs (with, e.g. x86 architecture), which are highly flexible in their ability to handle general tasks, and one or more accelerator chips. The general idea is that while complex operations are handled by the CPU, 'threadable' calculations can be offloaded to the accelerators.

A challenge with accelerator-based systems is that the established programming models typical for CFD codes, i.e. MPI communication on homogeneous-architecture multicore computers, will not translate effectively (or at all). Thus, CFD codes that perform well today under MPI-only communication will likely need substantial retooling to maintain performance on next-generation machines and to better exploit the increase in performance that will be enabled through extreme concurrency of exascale platforms [7].

There is the view that the next-generation of low-power supercomputing will bring us into a 'FLOPS are free' era (see, e.g. [7,38]), for which data movement and memory constraints are the bottlenecks rather than floating-point calculation speed. As described in [38], significant modifications are needed within the 'math stack', e.g. linear-system solvers, to be able to effectively exploit next-generation systems. This is true of the whole solver stack, and new algorithms that have lower communication and synchronization overheads will likely be key. For example, moving to high-order spatial discretization is of interest for next-generation systems as the computational work per degree of freedom is higher than that for low-order methods.

There are a number of additional challenges that will come with exascale wind energy simulations [7,38]. Model sizes will become huge, which presents challenges to creating and moving model files. Given the sheer number of connected hardware components in exascale systems, hardware faults are expected to occur more frequently than on petascale systems. CFD simulations are typically run with a number of 'checkpoints', when the full simulation state is written to disk, enabling simulation restart from that point. These checkpoint files can be enormous, and the input–output system may be challenged in synchronizing, gathering, and storing the full system data. Further, given the challenges of storing the full system state for later analysis, *in situ* data reduction and analysis will be the key to enabling insight into simulation results.

The computational architecture and programming models chosen for the first exascale systems are very important to the success of wind energy calculations. As described in the previous section, today's state-of-the-art CFD codes that are used for wind energy calculations are typically designed to perform well on massively parallel systems through MPI communication. Those CFD codes will need significant retooling to continue being the state-of-the-art on next-generation exascale machines.

Although this chapter is focused on supercomputing-type simulations, in the context of smaller multicore clusters, the days may be numbered for homogeneous CPU-only architecture. It is likely that the new power-efficient heterogeneous systems being developed for exascale-class systems will become the norm in smaller clusters as well. Thus, CFD developers should be aware of the coming changes and consider retooling codes earlier rather than later.

1.5 Concluding remarks

Predictive computational simulation of wind turbines and plants demands the resolution of spatial and temporal scales that span many orders of magnitude. Such simulations are critical to understanding the complex fluid and structural dynamics that directly drive the cost of wind energy. Better understanding of the system, coupled with the ability to predict system response to, e.g. various inflow configurations and control algorithms, will enable scientists and engineers to create the most efficacious pathways to reduce the cost of wind energy.

Single-turbine simulations with many of the features required for predictivity are only recently being realized. However, known and expected deficiencies in the underlying models (e.g. turbulence models) and insufficient space-time grid resolution in those simulations bring into question their accuracy and predictivity, especially for complex transient phenomena like boundary-layer separation and wake formation. Modelling deficiencies aside, the resolution requirements for predictivity will require petascale-class supercomputing.

Currently, strong-scaling performance limits the effective use of computational resources on modern architectures and impacts our ability to accomplish highly resolved simulations with practical wall-clock times. As described above, a well-resolved simulation can take weeks until statistically meaningful results are attained simply because the system cannot be fully exploited and the wall-clock time per time step is too great. The computational power of next-generation exascale-class supercomputers may provide the resources necessary to unlock predictive simulations of the dynamics of wind turbines and plants. However, the strong-scaling limit continues to present an obstacle to utilizing a significant portion of a petascale supercomputer, which, without significant preparation, could become worse with exascale-class computing. Complicating further is the knowledge that exascale machines will have greatly different, but power-efficient, architecture that may well degrade strong-scaling performance. These facts point to much needed effort in pushing the strong-scaling limit on existing and future architectures.

Apart from the capability-type ‘hero runs’ of models with extreme resolution in space and time, petascale and exascale computers provide the opportunity for scientific discovery through capacity-type simulations, where many simulations of lower fidelity but computationally efficient models are run. Capacity-type simulations are necessary for parameter sensitivity studies, design optimization, and uncertainty quantification.

The demand for HFM development and HPC resources for improved predictive modelling capability for wind energy will continue to grow as industry and government seek solutions to (1) mitigate risk and uncertainty of technology development at multimegawatt-turbine scales, (2) address new requirements driving renewable energy deployment and high-penetration scenarios, (3) understand the impacts of multiple gigawatt-scale plant deployments on microclimatology and grid architectures for both land-based and offshore installation, and (4) evolve the systems and processes needed for the dispatch of renewable energy generation on a national scale supporting a mix of power-generation technologies. When considering the cost of developing a new offshore multimegawatt commercial wind turbine product line or the financial risk of a large offshore wind plant deployment, predictive HPC and HFM simulations that incorporate integrated design, cost analysis, and O&M predictions will become an essential part of the early design and development process.

Acknowledgements

This work was authored in part by Alliance for Sustainable Energy, LLC, the manager and operator of the National Renewable Energy Laboratory for the U.S. Department of Energy under Contract No. DE-AC36-08GO28308. M.A. Sprague's contribution to this work was supported by the Exascale Computing Project (17-SC-20-SC), a collaborative effort of two U.S. Department of Energy organizations (Office of Science and the National Nuclear Security Administration) responsible for the planning and preparation of a capable exascale ecosystem, including software, applications, hardware, advanced system engineering, and early testbed platforms, in support of the nation's exascale computing imperative. M.C. Robinson is a National Renewable Energy Laboratory (NREL) senior technology advisor to the U.S. Department of Energy's Wind Energy Technologies Office (DOE/WETO) and Chief Engineer of the Atmosphere to Electrons Program sponsoring the HPC and HFM program activities in wind energy sciences. Funding for M.C. Robinson was provided by the U.S. Department of Energy Office of Energy Efficiency and Renewable Energy Wind Energy Technologies Office. The views expressed in the article do not necessarily represent the views of the DOE or the U.S. Government. The U.S. Government retains and the publisher, by accepting the article for publication, acknowledges that the U.S. Government retains a nonexclusive, paid-up, irrevocable, worldwide license to publish or reproduce the published form of this work, or allow others to do so, for U.S. Government purposes. The authors thank Matt Churchfield, Ray Grout, Shreyas Ananthan, and Ganesh Vijayakumar for useful recommendations.

References

[1] University of Virginia. 50 MW Segmented Ultralight Morphing Rotors for Wind Energy. Charlottesville, VA: University of Virginia; 2015. U.S. DOE ARPA-E Award # DE-AR0000667.

[2] Wiser R, Bolinger M. 2017 Wind Technologies Market Report. U.S. Department of Energy, Office of Energy Efficiency and Renewable Energy, Wind Energy Technologies Office; 2018. DOE/EE-1798. Available from: https://www.energy.gov/sites/prod/files/2018/08/f54/2017_wind_technologies_market_report_8.15.18.v2.pdf [cited 2018 Dec 21].

[3] Wind Vision: A New Era for Wind Power in the United States. U.S. Department of Energy; 2015. DOE/GO-102015-4557. Available from: https://www.energy.gov/eere/wind/maps/wind-vision [cited 2018 Nov 19].
[4] Dykes K, Hand M, Stehly T, *et al.* Enabling the SMART Wind Power Plant of the Future Through Science-Based Innovation. National Renewable Energy Laboratory; 2017. NREL/TP-5000-68123. Available from: https://www.nrel.gov/docs/fy17osti/68123.pdf [cited 2018 Nov 19].
[5] Larsen HH, Sønderberg Petersen L, editors. DTU International Energy Report 2014: Wind Energy Drivers and Barriers for Higher Shares of Wind in the Global Power Generation Mix. Technical University of Denmark (DTU); 2014.
[6] Churchfield M, Moriarty P. Modeling the flow through full wind plants: Wakes and wake interactions. In: Veers P, editor. Wind Energy Modelling and Simulation: Atmosphere and Plant. Institution of Engineering and Technology; 2018.
[7] Sprague MA, Boldyrev S, Fischer P, *et al.* Turbulent Flow Simulation at the Exascale: Opportunities and Challenges Workshop. U.S. Department of Energy, Office of Science, Advanced Scientific Computing Research; 2017. Published as Tech. Report NREL/TP-2C00-67648 by the National Renewable Energy Laboratory. Available from: https://www.nrel.gov/docs/fy17osti/67648.pdf [cited 2018 Nov 19].
[8] Fleming P, Annoni J, Churchfield M, *et al.* A simulation study demonstrating the importance of large-scale trailing vortices in wake steering. Wind Energy Sciences. 2018;3:243–255.
[9] Hammond SW, Sprague MA, Womble D, *et al.* A2e High Fidelity Modeling: Strategic Planning Meetings. National Renewable Energy Laboratory; 2015. NREL/TP-2C00-64697. Available from: https://www.nrel.gov/docs/fy16osti/64697.pdf [cited 2018 Nov 19].
[10] Guntur S, Jonkman J, Sievers R, *et al.* A validation and code-to-code verification of FAST for a megawatt-scale wind turbine with aeroelastically tailored blades. Wind Energy Science. 2017;2:443–468.
[11] Wang Q, Sprague MA, Jonkman J, *et al.* BeamDyn: An efficient high-fidelity wind turbine blade solver in the FAST modular framework. Wind Energy. 2017;20:1439–1462.
[12] Anderson JD. Fundamentals of Aerodynamics. 4th ed. New York: McGraw-Hill; 2007.
[13] Turkel E, Radespiel R, Kroll N. Assessment of preconditioning methods for multidimensional aerodynamics. Computers and Fluids. 1997;26:613–634.
[14] Guillard H, Viozat C. On the behaviour of upwind schemes in the low Mach number limit. Computers and Fluids. 1999;28:63–86.
[15] Turkel E. Preconditioning techniques in computational fluid dynamics. Annual Review of Fluid Mechanics. 1999;31:385–416.
[16] Vijayakumar G, Brasseur J. Blade-resolved modeling with fluid–structure interaction. In: Veers P, editor. Wind Energy Modelling and Simulation: Atmosphere and Plant. Institution of Engineering and Technology; 2019.
[17] Mittal R, Iaccarino G. Immersed boundary methods. Annual Review of Fluid Mechanics. 2005;37:239–261.

[18] Domino SP. Design-order, non-conformal low-Mach fluid algorithms using a hybrid CVFEM/DG approach. Journal of Computational Physics. 2018;359:331–351.

[19] Jonkman J, Butterfield S, Musial W, *et al.* Definition of a 5-MW Reference Wind Turbine for Offshore System Development. Golden, CO: National Renewable Energy Laboratory; 2009. NREL/TP-500-38060. Available from: https://www.nrel.gov/docs/fy09osti/38060.pdf [cited 2018 Nov 30].

[20] Lawson M, Melvin J, Ananthan S, *et al.* Blade-Resolved, Single-Turbine Simulations Under Atmospheric Flow. Golden, CO: National Renewable Energy Laboratory; 2018. NREL/TP-5000-72760. Available from: https://www.nrel.gov/docs/fy18osti/72760.

[21] Top500 List. Available from: https://www.top500.org/ [cited 2018 Oct 3].

[22] Churchfield MJ, Sprague MA. Exascale computing and wind energy. In: DOE ASCR Turbulent Flow Simulation at the Exascale: Opportunities and Challenges Workshop; 2015. White Paper.

[23] Fischer PF, Heisey K, Min M. Scaling limits for PDE-based simulation. In: Proceedings of the 22nd AIAA Computational Fluid Dynamics Conference, AIAA AVIATION Forum. 2015-3049; 2015. Available from: https://arc.aiaa.org/doi/10.2514/6.2015-3049 [cited 2018 Nov 19].

[24] Sprague MA, Churchfield MJ, Purkayastha A, *et al.* A comparison of Nek5000 and OpenFOAM for DNS of turbulent channel flow. In: Presentation at the Nek5000 Users Meeting, Argonne National Laboratory; 2010. Available from: https://www.mcs.anl.gov/~fischer/nek5000/sprague_nek5000_dec2010.pdf [cited 2018 Nov 19].

[25] OpenFOAM. Available from: https://openfoam.org/ [cited 2018 Nov 2].

[26] Nek5000, A Spectral Element code for CFD. Available from: https://nek5000.mcs.anl.gov/ [cited 2018 Nov 2].

[27] Moser RD, Kim J, Mansour NN. Direct numerical simulation of turbulent channel flow up to Re_tau=590. Physics of Fluids. 1999;11:943–945.

[28] Peherstorfer B, Wilcox K, Gunzburger M. Survey of multifidelity methods in uncertainty propagation, inference, and optimization. SIAM Review. 2018; 60:550–591.

[29] U.S. Department of Energy. Scientific Grand Challenges: Challenges in Climate Change Science and the Role of Computing at the Extreme Scale. U.S. Department of Energy Workshop Report; 2008. Published as Tech. Report PNNL-18362 by the Pacific Northwest National Laboratory. Available from: https://science.osti.gov/~/media/ascr/pdf/program-documents/docs/Climate_report.pdf [cited 2019 Aug 2].

[30] U.S. Department of Energy, Scientific Grand Challenges: Challenges for Understanding the Quantum Universe and the Role of Computing at the Extreme Scale. U.S. Department of Energy Workshop Report; 2008. Published as Tech. Report PNNL-18775 by the Pacific Northwest National Laboratory. Available from: https://science.osti.gov/~/media/ascr/pdf/program-documents/docs/Hep_report.pdf [cited 2019 Aug 2].

[31] U.S. Department of Energy. Scientific Grand Challenges: Forefront Questions in Nuclear Science and the Role of Computing at the Extreme Scale.

U.S. Department of Energy Workshop Report; 2009. Published as Tech. Report PNNL-18739 by the Pacific Northwest National Laboratory. Available from: https://science.osti.gov/~/media/ascr/pdf/program-documents/docs/Np_report.pdf [cited 2019 Aug 2].

[32] U.S. Department of Energy. Science Based Nuclear Energy Systems Enabled by Advanced Modeling and Simulation at the Extreme Scale. U.S. Department of Energy Workshop Report; 2009. Available from: https://science.osti.gov/~/media/ascr/pdf/program-documents/docs/Sc_nework_shop_report.pdf [cited 2019 Aug 2].

[33] U.S. Department of Energy. Scientific Grand Challenges: Fusion Energy Sciences and the Role of Computing at the Extreme Scale. U.S. Department of Energy Workshop Report; 2009. Published as Tech. Report PNNL-19404 by the Pacific Northwest National Laboratory. Available from: https://science.osti.gov/~/media/ascr/pdf/program-documents/docs/Fusion_report.pdf [cited 2019 Aug 2].

[34] U.S. Department of Energy. Scientific Grand Challenges: Opportunities in Biology at the Extreme Scale of Computing. U.S. Department of Energy Workshop Report; 2009. Available from: https://science.osti.gov/~/media/ascr/pdf/program-documents/docs/Biology_report.pdf [cited 2019 Aug 2].

[35] U.S. Department of Energy. Scientific Grand Challenges for National Security: The Role of Computing at the Extreme Scale. U.S. Department of Energy Workshop Report; 2009. Published as Tech. Report PNNL-19379 by the Pacific Northwest National Laboratory. Available from: https://science.osti.gov/~/media/ascr/pdf/program-documents/docs/NNSA_grand_challenges_report.pdf [cited 2019 Aug 2].

[36] U.S. Department of Energy. Scientific Grand Challenges: Discovery in Basic Energy Sciences: The Role of Computing at the Extreme Scale. U.S. Department of Energy Workshop Report; 2009. Published as Tech. Report PNNL-19902 by the Pacific Northwest National Laboratory. Available from: https://science.osti.gov/~/media/ascr/pdf/program-documents/docs/Bes_exascale_report.pdf [cited 2019 Aug 2].

[37] U.S. Department of Energy. ASCAC Subcommittee Report: The Opportunities and Challenges of Exascale Computing; 2010. Available from: https://science.osti.gov/-/media/ascr/ascac/pdf/reports/Exascale_subcommittee_report.pdf [cited 2019 Aug 2].

[38] Dongarra J, Hittinger J, Bell J, *et al.* Applied Mathematics Research for Exascale Computing. US Department of Energy; 2014. Available from: http://science.energy.gov/~/media/ascr/pdf/research/am/docs/EMWGreport.pdf [cited 2018 Nov 19].

[39] Messina P. Update on the Exascale Computing Project (ECP). Presentation at the HPC User Forum, Santa Fe, New Mexico, 18 April 2017. Available from: https://science.osti.gov/~/media/ascr/pdf/research/am/docs/EMWGreport.pdf [cited 2017 Nov 26].

Chapter 2
Blade-resolved modeling with fluid–structure interaction

Ganesh Vijayakumar[1] *and James G. Brasseur*[2]

Blade-resolved simulations are a high-fidelity modeling tool that helps understand the fundamental flow physics that create loads on wind turbines. Along with fluid–structure interaction (FSI), they provide a foundation to improve the state-of-the-art aerodynamic models of modern wind turbines with large flexible blades and thus the design of the next generation of wind turbines. One of the most important parts of wind-turbine simulation is modeling the conversion of kinetic energy in the wind into mechanical energy using principles of aerodynamics. The hierarchies of aerodynamic models for wind turbines in increasing the order of complexity, cost, and fidelity are blade element momentum (BEM) theory (BEMT), vortex methods, actuator disc/line methods, and blade-resolved modeling. As discussed in Chapter 1 (Volume 2), BEMT and vortex methods make various approximations regarding the nature of flow around the blade sections and the turbine wake. Specifically, they use a table lookup of pre-calculated/measured lift and drag polars for the two-dimensional (2D) flow around the airfoils of blade sections. The actuator methods relax many of the approximations of the BEMT/vortex methods as discussed in Chapter 6 (Volume 1) but still rely on the empirical 2D lift and drag polars at blade sections used in BEMT. The blade-resolved modeling approach relaxes all the assumptions of the lower fidelity models by resolving the boundary layers around the geometry of the turbine surface. The resolution of the boundary layers on the blades at high Reynolds numbers using computational fluid dynamics (CFD) and turbulence modeling comes at a significant increase in the computational cost compared to lower fidelity models. Figure 2.1 shows a comparison of the wake predicted by the blade-resolved and actuator models.

Blade-resolved modeling, in different forms and at various resolutions, is currently used for verification of a few design load cases of the result of multidisciplinary optimization process of wind-turbine design (see Chapter 3, Volume 2); the high-current computational cost of blade-resolved simulations prevents further integration into the design process. It is also used to tune/correct the lower fidelity aerodynamic models for specific applications: blade-root section design and corrections to

[1]National Renewable Energy Laboratory, Golden, CO, USA
[2]Aerospace Engineering Sciences, University of Colorado, Boulder, CO, USA

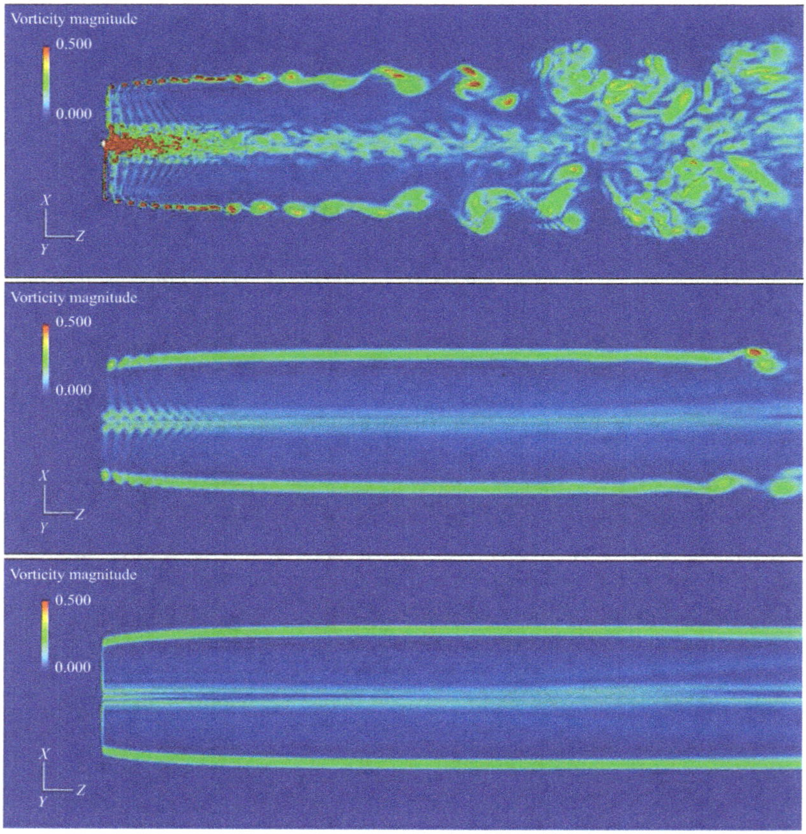

Figure 2.1 Comparison of wake of the NREL 5-MW turbine, using contours of vorticity magnitude, in uniform inflow predicted by blade-resolved (top), actuator line (middle), and actuator disk (bottom) simulations. © 2015. Reprinted, with permission, from [1]

induction calculations at high induction [2], trailing-edge noise [3], vortex-generator sizing and placement [4], design of other passive aerodynamic devices on the blade, and analysis of the effect of blade erosion [5–7]. Despite the increased computational cost, the following several trends in wind-turbine design increase the need for blade-resolved modeling in the future.

Focus on cost of energy: The focus on the levelized cost of energy (LCOE) drives the need for lower safety factors that arise from the inaccuracies in modeling during the design process. Empirical correction factors are currently used in lower fidelity aerodynamic models to correct the lift and drag polars of airfoils to account for the three-dimensional (3D) flow effects. Blade-resolved modeling allows for a physics-based prediction of rotational augmentation and stall delay

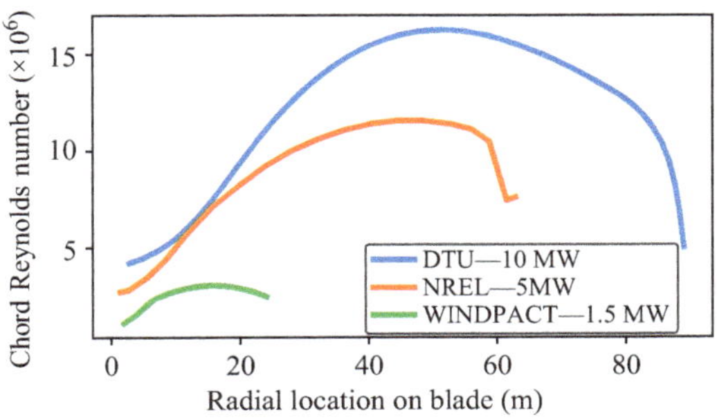

Figure 2.2 Variation of chord Reynolds number over the blade length at rated operating condition for three commercially relevant wind turbines: DTU 10 MW [8], NREL 5 MW [9], and WINDPACT 1.5 MW [10]

near the blade root [11], especially at off-design conditions. Blade-resolved simulations are also necessary to improve the calculation of induction in low-fidelity models at high induction and yawed conditions [12]. This will help improve the accuracy of aerodynamic models over the entire operating envelope of wind turbines to reduce the LCOE.

Increase in rotor size: The current trend of continuous increase in rotor size of wind turbines pushes the operating Reynolds number outside the range of validated aerodynamic models [12]. Figure 2.2 shows the variation of chord Reynolds number over the blade length of three commercially representative wind turbines, DTU 10 MW [8], NREL 5 MW [9], and WINDPACT 1.5 MW [10]. Measurement of airfoil lift and drag polars are very expensive and increasingly hard to obtain at the high Reynolds of the next-generation wind turbines [13]. Blade-resolved modeling will be required in the future to create a hierarchy of validated aerodynamic models [14].

Atmospheric turbulence: Modern wind turbines increasingly interact with a larger portion of the atmospheric boundary layer (ABL). The large range of length and time scales in atmospheric turbulence force load fluctuations that fatigue the blades and the drive train of wind turbines. The turbulence in the atmosphere is very different compared to turbulence in the wind tunnels (Chapter 5, Volume 1). Blade-resolved simulations will help improve the understanding of the interaction between atmospheric turbulence and blade aerodynamics. This will help drive the next round of reductions in LCOE.

FSI: Modern wind turbines increasingly use aeroelastically tailored lighter and more flexible blades to avoid the increase in weight due to upscaling to larger rotor sizes. Such blades use large deflections to alleviate load fluctuations and fatigue even under normal operating conditions. Blade-resolved modeling will be necessary to create the next generation of wind-turbine aeroelastic models to accurately simulate unsteady aerodynamic effects caused by bend-twist coupling and dynamic stall. Hybrid Reynolds-averaged Navier Stokes (RANS)/large Eddy simulations (LES) will be necessary to capture these effects along with a significant increase in spatiotemporal resolution and cost compared to typical RANS-based blade-resolved simulations. RANS, LES, and hybrid RANS/LES turbulence modeling strategies are described in more detail in Section 2.2.3.

Blade-resolved modeling of wind turbines along with FSI is under active development thanks to several well-funded projects like UPWIND [15], INNWIND.EU [16], AVATAR [12], Atmosphere to Electrons initiative [17], and the Exascale Computing Project [18]. In this chapter, we describe the numerical and modeling aspects of the state-of-the-art blade-resolved simulations of wind turbines. First, Section 2.1 describes the extraordinary range of length and time scales relevant to blade-resolved CFD of wind turbines operating in the atmosphere. Section 2.2 discusses the essential numerical and modeling elements in blade-resolved simulation of wind turbines. We discuss some of the practical issues in running blade-resolved wind-turbine simulations in Section 2.3. Finally, we conclude with a discussion of the impact of blade-resolved modeling on the wind-turbine industry so far along with an outlook on the challenges for the future in Section 2.4.

2.1 The extraordinary range of length and time scales relevant to wind-turbine operation

Utility-scale wind turbines in the field experience flow features across a range of length scales $\sim 10^8$ orders of magnitude apart, from the energy-containing atmospheric turbulence eddies to the rotor disk to the dynamics of the thin boundary layers on the blade. Conceptualization of blade-boundary-layer-resolved CFD at the highest levels of fidelity requires the modeled dynamical system—the physics-based submodels, numerical algorithms, and design of the grid—to capture all functionally relevant responses to flow features in the atmosphere and around the wind turbine and wind farm over an extraordinary range of length and time scales. In the following, we describe in more detail the range of spatial and temporal scales in the 3D nonsteady wind velocity field that force wind turbines individually and within wind farms. We begin with a description of the atmospheric "micro scales," the turbulence eddies associated with the ABL, the lowest 1,000–2,000 m of the atmosphere. We then describe the response of the rotor and blade boundary layer scales in response to spatiotemporal forcing by the atmospheric microscale motions. We end with a brief discussion of the role of "mesoscale" atmospheric motions, the predominantly horizontal winds associated with weather in the atmospheric free troposphere. Not considered here are additional scales associated with forcing by eddying motions in the atmosphere induced by nonplanar surface topography (hills and mountains), or scales associated with wave and platform motion in offshore operations of floating wind turbines.

2.1.1 Impacts of atmospheric "microscale" turbulence

Utility-scale wind turbines in the field experience the continual passage of energy-containing turbulence eddies through the disk that induce nonsteady responses in blade and rotor loads. The origin of the turbulence on operating utility-scale wind turbines is the lowest 10%–20% of the ABL, which itself is in the lower 10%–20% of the atmospheric troposphere; the troposphere is approximately the lowest 15 km of the atmosphere, the region where the weather events experienced by wind turbines reside. The wind-turbine rotor responds to variabilities in wind velocity along the blades and in time—the variabilities are created by the highly turbulent flow field during the daytime and the intermittency of turbulence during nighttime. These spatiotemporal variabilities can be sufficiently strong to move local blade aerodynamics to parameter spaces outside the blade-design envelope. At the atmospheric level, the 3D energy-dominant coherent structure of the velocity deviations from the mean atmospheric winds vary a great deal in relationship to

- fundamental differences between the unstable daytime versus stable nighttime boundary layer;
- the diurnal changes in the daytime in both turbulence eddy size and in the convective- versus shear-driven structure of the energy-dominant eddies as solar heating increases from sunrise, to peak surface heat flux midday, to sunset;
- the highly nonstationary collapse of the boundary layer in the transition from the daytime to nighttime stably stratified ABL with potential formation of coherent structures, such as low-level jets;
- other nighttime nonequilibrium transitions as turbulence is progressively suppressed by increasing stable stratification; and
- the modulation of the mean winds and turbulence eddy structure at the "microscale" (i.e., within the ABL) during the passage of weather events at the "mesoscale" (i.e., above the ABL in the free troposphere).

The canonical daytime ABL is capped by a local stable inversion layer that creates a "soft lid" to the highly 3D, convectively driven eddying motions below. This "capping inversion" typically initiates after sunrise at 200–300 m, then it rises during the day to reach peak apogee in the early afternoon until its collapse during sunset late in the day. The capping inversion depth is the agreed-upon measure of the boundary layer depth; therefore, the largest vertical extent of atmospheric eddies, driven by buoyancy, is of order 1,000–2,000 m, while the horizontal scale can be extended by mean shear [19,20].

Over land, the period of wind–turbine interaction that can be described as quasi-stationary occurs in the early afternoon, typically for a few hours after peak solar apogee. In this quasi-equilibrium state, the height of the atmospheric surface layer (ASL), the lowest part of the ABL, is 15%–20% of the boundary layer height. In the transition from the early morning ABL shortly after sunrise to peak solar apogee after noon, the ABL is out of equilibrium as the boundary layer grows from a couple hundred meters to a peak capping inversion height of 1,000–2,000 m. During this period, the surface layer forms and grows from 50–100 to 150–400 m at peak solar apogee. The transverse dimension of the energy-containing eddies that interact with the wind-turbine scale on the distance from the ground in the ASL. The streamwise dimensions of the energy-containing eddies, however, are stretched by mean shear to lengths much larger than the rotor disk diameter. The rotor disk of current 2–3-MW

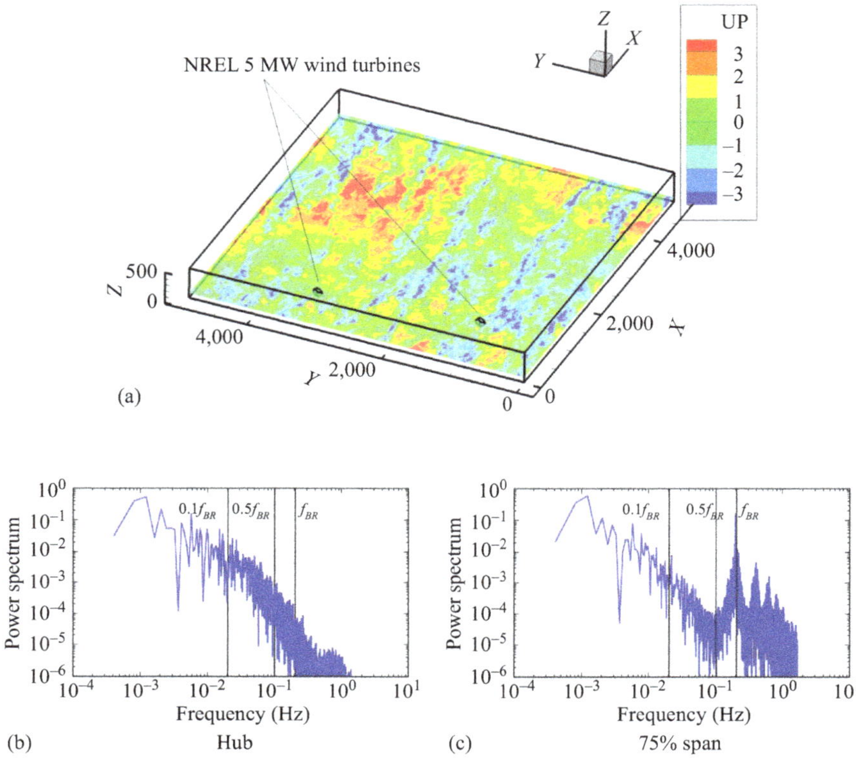

Figure 2.3 (a) Isocontours of streamwise velocity fluctuations at hub height (90 m) in an equilibrium daytime ABL along with two NREL 5-MW wind-turbine disks drawn to scale. (b) and (c) Power spectrum of the streamwise velocity sampled at two different points on a hypothetical NREL 5-MW turbine in a daytime ABL. The three black lines are at 0.1f_{BR}, 0.5f_{BR}, and f_{BR}, where f_{BR} is the blade rotation frequency. Figures from Vijayakumar [21]

onshore wind turbines has rotor diameters of ~100 m and hub heights of ~100 m. Thus, the largest spatial and temporal scales of turbulence eddy passage through the rotor disk grow through the morning until the surface layer fully covers the rotor disk midday.

Figure 2.3(a) shows contours of streamwise horizontal velocity fluctuations at the hub height of the NREL 5-MW turbine in a daytime ABL generated using LES; two NREL 5-MW turbines are shown to scale. The characteristic energy-dominant eddy size is roughly 40–140 m from the lower to upper (LU) margins of the rotor disk; that is, of order blade length to rotor disk diameter. The blade sectional chord lengths, however, are typically in the range of 2–4 m, 1–2 orders of magnitude below the energy-dominant eddy size in the afternoon ASL. Figure 2.3(b) and (c) shows

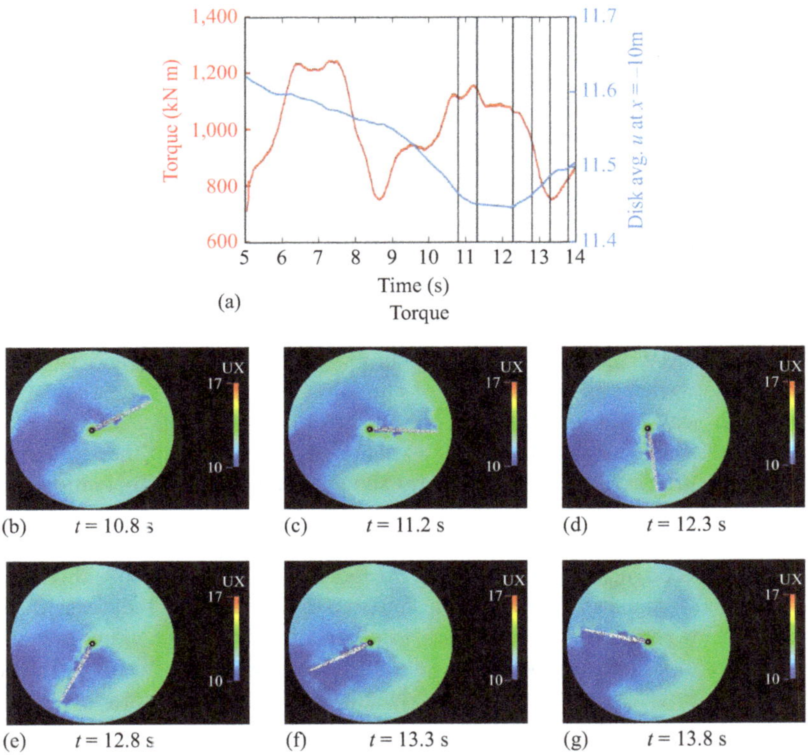

Figure 2.4 (a) Correlation of the downward ramp in torque and its recovery as the blade cuts through a low-speed streak from t = 10.8–13.8 s. Parts (b)–(g) show contours of the streamwise velocity at different times on a plane 2 m in front of the turbine. Data obtained from blade-resolved hybrid RANS/LES simulation of a single blade of the NREL 5-MW turbine in the atmosphere [28]

the spectrum of the streamwise velocity sampled at a point at hub height and at a point rotating at the 75% span location on a hypothetical NREL 5-MW wind turbine. Figure 2.3(b) shows that the turbulent energy in the ABL near the blade rotation timescales is four orders of magnitude smaller than at the integral scales. The physical rotation of the turbine through this turbulence generates significant energy at the rotation frequency and its harmonics—comparable to the energy level in the integral scales of the ABL. This energy experienced by the outer sections of the blade is likely to generate the dominant load fluctuations on wind turbines in the field.

Several authors have commented on the primary mechanisms underlying nonsteady load fluctuations induced by atmospheric turbulence in the daytime ABL, including the question, "what scales in atmospheric turbulence are the strongest contributors to nonsteady load fluctuations on a wind turbine in the field?"

Wächter *et al.* [22] show the non-Gaussian probability density function of the increments in the velocity and the loads separated by certain timescales; they claim that the similarity in the probability density functions between the two to suggest that the energy in the small scales of the ABL are responsible for the large load fluctuations. Berg *et al.* [23] show that the non-Gaussianity in the ABL turbulence at small scales compared to the integral scales has little effect on wind-turbine load fluctuations. Rai *et al.* [24] and Sim *et al.* [25] analyze the same issue through the determination of the required numerical resolution of ABL turbulence to capture wind-turbine loads but arrive at opposite conclusions. Lavely *et al.* [26] and Nandi *et al.* [27] use high-resolution LES of the ABL with actuator-line representations of utility-scale rotor blades, and Vijayakumar *et al.* [28], with blade-boundary-layer-resolved LES of a single NREL 5-MW blade rotating in the atmosphere, draw a different, more complex, conclusion. In addition to the obvious one/three-per-rev response, they uncover large peak-to-peak variations in blade and rotor loadings at two disparate timescales from two different, but related, sources associated with passage of the surface-layer integral-scale turbulence eddies. Figure 2.4 shows that the largest fluctuations in the torque occur at timescales less than the rotation timescale as the blade "cuts" through the integral-scale eddies in the atmosphere. These large, ramp-like peak-to-peak variations are an aerodynamic load response to individual blades passing through the internal spatial structure of atmospheric integral-scale eddies as they pass through the wind-turbine rotor disk. The passage of individual atmospheric eddies creates the second source of large peak-to-peak variations in loadings experienced by the wind-turbine blades and rotor, occurring at an advective timescale on the order of 30–60 s, two orders of magnitude longer than the small ramp-like events. The roles of these two large variability-inducing processes in wind-turbine function remain a current area of study.

Toward the end of the day, as the sun sets, solar heating turns off and surface heat flux reverses direction from heating to cooling, buoyancy production of turbulence rapidly changes to turbulence destruction near the surface, and a highly nonstationary collapse of the boundary layer takes place as the ABL transitions to its nighttime stability stratified state. Not only does ABL turbulence structure rapidly change along with the turbulence length and time scales, this nonequilibrium collapse is often accompanied by the formation of internal flow structures—some of which can impact wind turbines. For example, a common occurrence in the transition from unstable daytime to stable nighttime ABL in the central plains of the United States is the formation of "low-level jets," high-speed flow localized in vertically near the ground that often occurs within a couple hundred meters from the surface, impacting a wind-turbine rotor disk [29]. The relative fluctuation levels decrease during the evening because of stable stratification and turbulence destruction at the surface, creating a thinner internal boundary layer and a transition to a more shear-dominated turbulence with a 3D turbulence structure suppressed by increasing stable stratification, in comparison to the daytime ABL. The ABL just before sunrise, for example, is often thin, highly stratified, and quasi-laminar.

2.1.2 The rotor and blade-boundary-layer response length and time scales

The boundary layers on the blade surfaces contain important flow physics that generate the loads on the turbine. They respond to the turbulence in the atmosphere modified by the axial and angular induction upstream of the turbine and the turbine geometry and control actions. These boundary layers occur at peak sectional chord Reynolds numbers of $O(10^7)$ on modern utility-scale wind turbines, as shown in Figure 2.2. Figure 2.5 shows an estimate of the four primary length scales in the flow around commercial-scale wind turbines: the rotor diameter, D, the average blade chord, c, and the characteristic boundary layer thickness, δ, and viscous length scale, $l_\nu = \nu/u^*$, where ν is the kinematic viscosity of air and u^* is an average friction velocity over the blade surface. Similarly, Figure 2.6 shows an estimate of four primary timescales in the flow around turbines operating in uniform inflow: the eddy passage timescale, D/V_{rated}, where V_{rated} is the rated wind speed of the turbine and atmospheric eddies of order the rotor diameter considered, the rotation timescale, $1/\Omega_{rated}$, where Ω_{rated} is the rotation speed of the turbine, the local chord passage timescale, c/V_{loc}, where V_{loc} is the resultant local velocity at any blade section, and the boundary layer viscous scale, $\tau_\nu = l_\nu/u^*$. Figures 2.5 and 2.6 show that both the length and time scales of the flow in the blade boundary layers are separated from the rotor-wide scales by more than six orders of magnitude. It is for this reason that direct numerical simulation (DNS), a simulation that resolves all the turbulent length and time scales down to the Kolmogorov scale, is impractical—the requirements for the resolution of the extremely small length and time scales are very high. A RANS formulation for the surface boundary layers is computationally feasible because it resolves only the gradients of the ensemble-averaged quantities.

The energy in the different scales of atmospheric turbulence in the ASL forces the blade boundary layers to respond through fluctuations in the local angle of attack and velocity magnitude. The reduced frequency, $\omega = fc/2V_{loc}$, determines the type of response, where f is the forcing frequency and $2V_{loc}/c$ is half the chord passage timescale. When the reduced frequency is below ~0.05, the forcing is generally slow enough for the boundary layer to respond in a quasi-steady manner. According to Leishman [30], unsteady aerodynamic effects become important when the reduced frequency exceeds ~0.05 and are strong above 0.2. The streamwise and transverse integral scales of turbulence in the ASL force fundamentally different types of boundary-layer responses.

As mentioned in the previous section, the streamwise integral scales in the ASL are several times larger than the rotor disk with correspondingly large timescales. Nandi *et al.* [27] analyzed the data collected in the field on a GE 1.5-MW turbine in the daytime ABL. They showed that the longest timescale of blade response, associated with the passage of energy-dominant eddies, is ~30–60 s. Lavely *et al.* [26] showed similar results through an analysis of the daytime ABL using LES. These results are consistent with the findings of Wächter *et al.* [22], who observed the largest increments in streamwise velocity, $u(x, t + \Delta t) - u(x, t)$, at similar timescales. These

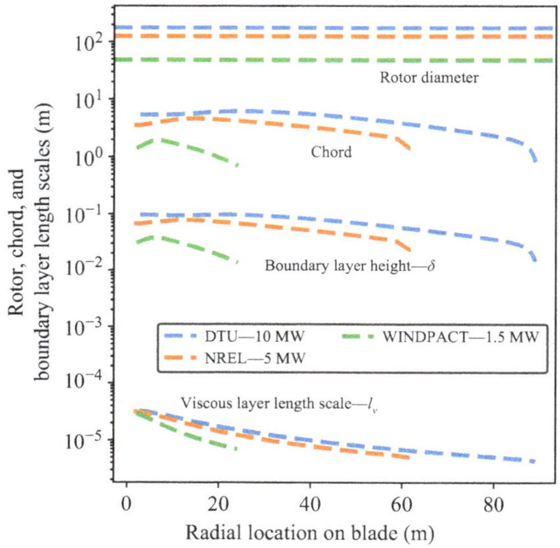

Figure 2.5 Variation of rotor, chord, and boundary-layer length scales over the blade length at rated operating condition for three commercially relevant wind turbines: DTU 10 MW [8], NREL 5 MW [9], and WINDPACT 1.5 MW [10]

timescales correspond to a reduced frequency much smaller than 0.05 and are hence classified as a quasi-steady rotor-response timescale.

The transverse integral scales in the ASL are approximately the same as the rotor disk. As shown in Figure 2.4, the rotation of the blade through the turbulence eddies creates a forcing at timescales smaller than the rotation timescale. Vijayakumar *et al.* [28] and Lavely *et al.* [26] have shown, with high-fidelity CFD of the utility-scale wind turbine within the daytime ABL, that the impact of the fluctuations in wind velocity vector on fluctuating sectional loads along wind-turbine blades is primarily through local changes in sectional angle of attack as the rotating blade cuts through energetic atmospheric eddies. Nandi *et al.* [27] found that the passage of the blades through the internal structure of the turbulence eddies creates large ramp-like excursions in sectional angle of attack, lift, and integrated blade torque at timescales below 1 s (~0.5–1 s). These same ramp-like events are observed in the rotor moments that pass through the drivetrain in Lavely *et al.* [26]. Figure 2.7 shows the variation of the reduced frequency over the blade at a forcing of 1 per revolution and 3 per revolution for three commercially relevant wind turbines, DTU 10 MW [8], NREL 5 MW [9], and WINDPACT 1.5 MW [10]. Thus, the inner 50% of the blade is likely influenced by unsteady aerodynamic effects when forced at the blade rotation frequency. The unsteady aerodynamic effects become particularly significant when the forcing frequency is three times the blade rotation frequency. This suggests that even

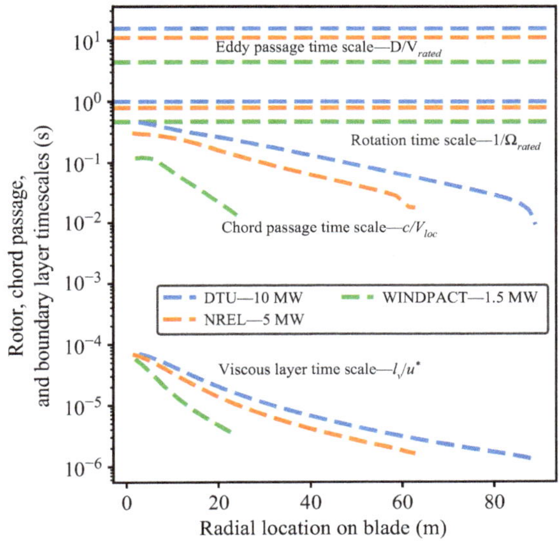

Figure 2.6 Variation of eddy, blade rotation, chord passage, and viscous boundary layer timescales over blade length at rated operating conditions for three commercially relevant wind turbines: DTU 10 MW [8], NREL 5 MW [9], WINDPACT 1.5 MW [10]

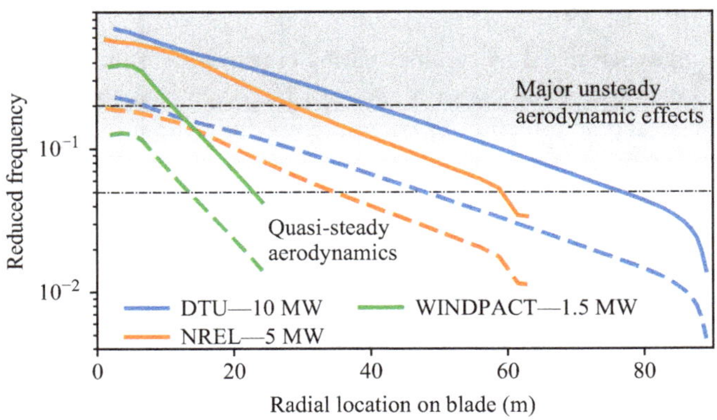

Figure 2.7 Variation of reduced frequency $\omega = f\,c/2\,V_{loc}$ over the blade length at a forcing of 1 per-rev (dashed lines) and 3 per-rev (solid lines) for three commercially relevant wind turbines: DTU 10 MW [8], NREL 5 MW [9], and WINDPACT 1.5 MW [10]

regular operation of a wind turbine in the atmosphere could lead to dynamic stall and boundary layer separation over significant portions of the blade. For example, the shear in the ABL alone will force the blade sections at 1 per revolution. When the rotor operates at off-design conditions like yaw misalignment, strong eddy passage, localized gusts, and so on, the blade sections may experience forcing at much higher than 1 per revolution.

2.1.3 The wake response length and time scales

Within a wind farm, the rotor experiences a mix of ABL turbulence and turbulence motions from the wakes of upstream wind turbines. Although the "front" rows of wind turbines within a wind farm respond directly to atmospheric turbulence, wind turbines further into the wind farm also respond to turbulent wakes from upstream wind turbines. The wake "meandering" (the time-varying transverse motions of the wake) is impacted by interactions between the rotor wakes and atmospheric turbulence eddies of comparable transverse scale. The spatial and temporal variability of the flow field in the turbine wakes generates an additional source of load transients on many turbines operating in a wind farm. At the scale of the rotor disk several diameters downstream of the upstream wind turbine, the large variances in the wake of the upstream wind turbine pass over the downstream rotor disk at the wake-meandering frequency. Wake meandering is very different in the daytime than the nighttime ABL. The generally lower turbulence intensity levels at night make "wake steering" control strategies more effective in the nighttime ABL. Churchfield and Moriarty discuss the effects of wakes on wind-turbine loads and performance in more detail in Chapter 6 (Volume 1).

2.1.4 Influences from atmospheric mesoscales and related weather events

The mean wind experienced by turbines varies in response to the diurnal cycle over timescales on the order of an hour (or greater). However, it is not unusual for mean wind magnitude and/or direction to change with the passage of common weather events at shorter timescales. These weather-induced changes also impact shorter timescale nonequilibrium changes in the energy-dominant atmospheric eddies that directly impact wind turbines driving turbine-blade aerodynamics off design. We have already shown a scale separation of eight orders of magnitude relevant to wind-turbine function in the atmosphere. However, there are an additional three orders of magnitude of relevant separation in spatial scale when one also considers potential significant roles of flow patterns at the mesoscale, largely horizontal flow patterns with scales of motion from hundreds of kilometers to tens of kilometers. The impacts of these large-scale weather patterns, largely horizontal and within the free troposphere, on the evolution of the ABL below, within which are embedded wind turbines and wind farms, are 2-fold: (1) The role of the passage of weather events on the modulation of the mean and fluctuating wind coherent structure in the ABL that interacts with wind turbines/farms, and (2) the role of mesoscale motions on the generation of power across large wind farms and the interactions among wind farms. Weather events can alter both the mean velocity field and the effective stability state of the ABL that affect the nonsteady forcing of wind turbines.

In addition to mesoscale-forced nonequilibrium effects on the ABL wind and wind-turbine-scale turbulence structure, mesoscale wind patterns can impact wind farms (e.g., groupings of wind turbines within very large wind farms, those approaching $O(100\,\text{km})$ in the horizontal direction, directly at the mesoscale). Coriolis acceleration, which turns the flow through the wind farm, can have an impact on the control of wind-turbine orientation within the wind farm as well as on the interaction between the wake of one wind farm and another located downstream. The Coriolis timescale is the inverse of the local projected rotation rate of the earth. At midlatitude, the Coriolis timescale is $\sim 10^{-4}$ rad/s $\sim$2.75 h. Haupt *et al.* discuss the effects of mesoscale motions on wind turbines in detail in Chapter 3 (Volume 1).

2.1.5 Concluding discussion

In conclusion, the mechanical responses in forces and moments to mean and nonsteady wind loadings that underlie both wind-turbine power generation and reliability concerns result from the interactions among motions over very large ranges in relevant length and time scales. For individual wind turbines, and for wind turbines within wind farms, the dominant wind forcings are a reaction to coherent turbulence motions embedded within the atmosphere and wind-turbine wakes. Part of this response is the creation of highly turbulent, extremely thin boundary layers on the blade surfaces. In summary, the scales of fluid motion relevant to wind-turbine function are as follows:

Smallest: The viscous layers within the blade boundary layers, $\sim$5–30 μm, in which the viscous stresses are produced on the blade surface and, more importantly, that interact with turbulence motions within the blade boundary layer mentioned earlier to either maintain attached blade boundary layers or allow the boundary layer to separate and dramatically lower sectional lift and increase sectional drag.

Mid (a): The internal structure of rotor wakes will generate smaller spatial-scale responses on downstream wind-turbine blades and rotors (the significance of which remains largely unexplored); furthermore, wake instabilities, and the impact of ABL microscale eddy structure on wake instabilities, create spatiotemporal oscillations in the wake ("meandering") that likely generate ranges of spatial and temporal force and moment timescales (also largely unexplored).

Mid (b): The internal structure of the daytime integral-scale eddies (those that contain the largest variances in wind velocity vector magnitude and direction), through which the rotating wind-turbine blades pass and as a result of which wind-turbine load responses at three ranges of characteristics time are generated: (1) integral-scale eddy passage time ($\sim$1 min), (2) blade-passage time ($\sim$4–7 s), depending on rotor rpm), and (3) a short important "ramp" timescale associated with the passage of rotor blades through the internal structure of the large coherent turbulence eddies ($<$1 s).

Mid (c): The blade chord scale, $\sim$2–4 m, over which sectional lift and drag are produced as the dominant forces involved in the generation of torque and power and that create nonsteady moments at the hub that pass through the drivetrain.

Largest (a): Internal flow structures developed in the nonequilibrium collapse of the unstable daytime ABL into the stable nighttime ABL (e.g., low-level jets of order rotor diameter in height).

Largest (b): The interaction of turbine rotor wakes of upstream wind turbines (of order the rotor diameter, ~100 m) with downstream wind turbines within wind farms.

Microscale largest: The energy-dominant integral-scale eddies of fully developed microscale ABL turbulence within the daytime ASL, ~40–140 m, of order the wind-turbine blade length and wind-turbine rotor diameter, which represents eight orders of magnitude in the range of important spatial scales that interact in the functions of wind turbines, both on and off design. However, at the wind farm level, and potentially even at the wind-turbine level, mesoscale (nonextreme) weather events create responses in function at scales of up to three orders of magnitude larger than the relevant ABL turbulence scales.

Mesoscale largest: Direct impact of mesoscale motions and weather events on large wind farm function as a unit at the mean flow scale for wind-turbine function, including Coriolis effects on mean wind direction.

Impacts of mesoscale weather on largest microscale motions: The time changes in the mean mesoscale wind vector (from the wind-turbine perspective) create deviations from equilibrium in the evolution of the ABL microscale motions that can impact the scale and structure of the energy-dominant turbulence motions that force the wind turbine to create nonsteady load variations.

From the previous discussion, we conclude that blade boundary-layer-resolved CFD should resolve and/or model at least eight orders of magnitude in spatial and temporal scale separation, from the viscous layers adjacent to blade surfaces to the rotor, wake, and ABL eddy scales. If one also includes atmospheric mesoscale motions and weather events as relevant to the response of both individual wind turbines within wind farms and the wind farm as a whole, then consideration extends to 11 orders of magnitude in relevant spatial scale separation to capture impacts of the atmospheric winds on wind-turbine and wind-farm function. Even so, eight orders of magnitude in the range of dynamically important spatial scales of motion for evaluating and improving LCOE in wind turbine and wind farm function is very large. Further, it points to the severe computational (and experimental) challenges in the evolution of the science of wind turbines and wind farms, which is needed to advance the technology to achieve more efficient production of energy from the wind.

2.2 Essential numerical and modeling elements in blade-resolved simulation of wind turbines

2.2.1 CAD model and mesh generation

The first step of performing a blade-resolved CFD simulation of wind turbines is to generate a computer-aided design (CAD) model that defines the surface that interacts with the air around the turbine. The required level of accuracy and detail in the

CAD model will depend on the sensitivity of the desired physics (e.g., the integrated loads, sectional loads, and acoustics) to the resolution of different features on the turbine surface. For instance, any errors in the modeling of the turbine hub or nacelle surface are unlikely to affect the integrated loads, whereas the same errors could affect the prediction of the local flow and sectional loads around the hub. The essential requirements of a CAD model for a successful blade-resolved simulation are a "water-tight" geometry, smooth interpolation between airfoil sections, and special care for add-on features and airfoil trailing edges. A "water-tight" geometry implies the lack of gaps up to a desired tolerance at any intersecting surfaces like the blade–nacelle interface, trailing edge of airfoil sections, and the blade tip. The first cell height of the mesh is likely to be $\sim$1–10 μm when performing blade-resolved simulations at Reynolds numbers of more than one million. One may need to set the tolerance in the CAD software appropriately to avoid significant errors in the definition of the blade shape. Wind-turbine design procedures (Chapter 3, Volume 2) mostly use BEM/vortex theory-based methods with airfoil shapes defined at many cross sections of the blades. To maintain consistency with the design procedure, the CAD geometry of a turbine used in blade-resolved simulations needs to smoothly interpolate between the airfoil sections. Special care must be taken to incorporate any add-on features like vortex generators or serrated trailing edges into the CAD model. Although some airfoil sections are defined with a specified thickness at the trailing edge, most (in the literature) are defined with a sharp trailing edge. While it is easy to generate 2D meshes around airfoils with a sharp trailing edge, it may not be simple to generate 3D meshes around blades with sharp trailing edges on sections. Trailing edges of airfoil sections on real wind-turbine blades are of a finite thickness as a result of manufacturing and noise constraints. One could truncate the airfoil at an arbitrary location close to the trailing edge or modify the shape of the airfoil near the trailing edge to include a finite thickness. When using the first option, it may be necessary to scale the airfoil shape by a small factor to retain the chord. The suggested modifications to the shape of the airfoil trailing edge are unlikely to affect the load predictions on the blade; it may be best to avoid any such ad-hoc modifications for aeroacoustics applications.

The generation of a mesh is one of the most time-consuming parts of the total process of performing blade-resolved simulations. The three main aspects of mesh generation are selection of domain size, choice of mesh resolution, and quality of discretized elements/cells. The choice of a mesh generation strategy also depends on the chosen CFD solver. Some CFD solvers can handle only block-structured meshes and/or hex/quad elements. Some CFD solvers may not be able to handle hanging nodes that are common at the interface between grid refinements. Most general-purpose CFD solvers can handle arbitrary types of elements/cells. Examples of grid-generation software used for blade-resolved CFD of wind turbines include GridPro [31], Pointwise [32], and CHIMERA [33]. Figure 2.8 shows a few examples of meshes used for blade-resolved simulations of wind turbines in the literature. Irrespective of the type of CFD solver and mesh generation procedure, the quality of the mesh impacts both the accuracy of the solution as well as the time to solution. The aspect ratio of elements/cells, the volume ratio of adjacent elements/cells, and the minimum and maximum included angles between faces of elements/cells are important metrics that

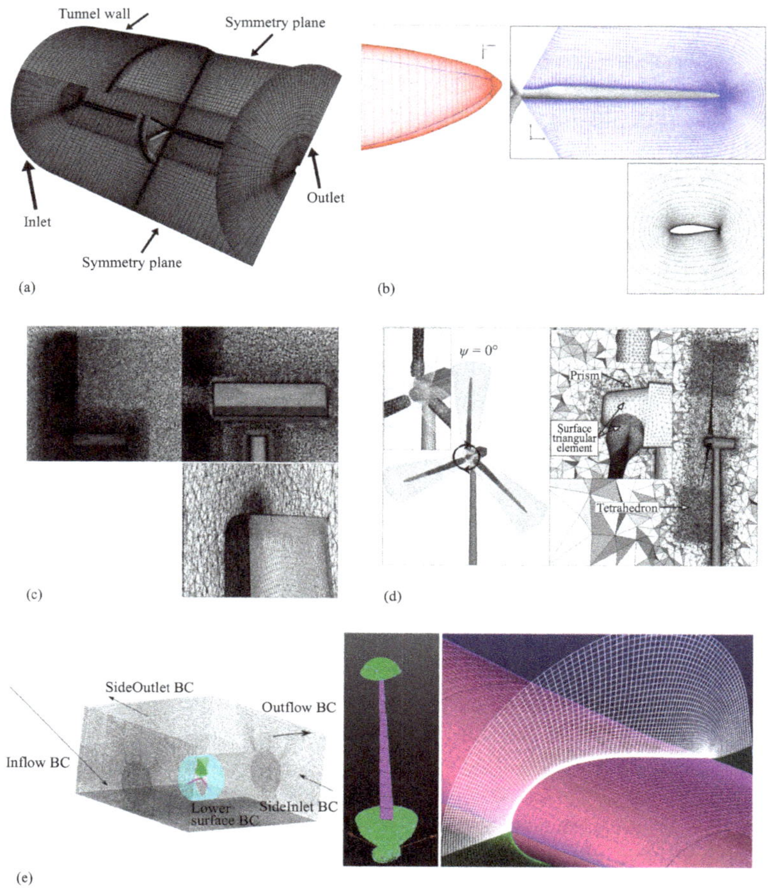

Figure 2.8 Examples of meshes used for blade-resolved CFD simulations in the literature. (a) Block-structured mesh used by Sørensen et al. for the NREL Phase VI rotor. © 2002. Reprinted, with permission, from [34]. (b) Block-structured mesh used by Sørensen and Johansen for the UPWIND rotor. © 2007. Reprinted, with permission, from [35]. (c) Overset mesh used by Lynch and Smith for the NREL Phase VI rotor. © 2013. Reprinted, with permission, from [36]. (d) Overset mesh used by Yu et al. for the NREL Phase VI rotor. © 2013. Reprinted, with permission, from [37]. (e) Unstructured mesh with a sliding interface used by Vijayakumar et al. [28] for the NREL 5-MW rotor.

quantify the quality of a mesh for blade-resolved simulations. The convergence and accuracy of typical finite volume solvers is improved by increasing the orthogonality of the control volume faces to the line-connecting nodes or cell centers and the local velocity vector.

The required mesh resolution at any given point in the computational domain depends on the local geometry, relevant length scales of the flow around the point (Section 2.1), and the turbulence and wall-modeling strategies. Most of the mesh resolution is typically concentrated near the blade surface to capture the large gradients of velocity in the blade boundary layer and the shear layers in the turbine wake. It may be necessary to compensate for low mesh quality with additional resolution in certain parts of the computational domain. Modeling the flow in the blade boundary layer using LES/DNS up to the wall without any special models requires the resolution of Kolmogorov scales near the turbine surface (see Section 2.2.3). Hence, LES/DNS modeling techniques are prohibitively expensive for any practical blade-resolved CFD simulations of wind turbines. Practical modeling strategies for blade-resolved CFD simulations of wind turbines in increasing order of required mesh resolution and complexity of mesh design are RANS, URANS, wall-modeled LES (WMLES), or hybrid URANS/LES. The mesh in the immediate vicinity of the turbine is typically generated by extruding the blade-surface mesh. The required resolution in the wall-normal direction can be very different from that along the blade surface depending on the turbulence modeling strategy. The resolution of the viscous layer ($0 \leq y^+ \leq 10$) requires the first grid point to satisfy $y^+ < 2$ [38]. $y^+ = y/\sqrt{(\nu/\left|<\boldsymbol{\tau}_{\mathbf{wall}} \cdot \hat{\boldsymbol{n}}>\right|)}$ is the wall-normal distance, y, nondimensionalized on the local viscous length scale, where ν is the kinematic viscosity, $\boldsymbol{\tau}_{\mathbf{wall}}$ is the viscous stress tensor and $\hat{\boldsymbol{n}}$ is a unit surface-normal vector. Both RANS and LES turbulence modeling strategies can be combined with wall models that allow the first grid point to be in the log layer $30 < y^+ < 100$. LES typically requires a mesh aspect ratio of $O(1)$ to resolve the isotropic eddies near the filter scale, whereas meshes for RANS near the wall require aspect ratios to be much larger than $O(1)$, commonly as high as 30,000. Hence, meshes for hybrid RANS/LES simulations require a complex design—the aspect ratio must transition from $\gg 1$ at the surface to $\sim O(1)$ in the region where the turbulence model transitions from RANS to LES [39].

2.2.2 CFD solver

A CFD solver solves the discretized Navier–Stokes equations on a mesh/grid described in Section 2.2.1 with appropriate boundary conditions. Table 2.1 lists a short and nonexhaustive selection of CFD solvers that have been used in the literature for blade-resolved simulations of wind turbines. The highest Mach number near the tip of a wind-turbine blade is sufficiently low, $M_{tip} \leq 0.3$, so that the incompressible flow approximation is reasonable to model the flow around a wind turbine. However, many authors choose to use fully compressible or artificial compressibility-based CFD solvers for blade-resolved simulations of wind turbines for other considerations including availability, history of use, and so on. In the near future, the offshore turbines may be able to increase the tip speed because of relaxed noise constraints. Compressible flow solvers may then become necessary to model the flow around the tip of large offshore wind turbines [40]. An increasingly common trend is to use a framework that allows the use of multiple solvers in different regions of the flow coupled using an overset methodology. Sitaraman *et al.* [41] used the CREATE-AV

Table 2.1 Overview of CFD solvers used in the literature for blade-resolved simulations around wind turbines

CFD Solver	**Description**
Ellipsys3D—DTU [13]	Incompressible, block structured, finite volume
OpenFOAM [28,44]	Incompressible, unstructured, finite volume, PISO/SIMPLE
ANSYS—FLUENT [45]	Incompressible, unstructured, finite volume, unsteady SIMPLE
THETA—DLR [46]	Incompressible, unstructured, finite volume, unsteady SIMPLE
ANSYS CFX [47]	Incompressible, unstructured, control-volume finite element (CVFEM), Unsteady SIMPLE
Tenasi—UTC [48]	Compressible, artificial compressibility Chorin's method
NSU-3D [43]	Compressible, unstructured, finite volume, AMR, overset
OverFLOW [49]	Compressible, overset, block-structured, implicit finite difference
elsA—ONERA [50]	Compressible, block-structured finite volume, Jameson scheme
Bazilevs [51,52]	Incompressible residual-based variational multiscale finite element
AcuSolve [4,53]	Incompressible Galerkin/least squares finite element
TAU—DLR [46]	Compressible, block-structured, overset, Jameson scheme
CREATE-AV HELIOS [41]	Compressible-solver framework
W^2A^2KE3D [42]	Compressible-solver framework

HELIOS framework and Kirby *et al.* [42] used the W^2A^2KE3D framework to combine the NSU-3D [43] solver near the wind turbines with different off-body solvers to simulate the Lillgrund wind farm.

As discussed in Section 2.1, the huge range of relevant length and time scales in the flow around wind turbines in the atmosphere makes DNS impossible for full-scale wind-turbine simulations. RANS models of the boundary layer integrated with resolution of the integral-scale eddies in the ABL and rotor wakes make hybrid URANS/LES the approach of choice for highest fidelity CFD of wind turbines in the atmosphere.

2.2.2.1 Incompressible-flow solvers

Incompressible solvers usually solve the filtered or averaged Navier–Stokes equations involving the continuity and the momentum equation (see Section 2.2.3). In atmospheric flows, a body force term in the momentum equation can include buoyancy force and the Coriolis acceleration. The addition of buoyancy force typically requires the solution of the temperature equation, usually written in terms of a potential temperature [54]. There are four major components of an incompressible CFD solver: the splitting scheme, discretization scheme, convection scheme, and the linear solver.

Discretization scheme: The discretization scheme decides the way in which the governing conservation equations are represented on the discretized mesh/grid. The three main discretization schemes used for wind-turbine CFD are finite difference, finite volume, and finite element schemes. The finite difference schemes enforce the governing conservation equations in the differential form at a finite number of points in the domain, and the finite volume-based schemes enforce the integral form of the

Navier–Stokes equations over small control volumes associated with the mesh/grid. The finite volume schemes are typically used in the wind-turbine CFD literature because of their better conservation properties compared to finite difference schemes. The Gauss theorem is used in finite volume schemes to evaluate as many terms as possible using surface integrals. The finite element schemes solve a "weak form" of the governing conservation equations integrated over the computational domain. They typically use Gauss–Legendre-type quadrature over small elements associated with the mesh/grid.

Convection scheme: The finite volume schemes require interpolation of nodal quantities to the surfaces of the discretized control volumes to compute fluxes. Finite volume CFD solvers typically use a nonlinear scheme to weight the interpolation toward the local upwind direction to maintain the boundedness and stability of the overall time-stepping scheme at the cost of some accuracy. A nonlinear interpolation scheme is usually formulated as a flux limiter [55], which is a function of the local gradients, or a hybrid scheme that is a function of the local cell Peclet number [56]. Different convection schemes may be necessary to maintain boundedness and accuracy for the momentum equation and for the prognostic equations for a turbulence model. The convection scheme for any CFD solver and application should be selected and tuned through an appropriate validation exercise.

Splitting scheme: The filtered/averaged incompressible conservation equations (2.14) and (2.15) are coupled, nonlinear partial differential equations involving multiple variables including velocity, pressure, and any variables for the prognostic equations of the turbulence model. "Colocated" incompressible solvers track the mass flow rate through the faces of the control volumes independent of the velocity at the cell centers/nodes—this mimics the "staggered" mesh approach and allows for direct implementation of discretized conservation of mass. The advective terms are linearized using the mass flow rate from the previous time step or "outer" iteration. The linearized, yet still coupled, equations are solved in a staggered manner instead of solving all the equations together. The splitting scheme decouples the partial differential equations through the introduction of splitting errors—which are then reduced through appropriate "outer" iterations. The unsteady SIMPLE/PISO scheme is one of the most common splitting schemes used for incompressible blade-resolved CFD [56], as they are based on an approximate LU factorization of the implicit discretized governing equations. After linearization and temporal and spatial discretization, the semi-implicit incompressible momentum and continuity equations are expressed as

$$\begin{bmatrix} A & G \\ D & 0 \end{bmatrix} \begin{bmatrix} u^{n+1} \\ p^{n+1} \end{bmatrix} = \begin{bmatrix} f \\ 0 \end{bmatrix}, \tag{2.1}$$

where D and G are the divergence and gradient operators, respectively, and A is the discretized operator for the momentum equation without the pressure-gradient term. If the discretized mesh contains unknowns at m nodes or cell centers, the A matrix is of size $3\,\text{m} \times 3\,\text{m}$, D of size $m \times 3m$, and G of size $3m \times m$. u^{n+1} of size $3m$ and p^{n+1} of size m are vectors containing velocity and pressure at the nodes or cell centers at the next time step, $n+1$. f is the right-hand side of the discretized momentum equation based on information from the velocity and pressure from time step n and any boundary conditions. An exact factorization of (2.1) is

$$\begin{bmatrix} A & 0 \\ D & -D\,A^{-1}\,G \end{bmatrix}\begin{bmatrix} I & A^{-1}\,G \\ 0 & I \end{bmatrix}\begin{bmatrix} u^{n+1} \\ p^{n+1} \end{bmatrix} = \begin{bmatrix} f \\ 0 \end{bmatrix},$$

$$\begin{bmatrix} A & 0 \\ D & -D\,A^{-1}\,G \end{bmatrix}\begin{bmatrix} u^{n+1} + A^{-1}\,G\,p^{n+1} \\ p^{n+1} \end{bmatrix} = \begin{bmatrix} f \\ 0 \end{bmatrix},$$

$$\begin{bmatrix} A & 0 \\ D & -D\,A^{-1}\,G \end{bmatrix}\begin{bmatrix} u^{*} \\ p^{n+1} \end{bmatrix} = \begin{bmatrix} f \\ 0 \end{bmatrix}. \tag{2.2}$$

where I is the identity matrix and u^* is and intermediate velocity. The exact inversion of A to compute the Schur complement $D\,A^{-1}\,G$ is prohibitively expensive. Hence, most solvers use the approximate factorization outlined below instead:

$$\begin{bmatrix} A & G \\ D & 0 \end{bmatrix}\begin{bmatrix} u^{n+1} \\ \Delta p^{n+1} \end{bmatrix} + \begin{bmatrix} I & G \\ D & 0 \end{bmatrix}\begin{bmatrix} 0 \\ p^{n} \end{bmatrix} = \begin{bmatrix} f \\ 0 \end{bmatrix},$$

$$\begin{bmatrix} A & 0 \\ D & -L_B \end{bmatrix}\begin{bmatrix} I & B\,G \\ 0 & I \end{bmatrix}\begin{bmatrix} u^{n+1} \\ \Delta p^{n+1} \end{bmatrix} + \begin{bmatrix} I & 0 \\ DB & -L_B \end{bmatrix}\begin{bmatrix} I & G \\ 0 & I \end{bmatrix}\begin{bmatrix} 0 \\ p^{n} \end{bmatrix} = \begin{bmatrix} f \\ 0 \end{bmatrix},$$

$$\begin{bmatrix} A & 0 \\ D & -L_B \end{bmatrix}\begin{bmatrix} u^{n+1} + B\,G\Delta p^{n+1} \\ \Delta p^{n+1} \end{bmatrix} + \begin{bmatrix} I & 0 \\ DB & -L_B \end{bmatrix}\begin{bmatrix} Gp^{n} \\ p^{n} \end{bmatrix} = \begin{bmatrix} f \\ 0 \end{bmatrix},$$

$$\begin{bmatrix} A & 0 \\ D & -L_B \end{bmatrix}\begin{bmatrix} u^{*} \\ \Delta p^{n+1} \end{bmatrix} + \begin{bmatrix} Gp^{n} \\ D\,B\,Gp^{n} - L_B p^{n} \end{bmatrix} = \begin{bmatrix} f \\ 0 \end{bmatrix}, \tag{2.3}$$

where B is an approximation to A^{-1} typically chosen as the inverse of only the diagonal part of A and $\Delta p^{n+1} = p^{n+1} - p^n$. L_B is the Laplacian approximation to $D\,B\,G$. While L_B and $D\,B\,G$ are mathematically equivalent, their numerical evaluation procedure is different. The term involving the difference between the two evaluations is referred to as Rhie–Chow interpolation and is essential for avoiding the "checkerboarding" problem in incompressible solvers. The final sets of equations typically solved at every time step are

$$Au^* + Gp^n = f, \tag{2.4}$$

$$L_B\Delta p^{n+1} = D\,(u^* + B\,Gp^n) - L_B p^n, \tag{2.5}$$

$$u^{n+1} = u^* - B\,G\Delta p^{n+1} \tag{2.6}$$

SIMPLE-based algorithms solve (2.4)–(2.6) with some relaxation factors for velocity and pressure to maintain stability and accelerate convergence, whereas PISO-based algorithms repeat the pressure solve step for the same reasons. The solution of (2.4)–(2.6) is repeated several times for each time step as "outer" iterations to reduce the splitting and linearization errors. Ferziger and Peric [56] detail the difference between various SIMPLE- and PISO-based algorithms.

Linear solver: The splitting and discretization steps results described earlier result in linear systems of equations corresponding to the momentum, continuity, temperature (when buoyancy is included), and any prognostic equations for the turbulence model. These linear systems of equations are typically of the form $P\mathbf{x} = q$, where P is an $m \times m$ matrix and $\mathbf{x}$ is a vector of m unknowns corresponding to the number of degrees of freedom in the meshing and discretization procedure. For typical simulations of flow past wind turbines that use meshes of the order of 1–100 million degrees of freedom, iterative solvers [57,58] are used to solve the linear system of equations to a desired tolerance. Solvers typically used for wind-turbine CFD include generalized minimum residual (GMRES), biconjugate gradient, conjugate gradient, and multigrid methods. In practice, most iterative solvers are used along with a preconditioner to reduce the stiffness of the matrix P. The choice of the preconditioner, solver, and solver tolerance greatly affects the total run time and feasibility of wind-turbine CFD simulations. Best practices on using/writing parallel iterative solvers for sparse linear systems on modern high-performance computing platforms can be found in open-source linear solver software packages like HYPRE [59] and PETSC [60]. Further discussion on this subject can be found in Chapter REFERENCE Mike Sprague CHAPTER HERE.

2.2.2.2 Compressible-flow solvers

As listed in Table 2.1, compressible CFD solvers are also used to perform blade-resolved CFD simulations of wind turbines. Compressible CFD solvers are widely used in the aerospace industry and have a long history of development along with verification and validation. Most compressible flow solvers solve the governing equations in the conservative form:

$$\frac{\partial \mathbf{q}}{\partial t} + \frac{\partial \mathbf{E}}{\partial x} + \frac{\partial \mathbf{F}}{\partial y} + \frac{\partial \mathbf{G}}{\partial z} = \mathbf{R}(\mathbf{q}) = 0 \tag{2.7}$$

$$\text{where } \mathbf{q} = \begin{bmatrix} \rho & \rho u & \rho v & \rho w & \rho e \end{bmatrix},$$

where u, v, and w are the filtered/averaged components of the velocity; e is the filtered/averaged internal energy; and $\mathbf{E}$, $\mathbf{F}$, and $\mathbf{G}$ are the conservative fluxes including the viscous and turbulent fluxes. Most applications of compressible flow solvers to wind energy applications extend the $\mathbf{q}$ to include the dependent variables in prognostic equations underlying the turbulence model. Applying Newton's method to (2.7) produces

$$\frac{\partial \mathbf{R}}{\partial \mathbf{q}} \Delta \mathbf{q}^{k+1} = -\mathbf{R}(\mathbf{q}^k) \tag{2.8}$$

$$\text{where } \Delta \mathbf{q}^{k+1} = \mathbf{q}^{k+1} - \mathbf{q}^k. \tag{2.9}$$

Equation (2.8) is typically solved ~20–100 times per time step to obtain a solution to the desired tolerance in the residual $\mathbf{R}(\mathbf{q})$. Krylov-based solvers like GMRES are typically used to solve the resulting system of equations. The low Mach numbers typical of wind energy applications ($M_{max} \leq 0.3$) make the Jacobian matrix, $\partial \mathbf{R}/\partial \mathbf{q}$, very

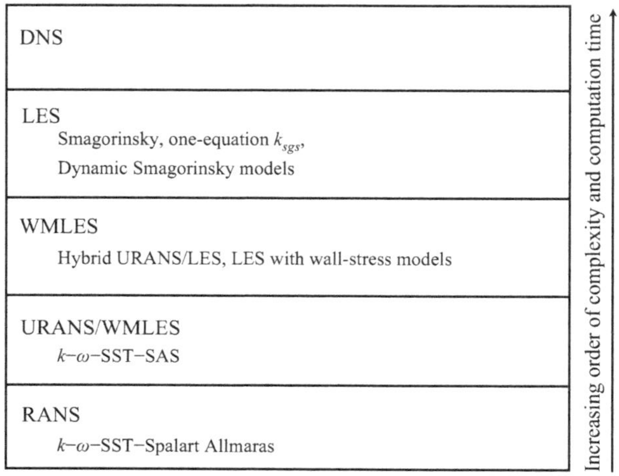

Figure 2.9 Turbulence modeling hierarchy ordered by complexity and computation time

stiff and ill-conditioned. Low-Mach preconditioning is typically necessary to speed up convergence of the solution of (2.8) for wind-energy applications. The National Atmospheric and Space Administration's OVERFLOW manual [61] describes other details of compressible flow solvers used in production level codes.

Most CFD solvers used for blade-resolved simulations of wind turbines are formulated to allow for rigid body mesh motion corresponding to the rotation of the rigid blades. Some solvers that incorporate FSI are formulated using the arbitrary Lagrangian–Eulerian method [62] to solve for the mesh displacement and velocity caused by the deflection of the wind-turbine components. Many authors use an overset approach [63] or a sliding mesh approach [64] to allow for motion of a grid attached to the rotating blade relative to a fixed far-field mesh. The overset approach typically allows for a greater range of motion, such as yaw and pitch, without considerable overhead during the mesh-generation procedure.

2.2.3 Turbulence modeling

As discussed in Section 2.1, the flow around a wind turbine is highly turbulent because the large Reynolds number and velocity field contains fluctuations over a wide range of length and time scales. According to Spalart [65], a DNS of flow over an aircraft wing is not feasible in the present century; a similar statement can be made for the simulation of flow around a wind turbine as well. Also, DNS will generate data sets so large as to be difficult to manipulate. The main purpose of blade-resolved CFD of the flow around wind turbines is the prediction of the loads on the turbine and the structure of the wake relevant to any wind turbines present downstream. Hence, modeling is necessary to reduce the number of degrees of freedom for practical blade-resolved CFD of wind turbines.

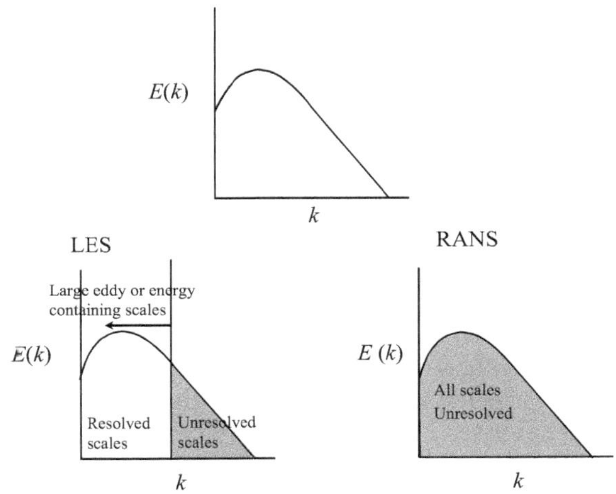

Figure 2.10 Illustration of modeling strategy in LES and RANS shown using the turbulent kinetic energy spectrum

We summarize the primary modeling strategies in Figure 2.9. The two main turbulence modeling techniques available for CFD are RANS and LES. Other modeling techniques, like WMLES, borrow or blend concepts from RANS and LES. Consider the energy spectrum schematic in Figure 2.10 of a turbulent flow region in the vicinity of a wind-turbine blade. The RANS approach applies an ensemble averaging to separate the ensemble average velocity (U_i) from all the turbulent fluctuations (u_i') as

$$u_i(\mathbf{x}, t) = U_i(\mathbf{x}, t) + u_i'(\mathbf{x}, t), \tag{2.10}$$

$$\text{where } U_i(\mathbf{x}, t) = \langle u_i(\mathbf{x}, t) \rangle . \tag{2.11}$$

The effect of all the scales containing the turbulent fluctuations, u_i', on the evolution of the ensemble-averaged velocity field, U_i, is modeled using the Reynolds stress tensor, T_{ij}, in (2.14). Thus, a RANS simulation only requires the resolution of the spatial and temporal gradients of the ensemble-averaged quantities, U_i, P, and T_{ij}, which greatly reduces the required number of degrees of freedom compared to DNS. The use of the term RANS in the wind-turbine community refers to the generalized case in which the ensemble average velocity field $U_i(\mathbf{x}, t)$ is a function of both space and time. Some authors choose to highlight the temporal variation of the ensemble average velocity field as unsteady RANS (URANS). The URANS approach is adopted for simulations of unsteady flows when turbulence is not the driving phenomenon of temporal variations in U_i [66]. Both steady-state RANS and URANS techniques have been used with great success to predict the loads on turbines under nominal operating conditions. Unlike hybrid URANS/LES simulations, no new physics are added to the turbulence model in the URANS framework.

The LES approach filters the velocity field up to a scale supported by the grid resolution as

$$u_i = \widetilde{u}_i + u_i^s, \tag{2.12}$$

$$\text{where } \widetilde{u}_i = \int G_\Delta(\mathbf{x} - \mathbf{y})\, u_i(\mathbf{y})\mathrm{d}\mathbf{y}. \tag{2.13}$$

G_Δ is a spatial filter with a cutoff scale that is proportional to the grid scale, Δ. Although (2.12) contains an explicit form of this filtering operator, the precise form of G_Δ is usually unknown in practice and heavily depends on the numerical discretization. The LES approach expects to resolve the energy-containing scales of turbulence, up to the inertial range as shown in Figure 2.10, on the grid. The effect of the unresolved/subfilter/subgrid scales on the evolution of the resolved scales is modeled through the subfilter stress tensor, T_{ij}, in (2.14). The resolution of a significant portion of the turbulent kinetic-energy spectrum in LES requires a much greater mesh resolution compared to RANS. However, the modeling of the subfilter-scale eddies in LES is expected to be more accurate and universal compared to the modeling of the entire turbulent kinetic energy spectrum in RANS. The integral scales are known to decrease with distance from the wall in high-Reynolds-number boundary layers [38]. Resolution of the integral scales in surface viscous layers or roughness elements with LES requires the same resolution as DNS—this is impossible or impractical at the high Reynolds numbers of commercial wind turbine and atmospheric flows [65]. The practical solution used in the external aerodynamics community is to bypass the wall-resolution requirement of LES using hybrid URANS/LES or detached-eddy simulation (DES). The hybrid URANS/LES approach models the boundary layer adjacent to the turbine surface using URANS and transitions to LES away from the surface.

Both the Reynolds-averaged and filtered incompressible Navier–Stokes equations look very similar to

$$\frac{\partial \overline{u}_i}{\partial t} + \frac{\partial \overline{u}_i \overline{u}_j}{\partial t} = -\frac{1}{\rho}\frac{\partial \overline{p}}{\partial x_i} - \frac{\partial \overline{\tau}_{ij}}{\partial x_j} - \frac{\partial T_{ij}}{\partial x_j}, \tag{2.14}$$

$$\frac{\partial \overline{u}_k}{\partial x_k} = 0 \tag{2.15}$$

$$\text{where } T_{ij}^h = \overline{(u_i u_j)} - \overline{u}_i \overline{u}_j, \tag{2.16}$$

$$\overline{u}_i = \widetilde{u}_i \text{ (filtered LES) or } U_i \text{ (averaged RANS)}, \tag{2.17}$$

$$\overline{(u_i u_j)} = \widetilde{(u_i u_j)} \text{ (filtered LES) or } \langle (u_i u_j) \rangle \text{ (averaged RANS)}, \tag{2.18}$$

$$\overline{\tau}_{ij} = \widetilde{\tau}_{ij} \text{ (LES) or } \langle \tau_{ij} \rangle \text{ (RANS)}. \tag{2.19}$$

While the filtered and RANS equations may look very similar, they have very different physical meaning and interpretations as explained in the preceding paragraphs. The similarity between the two equations is used heavily in several CFD solvers to streamline the implementation of the different turbulence modeling approaches, as well as in the implementation hybrid URANS/LES turbulence models. Most applications of CFD to wind-turbine simulations use the eddy viscosity representation for the deviatoric part of the modeled "turbulent stress" tensor τ^h_{ij}:

$$T_{ij} - \frac{1}{3}T_{kk}\delta_{ij} = -2\nu_t\overline{S}_{ij}, \tag{2.20}$$

$$\text{where } \overline{S}_{ij} = \frac{1}{2}\left(\frac{\partial U_i}{\partial x_j} + \frac{\partial U_j}{\partial x_i}\right) \text{ RANS,}$$

$$\overline{S}_{ij} = \frac{1}{2}\left(\frac{\partial \widetilde{u}_i}{\partial x_j} + \frac{\partial \widetilde{u}_j}{\partial x_i}\right) \text{ LES.}$$

RANS is commonly used in wind-turbine CFD to simulate the flow past a rigid turbine rotating under uniform inflow conditions. The boundary layer on the blade is typically assumed to be fully turbulent even though the flow may undergo transition to turbulence over the blade chord. The Spalart–Allmaras [67] and $k - \omega - SST$ [68] turbulence models are the most commonly used models for blade-resolved CFD of wind turbines.

Some RANS simulations of wind turbines are performed in a frame of reference rotating with the blade(s) with added source terms for Coriolis and centrifugal forces. These simulations do not include the effects of rotor–tower interactions and occasionally ignore the effects of the nacelle for larger turbines. In the rotating frame of reference approach, the RANS equations can be solved with a steady-state formulation. Sørensen *et al.* [34], Duque *et al.* [69], Le Pape *et al.* [50], and Chao and van Dam [70] have used this approach to predict the average sectional and integrated loads at different wind speeds on the NREL Phase VI rotor [71]. Simulation of a rotor in an inertial frame of reference with URANS requires an unsteady simulation over several rotations of the turbine to achieve convergence of loads. Potsdam and Mavriplis [43], Lynch and Smith [36], and Länger-Möller *et al.* [46] have used this approach to achieve similar results on the NREL Phase VI rotor. Despite using different solvers, formulations, meshes, and turbulence models, almost all of these studies capture the integrated and sectional loads under fully attached flow conditions ($V_{in} = 7$ m/s), as shown in Figure 2.11, with many of them capturing the stall region only qualitatively ($V_{in} = 10 - 18$ m/s). Mittal *et al.* [48] achieve good prediction of integrated loads on the BT1 rotor [72] operating at both design and stalled conditions using the inertial frame of reference approach. Sørensen *et al.* [13] also report good predictions of the sectional loads on the NEW-MEXICO rotor [73] under different flow conditions using the rotating frame of reference approach. More recent efforts have focused on RANS computations of megawatt-scale wind turbines: Sørensen and Johansen [35], Bazilevs *et al.* [51,52], and Chow and van Dam [49]. Because of the lack of publicly available experimental data to validate predictions of full-scale rotors, authors choose to perform extensive grid and time-step refinement studies to verify their results. Using a comprehensive off-body grid refinement study, Chow and van Dam [49] found that a domain size of 10 diameters (D) upstream and 20D downstream of the rotor is necessary to accurately predict the loads on the NREL

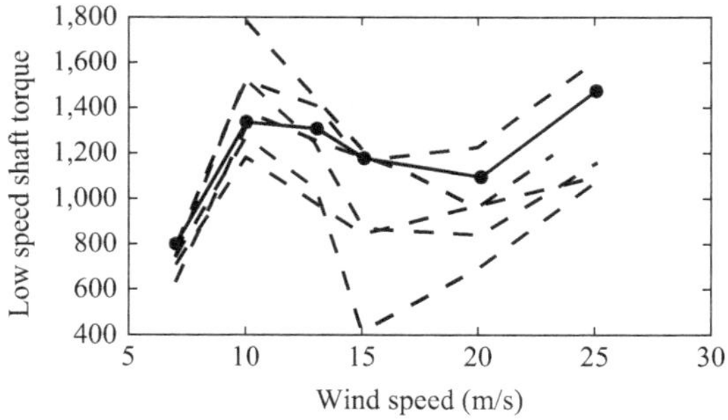

Figure 2.11 Comparison of low-speed shaft torque predicted by different blade-resolved CFD computations (dashed lines) [34,36,43,46,69] of the NREL Phase VI rotor under uniform inflow conditions. The dark black line with dots represents the measurements from the NREL Phase VI experiment [71]

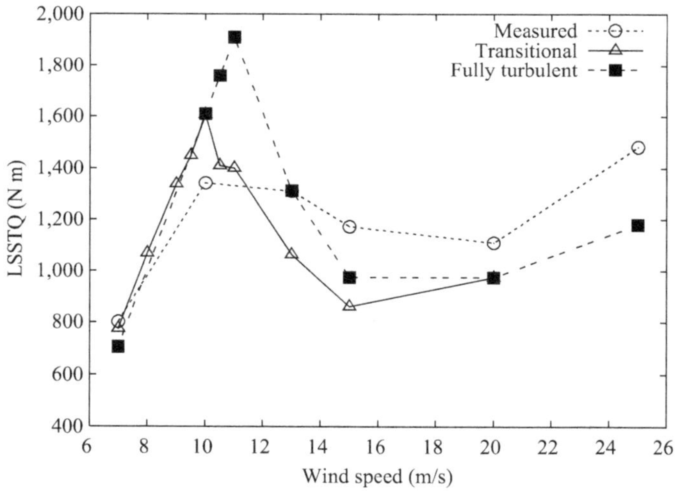

Figure 2.12 Comparison of Low Speed Shaft Rotor Torque (LSSTQ) predicted using blade-resolved CFD computations of the NREL Phase VI rotor in uniform inflow with and without the $\gamma - Re_{\theta_t}$ transition model with experiment. © 2009. Reprinted, with permission, from [74]

5-MW turbine along with a resolution of 1 m in the near-wake grid extending up to 0.5D upstream and 3D downstream of the rotor.

Several authors have attributed the inability to predict rotor stall in CFD computations of the NREL Phase VI rotor to the lack of a transition model in the turbulence model; many of them have started to include a transition model when simulating a rotor in a wind tunnel with mixed results. Sørensen [74] found a slight improvement in

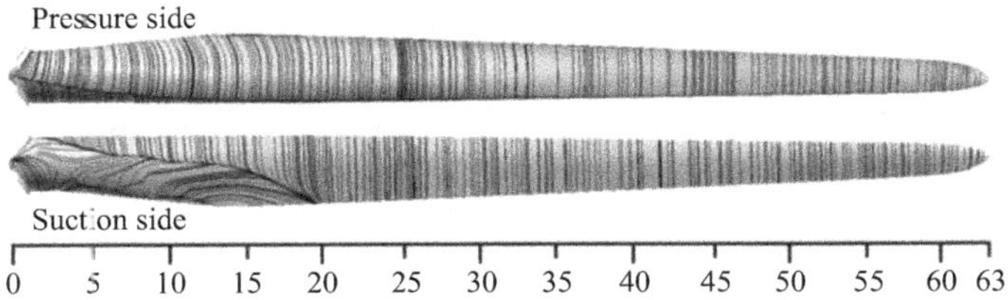

Figure 2.13 Limit surface streaklines for the baseline NREL 5-MW rotor in uniform inflow of 8 m/s. © 2012. Reprinted, with permission, from [76]

the prediction of stall of the NREL Phase VI rotor when using the $\gamma - Re_{\theta_t}$ turbulence model as shown in Figure 2.12. The presence and importance of modeling laminar to turbulent transition in the boundary layer on rotors operating in the real atmosphere have not been established and are still a research topic as shown by Schaffarczyk *et al.* [75]. In addition, Sørensen *et al.* [13] have found that the correlation-based $\gamma - Re_{\theta_t}$ transition model fails to correctly predict the natural transition behavior at high Reynolds numbers relevant to full-scale turbines in the field.

URANS simulations are required to simulate the unsteady flow field around the rotor operating at a nonzero yaw angle. Yu *et al.* [37] use the URANS approach to simulate the flow around the NREL Phase VI rotor operating at 30° and 60° yaw.

Figure 2.13 shows the surface streamlines on an NREL 5-MW blade operating well within the design envelope in uniform inflow of 8 m/s by Chow and van Dam [76]. The flow is separated in the first 33% of the blade. This is a particularly challenging problem for RANS models, as they are expected to accurately capture the mean load and flow characteristics in the separated regions without triggering any numerical/physical flow instabilities in attached flow regions. The challenges only become harder under off-design conditions. A major limitation of RANS/URANS is the poor prediction of time-varying loads caused by separation in off-design conditions. Accurate prediction of unsteady separation that impacts achievable performance and life-cycle costs will be increasingly required earlier in the design cycle of wind turbines. This will drive the next round of reductions in the levelized cost of wind energy. The most common approach to predict unsteady loads on wind turbines under separated flow conditions is hybrid URANS/LES. Spalart *et al.* [65] motivate the need for hybrid URANS/LES or DES as

> "Separation is the essential flow feature motivating DES, with the expectation that the boundary layer is treated with RANS and is quasi-steady, but the free shear layer it feeds develops LES content."

Johansen *et al.* [77] and Lynch and Smith [36] observe better load predictions at stall using hybrid URANS/LES [78] computations of the NREL Phase VI rotor. Lynch and Smith [36] also simulated the NREL Phase VI rotor in uniform flow at 30° yaw; they were able to qualitatively capture the hysteresis loops of section-wise loads, albeit with a phase lag. Sørensen and Schreck [79] were also able to reproduce the qualitative behavior of the hysteresis loops of section-wise loads on the NREL Phase

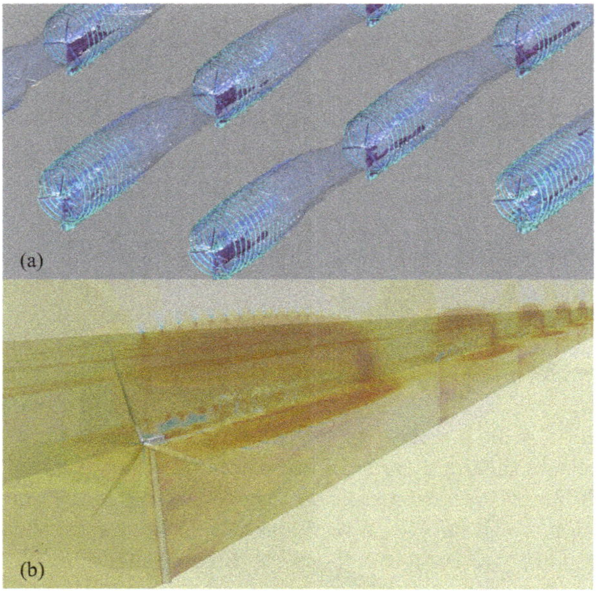

Figure 2.14 (a) Iso-surfaces and (b) isocontours of velocity magnitude from a blade-resolved simulation of the Lillgrund Wind Farm in uniform inflow. © 2017. Reprinted with permission from [42]

VI rotor in pitch motion during stand still. They discuss the technical difficulties of obtaining grid-independent and periodic results using hybrid RANS/LES methods under highly stalled conditions with a time-varying angle of attack using the scatter (not uncertainty) in the observed pressure profiles on the turbine blades. Mittal *et al.* [48] highlight the need for using hybrid URANS/LES models to accurately predict the structure of the wake behind turbines. Most of the current hybrid URANS/LES models are designed to model the flow past aerodynamic surfaces in relatively uniform inflow with some background turbulence. However, the length scales of the flow features near the blade (e.g., boundary layer dynamics, separation, and vortex shedding) are orders of magnitude smaller than the energy-containing eddies in atmospheric turbulence. The ability to capture the interaction between eddies of such disparate length scales is missing in many of the current hybrid URANS/LES methods. Vijayakumar *et al.* [28] developed a new hybrid URANS/LES model for application to wind-turbine blades operating in atmospheric turbulence. Sitaraman *et al.* [41] and Kirby *et al.* [42] have performed a simulation of an entire wind farm in uniform inflow with rigid blades using hybrid URANS/LES. Figure 2.14 shows isosurfaces and isocontours of velocity magnitude on a small portion of the Lillgrund wind farm by Kirby *et al.* [42].

The state-of-the-art hybrid URANS/LES methods have several limitations, such as, modeled stress depletion in the gray area between LES and RANS, log-layer mismatch or overshoot, and grid-induced separation [66]. Several of these limitations can be attributed to the semiheuristic construction of hybrid URANS/LES models without any reference to a generalized scale-separation procedure [66]. The same limitations of hybrid URANS/LES models causes inconsistent prediction of loads on wind turbines under separated flow conditions. The development of hybrid URANS/LES turbulence models and their application to wind-turbine applications continues to be an area of active research.

2.2.4 Fluid–structure interaction

Most of the blade-resolved CFD simulations of wind turbines in the literature have used rigid blades rotating at a specified rate. However, modern turbines use highly flexible blades with a controller that varies the rotation rate and the blade pitch. Many modern turbines also use aeroelastically tailored blades with bend-twist coupling that deform in a specific manner when the blade is rotating. Blade-resolved CFD simulations must be coupled to a computational structural dynamics (CSD) model of the turbine components and a controller to accurately predict the FSI under all operating conditions. We use the term "blade-resolved FSI" simulations to refer to simulations in which the CFD solver resolves the geometry and the boundary layer adjacent to the solid–fluid interface. Only a few authors have performed blade-resolved FSI simulations of wind turbines so far: Bazilevs *et al.* [52], Heinz *et al.* [80], and Yu and Kwon [37] simulate the flow past the NREL 5-MW rotor under different inflow conditions; Sayed *et al.* [81] simulate the flow past the large InnWind rotor; Carrión *et al.* [82] analyze the effect of varying torsional stiffness of the blade on the MEXICO and NREL Phase VI rotors using blade-resolved FSI simulations.

Many blade-resolved FSI simulations of upwind wind-turbines model only the interaction of the fluid with a structural model of the blades and ignore the hub, nacelle, and tower. Most blade-resolved FSI computations of wind-turbines model the blades as cantilevered beams with rigid cross sections. Carrión *et al.* [82] use the beam representation with precomputed responses to loads in terms of blade modes and shapes. Yu and Kwon [37] model the blade using Euler–Bernoulli beam theory, whereas Heinz *et al.* [80] and Sayed *et al.* [81] use a geometrically exact beam theory. Bazilevs *et al.* [52] chose to model the blade using the full geometry with the composite laminate layout using shell elements.

The FSI coupling algorithm typically involves the transfer of loads from the CFD to the CSD solver and the transfer of structural deformations and deformation rate from the CSD to the CFD solver. In any FSI simulation, the spatial and temporal discretization of the fluid (CFD) and structural (CSD) models may not necessarily match and yet have to be coupled to each other. The three main aspects of blade-resolved FSI coupling algorithm are as follows:

1. The coupling strategy should be consistent with the time-stepping algorithm in the CFD and CSD model and retain the desired order of accuracy.
2. The mesh motion algorithm of the fluid mesh should satisfy the geometric conservation law [83].
3. The transfer of loads and deflections from between the fluid and structural model should not introduce any spurious new energy.

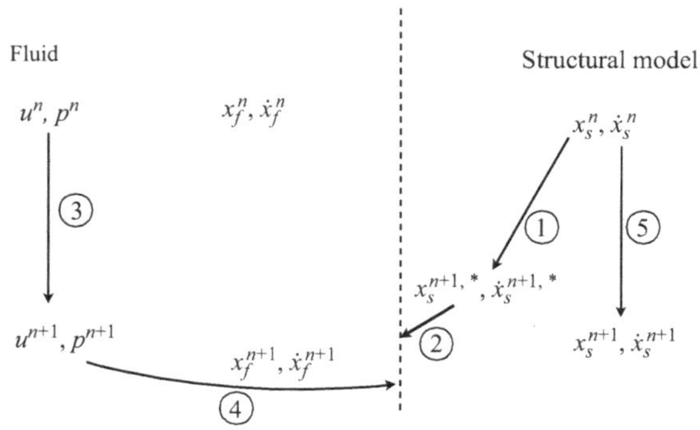

Figure 2.15 Illustration of the conventional serial staggered scheme [84] commonly used in blade-resolved FSI simulations. The dotted line represents the transfer of information at the solid–fluid interface

Depending on their background, authors use different terminology to describe FSI coupling algorithms including tight, strong, and loose coupling. In this chapter, we redefine these terms for simplicity. A tightly coupled FSI simulation uses a monolithic solver and a matching temporal discretization for the fluid and structural models. A strongly coupled FSI simulation uses partitioned/segregated solvers for the fluid and structural models, with several outer iterations between the two at every time step (e.g., Bazilevs *et al.* [52]). A loosely coupled FSI simulation uses partitioned/segregated solvers for the fluid and structural models with no outer iterations between the two at every time step (e.g., Heinz *et al.* [80]). Other FSI coupling approaches exploit periodicity under uniform inflow conditions to exchange information between the fluid and structural models only every revolution, as is typical of helicopter modeling approaches (e.g., Yu and Kwon [37]).

A strong or loose coupling approach is the most reasonable strategy for blade-resolved FSI simulations of wind turbines under generic inflow conditions. The most common loose coupling FSI algorithm is the conventional serial staggered (CSS) scheme [84] shown in Figure 2.15. x_s and $\dot{x}_s$ represent the position and the velocity of the structure on the solid–fluid interface, respectively; x_f and $\dot{x}_f$ represent the same information after mapping to the fluid side. Lesoinne, Farhat, and Piperno *et al.* have shown in several papers that [84–88] the CSS procedure is "only first-order time accurate," even when the underlying flow and structural solvers are second-order time accurate. Heinz *et al.* [80] observe almost no difference between strong and loose coupling approaches in their blade-resolved FSI simulations of the NREL 5-MW turbine under uniform inflow conditions. The importance of strong coupling in blade-resolved FSI simulations of wind turbines under generic inflow conditions is still a research topic.

Blade-resolved FSI simulations are likely to provide more accurate prediction of the chord-wise bending moment and corresponding torsional deflection in the outer regions of the blade compared to rigid blade CFD simulations. Blade-resolved

FSI simulations are currently the only way to predict the effect of vortex-induced vibrations on the rotor at standstill. The operation of the next generation of increasingly larger and slender turbines over a large wind shear will increase the need for blade-resolved FSI simulations. Availability of validation data sets under varying inflow conditions that include the details of the structural properties of the blade and tower will spur further research on blade-resolved FSI simulation technology.

2.3 Practical issues in performing blade-boundary-layer-resolved simulations

Several issues arise when all the tools from the previous section are put together to perform blade-resolved simulations around wind turbines. In this section, we discuss some of the practical issues that arise when performing a real blade-resolved CFD simulation.

2.3.1 Mesh generation

The size of meshes used for blade-resolved CFD of wind turbines has increased from $O(1-10\text{ M})$ cells to $O(50-100\text{ M})$ cells over the last decade, in large part a result of increased availability of computational power. However, the mesh generation procedures continue to be based on serial/single processor codes. The mesh generation step and the decomposition of the mesh to all processors usually require a computational node with large amounts of memory.

2.3.2 Mesh quality

The quality of the mesh greatly affects the rate of convergence of the CFD solver at every time step. The quality of the mesh also controls the maximum allowable time step for a stable and accurate solution. On a mesh with severely deformed cells, non-orthogonal correctors [89], or other approximate methods of computing the Laplacian for diffusion terms may be required to achieve convergence.

2.3.3 Convergence and time step

The convergence of the CFD solution as a whole is typically determined by the drop in the norm of the residual of the several linear systems of equations $|Ax-b|$ in the solution procedure at every time step or the nonlinear residual, $|b|$, if b represents the discretized Navier–Stokes equations itself. Several authors report using a time step corresponding to $0.5^\circ - 2.0^\circ$ of rotation of the turbine. Such a time step will be necessary to complete a reasonable number of rotations of the turbine to get statistically relevant results. On any reasonable mesh, such a time step will correspond to a Courant–Friedrichs–Lewy (CFL) number [56] that is much higher than one. While this is counter-intuitive to the typical requirement of $CFL < 1$, the regions of high CFL usually occur close to the leading and trailing edge of the airfoil sections near the tip of the blades. Figure 2.16 shows the contours of element CFL number on a

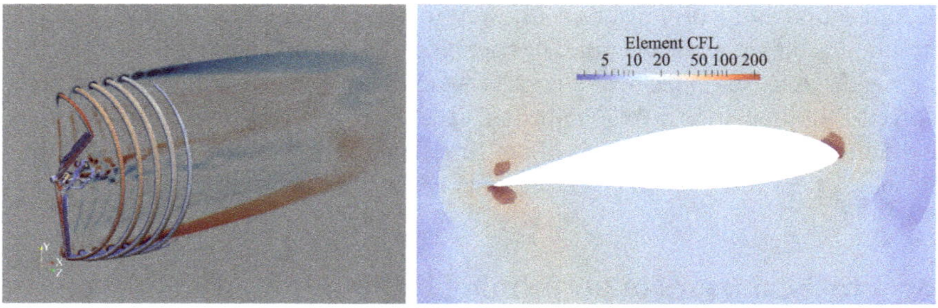

(a) (b)

Figure 2.16 Overset simulation of the NREL 5-MW turbine in uniform inflow of 8 m/s at Δ t corresponding to 0.25° rotation using Nalu-wind. (a) Isosurfaces of q criterion along with isocontours of vorticity on a vertical plane through the hub. (b) Contours of element CFL number on a 2D cross section close to the blade tip. Shaded areas with the mesh show elements with CFL > 50. Nalu-wind is the solver used in the Atmosphere to Electrons project [17], available from https://github.com/Exawind/nalu-wind. © 2019. Reprinted with permission from [90]

2D cross section close to the blade tip from a simulation of an NREL 5-MW turbine in uniform inflow at 8 m/s at a Δt corresponding to 0.25° rotation. The regions close to the leading edge and the trailing edge near the tip of the blade have the largest CFL number, while the rest of the domain operates closer to a CFL number of 1. The CFD solver must be capable of producing stable and accurate results at high local CFL numbers. For hybrid URANS/LES simulations, the $CFL < 1$ constraint probably needs to be satisfied in the LES region. The time step and the number of Newton iterations per time step are usually adjusted to optimize wall clock time for each specific problem (e.g., as shown by Mittal *et al.* [48]).

2.3.4 Verification

Code verification and validation are essential to performing predictive simulations with any CFD solver [91]. Verification shows that the CFD code does solve the intended conservation equations to the desired order of accuracy through a process of mesh refinement. Code verification should be typically performed using known analytical solutions or the method of manufactured solutions as shown by Roache [91]. However, CFD codes in the literature usually do not explicitly show verification studies. Verification of CFD codes is usually complicated through addition of runtime features like relaxation and turbulence modeling. Domino [92] shows verification studies of non-conformal low-Mach fluid algorithms for applications to blade-resolved CFD of wind turbines.

2.3.5 Validation

As described in Section 2.2, a blade-resolved simulation of the flow past a wind turbine involves several components acting together. Validation of CFD codes requires a comparison to experimental data under controlled inflow conditions. There are three main data sets available for validation of blade-resolved CFD solvers, including NREL Phase VI [71], NEW-MEXICO [73], and BT data sets [72]. All data sets provide sectional and integrated loads on the turbine over a wide range of canonical operating conditions. The NEW-MEXICO and BT data sets also provide measurements in the wake region for validation purposes. Finally, the BT data set also includes cases with multiple turbines with wake interaction. Transition modeling may be necessary when performing blade-resolved computations of these turbines for comparison to experiments. Modeling the nacelle and blade root geometry may also be required, as the ratio of nacelle to blade diameter is usually larger in model wind turbines than wind turbines in the field. Miller *et al.* [93] have recently claimed experiments on a model wind turbine in a wind tunnel that have full-scale similarity with $O(100\,\mathrm{kW})$ scale wind turbines in the field. The adoption of this data set for validation of blade-resolved wind-turbine simulations could be interesting.

2.4 Conclusions and challenges for future advancement in the state-of-the-art

The wind-turbine CFD literature points to the success of full Navier–Stokes computations in predicting wind-turbine loads under attached flow conditions and indicates a need for using hybrid URANS/LES methods to capture stalled flow regimes accurately. A major impact of blade-resolved CFD has been the successful adaptation of the blade geometry to changing design trends and reducing the cost of wind energy over the last two decades. Wind-turbine manufacturers have successfully incorporated blade-resolved CFD to tune their lower fidelity models based on BEM theory and design of increasingly larger blades. Engineering stall delay models like those by Du and Selig [94] are typically used to adjust the lookup tables for aerodynamic coefficients to allow the use of BEM-theory-based models for design. The use of blade-resolved CFD allows for better prediction of the effects of 3D flow, rotational augmentation, and stall delay and tip loss on the aerodynamic performance of the turbine [11]. Blade-resolved CFD has also played a part in the proliferation of the use of thicker airfoils with flat-back profiles near the blade root of wind turbines without the loss of aerodynamic performance through reduction of "root leakage" [95]. The use of thicker airfoils near the blade root has helped reduce the structural weight for increasingly larger turbines. Many wind-turbine manufacturers have adapted the use of vortex generators on the suction side of the blade over the past decade. Blade-resolved CFD plays an important role in the optimization of the size, configuration, and placement of the vortex generators on the blades [4,96]. Many wind-turbine manufacturers are actively using blade-resolved CFD to help design devices like flaps, DinoShells and DinoTails [95] to achieve the next round of improvements in aerodynamic performance and control of wind-turbine blades.

Because of the highly nonlinear relationship between mechanical power and rotor diameter, the trend in wind-turbine design is moving toward ever-larger wind-turbine rotors, requiring more flexible blades to minimize weight. The corresponding increase in blade elastic deformation leads to greater response between the spatially and temporally varying turbulent winds and the spatially and temporally varying deformations of the blade, especially in the daytime ABL when 3D turbulent motions are dominant. The response of the blade boundary layer to these stronger spatiotemporal variations in blade geometry will progressively play a more important role in aerodynamic response and temporal variations in integrated loads (moments and forces at the blade hub) and on the release of vorticity into the rotor wake. Consequently, blade-resolved CFD will play an increasingly important role in blade and rotor design of increasingly larger wind turbines, especially in offshore applications.

In the future, blade-resolved CFD of wind turbines will continue to play a major role in the design of larger wind turbines for offshore applications. Adjoint-based optimization of turbine blades, as shown by Dhert *et al.* [97] and Mishra *et al.* [98], will likely become more mainstream. The large turbines of the future will require aerodynamic data for airfoils at larger Reynolds numbers. Sørensen *et al.* [13] discuss the difficulty of obtaining experimental data for airfoils at high Reynolds numbers due to the expensive nature of the experiments. Thus, blade-resolved CFD may be required to generate aerodynamic characteristics of airfoils for the design of wind turbines in the future. These future wind turbines will also operate over a larger shear profile of the ABL, thereby increasing the likelihood of off-design blade operation in addition to variations associated with large-eddy wind fluctuations. Without any new aerodynamic control devices, the turbine blade will experience increased dynamic boundary layer separation and unsteady aerodynamic effects. According to Gupta [99], using a single degree of freedom blade pitch control for a rotor larger than 60 m in length is not optimal. The design of new local aerodynamic control devices like flaps [100] will require a tighter integration of blade-resolved CFD into the wind-turbine design process.

Current wind-turbine drivetrains favor the use of gearboxes to increase the rpm between the main shaft and the high-speed shaft that drives the generator. The mechanical transfer of torque through the gearbox changes the characteristics of torque fluctuations at the generator relative to the main shaft [27]. However, one increasingly used approach to avoid problems with premature gearbox failure is a direct-drive system [101], for which fluctuations entering the drivetrain from the rotor hub will more directly be passed to the generator. Thus, the more precise predictions of the fluctuating load transients afforded by blade-resolved CFD can potentially be important for generator design in direct-drive wind-turbine systems.

Significant challenges need to be resolved to enable blade-resolved modeling to improve the next generation of aeroelastic models of wind turbines. Resolution of these challenges is required to develop feasible, perhaps eventually even practical, blade-resolved CFD of individual wind turbines in true atmospheric turbulence. These challenges fall roughly into two interdependent groupings: representations of boundary layer physics and computational efficiency.

Key physics-based modeling improvements are needed for the development of validated hybrid URANS/LES formulations that predict the turbine performance over the entire operating envelope. Blade-boundary-layer-resolved simulations of wind-turbine rotors resolve and model the flow physics across scales separated by several orders of magnitude including the $\sim 10\ \mu m$ viscous layers adjacent to the blade surface as shown in Figures 2.5 and 2.6. Turbine blades operating within their design envelope are usually modeled using RANS as the blade boundary layers are generally attached, with the exception of the near-hub region with thickened quasi-circular sections that do not contribute significantly to blade torque. Hybrid RANS/LES models may be required to capture subtle viscous-layer instability and 3D separation dynamics under off-design conditions, along with significantly increased resolution in the chord and span-wise directions and in the outer layer of the boundary layer in the surface-normal direction. An open question is the extent to which these subtle viscous-layer dynamics are important to wind-turbine function—that is, to what extent do subtle boundary layer separation dynamics impact wind-turbine torque and power or deleterious non-steady response dynamics that impact fatigue and component failure? These more subtle blade boundary layer dynamics may be functionally significant when the blade is operating outside its design envelope, which can occur with sufficient rotor yaw error or dynamic pitching of individual blades. We may anticipate, therefore, that separation associated with viscous-layer instability will likely become progressively more important as wind-turbine technology advances, and more sophisticated control strategies are implemented that move the blade flows further outside their design envelope for longer periods of time during normal wind-turbine operation. The most severe modeling requirements are adjacent to the blade surfaces wherein turbulence models are needed that predict the turbine performance over the entire operating envelope including time-varying loads under dynamic stall and highly stalled regions. To this end, improvements are needed to better represent the transition between the RANS-like resolutions of high-aspect-ratio grid cells near the surface, and $O(1)$ aspect-ratio grid cells required for LES at some distance from the blade surfaces. These models should be developed alongside appropriate wind-turbine-specific experiments that bridge the divide between disparate scales in the atmosphere and blade boundary layer.

Computational efficiency challenges couple directly with the development of physically realistic model predictions because local model fidelities and global computational efficiency are both directly coupled to the resolution and design of the computational grid. Indeed, from a practical perspective, grid development is often a weak link, as it requires extensive effort with major implications to predictive accuracy, numerical instability and fidelity, and, importantly, computational load. The computational cost of blade-resolved simulations of wind turbines must come down significantly to allow them to be integrated tightly into the design process (Chapter 3, Volume 2). For example, blade-resolved simulations of specific design load conditions should be completed within 1 day. This will likely be driven by advances in the next generation of high-performance parallel computing platforms in which "flops are free" and data movement will become the primary bottleneck (Chapter 1, Volume 1). Parallel mesh generation strategies must be developed with better integration with

CAD and decomposition to very large numbers of processors. Scalable and robust CFD solvers that incorporate submodels with sufficient physical accuracy would also need to be developed that target the next generation of computational platforms.

Acknowledgments

This work was authored, in part, by the National Renewable Energy Laboratory, operated by Alliance for Sustainable Energy, LLC, for the U.S. Department of Energy (DOE) under Contract No. DE-AC36-08GO28308. Funding provided by the U.S. Department of Energy Office of Energy Efficiency and Renewable Energy Wind Energy Technologies Office. G. Vijayakumar's contribution to this work was supported by U.S. DOE Energy Efficiency and Renewable Energy Wind Energy Technologies Office. The views expressed in the article do not necessarily represent the views of the DOE or the U.S. Government. The U.S. Government retains and the publisher, by accepting the article for publication, acknowledges that the U.S. Government retains a nonexclusive, paid-up, irrevocable, worldwide license to publish or reproduce the published form of this work, or allow others to do so, for U.S. Government purposes.

References

[1] Troldborg N, Zahle F, Réthoré PE, *et al.* Comparison of wind turbine wake properties in non-sheared inflow predicted by different computational fluid dynamics rotor models. Wind Energy. 2015;18(7):1239–1250.

[2] Laursen J, Enevoldsen P, Hjort S. 3D CFD quantification of the performance of a multi-megawatt wind turbine. Journal of Physics: Conference Series. 2007;75(1):012007.

[3] Wolf W, Lele S, Jothiprasad G, *et al.* Investigation of Noise Generated by a DU96 Airfoil. In: 18th AIAA/CEAS Aeroacoustics Conference (33rd AIAA Aeroacoustics Conference). American Institute of Aeronautics and Astronautics; 2012.

[4] Tian QQ, Corson D, Baker JP. Application of Vortex Generators to Wind Turbine Blades. In: 34th Wind Energy Symposium. American Institute of Aeronautics and Astronautics; 2016.

[5] Schramm M, Rahimi H, Stoevesandt B, *et al.* The influence of eroded blades on wind turbine performance using numerical simulations. Energies. 2017;10(9).

[6] Sareen A, Sapre CA, Selig MS. Effects of leading edge erosion on wind turbine blade performance. Wind Energy. 2014;17(10):1531–1542.

[7] Gaudern N. A practical study of the aerodynamic impact of wind turbine blade leading edge erosion. Journal of Physics: Conference Series. 2014;524(1):012031.

[8] Bak C, Zahle F, Bitsche R, *et al.* The DTU 10-MW Reference Wind Turbine; 2013.

[9] Jonkman J, Butterfield S, Musial W, *et al.* Definition of a 5-MW Reference Wind Turbine for Offshore System Development. National Renewable Energy Laboratory; 2009. NREL/TP-500-38060.

[10] Rinker J, Dykes K. WindPACT Reference Wind Turbines. National Renewable Energy Laboratory; 2018. NREL/TP-5000-67667.

[11] Guntur S, Sørensen NN. A study on rotational augmentation using CFD analysis of flow in the inboard region of the MEXICO rotor blades. Wind Energy. 2015;18(4):745–756.

[12] AVATAR – Advanced Aerodynamic Modelling, Design and Testing for Large Rotor Blades. European Energy Research Alliance; 2018.

[13] Sørensen NN, Zahle F, Boorsma K, *et al.* CFD computations of the second round of MEXICO rotor measurements. Journal of Physics: Conference Series. 2016;753(2):022054.

[14] Ananthan S. Atmosphere to Electrons (A2e) Overview. In: The 3rd Wind Energy Systems Engineering Workshop; 2015.

[15] UPWIND – Design Limits and Solutions for Very Large Wind Turbines. European Community – Sixth Framework Programme; 2011.

[16] LCOE Reduction for the Next Generation Offshore Wind Turbines – Outcomes From the INNWIND.EU Project. European Community - Seventh Framework Programme; 2017.

[17] Atmosphere to Electrons (A2e) Overview. US Department of Energy – Office of Energy Efficiency and Renewable Energy; 2015.

[18] Lawson M, Melvin J, Ananthan S, *et al.* Blade-Resolved, Single-Turbine Simulations Under Atmospheric Flow. Golden, CO: National Renewable Energy Laboratory; 2018. NREL/TP-5000-72760.

[19] Sullivan PP, McWilliams JC, Moeng CH. A subgrid-scale model for large-eddy simulation of planetary boundary-layer flows. Boundary-Layer Meteorology. 1994;71(3):247–276.

[20] Khanna S, Brasseur JG. Three-dimensional buoyancy- and shear-induced local structure of the atmospheric boundary layer. Journal of the Atmospheric Sciences. 1998;55(5):710–743.

[21] Vijayakumar G. Non-Steady Dynamics of Atmospheric Turbulence Interaction with Wind Turbine Loadings Through Blade-Boundary-Layer-Resolved CFD. The Pennsylvania State University, University Park, PA; 2015.

[22] Wächter M, Heißelmann H, Hölling M, *et al.* The turbulent nature of the atmospheric boundary layer and its impact on the wind energy conversion process. Journal of Turbulence. 2012;13:N26.

[23] Berg J, Natarajan A, Mann J, *et al.* Gaussian vs non-Gaussian turbulence: impact on wind turbine loads. Wind Energy. 2016;19(11): 1975–1989.

[24] Rai RK, Gopalan H, Naughton JW. Effects of spatial and temporal resolution of the turbulent inflow on wind turbine performance estimation. Wind Energy. 2016;19(7):1341–1354.

[25] Sim C, Basu S, Manuel L. On space-time resolution of inflow representations for wind turbine loads analysis. Energies. 2012;5(7):2071–2092.

[26] Lavely AW. Effects of Daytime Atmospheric Boundary Layer Turbulence on the Generation of Nonsteady Wind Turbine Loadings and Predictive Accuracy of Lower Order Models. The Pennsylvania State University, University Park, PA; 2017.

[27] Nandi TN, Herrig A, Brasseur JG. Non-steady wind turbine response to daytime atmospheric turbulence. Philosophical Transactions of the Royal Society of London A: Mathematical, Physical and Engineering Sciences. 2017;375(2091).

[28] Vijayakumar G, Brasseur J, Lavely AW, *et al.* Interaction of Atmospheric Turbulence With Blade Boundary Layer Dynamics on a 5MW Wind Turbine Using Blade-Boundary-Layer-Resolved CFD With Hybrid URANS-LES. In: 34th Wind Energy Symposium. American Institute of Aeronautics and Astronautics; 2016.

[29] Kelly ND, Jonkman BJ, Scott GN. The Great Plains Turbulence Environment: Its Origins, Impact and Simulation. National Renewable Energy Laboratory; 2006. NREL/CP-500-40176.

[30] Leishman GJ. Principles of Helicopter Aerodynamics. Cambridge Aerospace Series, Cambridge University Press, Cambridge; 2006.

[31] GridPro – Automatic Multiblock Grid Generation Software. http://www.gridpro.com/.

[32] Pointwise – Mesh Generation Software. Available from: http://www.pointwise.com/.

[33] Chan WM, Pandya SA, Rogers SE, *et al.* Chimera Grid Tools User's Manual, Version 2.2. NASA Advanced Supercomputing Division; 2018.

[34] Sørensen NN, Michelsen JA, Schreck S. Navier–Stokes predictions of the NREL phase VI rotor in the NASA Ames 80 ft × 120 ft wind tunnel. Wind Energy. 2002;5(2–3):151–169.

[35] Sørensen NN, Johansen J. UPWIND, Aerodynamics and Aero-elasticity, Rotor Aerodynamics in Atmospheric Shear Flow. Technical University of Denmark, Lyngby; 2007.

[36] Lynch CE, Smith MJ. Unstructured overset incompressible computational fluid dynamics for unsteady wind turbine simulations. Wind Energy. 2013;16(7):1033–1048.

[37] Yu DO, You JY, Kwon OJ. Numerical investigation of unsteady aerodynamics of a Horizontal-axis wind turbine under yawed flow conditions. Wind Energy. 2013;16(5):711–727.

[38] Pope SB. Turbulent Flows. Cambridge University Press, Cambridge; 2005.

[39] Spalart PR, Strelets MK, Travin A. Expectations in the Wall Region of a Large-Eddy Simulation. In: Meyers J, Geurts BJ, Sagaut P, editors. Quality and Reliability of Large-Eddy Simulations. vol. 12 of ERCOFTAC Series. Springer Netherlands; 2008. pp. 181–191.

[40] Sørensen NN, Bertagnolio F, Jost E, *et al.* Aerodynamic effects of compressibility for wind turbines at high tip speeds. Journal of Physics: Conference Series. 2018;1037(2):022003.

[41] Sitaraman J, Mavriplis DJ, Duque EP. Wind Farm Simulations Using a Full Rotor Model for Wind Turbines. In: 32nd ASME Wind Energy Symposium. American Institute of Aeronautics and Astronautics; 2014.

[42] Kirby AC, Brazell M, Yang Z, *et al.* Wind Farm Simulations Using an Overset hp-Adaptive Approach with Blade-Resolved Turbine Models. In: 23rd AIAA Computational Fluid Dynamics Conference. American Institute of Aeronautics and Astronautics; 2017.

[43] Potsdam M, Mavriplis D. Unstructured Mesh CFD Aerodynamic Analysis of the NREL Phase VI Rotor. In: Aerospace Sciences Meetings. American Institute of Aeronautics and Astronautics; 2009.

[44] Dose B, Rahimi H, Herráez I, *et al.* Fluid–structure coupled computations of the NREL 5MW wind turbine blade during standstill. Journal of Physics: Conference Series. 2016;753(2):022034.

[45] van Rooij RPJOM, Arens EA. Analysis of the experimental and computational flow characteristics with respect to the augmented lift phenomenon caused by blade rotation. Journal of Physics: Conference Series. 2007;75(1):012021.

[46] Länger-Möller A, Löwe J, Kessler R. Investigation of the NREL phase VI experiment with the incompressible CFD solver THETA. Wind Energy. 2017;20(9):1529–1549.

[47] Yelmule MM, Anjuri EVSJ. CFD predictions of NREL phase VI rotor Experiments in NASA/AMES wind tunnel. International Journal of Renewable Energy Research. 2013;3(2):261–269.

[48] Mittal A, Sreenivas K, Taylor LK, *et al.* Blade-resolved simulations of a model wind turbine: effect of temporal convergence. Wind Energy. 2016;19(10):1761–1783.

[49] Chow R, van Dam CP. Computational investigations of blunt trailing-edge and twist modifications to the inboard region of the NREL-5MW rotor. Wind Energy. 2013;16(3):445–458.

[50] Le Pape A, Lecanu J. 3D Navier–Stokes computations of a stall-regulated wind turbine. Wind Energy. 2004;7(4):309–324.

[51] Bazilevs Y, Hsu MC, Kiendl J, *et al.* 3D simulation of wind turbine rotors at full scale. Part II: Fluid–structure interaction modeling with composite blades. International Journal for Numerical Methods in Fluids. 2011;65(1–3): 236–253.

[52] Bazilevs Y, Hsu MC, Akkerman I, *et al.* 3D simulation of wind turbine rotors at full scale. Part I: Geometry modeling and aerodynamics. International Journal for Numerical Methods in Fluids. 2011;65(1–3):207–235.

[53] Corson D, Griffith DT, Ashwill T, *et al.* Investigating Aeroelastic Performance of Multi-Mega Watt Wind Turbine Rotors Using CFD. In: 53rd AIAA/ASME/ASCE/AHS/ASC Structures, Structural Dynamics and Materials Conference. American Institute of Aeronautics and Astronautics; 2012.

[54] Wyngaard JC. Turbulence in the Atmosphere. Cambridge University Press, Cambridge; 2010.

[55] Waterson NP, Deconinck H. Design principles for bounded higher-order convection schemes – a unified approach. Journal of Computational Physics. 2007;224(1):182–207. Special Issue Dedicated to Professor Piet Wesseling on the occasion of his retirement from Delft University of Technology.

[56] Ferziger JH, Peric M. Computational Methods for Fluid Dynamics. 3rd ed. Springer, Berlin; 2002.

[57] Golub GH, Loan CFV. Matrix Computations. 3rd ed. The Johns Hopkins University Press, Baltimore, MD; 1996.

[58] Saad Y. Iterative Methods for Sparse Linear Systems. 2nd ed. Society for Industrial and Applied Mathematics, Philadelphia, PA; 2003.

[59] HYPRE; 2015. https://computation.llnl.gov/projects/hypre-scalable-linear-solvers-multigrid-methods/software.

[60] Balay S, Abhyankar S, Adams MF, *et al.* PETSc Web Page; 2015. http://www.mcs.anl.gov/petsc. Available from: http://www.mcs.anl.gov/petsc.

[61] Nichols RH, Buning PG. User's Manual for OVERFLOW 2.2; 2017. https://overflow.larc.nasa.gov/home/users-manual-for-overflow-2-2/.

[62] Donea J, Huerta A, Ponthot JP. 14: Arbitrary Lagrangian–Eulerian Methods. In: Encyclopedia of Computational Mechanics. John Wiley and Sons Ltd., Hoboken, NJ; 2004.

[63] Chan WM, Gomez RJ, Rogers SE, *et al.* Best Practices in Overset Grid Generation. In: 32nd AIAA Fluid Dynamics Conference. AIAA 2002-3191; 2002.

[64] Farrell PE, Maddison JR. Conservative interpolation between volume meshes by local Galerkin projection. Computer Methods in Applied Mechanics and Engineering. 2011;200:89–100.

[65] Spalart PR. Detached-Eddy simulation. Annual Review of Fluid Mechanics. 2009;41(1):181–202.

[66] Sagaut P, Deck S. Multiscale and Multiresolution Approaches in Turbulence: LES, DES and Hybrid RANS/LES Methods: Applications and Guidelines. 2nd ed. Imperial College Press, London; 2013.

[67] Spalart PR. A one-equation turbulence model for aerodynamic flows. La Recherche Aerospatiale. 1994;1:5–21.

[68] Menter F, Kuntz M, Langtry R. Ten years of industrial experience with the SST turbulence model. In: International Symposium on turbulence, heat and mass transfer; 2003. vol. 4, pp. 625–632.

[69] Duque EPN, Burklund MD, Johnson W. Navier–Stokes and comprehensive analysis performance predictions of the NREL phase VI experiment. Journal of Solar Energy Engineering. 2003;125(4):457–467.

[70] Chao DD, van Dam CP. Computational aerodynamic analysis of a blunt trailing-edge airfoil modification to the NREL Phase VI rotor. Wind Energy. 2007;10(6):529–550.

[71] Hand M, Simms D, Fingersh LJ, *et al.* Unsteady Aerodynamics Experiment Phases VI: Wind Tunnel Test Configurations and Available Data Campaigns; 2001. NREL/TP-500-29955.

[72] Krogstad PÅ, Sætran L. Wind turbine wake interactions; results from blind tests. Journal of Physics: Conference Series. 2015;625(1):012043.

[73] Boorsma K, Schepers JG. Rotor experiments in controlled conditions continued: New Mexico. Journal of Physics: Conference Series. 2016;753(2):022004.

[74] Sørensen NN. CFD modelling of laminar-turbulent transition for airfoils and rotors using the $\gamma - Re_{\Theta_t}$ model. Wind Energy. 2009;12(8):715–733.

[75] Schaffarczyk AP, Schwab D, Breuer M. Experimental detection of laminar-turbulent transition on a rotating wind turbine blade in the free atmosphere. Wind Energy. 2016;20(2):211–220.

[76] Chow R, van Dam CP. Verification of computational simulations of the NREL 5 MW rotor with a focus on inboard flow separation. Wind Energy. 2012;15(8):967–981.

[77] Johansen J, Sørensen NN, Michelsen JA, *et al.* Detached-eddy simulation of flow around the NREL Phase VI blade. Wind Energy. 2002;5(2–3): 185–197.

[78] Travin A, Shur M, Strelets M, *et al.* Physical and Numerical Upgrades in the Detached-Eddy Simulation of Complex Turbulent Flows. In: Friedrich R, Rodi W, editors. Advances in LES of Complex Flows. vol. 65 of Fluid Mechanics and Its Applications. Springer Netherlands; 2004. pp. 239–254.

[79] Sørensen NN, Schreck S. Computation of the National Renewable Energy Laboratory Phase-VI rotor in pitch motion during standstill. Wind Energy. 2012;15(3):425–442.

[80] Heinz JC, Sørensen NN, Zahle F. Fluid–structure interaction computations for geometrically resolved rotor simulations using CFD. Wind Energy. 2016;19(12):2205–2221.

[81] Sayed M, Lutz T, Krämer E, *et al.* High fidelity CFD-CSD aeroelastic analysis of slender bladed horizontal-axis wind turbine. Journal of Physics: Conference Series. 2016;753(4):042009.

[82] Carrión M, Steijl R, Woodgate M, *et al.* Aeroelastic analysis of wind turbines using a tightly coupled CFD–CSD method. Journal of Fluids and Structures. 2014;50:392–415.

[83] Lesoinne M, Farhat C. Geometric conservation laws for flow problems with moving boundaries and deformable meshes, and their impact on aeroelastic computations. Computer Methods in Applied Mechanics and Engineering. 1996;134(1):71–90.

[84] Farhat C, Lesoinne M, Tallec PL. Load and motion transfer algorithms for fluid/structure interaction problems with non-matching discrete interfaces: Momentum and energy conservation, optimal discretization and application to aeroelasticity. Computer Methods in Applied Mechanics and Engineering. 1998;157(1):95–114.

[85] Farhat C, Lesoinne M, Maman N. Mixed explicit/implicit time integration of coupled aeroelastic problems: three-field formulation, geometric conservation and distributed solution. International Journal for Numerical Methods in Fluids. 1995;21(10):807–835.

[86] Farhat C, Lesoinne M. Two efficient staggered algorithms for the serial and parallel solution of three-dimensional nonlinear transient aeroelastic problems. Computer Methods in Applied Mechanics and Engineering. 2000;182(3–4):499–515.
[87] Piperno S, Farhat C. Design of efficient partitioned procedures for the transient solution of aeroelastic problems. Revue Européenne des Éléments Finis. 2000;9(6–7):655–680.
[88] Farhat C, Rallu A, Wang K, *et al.* Robust and provably second-order explicit–explicit and implicit–explicit staggered time-integrators for highly non-linear compressible fluid–structure interaction problems. International Journal for Numerical Methods in Engineering. 2010;84(1):73–107.
[89] Jasak H. Error Analysis and Estimation for the Finite Volume Method with Applications to Fluid Flows. Department of Mechanical Engineering Imperial College of Science, Technology and Medicine, London; 1996.
[90] Ananthan S, Vijayakumar G, Sitaraman J, Sharma A, Sprague MA. Blade-resolved, overset-mesh turbine simulations using Nalu-Wind. In: Wind Energy Science Conference, Cork, Ireland; 2019.
[91] Roache PJ. Verification and Validation in Computational Science and Engineering. Hermosa Publishers, Socorro, NM; 1998.
[92] Domino SP. Design-order, non-conformal low-Mach fluid algorithms using a hybrid CVFEM/DG approach. Journal of Computational Physics. 2018;359:331–351.
[93] Miller MA, Kiefer J, Westergaard C, *et al.* Model wind turbines tested at full-scale similarity. Journal of Physics: Conference Series. 2016;753(3):032018.
[94] Du Z, Selig M. A 3-D Stall-Delay Model for Horizontal Axis Wind Turbine Performance Prediction. In: 1998 ASME Wind Energy Symposium. American Institute of Aeronautics and Astronautics; 1998.
[95] Siemens. Fact Sheet: Rotor Blades; 2012. Available from: https://w5.siemens.com/italy/web/pw/press/News/Documents/Fact%20Sheet%20Blade%20Technology.pdf.
[96] Manolesos M, Sørensen NN, Troldborg N, *et al.* Computing the flow past vortex generators: comparison between RANS simulations and experiments. Journal of Physics: Conference Series. 2016;753(2):022014.
[97] Dhert T, Ashuri T, Martins JRRA. Aerodynamic shape optimization of wind turbine blades using a Reynolds-averaged Navier–Stokes model and an adjoint method. Wind Energy. 2017;20(5):909–926.
[98] Mishra A, Mani K, Mavriplis D, *et al.* Time dependent adjoint-based optimization for coupled fluid–structure problems. Journal of Computational Physics. 2015;292(Supplement C):253–271.
[99] Gupta A. Trends and Changing R&D Needs in Blade Technology. In: 2016 Sandia Wind Turbine Blade Workshop; 2016.
[100] Nader J. Large Wind Turbine Blade Design Challenges and R&D Needs. In: 2016 Sandia Wind Turbine Blade Workshop; 2016.
[101] Siemens. Fact Sheet: The New SWT-6.0; 2013. Available from: https://www.siemens.com/press/pool/de/feature/2013/energy/2013-01-rotorblade/factsheet-6mw-e.pdf.

Chapter 3
Mesoscale modeling of the atmosphere

Sue Ellen Haupt[1], *Branko Kosović*[1], *Jared A. Lee*[1] *and Pedro Jiménez*[1]

3.1 Introduction to meteorology for wind energy modeling

Understanding the full range of scales and motions in atmospheric flow is the basis for modeling for wind plants. Because the flow field of the wind plant is embedded in a large-scale atmospheric flow, it is imperative to correctly represent all appropriate forcing scales to accurately simulate wind park flow. This atmospheric forcing is not merely an inflow to wind plant simulations but rather a constantly changing flow of energy at various scales. This inherent variability of the atmosphere on a range of temporal and spatial scales causes variability in conditions for wind energy generation that must be understood and modeled well before applying wind plant-scale simulations.

This chapter discusses the mechanisms responsible for generating atmospheric motion, scales of the various phenomena of the atmosphere relevant to modeling the wind plant, atmospheric energetics, and introduces the chaotic nature of the flow. Then the equations of motion and their discretization are presented, along with the discussion of typical boundary and initial conditions, including a discussion of data assimilation (DA) as a way to assure that the initial conditions best incorporate the observations. We then discuss how the atmospheric models are applied to wind-resource assessment and forecasting, and how the wind farm wakes are currently parameterized in the atmospheric models. For completeness, we introduce methods for assessment, postprocessing, and uncertainty quantification. As the field is rapidly evolving, we finish with discussion of where the field is likely to go in the next decade.

3.1.1 Forces and the general circulation of the atmosphere

The primary forcing mechanism for atmospheric flow is differential heating of the planet, modified by its rotation. Because the Earth is nearly spherical and rotates on an axis with a 23.4° tilt as it circles the sun, the solar heating is stronger at the equator on average and lesser at the poles. This differential heating creates a pressure gradient force that acts to generate poleward motion at the Earth's surface. In the

[1]National Center for Atmospheric Research, Boulder, CO, USA

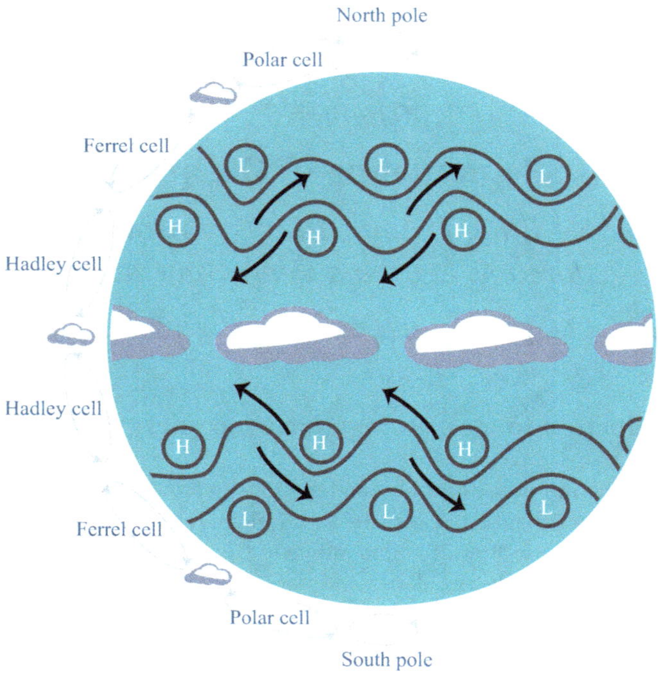

Figure 3.1 Depiction of the general circulation of the atmosphere

tropics, air rises, forming the Inter-Tropical Convergence Zone (ITCZ). A return flow aloft maintains continuity, creating a Hadley cell [1]. The rotation of the Earth causes an apparent Coriolis force to deflect the flow to the right in the Northern Hemisphere (left in the Southern Hemisphere). If these were the only forces acting, we would thus have surface easterly* flow in the Northern Hemisphere, which indeed is what is observed in the tropics and subarctic regions. Once again, continuity demands a westerly return flow aloft. The flow is actually complicated by instabilities that force a three-cell circulation system [2], as displayed in Figure 3.1. The counter-rotating cell in the mid-latitudes is known as the Ferrel Cell and the Polar Cell near each pole completes the general circulation.

Where the cells meet, semipermanent atmospheric features are created that ebb and flow seasonally as the Earth circles the sun. The ITCZ is associated with low pressures at the surface, causing the region of low wind near the thermal equator known as the doldrums, as air is mostly ascending due to convergence. Away from the equator, the Hadley cells create the northeast and southeast trade winds at the surface. Where the Hadley and Ferrel cells meet, the downward motion generates semipermanent higher pressure regions, resulting in the Bermuda-Azores High and

*Meteorologists refer to easterly flow as originating from the east, and similarly for other flow directions.

North Pacific High, with similar features in the southern hemisphere, the Kalahari High and South Pacific High; these are also popularly known as the "horse latitudes" due to the typically light and variable surface winds. The downward motion where the Ferrel Cell meets the Polar Cell creates a continuous belt of low pressure in the Southern Ocean and in the Northern Hemisphere creates the Aleutian Low and Icelandic Low features in winter. Each of these semipermanent low-pressure systems impacts the general circulation and the climatology of the Earth's regions. The details of these features are forced by the position of the major continents, which generate differences in surface heating and heat capacity as well as differences in the surface roughness that modify the general circulation.

The 23.4° tilt to the Earth's axis generates the seasonality as the Earth circles the sun. The features of the general circulation move northward and southward seasonally, modifying the energetics of the flow system.

The vertical structure of the atmosphere is also due to radiative forcing modified by dynamic factors. The lowest layer of the atmosphere, the troposphere, which is roughly 11 km deep at midlatitudes, is the one most relevant for harvesting wind energy. Note, however, that the forcing from the stratosphere above alters the flow patterns in the troposphere, and global modeling should be sufficiently deep to include these effects. Within the troposphere, the atmospheric boundary layer (ABL) represents the lowest layer that is well mixed, which ranges from very shallow depths (tens of meters) at night to depths of 1–3 km during the day when convection is occurring.

3.1.2 Scales and phenomena in the atmosphere

The atmospheric flow is further complicated by instabilities and smaller scale forcing, which generate secondary and tertiary circulations. As the nonlinear fluid flow evolves, instabilities are generated, resulting in waves, which are manifested as the passing weather systems. These weather systems are essentially ridges and depressions in the flow that are observed as fronts separating the warm air typically to the south (in the northern hemisphere) from the colder air to the north. Figure 3.2 is a diagram of a typical weather system, named the Norwegian cyclone model after the original discoverers [3]. The fronts are typically slanted vertically, with the warm front having a shallow slope as the warm air overruns the cooler air, while the cold front tends to be steeper as the colder air at the surface pushes through, displacing the warmer air upward as seen in Figure 3.2. These passing weather systems are important for wind energy in that they can cause rapid changes in wind speed and direction that lead to wind ramps and force changes in turbine orientation. Other phenomena that can cause such ramps include thunderstorm outflows, land and sea breezes, low-level jets, and more.

Secondary forcing mechanisms further alter the flow, including terrain, differential heating and stresses between the land and water regions of the Earth, surface canopy differences, and beyond. These effects cause local flow situations that can be modeled using numerical weather prediction (NWP) models that resolve these features. But some local features, such as mesoscale convective systems, can cause

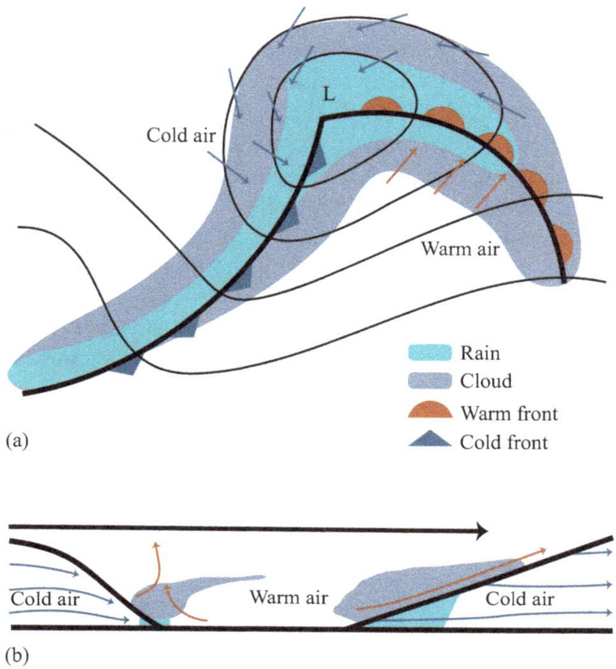

Figure 3.2 Schematic of the idealized Norwegian cyclone model. Part (a) shows a planar view looking down on the system while part (b) shows a vertical cross section

outflows that result in rapid changes in wind speed and direction, are often modeled more stochastically for forecasts beyond a couple hours, and real-time local observations are required to correctly capture the specific location and timing of such events.

Another predominant cause of atmospheric variability is the diurnal cycle of heating due to the Earth's rotation. The ABL evolves over the course of the day (Figure 3.3), with surface heating in the morning leading to vertical motion (convection), which causes rising air, increasing turbulence, and instability [4]. As the heating continues, the boundary layer enlarges and instability increases. At the same time, the rising air cools, the water vapor condenses, clouds form, and the heating becomes self-limited by the cloud cover. As evening arrives, heating subsides and so does the upward motion. The heat flux becomes negative as the sun sets, the top of the ABL collapses, and the atmosphere becomes stable. It is not unusual for a low-level jet to form under these stable conditions, causing a nocturnal increase in the wind speed within the ABL (Figure 3.4). Such conditions can cause wind shear across the rotor diameter and lead to differences in wind-power output, as well as to generate a vertical component of torque on the turbine.

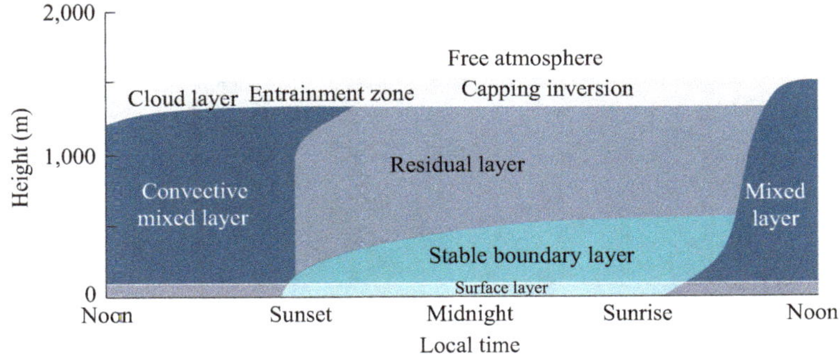

Figure 3.3 Schematic of the diurnal evolution of the atmospheric boundary layer

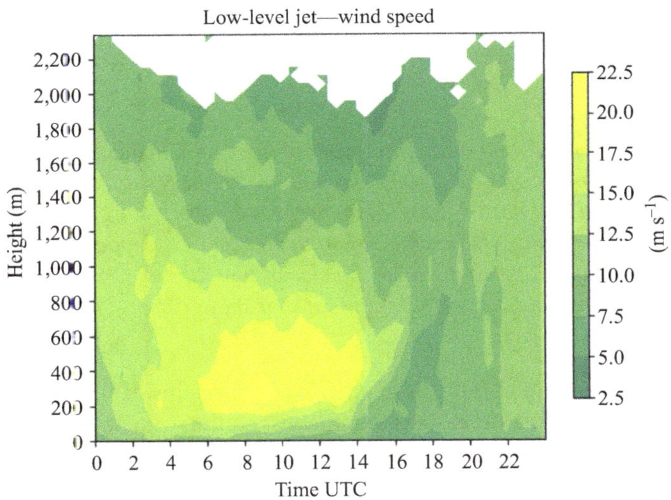

Figure 3.4 The diurnal pattern of wind speed at the CEWEX13 field study site on August 26, 2013. The ABL is well mixed during the daytime and has high shear during nighttime (from 06 to 14 UTC). The wind-speed profile was measured using a scanning Leosphere Windcube 200-S lidar [5]

3.1.3 Atmospheric energetics

All of these forces and resulting motions act together to redistribute the atmosphere's energy. Rather than having the energy from the sun focused near the equator, these motions redistribute it throughout the atmosphere. Conservation of angular momentum plays a role in the ebb and flow of the weather systems. Because the Coriolis parameter ($f = 2\Omega \sin \phi$, where Ω is the rotation rate of the Earth and ϕ is the latitude) that denotes the multiplicative effect of the Coriolis force disappears at the

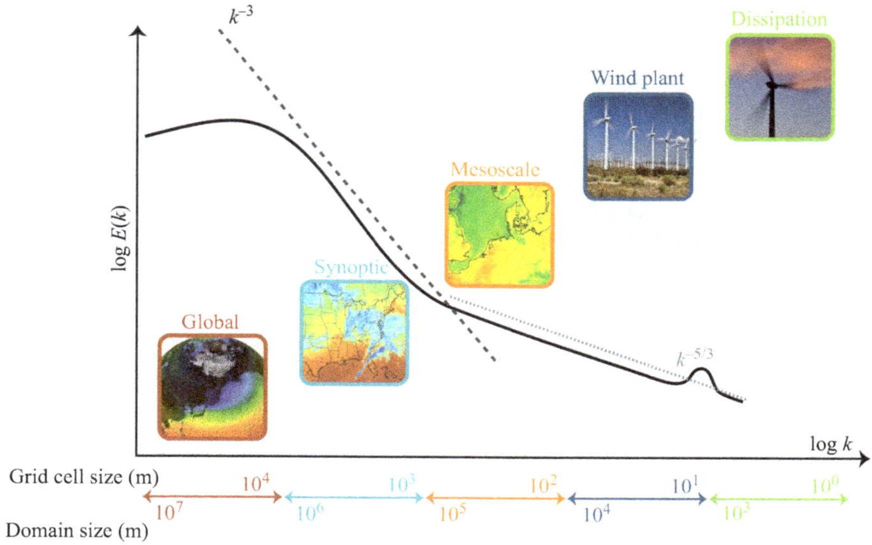

Figure 3.5 Scales of motion in the atmosphere interact. This energy spectrum (solid black line) indicates the amount of energy at each scale. The light dashed and dotted lines indicate the slopes of the different portions of the spectrum. Also included is a depiction of the type of motion represented by each scale, the size of typical domains, and the typical grid cell spacing for modeling at that scale (After [7])

equator and is maximum at the poles,† the rotating systems of the atmosphere seek to equalize the angular momentum through conservation of planetary vorticity. The energy and vorticity conservation interact to produce a cascade of energy from the large planetary scales to the smaller scales. The small-scale turbulence in the ABL further acts to redistribute energy on the shorter time scales/smaller spatial scales.

Figure 3.5 displays an average energy spectrum for the atmosphere [6]. The nonlinearly interacting waves of the atmosphere span a large range of scales. The weather systems that are most simply represented by the Norwegian cyclone model discussed above represent the largest scale Rossby waves (lows followed by highs) that circle the globe in the mid-latitudes (see Figure 3.1). These weather systems have wavelengths on the order of about 10,000 km, and their influence persists over a given region for a few days. The secondary circulations develop at the next smaller scales. Mesoscale circulations include the diurnal cycles described above as well as those features forced by surface heterogeneities, such as air–sea temperature differences that force land and sea breezes. Smaller waves include gravity waves (analogous to waves due to perturbations on the surface of a water body), inertial waves traveling

†This differential value of the Coriolis parameter with latitude is known as the beta effect.

through the interior of the rotating atmosphere, and sound waves due to atmospheric compressibility. The smallest eddies in the boundary layer have wavelengths on the order of centimeters. Thus, the range of scales of atmospheric motion encompasses about eight orders of magnitude. This fact implies that one cannot resolve all of these scales numerically with current computing capabilities. Note that energy in the smallest scales also propagates upscale. Thus, careful thought must be given to filtering to avoid having numerical simulations become unstable.

Thus, the atmosphere is an undulating structure with dynamic interactions between scales that drive the energy flow, including transfers between potential and kinetic energy, in an attempt to redistribute the global energy difference driven by incoming solar radiation and outgoing surface radiance. Although temporal and spatial averages and medians can be defined, they are seldom actually achieved. Thus, it is critical to model this variability and also to understand the limits of our ability to predict this variability.

3.1.4 The chaotic nature of atmospheric flow

In his seminal 1963 paper, Edward Lorenz demonstrated that the atmospheric equations of motion are quite sensitive to initial conditions. Extreme sensitivity implies that even tiny perturbations to the initial (or boundary) conditions can lead to large differences in the flow field at a future time [8].

Meteorologists deal with this chaotic nature of the atmosphere in two primary ways. First, they work to obtain the best possible initial and boundary conditions with which to initialize the models. This is accomplished by assimilating as many observations as possible into the model as an initial condition. This topic is covered in Section 3.3. The second way that this inherent chaotic nature is embraced is in quantifying the uncertainty in the flow by running ensembles of model simulations, each with varied initial or boundary conditions or with different model physics parameterizations. This uncertainty quantification is the topic of Section 3.6.

However, we can never and will never have a precisely accurate snapshot of the full atmospheric state globally; the resultant uncertainty in the initial state leads to inherent limits on the predictability of the future atmospheric state. This inherent predictability limit of the atmosphere, beyond which NWP models cannot have any skill compared to climatology (the 30-year average weather conditions for a particular date and location), was once thought to be about 15 days [9], but recent advances in modeling have suggested the actual limit for synoptic-scale phenomena is somewhere closer to 3 weeks, with shorter limits of predictability for smaller scale phenomena (e.g., [10–12]).

3.2 Basics of atmospheric modeling

3.2.1 Historical perspective

The history of simulations of atmospheric flows could be traced back to the work of Norwegian physicist Vilhelm Bjerknes. In 1904, he published a paper suggesting that it should be possible to forecast the weather by solving a system of nonlinear partial differential equations [13]. This idea was pursued by Lewis Fry Richardson. Richardson was an English mathematician and physicist after whom the Richardson number, the nondimensional ratio of the temperature gradient to the shear, is named. Richardson had an interest in meteorology. In his 1922 paper "Weather Prediction by Numerical Process," he envisioned NWP as using a massively parallel computer consisting of human "computers," or as we would say today, "processor cores":

> After so much hard reasoning, may one play with a fantasy? Imagine a large hall like a theatre, except that the circles and galleries go right round through the space usually occupied by the stage. The walls of this chamber are painted to form a map of the globe. The ceiling represents the north polar regions, England is in the gallery, the tropics in the upper circle, Australia on the dress circle and the Antarctic in the pit.
>
> A myriad of computers are at work upon the weather of the part of the map where each sits, but each computer attends only to one equation or part of an equation. The work of each region is coordinated by an official of higher rank. Numerous little "night signs" display the instantaneous values so that neighbouring computers can read them. Each number is thus displayed in three adjacent zones so as to maintain communication to the North and South on the map…[14]

Richardson's idea, described in Peter Lynch's book *The Emergence of Numerical Weather Prediction: Richardson's Dream* [15], was ahead of its time. Richardson's own attempt at NWP was not successful. Richardson attempted to predict the change of pressure at a single point 6 h ahead using only a slide rule and a table of logarithms. The calculation took Richardson 6 weeks. However, due to using a large time step, numerical instability developed and the prediction was not realistic. After World War II, development of communication technology and the first electronic computer enabled the development of NWP [16].

John von Neumann was a driving force behind the development of the first computer, the Electronic Numerical Integrator and Computer, and for computational science and numerical methods for weather forecasting and turbulence simulations [17]. In 1948, he assembled a group of atmospheric scientists at the Institute for Advanced Study in Princeton, New Jersey. Jule Charney headed the effort to construct a mathematical model based on the simplified equations of atmospheric dynamics and to demonstrate the feasibility of NWP. The team made the first one-day, nonlinear weather prediction in April 1950. Four years later, advances in computer technology and modeling enabled real-time operational NWP, and the Joint Numerical Weather Prediction Unit was formed, funded by the US Weather Bureau, the US Air Force, and the US Navy.

Early NWP models focused on simulation of atmospheric dynamics. However, by the late 1960s, a steady development of computational facilities enabled the use of higher and higher spatial resolutions. This in turn prompted researchers to develop new limited area models (LAMs) that focus on resolving atmospheric motions on regional scale and depend on global models for initial and boundary conditions. LAMs also include parameterizations of other atmospheric processes such as cloud microphysics and radiative transfer. Here is Richard Anthes' recollection of the early days in the development of one of the first LAMs:

> While I was an assistant professor at Pennsylvania State University in the early 1970s, my first Ph.D. student, Thomas Warner, and I began converting the 3-D hurricane model I had developed in Miami to a general limited-area mesoscale model. We called it the "Meso Monster," or MM, because of its size and complexity. This work, aided by many graduate students and collaborators, eventually led to the fifth generation Penn State-NCAR mesoscale model MM5, which was widely used by scientists around the world. [18]

3.2.2 Governing equations for flows in the atmosphere

The fundamental equations describing fluid motion of a gas based on conservation of mass, momentum, and energy are the compressible Navier–Stokes equations. In the inertial frame, coordinate independent notation, the compressible Navier–Stokes equations take the following form:

$$\frac{D\rho}{Dt} = -\rho\nabla \cdot \mathbf{v} \tag{3.1}$$

$$\rho\frac{D\mathbf{v}}{Dt} = -\nabla p + \nabla \cdot \boldsymbol{\sigma} + \mathbf{F} \tag{3.2}$$

$$\rho\frac{De}{Dt} = -p\nabla \cdot \mathbf{v} + \nabla \cdot (k\nabla T) + \nabla \cdot (\boldsymbol{\sigma} \cdot \mathbf{v}) + \mathbf{F} \cdot \mathbf{v} \tag{3.3}$$

The density of a fluid is denoted by ρ; $\mathbf{v}$ is the velocity vector; p denotes pressure; $\boldsymbol{\sigma}_{ij}$ viscous stress; T temperature; k molecular diffusivity; $\mathbf{F}$ external forcing; and e total energy, a sum of kinetic, potential, and internal energy (here, bold face characters denote vectors ($\mathbf{v}$ and $\mathbf{F}$) or tensors ($\boldsymbol{\sigma}$)) defined as

$$e = \frac{1}{2}(\mathbf{v} \cdot \mathbf{v}) + \int_0^z g(z)dz + c_v T \tag{3.4}$$

where $g(z)$ is gravitational acceleration, and internal energy is a product of specific heat capacity at constant volume, c_v, and temperature, while the material derivative of any quantity, φ, is defined as

$$\frac{D\varphi}{Dt} = \frac{\partial\varphi}{\partial t} + (\mathbf{v} \cdot \nabla)\varphi \tag{3.5}$$

In general, pressure can be determined using the ideal gas law. The ideal gas law is a good approximation of the behavior of many gases, including air, under a wide range of conditions.

$$p = \rho RT \tag{3.6}$$

The viscous stress is modeled using a phenomenological model that relates it to the divergence and the rate of strain with bulk and dynamic viscosity:

$$\boldsymbol{\sigma} = \lambda\nabla \cdot \mathbf{v} + \mu_{\nu}\left[\nabla\mathbf{v} + (\nabla\mathbf{v})^{\mathrm{T}}\right] \tag{3.7}$$

where λ is the bulk viscosity and μ is the dynamic viscosity. Dynamic viscosity is $\mu = \rho\nu$, where ν is the kinematic viscosity. For air, the kinematic viscosity is $\sim 10^{-5}\mathrm{m^2 s^{-1}}$.

The compressible Navier–Stokes equations, (3.1)–(3.3), are the mathematical representation of a range of scales of motion in the atmosphere from large scale atmospheric cells and waves (e.g., Hadley, Ferrel, and polar cells and Rossby waves) to synoptic-scale motions, cyclones, mesoscale circulations, storms and convective circulations, to boundary layer flows, shear, all the way to three-dimensional turbulence eddies including dissipative scales and sound waves. The energy spectrum appears as Figure 3.5.

3.2.3 Numerical resolution requirements

A numerical approach that resolves all of the scales of fluid motion down to the dissipation is called direct numerical simulation (DNS). It can be shown that the number of grid points required to fully resolve a flow depends on the Reynolds number of the flow, defined as

$$Re = \frac{u\ell}{\nu} \tag{3.8}$$

where u is a characteristic velocity and l is a characteristic length scale.

When the production of kinetic energy is in balance with the dissipation, the dissipation can be estimated as follows:

$$\epsilon = \frac{u^3}{\ell} \tag{3.9}$$

Based on dimensional analysis, the Kolmogorov or dissipative length scale is defined as

$$\eta = \left(\frac{\nu^3}{\epsilon}\right)^{1/4} \tag{3.10}$$

Using expressions (3.8)–(3.10), we can estimate the number of grid points in one direction required to resolve all of the relevant scales of motion, from integral scales, ℓ, to dissipative scales, η, as

$$N = \frac{\ell}{\eta} = \left(\frac{u\ell}{\nu}\right)^{3/4} = Re^{3/4} \tag{3.11}$$

This means that in three dimensions, the number of grid points required for a DNS is

$$N^3 = Re^{9/4} \tag{3.12}$$

Considering that the characteristic length scale of energy containing synoptic atmospheric motions could be a few thousand kilometers, that the wind speeds are on the order of 10 m s^{-1}, and that the kinematic viscosity of the air is 1.5×10^{-5} m s^{-2}, the Reynolds number of large-scale atmospheric flows is on the order of 10^{11}. This means that one would need more than 10^{24} grid cells in order to fully resolve atmospheric flows. At present, this is far beyond the capabilities of even the most advanced high-performance computing platforms.

To simulate atmospheric flows on a computer, it is therefore necessary to make approximations based on physically justifiable arguments and thus reduce the range of scales of atmospheric motion that need to be resolved to capture the physical processes of interest. Atmospheric flow models therefore represent a subset of fluid flow models that include forcings characteristic of the Earth's atmosphere as well as approximations and parameterizations of physical processes that cannot be fully resolved. Since the inception of NWP, a hierarchy of atmospheric models has been developed. These models can be divided in three general groups: global circulation models, LAMs, and microscale models. Global circulation models solve the governing equations on a sphere. LAMs use different map projections to represent a region on the Earth's surface and solve the governing equations on a Cartesian computational grid. Finally, microscale models represent relatively small areas varying in size from a few hundred meters to tens of kilometers where the effects of the curvature of the Earth can be neglected.

"Weather patterns" evolve within the troposphere, which on average extends from the Earth's surface to on average about 11 km above the surface in midlatitudes. Considering that the Earth's radius at the equator is about 6, 371 km, Earth's atmosphere is relatively shallow. Most global circulation models therefore solve the primitive equations that are based on the hydrostatic assumption in a shallow atmosphere. If the grid spacing in atmospheric models is ~10 km or larger, the resolved vertical motions are relatively small. In that case, we can make the assumption that the gravitational acceleration is constant and the vertical pressure gradient is balanced by the buoyancy force:

$$\frac{\partial p}{\partial z} = -\rho g \tag{3.13}$$

It can be shown that this hydrostatic approximation implies that the terms eliminated from the vertical momentum conservation equation must be at least an order-of-magnitude smaller than the terms that are retained, i.e.,

$$\frac{dw}{dt} \ll g \tag{3.14}$$

3.2.4 Reynolds averaged Navier–Stokes simulation methodology

Since at present, DNS of atmospheric flows is not possible, numerical simulations must be based on resolving flow scales of interest and parameterizing the effect of the unresolved motions. This can be achieved using the Reynolds Averaging approach. Formally, the Reynolds Averaged Navier–Stokes (RANS) equations are derived by assuming multiple independent realizations of a flow governed by the Navier–Stokes equations and taking an ensemble average of these equations. When it is not feasible or meaningful to assume existence of an ensemble of independent flow realizations, the ergodic hypothesis is invoked and an ensemble average is replaced by a time average. Invoking the ergodic hypothesis implies that within the time span over which time averaging is carried out, the dynamical system explores all of the available region of the phase space, which is the space containing all possible solutions under the given conditions. Assuming validity of the ergodic hypothesis, fluid flow models based on the RANS equations were initially used to solve for steady-state solutions, i.e., long-time averages of prognostic variables. Although the atmosphere is obviously not in a steady state, RANS parameterizations of turbulence were also employed to reduce the computational complexity compared to DNS and to parameterize the effects of three-dimensional turbulent eddies in large-scale models. Most applications of NWP, including wind power integration, require forecasts or simulations at a wide range of time scales from a few minutes to several days. However, it is not justified to assume that conditions under which the ergodic hypothesis is valid are satisfied for these longer time spans. To derive the governing equations for a hierarchy of atmospheric models focusing on different ranges of scales of motion, it is therefore more appropriate to average, or more generally filter, the compressible Navier–Stokes equations over a volume representing the smallest length scales of relevance for a particular application [16,26]. A volume average of a quantity φ is defined as

$$\bar{\varphi} = \int_V G(\mathbf{x})\varphi dV \tag{3.15}$$

where $G(\mathbf{x})$ is a kernel function. Then the fluctuating component of φ can be defined as

$$\varphi' = \varphi - \bar{\varphi} \tag{3.16}$$

The RANS equations governing atmospheric flows can take different forms depending on the coordinate system used. Although the governing equations for global circulation models are expressed in spherical coordinates, these coordinates are avoided in numerical simulations due to the problems with a singularity at each pole where meridians intersect. Instead of spherical coordinates, global circulation models use cubed sphere, yin–yang, or icosahedral grids, while LAMs use coordinate systems that map the Earth's surface onto a plane using one of the conformal projections. Conformal projection preserves angles between intersecting curves, e.g., right angles between meridians and parallels. While these mappings are conformal, they are not isometric, i.e., the distances are not preserved, and, therefore, map factors must be introduced. However, map factors are locally the same in both horizontal directions. The primary conformal projections used in LAMs are Lambert, Mercator, and polar stereographic projection. Rotated spherical coordinates are also commonly used in LAMs. The advantage of this system is that the convergence of the meridians is minimized by tilting the pole, so that the equator of the sphere runs through the center of the model domain. In the case of small domains, the impact of the Earth's curvature is small, so that the equations are essentially identical to those for a tangential Cartesian coordinate system.

$$\frac{\partial \bar{\rho}}{\partial t} + \nabla \cdot (\bar{\rho}\langle \mathbf{v} \rangle) = 0 \tag{3.17}$$

$$\frac{\partial \bar{\rho}\langle u \rangle}{\partial t} + \langle \mathbf{v} \rangle \cdot \nabla(\bar{\rho}\langle u \rangle) - \bar{\rho}\frac{\langle u \rangle \langle v \rangle}{a} \tan\phi - f\bar{\rho}\langle v \rangle$$
$$= -\frac{1}{a\cos\phi}\frac{\partial \bar{p}}{\partial \vartheta} - (\nabla \cdot \boldsymbol{\sigma}) \cdot \mathbf{e}_1 \tag{3.18}$$

$$\frac{\partial \bar{\rho}\langle v \rangle}{\partial t} + \langle \mathbf{v} \rangle \cdot \nabla(\bar{\rho}\langle v \rangle) - \bar{\rho}\frac{\langle u \rangle^2}{a} \tan\phi + f\bar{\rho}\langle u \rangle$$
$$= -\frac{1}{a}\frac{\partial \bar{p}}{\partial \phi} - (\nabla \cdot \boldsymbol{\sigma}) \cdot \mathbf{e}_2 \tag{3.19}$$

$$\frac{\partial \bar{\rho}\langle w \rangle}{\partial t} + \langle \mathbf{v} \rangle \cdot \nabla(\bar{\rho}\langle w \rangle) = -\frac{\partial \langle p \rangle}{\partial z} - (\nabla \cdot \boldsymbol{\sigma}) \cdot \mathbf{e}_3 - \bar{\rho}g \tag{3.20}$$

$$\frac{\partial \bar{p}}{\partial t} + \langle \mathbf{v} \rangle \cdot \nabla \bar{p} = -\frac{c_{pd}}{c_{vd}}pD + \left(\frac{c_{pd}}{c_{vd}} - 1\right)\bar{\rho}c_{pd}Q_T \tag{3.21}$$

$$\frac{\partial \bar{\rho}\langle \theta \rangle}{\partial t} + \langle \mathbf{v} \rangle \cdot \nabla \bar{\rho}\langle \theta \rangle = \frac{1}{c_p}\left(\frac{\partial \bar{p}}{\partial t} + \langle \mathbf{v} \rangle \cdot \nabla \bar{p}\right) + Q_T \tag{3.22}$$

$$\frac{\partial \overline{q_v}}{\partial t} + \langle \mathbf{v} \rangle \cdot \nabla \overline{q_v} = -(S_l + S_f) + M_{qv} \tag{3.23}$$

where

$$\langle v \rangle \cdot \nabla \frac{1}{a\cos\phi}\left(\langle u \rangle \frac{\partial}{\partial \vartheta} + \langle v \rangle \cos\phi \frac{\partial}{\partial \phi}\right) + \langle w \rangle \frac{\partial}{\partial z} \tag{3.24}$$

Here, angle brackets define the Favre average of a quantity, φ, defined as

$$\langle \varphi \rangle = \frac{\overline{\rho\varphi}}{\bar{\rho}} \tag{3.25}$$

In (3.18)–(3.20) $\mathbf{e}_1, \mathbf{e}_2$, and $\mathbf{e}_3$ are unit vectors, ϕ and ϑ are angles of latitude and longitude, respectively, and g is gravitational acceleration. In (3.21), c_{pd} and c_{vd} are specific heat capacity of dry air at constant pressure and constant volume, respectively. In (3.23) $\overline{q_v}$ is specific humidity, S_l and S_f are sinks of water vapor due to condensation and freezing, respectively, while M_{qv} is a source of water vapor. In (3.24) u, v, and w are velocity vector components. In the vertical direction, some form of terrain-following coordinates is commonly used.

Since domains of microscale models are usually relatively small, effective coupling of mesoscale and microscale simulations requires relatively small mesoscale grid cell sizes, on the order of 1 km or less. As mentioned earlier, for domain sizes characteristic of microscale simulations with grid cell sizes on the order of 100 m or less, map factor effects can be neglected. For simplicity, we will therefore present model equations in Cartesian form where repeated indices denote summation. The mass conservation and the prognostic equation for momentum in atmospheric models take the following form:

$$\frac{\partial \bar{\rho}}{\partial t} + \frac{\partial \bar{\rho} \langle u_i \rangle}{dx_i} = 0 \tag{3.26}$$

$$\frac{\partial \bar{\rho} \langle u_i \rangle}{\partial t} + \frac{\partial \bar{\rho} \langle u_i \rangle \langle u_j \rangle}{\partial x_j} = -\frac{\partial \langle p \rangle}{\partial x_i} - \frac{\partial \tau_{ij}}{\partial x_j} - 2\bar{\rho}\varepsilon_{ijk}\zeta_j \langle u_k \rangle - \bar{\rho} g \delta_{i3} \tag{3.27}$$

The last two terms on the right-hand side of the prognostic equation for momentum (3.27) represent the Coriolis force and buoyancy, where ζ_j is the Earth's rotation rate vector and τ_{ij} is turbulent stress tensor, while ε_{ijk} and δ_{i3} are antisymmetric tensor and Kronecker delta tensor, respectively.

For atmospheric flow simulations, the energy equation is commonly replaced by the prognostic equation for potential temperature:

$$\frac{\partial \bar{\rho} \langle \theta \rangle}{\partial t} + \frac{\partial \bar{\rho} \langle u_j \rangle \langle \theta \rangle}{\partial x_j} = \frac{\partial H_j}{\partial x_j} + Q_h \tag{3.28}$$

Potential temperature, θ, is defined as the temperature a parcel of air would attain if brought adiabatically to a reference pressure, p_0 (e.g., surface pressure), neglecting water vapor effects:

$$\theta = T\left(\frac{p_0}{p}\right)^{R/c_p} \tag{3.29}$$

The reference pressure, p_0, is a commonly set to 10^5Pa, R is the gas constant for dry air, and c_p is the specific heat capacity at constant pressure. In the prognostic equation for potential temperature, H_j is the turbulent sensible heat flux and Q_h is the source term.

In atmospheric models, we must account for moist processes, which significantly impact the buoyancy, including a prognostic equation for water vapor, liquid water, and ice. All of these prognostic equations have a similar form, so a prognostic equation for water vapor is

$$\frac{\partial \bar{\rho} \langle q_v \rangle}{\partial t} + \frac{\partial \bar{\rho} \langle u_j \rangle \langle q_v \rangle}{\partial x_j} = \frac{\partial I_j}{\partial x_j} + Q_v \tag{3.30}$$

where I_j is the turbulent flux of water vapor and Q_v is the source term. Prognostic equations for liquid water, q_l, and ice, q_f, are similar to the prognostic equation for water vapor and can be obtained by replacing subscript v in the above equation by subscripts l and f, respectively.

In the prognostic equations for momentum (3.27), potential temperature (3.28), and water vapor (3.30) the turbulent stress, τ_{ij}, turbulent heat flux, H_j, and turbulent water vapor flux, I_j, are introduced. These three new terms are the consequence of the filtering of nonlinear terms, which are defined as the (subfilter) turbulent stress,

$$\tau_{ij} = \bar{\rho}(\langle u_i \rangle \langle u_j \rangle - \langle u_i \rangle \langle u_j \rangle) \tag{3.31}$$

and (subfilter) turbulent fluxes of sensible heat, H_j, and water vapor, I_j,

$$H_j = \bar{\rho}(\langle u_j\theta\rangle - \langle u_j\rangle\langle\theta\rangle) \tag{3.32}$$

$$I_j = \bar{\rho}(\langle u_j q_v\theta\rangle - \langle u_j\rangle\langle q_v\rangle) \tag{3.33}$$

The divergences of turbulent stress and fluxes represent the effect of the filtered scales on the resolved scales. The turbulent stress and fluxes must be parameterized. The simplest models for turbulent stress and heat flux are similar to the constitutive relation for viscous stresses and molecular flux, except that instead of molecular viscosity and diffusivity coefficient, a flow-dependent turbulent eddy viscosity and diffusivities, K_m, and, K_h are introduced so that

$$\tau_{ij} = \bar{\rho}K_m\left(\frac{\partial\langle u_i\rangle}{\partial x_j} + \frac{\partial\langle u_j\rangle}{\partial x_i}\right) \tag{3.34}$$

$$H_j = \bar{\rho}K_m\frac{\partial\langle\theta\rangle}{\partial x_j} \tag{3.35}$$

In the atmosphere, eddy viscosity and diffusivity are on the order of 0.1–1 m^2s^{-1}.

Notice that viscous stress, σ_{ij}, and molecular heat flux are not included in the prognostic equations for momentum and potential temperature. Since the kinematic viscosity of air is $10^{-4}\,m^2s^{-1}$ and the molecular thermal diffusion of air is $1.9 \times 10^{-4}m^2s^{-1}$ at 300K and standard surface pressure, viscous stress and molecular heat diffusion are orders of magnitude smaller than turbulent stress and heat diffusion, and, therefore, they are neglected in atmospheric models except for the diffusion fluxes of the liquid and solid forms of water.

As we have seen, resolving all the scales of atmospheric motion requires computational resources beyond the current capabilities of high-performance computing. While resolving dissipative scales requires small grid cell size, resolving relatively fast-propagating sound waves would require prohibitively small time steps. However, for weather-forecasting applications, we are not interested in forecasting sound-wave propagation. We can therefore introduce approximations that in essence filter sound waves and thus allow the use of larger time steps when integrating the equations of motion. One such approximation includes the derivation of a prognostic equation for pressure in the following form:

$$\frac{\partial\bar{p}}{\partial t} + \langle u_j\rangle\frac{\partial\bar{p}}{\partial x_j} = -\frac{c_p}{c_v}\bar{p}\frac{\partial\langle u_j\rangle}{\partial x_j} + \left(\frac{c_p}{c_v} - 1\right)Q_h - \frac{c_p}{c_v}Q_m \tag{3.36}$$

where c_v is specific heat capacity at constant volume, Q_h and Q_m. The density is computed using the equation of state:

$$\bar{\rho} = \frac{\bar{p}}{R_d(1-\alpha)\bar{T}} \tag{3.37}$$

where R_d is the gas constant for dry air and α is defined as

$$\alpha = \left(\frac{R_d}{R_d} - 1\right)\langle q_v\rangle - \langle q_l\rangle - \langle q_f\rangle \tag{3.38}$$

The base state of the atmosphere is defined as hydrostatically balanced so that

$$p_0 = \rho_0 R_d T_0 \tag{3.39}$$

$$\frac{\partial p_0}{\partial z} = -g\rho_0 = -\frac{g p_0}{R_d T_0} \tag{3.40}$$

Dudhia [20] assumed a constant rate of increase in temperature with the logarithm of pressure

$$\frac{\partial T_0}{\partial \ln p_0} = \beta \tag{3.41}$$

where p_0, T_0, and ρ_0 are reference surface pressure, temperature, and density, respectively.

Another approach to filtering sound waves is through the anelastic approximation. The simplest form of the anelastic approximation involves the assumption that the density does not vary in time. By taking the divergence of the momentum equation and requiring that the mass is conserved, we can obtain a Poisson equation for pressure.

$$\frac{\partial^2 \langle p \rangle}{\partial x_i \partial x_i} = -\frac{\partial \bar{\rho} \langle u_j \rangle \langle u_i \rangle}{\partial x_i \partial x_j} - \frac{\partial \tau_{ij}}{\partial x_i \partial x_j} - \frac{g}{T_0}\frac{\partial \bar{\rho}\theta}{\partial x_i} - 2\varepsilon_{ijk}\frac{\partial \bar{\rho}\zeta_j \langle u_k \rangle}{\partial x_i} \tag{3.42}$$

Figure 3.6 displays the family of types of approximations that are made in modeling the atmosphere. One must carefully choose the right level of detail for the needs of the modeling application.

3.2.5 Discretizations

Since the inception of the idea of NWP, it has been recognized that integrating equations governing atmospheric flows would require significant computational resources. Over the last 60 years, rapid development of computer technology enabled NWP at ever-increasing complexity and spatiotemporal resolution. To integrate the governing equations, NWP models employ spatial and temporal discretization approaches. While time discretization is commonly accomplished using a finite differencing approach, spatial discretization in addition to different finite differencing schemes often involves spectral and finite volume schemes.

3.2.6 Forcing physics and parameterizations

The parameterizations, or physics, account for physical processes that are not explicitly modeled but whose effects are relevant and should be included in the simulation. These processes may not be simulated explicitly either because we do not understand them or because an explicit representation would be prohibitive in terms of the computational cost. A typical example of the former is the atmospheric turbulent mixing, whereas an example of the later is radiation. Other processes that are typically parameterized are the effects of convection when the grid spacing is coarse, cloud microphysics, and land-atmosphere interactions including soil processes.

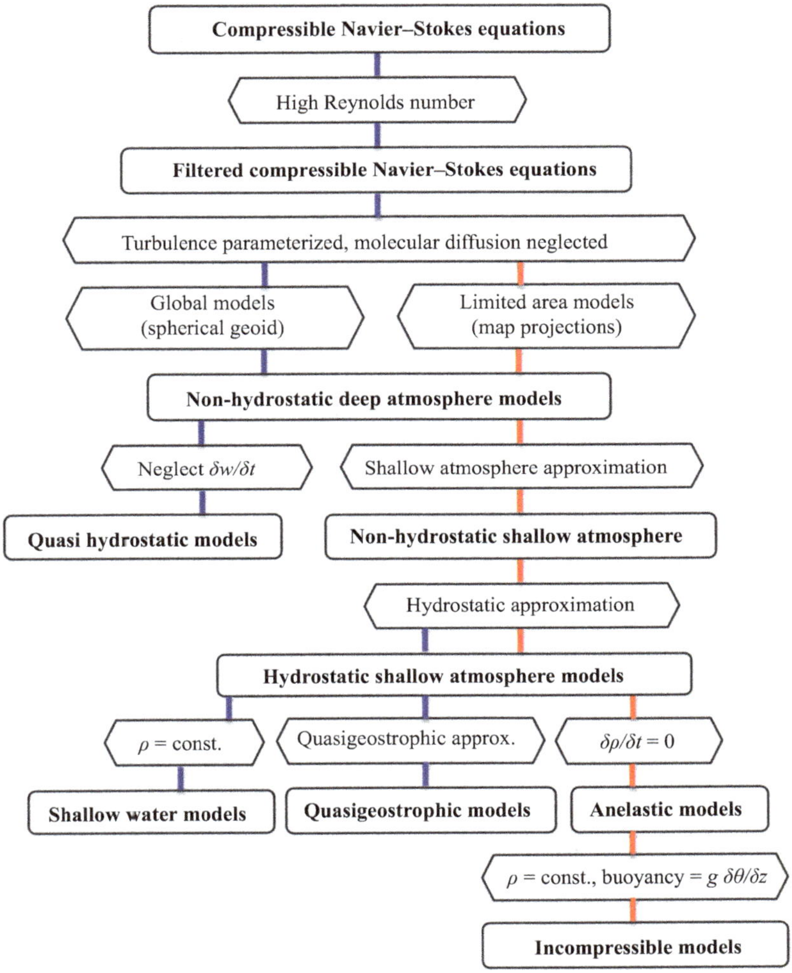

Figure 3.6 Family of approximations to the Navier–Stokes equations commonly used in modeling for renewable energy

Because parameterizations are mere approximations, they typically introduce larger errors than those introduced by the discretization of the atmospheric equations of motion. Furthermore, the parameterization's performance may depend on the atmospheric state or even on the geographical region. Therefore, one should be aware of what kinds of approximations are introduced by the parameterizations and how the approximations may affect the realism of a particular simulation.

A key parameterization is that of the radiative transfer process, which provides the heat balance of the system modeled through the incoming downward irradiance at the surface—the shortwave component—and the outgoing longwave component. The longwave radiation scheme accounts for the infrared radiation absorbed and emitted by atmospheric gases, clouds, and the Earth's surface. The shortwave and longwave radiative transfers are both determined by the modeled water vapor, clouds, and gases such as ozone, oxygen, and carbon dioxide. The most advanced radiation schemes model the interaction of the radiation with various species of modeled atmospheric aerosols.

Land surface models (LSMs) are used to represent the interactions between the land or water and the atmosphere. They quantify soil properties (soil type, vegetation type, soil moisture, and soil temperature) plus the surface turbulent fluxes of heat, moisture, and (sometimes) momentum. LSMs typically represent moisture and heat fluxes in several soil layers. Some advanced LSMs include the effects of vegetation, urban heat islands, snow cover, and hydrology. These processes determine the ground temperature, which is used to compute the emitted longwave radiation, which is passed to the radiation parameterization. LSMs must also account for surface albedo to provide feedback to the shortwave radiation scheme. Moreover, the turbulent fluxes provide the lower boundary condition to the planetary boundary layer (PBL) parameterization [21].

Parameterizations of the ABL, typically known as PBL schemes, account for the vertical mixing associated with atmospheric turbulence in the lowest layers of the atmosphere (the lowest couple thousand meters). As discussed in Section 3.1.2, this layer is quite dynamic and varies considerably over the course of a day (see Figure 3.3). The turbulent mixing impacts the horizontal and vertical distribution of temperature, water vapor, horizontal momentum, and trace gases, which in turn leads to redistribution by changes in the resolved winds. These turbulent fluxes are typically modeled by both local and nonlocal mixing. The various parameterization schemes require a surface turbulent flux. These are often models based on Monin–Obukhov (M–O) similarity theory [22].

Unresolved deep convective and/or shallow convective clouds are modeled via cumulus parameterizations to account for the effects on temperature and moisture profiles associated with cloud development. The schemes additionally model precipitation, which feeds back into the LSM. Deep convective schemes, often based on closure assumptions valid for horizontal grid spacing higher than 10 km, are less accurate when the model is run at finer grid spacing. In this case, the model begins to resolve the convective updrafts/downdrafts that are parameterized, providing dual mechanisms for the same process, Thus, it is not usually appropriate to employ deep convective schemes when the grid spacing is finer than about 5 km. Shallow convection must still be parameterized at grid spacings down to about 1 km.

Processes involving cloud droplets and/or ice crystals and the effects of their life cycle on the temperature and moisture profiles, precipitation, and cloud radiative properties are parameterized through microphysics parameterization schemes. These schemes account for processes in both liquid water clouds (warm clouds) and in clouds consisting of water and ice hydrometeors (mixed phase clouds). Most microphysics parameterizations are based on a bulk approach that uses functional forms for the size distribution of hydrometeors, and parameters of the distribution (e.g., mixing ratios) are predicted by the scheme. Other microphysics schemes use a spectral (bin) approach that divides microphysical particles into numerous mass bins, computing the size distribution and evolution of each bin independently. While bin schemes can be more accurate, they are also far more computationally intensive and thus are not typically used in operational NWP models (e.g., [23]).

Many models provide choices for parameterization schemes, and those choices will lead to different model performance. For instance, the Weather Research and Forecasting (WRF) model [24,25] allows a plethora of choices of parameterizations for each of the above-listed processes. One must be careful that the choices for one process are compatible with those of the other processes. There have been many studies of the relative merits of different parameterizations, but their success largely depends on the usage and the characteristics of the region modeled [21,26].

3.3 Initial conditions and data assimilation

Achieving more accurate initial states (or analyses, which are the best estimates of the atmospheric state at a particular time for a particular model) will lead, in general, to more accurate NWP forecasts. The primary way of improving the accuracy of analyses in modern NWP is by using DA. DA is the "process in which observations distributed in time are merged together with a dynamical model of the flow in order to determine as accurately as possible the state of the atmosphere" [27]. Several classes of DA algorithms exist and are discussed below, each of which ingests observations from a diverse array of platforms to update a background (prior) state to an analysis (posterior) state, which then can be used as an initial condition state for an NWP forecast. A schematic of this basic outline of DA procedures is shown in Figure 3.7.

There are numerous sources for the meteorological observations ingested by DA methods. All of the schemes discussed below can ingest observations from surface-based, sonde-based (e.g., weather balloons), aircraft-based, and satellite-based platforms. Many DA algorithms can also assimilate radial velocity and reflectivity observations from Doppler radars. In addition to the basic temperature, humidity, and wind observations that all DA algorithms assimilate, many DA algorithms also can assimilate barometric pressure, satellite radiances, chemical species concentrations, and other special observations if forward operators are written to map the observations onto model variable space. Observation density tends to be higher in the United States, Europe, China, and India than in most of the rest of the world, with observation density being especially low over the oceans. This disparity in observation network density affects the quality of analyses in different regions, and the greatest differences between DA algorithms can often be found in regions with sparse observations [28].

3.3.1 Nudging

Perhaps the simplest DA method is Newtonian Relaxation, also known as nudging. There are two main types of nudging schemes: observation nudging and analysis nudging. In both of these types of optimal interpolation schemes, the model forecast

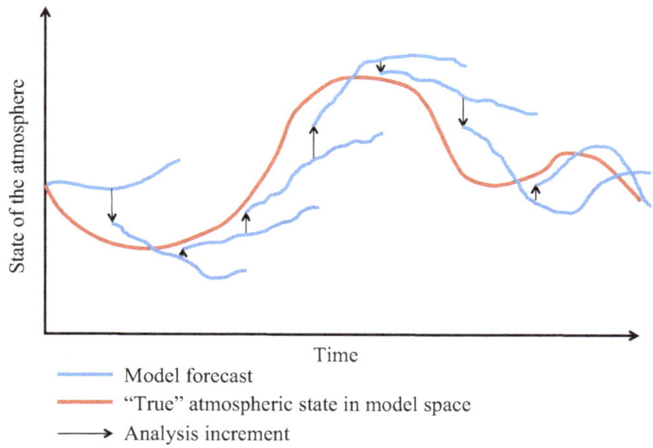

Figure 3.7 Schematic of the basic data assimilation procedure of correcting a short-range model forecast with observations to create a new analysis state from which a new forecast is initialized. The blue lines are the model forecasts, the red line is the "true" atmospheric state in model space (within the limitations of model resolution and imperfect physics), and the upward/downward arrows are the analysis increments from the background (prior) state to the new analysis (posterior) state

state is nudged or relaxed toward individual observations or a gridded analysis, respectively, through the addition of a small artificial term to the model prognostic tendency equations (e.g., [29–34]). In addition to weather-forecasting applications (e.g., [35]), analysis nudging has also been explored for regional climate modeling applications recently (e.g., [36]).

Observation nudging is an example of four-dimensional DA, as it assimilates observations continuously at every time step in three-dimensional space, rather than only at specific analysis times like intermittent DA schemes do. In the WRF model, for example, observation nudging for a quantity q at a model grid point (x, y, z) at time t is accomplished by

$$\frac{\partial q\mu}{\partial t}(x,y,z,t) = F_q(x,y,z,t) + \mu G_q \frac{\sum_{i=1}^{N} W_q^2(i,x,y,z,t)\left[q_0(i) - q_m(x_i,y_i,z_i,t)\right]}{\sum_{i=1}^{N} W_q(i,x,y,z,t)} \tag{3.43}$$

where μ is the dry hydrostatic pressure, F is the physical tendency terms of q, G is the nudging strength, i is the observation index, N is the total number of observations, W is the weighting function based on the separation in time and space between the observation and model grid point, q_o is the observed quantity, and q_m is the modeled quantity interpolated to the location of the observation. For each observation, the innovation, $q_o - q_m$, changes with time as the model evolves, and so does the magnitude of the nudging term. The innovation is spread spatially according to the specified radius of influence, and the temporal weight varies with time, having its maximum value close to the time of the observation, and gradually decreasing to zero at either edge of the nudging time window. This weighting, combined with constraining the magnitude of the nudging term to be small compared to the forcing terms, helps avoid creating artificial gravity waves in the model.

While the simplicity of observation nudging is certainly an advantage of the scheme, a notable disadvantage is that only observations that directly correspond to model prognostic variables can be assimilated. Thus, only temperature, humidity/moisture, and Cartesian wind components can be used with observation nudging. Other assimilation techniques use forward operators to map an observation of a non-prognostic quantity onto the model prognostic variables.

3.3.2 Variational DA

In contrast to nudging and optimal interpolation techniques, the other main class is statistical interpolation techniques. Statistical interpolation techniques require an estimate of the background error covariance between variables in the background state.

One category of statistical interpolation is variational DA ("Var") approaches. In pure variational DA [37,38], observations are combined with a single initial state and model trajectory to obtain an analysis. This analysis state is determined by minimizing a global cost function that takes into account the distance between the analysis and observations that are within the time window for assimilation; all the observations within this assimilation window are assimilated simultaneously to minimize the global cost function. In addition to the background error covariance matrix, this approach also requires the observation error covariance matrix, the accurate computation of both of which can often be difficult (e.g., [27,39]).

In three-dimensional variational (3D-Var) techniques, the background error covariance matrix is specified from a global climatology. Therefore, 3D-Var does not represent the "errors of the day," and the analysis cost function is minimized at the analysis time. An example of a 3D-Var cost function J is

$$\begin{aligned} J(x_a) &= (x_a - x_b)^T \mathbf{B}^{-1} (x_a - x_b) + (y - H[x_a])^T \mathbf{R}^{-1} (y - H[x_a]) \\ &= J_b + J_o \end{aligned} \tag{3.44}$$

where x_t is the true model state, x_a is the analysis model state, x_b is the background model state, y is the observation vector, H is the observation operator to map the observations onto model state variables, $\mathbf{B}$ is the background error covariance matrix ($x_b - x_t$), $\mathbf{R}$ is the observation error covariance matrix ($y - H[x]$), J_b is the background term, and J_o is the observation term. In order to solve this cost function, some assumptions are made: H is a linear operator; **B** and **R** are positive definite (i.e., non-trivial) matrices; the observation and background errors are unbiased and mutually uncorrelated; and the analysis is as close as possible to the true state in a root mean square sense.

In four-dimensional variational (4D-Var) techniques, however, the error matrices are flow-dependent, and the cost function is minimized to find the best NWP model trajectory through the entire assimilation window, rather than just at the analysis time. These important differences mean that 3D-Var techniques, while significantly less computationally expensive than 4D-Var, are also generally less accurate than 4D-Var (e.g., [40,47]).

3.3.3 Ensemble Kalman filters

A second category of statistical interpolation is the Kalman filter (KF; [41,42]). As with variational techniques, a wide array of observation types can be assimilated into the model via forward operators that map the observations onto model variables. In contrast to Var techniques, KF algorithms are a form of sequential DA, in that through the assimilation window, each observation is assimilated sequentially within a time window considered valid at the analysis time, thereby updating the background state. In KF algorithms, the background error covariance matrix is flow-dependent rather than a static climatology. This flow-dependent background error covariance matrix allows for improved accuracy of the analysis compared to 3D-Var, though it also is the primary reason why true KFs are intractable for NWP or other geophysical models with a large state space.

To address this issue of high computational cost, Evensen [43] developed the ensemble KF (EnKF). An EnKF uses a finite ensemble of simulations from a nonlinear model (such as an NWP model) and treats the ensemble as a random sample of the flow-dependent probability distribution of the model state. Each ensemble member is assumed to be an equally likely solution for the model state, and each is perturbed with different initial and boundary conditions to force a spread among the member solutions. Several types of EnKFs are in use by the NWP community currently (e.g., [44–50]), and they all solve these statistical analysis equations (though some solve a somewhat differently formed (3.47):

$$\mathbf{x}^a = \mathbf{x}^f + \mathbf{K}\left(\mathbf{y}^o - \mathbf{H}\mathbf{x}^f\right) \tag{3.45}$$

$$\mathbf{K} = \mathbf{P}^b\mathbf{H}^T\left(\mathbf{H}\mathbf{P}^b\mathbf{H}^T + \mathbf{R}\right)^{-1} \tag{3.46}$$

$$\mathbf{P}^a = (\mathbf{I} - \mathbf{K}\mathbf{H})\,\mathbf{P}^b \tag{3.47}$$

where $\mathbf{x}^a$ is the analysis state, $\mathbf{x}^f$ is the forecast (background) state, $\mathbf{y}^o$ is the observation vector, **H** is the linear observation operator, **K** is the optimal Kalman gain matrix, **R** is the (diagonal) observation error covariance matrix, $\mathbf{P}^a$ is the analysis error covariance matrix, $\mathbf{P}^b$ is the sample forecast (background) error covariance matrix, and **I** is the identity matrix. The forecast (background) error covariance matrix is estimated from the ensemble itself, from the outer product of the ensemble of perturbations, which is how it is flow-dependent [51]. Also note that $\mathbf{P}^b$ here in the EnKF equations and $\mathbf{B}$ in Var equations (Equation (3.44)) are both unique estimates of the same full-rank background error covariance matrix [52].

The success of an EnKF DA system lies in how well it can sample the distribution of the true atmospheric probability density function (pdf). The true atmospheric pdf can be thought of as the distribution that results from an infinite number of ensemble members with perturbations that fully span the uncertainty in the analysis [39,53]. Sampling error leads to noisy background error covariances, which can lead to spurious innovations, and insufficiently large variances across the ensemble, which can lead to overconfidence in the ensemble forecast compared to the observations (and eventually leads to "filter divergence"). In order to avoid sampling error, ensemble sizes with EnKF generally ought to be at least several dozen members. Even then, covariance localization (e.g., [54]) and ensemble variance inflation (e.g., [55,56]) are generally required to keep the cycling DA system stable and with sufficient spread.

A helpful review and summary of the myriad EnKF approaches that currently exist in the atmospheric sciences, and a discussion of several key issues relevant to improving the quality of EnKF analyses in various operational, quasi-operational, and experimental systems can be found in Houtekamer and Zhang [50]. The interested reader is referred there for more in-depth information on EnKFs.

3.3.4 EnVar and hybrid DA

Several research groups and operational centers have worked to combine the best portions of more than one of the approaches outlined above in order to yield an even more accurate analysis state. For example, Lei *et al.* [57] combined an EnKF with observation nudging, to leverage the benefits of those two approaches, and Schraff *et al.* [58] combines a local ensemble transform KF [59] with latent heat nudging, a technique to assimilate radar reflectivity data. Most of these hybrid efforts, however, combine EnKF with Var (e.g., [60–63]), and that is the approach that most NWP national operational centers currently take for DA. Developing these pure and hybrid "EnVar" techniques takes advantage of the flow-dependence of EnKF approaches and the often more robust background error statistics of Var (by avoiding some of the under-sampling issues with which EnKF typically contends). Bannister [52] provides a thorough, helpful review of the various pure and hybrid "EnVar" techniques in existence (e.g., 4DEnVar, En4DVar, and more), including those in use currently at operational centers (see Table 3.1), and proposes a harmonized, consistent terminology for these many approaches that have subtly different names but are substantially different algorithms in many key aspects. The interested reader is referred there for more details on those algorithms that are beyond the scope of this overview chapter.

Table 3.1 Modeling and DA systems used at various operational centers and groups around the world, including ensemble size N. Much of the information here is from Table 5 in [52]

Group	Model	DA setup	*N*	DA key references
CAWCR	ACCESS/UM (global)	4D-Var	24	Fraser *et al.* [64; Puri *et al.* [65]; Tripathi *et al.* [66]
DWD	ICON (global)	Hybrid 3DEnVar	40	Rhodin [67]
DWD	COSMO (regional)	LETKF with LHN	40	Schraff *et al.* [58]
EC	GEM (global)	Hybrid 4DEnVar	256	Buehner *et al.* [68]
EC	GEM (regional)	Hybrid 4DEnVar	256	Caron *et al.* [69]
ECMWF	IFS (global)	Recalibration of variables and errors in 4D-Var	25	Bonavita *et al.* [70]
HIRLAM	HIRLAM (regional)	Hybrid En4DVar, Hybrid 4DEnVar	40	Gustafsson *et al.* [71]; Gustafsson and Bojarova [72]
JMA	GSM (global)	4D-Var	27	Japan Meteorological Agency [73]
Météo-France	Arpège (global)	Recalibration of variables in 4D-Var	25	Berre *et al.* [74,75]; Raynaud *et al.* [76,77]
NCAR	WRF (regional)	EAKF	80	Schwartz *et al.* [49]
NCEP	GFS (global)	Hybrid 4DEnVar	80	Wang and Lei [78]; [79,80]
NCEP	WRF (regional)	4DEnVar	50	Wang *et al.* [60,81]; Liu *et al.* [82]; Wang [83]
NCEP	WRF-Chem (regional)	Hybrid 3D-Var	37	Schwartz *et al.* [84]
NRL	NOGAPS (global)	Hybrid En4DVar	80	Bishop *et al.* [85]; Kuhl *et al.* [86]
NRL	COAMPS (regional)	EnSRF	11	Zhao *et al.* [87]
PSU	WRF (regional)	Hybrid En4DVar	40	Zhang and Zhang [88]
UK Met Office	UM (global)	Hybrid En4DVar	23	Clayton *et al.* [89]

ACCESS, Australian Community Climate and Earth System; Arpège, Action de Recherche Petite Echelle Grande Echelle; CAWCR, Centre for Australian Weather and Climate Research; COAMPS, coupled ocean atmosphere mesoscale prediction system; COSMO, Consortium for Small-Scale Modeling; DWD, Deutscher Wetterdienst; EAKF, ensemble adjustment Kalman filter; EC, Environment Canada; ECMWF, European Centre for Medium Range Weather Forecasting; EnSRF, ensemble square root filter; GEM, Global Environment Multi-scale; GFS, Global Forecast System; GSM, Global Spectral Model; HIRLAM, High Resolution Local Area Modeling; ICON, ICOsahedral Non-hydrostatic model; IFS, Integrated Forecast System; JMA, Japan Meteorological Agency; LETKF, local ensemble transform Kalman filter; LHN, latent heat nudging; NCAR, National Center for Atmospheric Research; NCEP, National Centers for Environmental Prediction; NOGAPS, Navy Operational Global Atmospheric Prediction System; NRL, US Naval Research Laboratory; PSU, The Pennsylvania State University; UM, Unified Model; WRF, Weather Research and Forecasting model.

3.4 Boundary conditions

3.4.1 Forcing from global models

Unlike global NWP models, regional models (also called LAMs) by definition must have lateral boundary conditions (LBCs) specified from some other source. Those LBCs come from either a larger regional model or from a global model. For example, the National Centers for Environmental Prediction (NCEP) High Resolution Rapid Refresh model [90] obtains its boundary conditions from the Rapid Refresh (RAP) model [90], while the RAP model boundary conditions come from the NCEP Global Forecast System model [91]. These LBCs are generally applied on an hourly, 3-hourly, or 6-hourly basis, depending on the output frequency of the forcing model. Interpolating the forcing model's forecasts or analyses in space to the regional model's grid and in time between the LBC updates necessarily introduces errors that will be advected into the domain and negatively impact the quality of the forecast (e.g., [23,92]).

Many NWP models, such as WRF, allow for a higher resolution domain to be nested within a coarse-resolution parent domain. Even if the parent domain is global, the nested domain also suffers from some artificial errors introduced by interpolating coarser resolution model data to the finer resolution nest at the boundaries.

3.4.2 Top boundary

All NWP models require artificial upper boundary conditions to be imposed. There are different methods to implement an upper boundary, which are nicely summarized in Warner [26] and briefly introduced here. Both the altitude of this upper boundary and the type of boundary that is used can have significant impacts on the model solution.

For example, a "rigid lid" upper boundary will artificially reflect all upward-propagating internal gravity waves that are generated by flow over topography, convective storms, or other processes. This reflection of gravity waves does not happen in nature and contaminates the model state all the way down to the surface with artificial noise. Various types of "sponge layer" or "damping layer" upper boundaries attempt to absorb the gravity wave energy to varying degrees of effectiveness and computational overhead, depending on the thickness and other properties of the absorbing layer. Alternatively, "radiation" upper boundaries adjust variable values at the model top to facilitate gravity waves radiating through the upper boundary, neither being artificially reflected nor absorbed.

3.4.3 Bottom boundary

In addition to solar forcing and the effects of the Earth's rotation, exchanges of heat, moisture, gases, and momentum between the surface of the Earth and the atmosphere determine the state of the atmosphere and its evolution. A surface energy budget accounts for exchanges of heat and moisture among several (4–5) layers within the top few meters at the surface of the Earth. Surface energy budgets are parameterized using the land and sea surface models of varying complexity coupled with surface layer parameterizations and radiative transfer parameterizations. LSMs account for the effects of soil type, vegetation type, and state. In addition, surface snow or ice coverage affects the surface albedo and therefore radiative transfer. Surface layer parameterizations provide a link between the state of Earth's surface and the atmosphere. A comprehensive review of surface energy budget parameterizations in NWP models is given by Stensrud [21].

Considering that it is impossible to fully resolve boundary layer flows in rough boundary layers at the Earth's surface and that turbulent fluxes of heat, momentum, and moisture dominate molecular fluxes, it is necessary to parameterize these fluxes. M–O surface layer similarity theory provides a link between ensemble mean profiles of wind velocity and potential temperature to the turbulent fluxes of momentum and sensible heat at the surface. Considering the turbulent momentum flux, M–O similarity theory [22] represents an empirical extension of a logarithmic profile of a neutrally stratified ABL accounting for the effects of atmospheric stability. A nondimensional parameter representing stability effects is a ratio of distance from the surface, z, and Obukhov length, L, defined as [93]

$$L = -\frac{u_*^3}{\kappa(g/T)\langle w\theta_v\rangle} \tag{3.48}$$

where κ is the von Kármán constant, g is gravitational acceleration, T is the reference temperature, $\langle w\theta_v\rangle$ is the turbulent heat flux at the surface, and u_* is surface friction velocity. Fluctuating components of the vertical velocity and virtual potential temperature are denoted with w and θ_v, respectively. According to M–O similarity, the parameters controlling the structure of a surface layer are surface momentum flux (or surface friction velocity, u_*), wind shear $\partial U/\partial z$, distance from the surface, z, and the nondimensional atmospheric stability parameter, z/L. Then nondimensional shear can be expressed as a function of the nondimensional stability parameter, so that

$$\frac{\kappa z}{u_*}\frac{\partial\langle u\rangle}{\partial z} = \phi_m\left(\frac{z}{L}\right) \tag{3.49}$$

where ϕ_m is an experimentally determined stability function for momentum. Similarly, the nondimensional potential temperature gradient is

$$\frac{\kappa z}{\theta_*}\frac{\partial\langle\theta\rangle}{\partial z} = \phi_h\left(\frac{z}{L}\right) \tag{3.50}$$

where θ_* is the characteristic dynamic surface temperature defined as

$$\theta_* = -\frac{\langle w\theta_v\rangle}{u_*} \tag{3.51}$$

and ϕ_h is an experimentally determined stability function for heat. Different experiments provide different values of the von Kármán constant and different forms of universal stability functions (e.g., [94]). More widely accepted forms of the universal function are given by Dyer and Hicks [95] (for a detailed discussion see also [96]) together with the von Kármán constant, $\kappa = 0.41$:

$$\begin{aligned}
\phi_m\left(\frac{z}{L}\right) &= \left(1 - 15\frac{z}{L}\right)^{-1/4} \text{ for } -2 < \frac{z}{L} \le 0 \\
\phi_m\left(\frac{z}{L}\right) &= \left(1 + 5\frac{z}{L}\right) \text{ for } 0 \le \frac{z}{L} < 1 \\
\phi_h\left(\frac{z}{L}\right) &= \left(1 - 15\frac{z}{L}\right)^{-1/2} \text{ for } -2 < \frac{z}{L} \le 0 \\
\phi_h\left(\frac{z}{L}\right) &= \left(1 + 5\frac{z}{L}\right) \text{ for } 0 \le \frac{z}{L} < 1
\end{aligned} \tag{3.52}$$

Notice that eddy diffusivity coefficients for momentum and heat fluxes can be expressed as

$$K_m = \kappa z \frac{u_*}{\phi_m (z/L)} \text{ and } K_h = \kappa z \frac{u_*}{\phi_h (z/L)} \tag{3.53}$$

so that the turbulent Prandtl number is

$$Pr_t = \frac{K_m}{K_h} \tag{3.54}$$

When Dyer–Hicks universal functions are used, the turbulent Prandtl number is unity.

Expressions for the nondimensional shear and nondimensional potential temperature gradient can be integrated to obtain the following relationships:

$$\langle u \rangle = \frac{u_*}{\kappa}\left[\ln \frac{z}{z_0} + \psi_m\left(\frac{z}{L}\right)\right] \tag{3.55}$$

and

$$\langle \theta \rangle - \langle \theta_s \rangle = \frac{\theta_*}{\kappa}\left[\ln \frac{z}{z_0} + \psi_h\left(\frac{z}{L}\right)\right] \tag{3.56}$$

where stability functions ψ_m and ψ_h are integrals of universal functions ϕ_m and ϕ_h (cf., [97]) and $\langle \theta_s \rangle$ is land/sea surface skin temperature.

M–O similarity was developed under the assumption of horizontal homogeneity of a surface layer and using the ergodic hypothesis, which allows replacing ensemble averages with time averages. Nevertheless, M–O is widely used in atmospheric flow models at different scales to provide the lower boundary condition, i.e., surface momentum and heat fluxes. Most often M–O similarity theory is imposed indirectly through imposition of surface drag of momentum or heat where

$$u_*^2 = C_D U_V^2 \tag{3.57}$$

where U_N is the wind speed at some distance N meters above the surface. The momentum drag coefficient is then defined as

$$C_D = \frac{\kappa^2}{[\ln(z/z_0) + \psi_m(z/L)]^2} \tag{3.58}$$

Similarly, for heat flux

$$\langle w\theta_v \rangle = C_H U_N \Theta_N \tag{3.59}$$

with

$$C_H = \frac{\kappa^2}{[\ln(z/z_0) + \psi_m(z/L)][\ln(z/z_0) + \psi_h(z/L)]} \tag{3.60}$$

This approach is regularly used even for large-eddy simulations with grid cell sizes of a few meters. However, as the resolution of numerical simulations of atmospheric flows over heterogeneous terrain increases and the effects of grid cell size heterogeneities cannot be neglected, it will be necessary to revisit the accuracy or even applicability of M–O similarity theory for determining lower boundary conditions in atmospheric models.

3.4.4 Coupled models

To further enhance comprehensive prediction of the Earth system, NWP models can be coupled with models that simulate other parts of the system. For example, two-way coupled atmosphere-wave models (e.g., [98–100]) better simulate the marine ABL and the mutual interaction between wind and waves therein [101], yielding improved forecasts for offshore wind energy and other applications. Wind energy is among several applications that benefit from coupled atmosphere-hydrology models (e.g., [102]) as well, owing to the improved simulation of soil moisture, which impacts the growth and decay of the ABL, and thus low-level winds [103].

3.5 Using NWP for wind power

Mesoscale modeling is important for predicting wind-power production at several scales as described in more detail in this section. First, resource assessment requires modeling at these scales to accurately incorporate knowledge of the local impacts and expected variability over time. Second, wind-power forecasting relies on modeling the upstream flow accurately to be able to forecast changes in wind-power production. Third, accurate NWP model results are used to force the flow at finer scales in modeling the wind plants and turbine wakes. This section also discusses how postprocessing can improve the accuracy and some typical methods of assessing the accuracy of the mesoscale model output.

3.5.1 Resource assessment

Ideally, observations provide the most accurate information for wind-resource evaluation. However, observations have gaps in both spatial and temporal coverage. In addition, wind is usually measured near the surface (10 m above ground level), whereas the primary interest in wind-resource evaluations is to quantify the wind at hub height (~80 m and higher above ground level). In the absence of sufficient observations, which is too often the case, one can use NWP models to estimate the wind resource over the region of interest [104].

First, one should identify any previous relevant wind-resource assessment based on NWP models. Models have been used to estimate the wind resource over different regions, and different strategies have been used (e.g., [105–108] for the United States). Being familiar with previous efforts guides determining an appropriate strategy to simulate the wind resource over the region of interest.

An adequate use of NWP models requires careful consideration of the relevant physical mechanisms determining the wind resource over the region of interest in order to identify a suitable model configuration. For example, knowing the relevant physical phenomena helps to determine appropriate parameterizations. Theoretically, parameterizations are designed to work under all atmospheric conditions. In practice, performance depends on the weather situation. For example, the performance of parameterizations of the turbulent mixing within the ABL exhibits a dependence on atmospheric stability. In addition, these preliminary considerations should help one identify relevant surface heterogeneities. Complex topography or coastal regions should be properly represented in the model using an adequate horizontal grid spacing. Nesting is usually applied to progressively reach the desired horizontal grid spacing.

Another important decision is the data selected to create the initial and boundary conditions. One can use existing analysis or reanalysis datasets. Such datasets have been incorporated in a DA process to improve the estimations of the atmospheric state subject to satisfying the equations of the model. Reanalyses provide the advantage of using the same model to generate the atmospheric analysis. This could be important since wind-resource evaluations should span multiyear periods. Independent of using analysis or reanalysis datasets, one can apply DA to further improve the atmospheric analysis. This is more relevant if there are observations that were not assimilated in the (re)analysis dataset, such as surface observations from mesonets, or if the target grid spacing is much finer than the grid spacing of the (re)analysis.

After configuring the model and the data to create initial and boundary conditions, one must decide the strategy to simulate the target temporal period. One can run the model continuously or as a set of short simulations that span the complete target period. Performing the simulation as a set of short runs ensures that the synoptic scale is close to the observed one due to the frequent initializations. It also allows for an efficient parallelization of the simulation since the short simulations can be run concurrently. This is not the case for a single run of the model; thus, this kind of simulation will take a longer time to be completed. The potential drift of the synoptic scale due to the lack of re-initializations could be mitigated by nudging the coarsest domain(s) to the (re)analysis data. Analysis nudging [30,109], as described above, forces the model to closely follow the (re)analysis. This has the drawback of removing internal variability, however. A compromise is spectral nudging [110,111], which only nudges the lowest frequencies to the analysis fields, allowing the NWP model to develop internal variability in the higher frequencies. Nudging also has the advantage of reducing errors in the LBCs inherent to regional NWP models [92].

The selected strategy to perform the wind-resource evaluation should be tested, and potentially refined, before producing the final wind-resource evaluation. The period covered by this simulation should span at least a year to quantify the performance during different synoptic regimes through the annual cycle. This initial assessment may identify limitations in reproducing the wind resource. The origin of the limitations and possible refinements should be identified. These refinements may include changes to the model configuration to better represent relevant physical processes. The refined strategy should be tested during the same test period to quantify any potential improvement. This process may be repeated to improve the accuracy of the simulation until reaching acceptable levels of performance.

The selected configuration can be finally used to simulate the complete target period. Available observations over the region should be used to confirm that the evaluation produces acceptable estimations or to identify any potential deficiency of the resource assessment. Naturally, multiple years of simulations will also provide a more robust wind-resource assessment than will a single year.

3.5.2 Forecasting

Some of the challenges of wind forecasting are similar to the ones previously mentioned for the wind-resource evaluation. We need to understand the relevant physical processes affecting the wind behavior over the target region to configure the model. This includes physical settings or parameterizations, the spatial discretization, and dynamical aspects. The initial configuration can be refined over time to improve the forecast performance in a similar fashion as for resource assessment.

However, new challenges emerge primarily related to the latency of the forecast. If the NWP is a regional model, we must wait until the forecast becomes available to create the initial and boundary conditions. This latency cannot be avoided and has to be added to the time required to perform DA, if necessary, and to the time to run the NWP model. The latency of the DA may also include waiting for observations that are going to be assimilated.

Ultimately, there is a trade-off between the accuracy and the latency of the forecast. The configuration providing the best results may not be optimal for operational predictions. The gain in performance may not justify the increase in computational cost or any additional latency introduced into the forecasts. Hence, to improve operational forecasts, we need both accurate and efficient algorithms.

3.5.3 Turbine wake parameterization

If the region of interest is surrounded by wind farms already in operation, one should consider simulating the impacts that the wind turbines exert on the atmosphere. The parameterization of this effect has been implemented in mesoscale models (e.g., [112,113]). For example, the WRF model represents the turbines as an elevated sink of momentum and a source of turbulent kinetic energy (TKE) [112]. Recent evaluations point to acceptable performance of these types of parameterizations to reproduce the wind turbine wakes [114]. Additionally, modeling studies have pointed out that the wakes generated by wind farms have certain memory and thus affect downstream farms [115]. Hence, it is desirable to account for the effects of the wind turbines on the atmospheric flow in both wind forecasting and assessment studies.

The necessity to account for the effects of surrounding wind farms is illustrated with simulations performed in the European Energy Research Alliance—Design Tool for Offshore Wind Farm Clusters project. The simulations were performed with the WRF model to analyze any potential impact of the wind turbines over a specific wind farm (Figure 3.8). A couple of simulations are presented. The first one (WRF-default) does not account for the effects of any wind turbines, whereas the second one (WRF-cluster) includes the wind turbines of the wind farms surrounding the target one. By comparing both simulations, one can quantify the impact of the surrounding wind farms in the wind resource at the target location. The hub height winds during a day with NW/W winds near the center of the target location (center of black polygon in Figure 3.8) are shown in Figure 3.9. The simulations show small differences in the wind direction (Figure 3.9, top) with NW winds gradually veering to the west. On the contrary, the wind speeds have noticeable differences (Figure 3.9, bottom). These differences are small at the beginning of the simulated period that has NW winds, and thus the target location is not affected by upstream wind turbines. However, the simulations differ by more than $2\,\mathrm{m\,s^{-1}}$ during the second half of the simulation. These differences are a consequence of the upstream turbines during westerly winds (Figure 3.8). Hence, the simulations suggest a non-negligible impact of the surrounding wind farms on the wind resource at the target location.

3.5.4 *Postprocessing*

Because of the chaotic nature of the atmosphere and the differences in the model formulations, NWP model forecasts will differ. There are various types of error that, if systematic, can be corrected by postprocessing methods. For a single model, a typical approach is to apply model output statistics (MOS; [116]). MOS uses multivariate regression to determine and correct for those systematic errors. More complex methods use artificial intelligence methods, such as artificial neural networks, support vector machines, decision tree-based methods, and regime-dependent techniques.

More complex methods blend the results of multiple models, producing consensus forecasts. Woodcock and Engel [117] showed that blending such ensembles of forecasts can produce improved forecasts. Other systems, such as the Dynamical Integrated forecast (DICast®), combine the best of a dynamic version of MOS on each model with optimized consensus forecasting methods [118]. Such methods have been demonstrated to substantially improve wind forecasts over the raw model output [119–121].

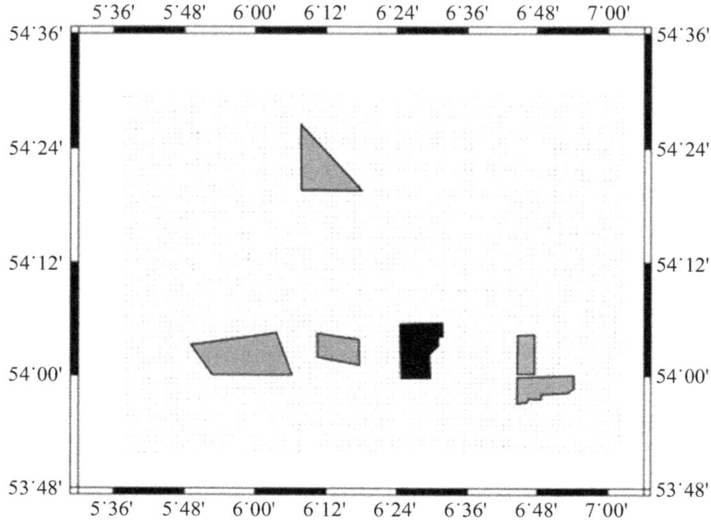

Figure 3.8 Wind farms (polygons) in the cluster of wind farms. The black polygon highlights the wind farm of interest, and the black dots represent the WRF grid.

3.5.5 Assessment

As with any prediction, NWP forecasts must be assessed for their quality. In some disciplines, this process is called assessment, validation, or evaluation, but in the atmospheric sciences, this process is called verification. Forecast verification is the process by which a forecast or set of forecasts are compared against the corresponding observation(s) of the predicted variable, to quantify some aspect of the relationship between the forecast(s) and observation(s). Numerous types of verification metrics have been developed, and for more comprehensive treatment of verification metrics and principles than space allows here, we refer the reader to Wilks [122].

For discrete predictands, such as the occurrence/nonoccurrence of cut-in wind speed, 2×2 contingency tables can be constructed (Figure 3.10). From these contingency tables, numerous metrics can be calculated, including the hit rate ($H = a/[a+c]$), false alarm ratio ($\text{FAR} = b/[a+b]$), and threat score, otherwise known as the critical success index ($\text{TS} = \text{CSI} = a/[a+b+c]$).

For continuous predictands, such as temperature or wind speed, mean absolute error (MAE) and root mean squared error (RMSE) are the two scalar metrics of forecast accuracy that are essentially ubiquitous in their use. RMSE is more sensitive to forecast outliers than MAE, and so they provide complementary information. Mean error (ME), also called bias, is another ubiquitous scalar metric for continuous predictands.

For probabilistic forecasts of dichotomous events, such as probability of precipitation, a commonly used scalar metric is the Brier score (BS), which is the average

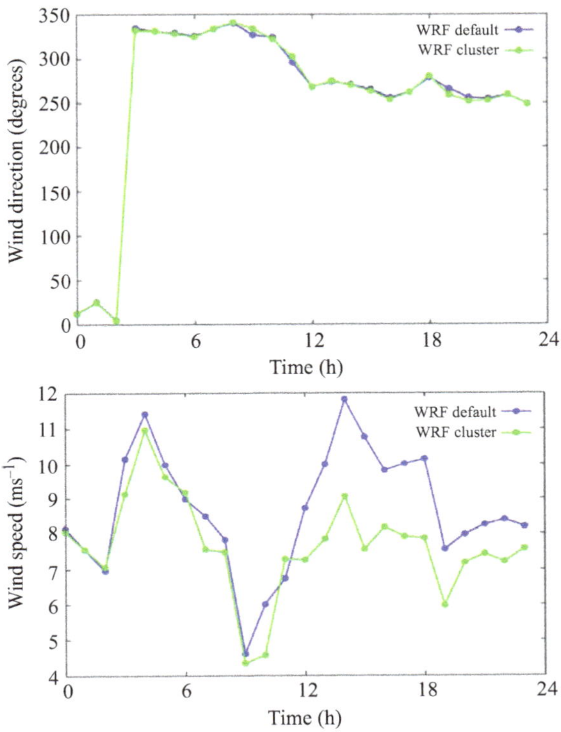

Figure 3.9 Wind direction (top) and wind speed (bottom) near the center of the wind farm under study (black polygon in Figure 3.8) from both WRF simulations. The blue line is for "WRF-default" and the green line is for "WRF-cluster" (see text for description of those experiments)

over n forecast-observation pairs of the squared difference of forecast probabilities y and observed probabilities o (1 if the event occurred, 0 otherwise)

$$BS = \frac{1}{n}\sum_{k=1}^{n}(y_k - o_k)^2 \tag{3.61}$$

The BS can be further decomposed into components denoting reliability, resolution, and uncertainty [122]. Reliability represents the calibration of the forecast probability (i.e., if the forecast probability for an event is 40%, it should verify about 40% of the time). Resolution represents the ability of a forecast to resolve different events (i.e., how similar or different are the average observational outcomes different for disparate forecasts?). The uncertainty term depends on observational error, and forecasts cannot reduce this component.

		Observed		
		Yes	**No**	Marginal totals
Forecast	**Yes**	***a***	***b***	$a + b$
	No	c	***d***	$c + d$
	Marginal totals	$a + c$	$b + d$	$n = a + b + c + d$

Figure 3.10 A 2 × 2 contingency table, where the letters a–d represent the counts of dichotomous events that were/were not forecast and also were/were not observed. The marginal totals are also included, as is the total sample size n

For probabilistic forecasts of continuous predictands, the continuous ranked probability score (CRPS; [123]) is now commonly used:

$$CRPS = \int_{-\infty}^{\infty} [F(y) - F_o(y)]^2 \, dy$$
$$F_o(y) = \begin{cases} 0, & y < \text{observed value} \\ 1, & y \geq \text{observed value} \end{cases} \tag{3.62}$$

where $F(y)$ is the cumulative distribution function (cdf) of the forecast y, and $F_o(y)$ is a step-function cdf that steps from 0 to 1 where the forecast variable y equals the observation. The CRPS is an extension of the BS and is an attractive metric because it assesses both the accuracy and sharpness of the forecast distribution. Sharpness is another term for the spread of a forecast distribution; sharper distributions by definition have less spread. Thus, if a forecast distribution is accurate but not sharp, or sharp but not accurate, it will be penalized with a higher CRPS. The CRPS is also a strictly proper scoring metric, meaning that the uniquely best score for the metric can only be achieved by forecasting for the same distribution from which the observation is drawn [124,125].

The selected few metrics that we highlighted above are all used for the verification of forecasts at a specific point and time, or averaged point-by-point over an entire model domain. However, one can easily envision scenarios in which point-based metrics in effect doubly penalize a forecast model for a spatial or temporal displacement of the prediction of some observed phenomenon (e.g., wind speed ramps from a frontal passage or from the outflow of a thunderstorm, precipitation fields, and cloud fields). To remedy this, an entire class of object-based verification methods has been developed (e.g., [126–129]). Objects in both the forecast and observed fields can be classified by attributes such as intensity/magnitude, area, centroid location, axis angle, aspect ratio, and curvature, with various summary metrics designed to assess the quality of match between forecast and observed objects. A popular tool called the method for object-based diagnostic evaluation (MODE; [129]) performs this object-based verification and is included in the Model Evaluation Tools (MET) verification package [130]. In addition to MODE, MET calculates a host of other statistics and is becoming the standard verification package in both the operational forecasting and atmospheric science research communities in the United States.

3.6 Uncertainty quantification

As mentioned in the introductory section of this chapter, meteorologists acknowledge the inherent limits to predictability of atmospheric flow and seek to quantify the uncertainty in NWP forecasts. In addition, model configurations and parameterizations are not perfect. As given in Section 3.2, there are multiple physics parameterizations, and Section 3.5 discussed the necessity of choosing those configurations carefully to assure best results for the specific use. Thus, for wind energy purposes, we often distinguish two sources of uncertainty, each with its own method of quantification. Structural uncertainty refers to that uncertainty due to the inherent chaotic fluctuations of the atmosphere. Parametric uncertainty derives from the differences that occur due to the application of differing parameterization schemes in model configuration. We treat these separately here. Both can contribute large portions to modeling uncertainty.

3.6.1 Quantifying parametric uncertainty

Because the choice of physics packages and parameter settings can influence the results of the model simulations, it is often useful to determine the extent of the resulting variations. It is typically prohibitive to test every possible combination of model configuration and parameter setting, so the process is often accomplished via a sampling process. Some studies have followed a method of Quasi Monte Carlo sampling [131,132]. Yang *et al.* [131] applied the method specifically to study sensitivity of hub-height wind speed (and resulting power) to values of the parameters used in the ABL and surface parameterization schemes in complex terrain, considering the impact on the variables most relevant to producing wind power, such as hub-height wind speed, shear across the rotor diameter, the vertical wind profile, and turbulent mixing parameters. They attributed the largest variances to perturbations in the parameters related to the TKE dissipation rate, the turbulent Prandtl number, the turbulent length scales in the Mellor–Yamada–Nakanishi–Nino ABL scheme [133], and surface roughness and von Kármán constant in the revised MM5 surface-layer scheme. The contributions of these different parameters depend on the terrain slope and the stability. They showed examples of where uncertainty in the wind-power production due to different parameter values can vary from near zero to full-rated power.

3.6.2 Quantifying structural uncertainty—ensemble methods

It is also important to quantify the inherent structural uncertainty due to the chaotic nature of atmospheric flow. The traditional method to accomplish this goal is to run ensembles of forecasts, each with a slightly different initial condition, boundary condition, parameterization configuration, and other types of small differences. Any difference at initialization will result in larger differences at a later time in the simulation. Careful construction of such ensembles can be effective in estimating the uncertainty of a forecast by quantifying the resulting spread of the members. For a well-constructed ensemble, on average, the spread will match the error, typically in terms of a mean squared error or variance. Many national centers and some wind-power forecasters routinely run ensembles. Although the number of members required to produce a naturally well calibrated ensemble is theoretically in the hundreds [134], methods to down-select to a smaller number for ensembles with distinguishable members have shown that on the order of seven to ten members may be sufficient if the ensemble is to be calibrated [135,136].

3.6.3 Calibrating ensembles

Seldom are raw ensembles sufficiently calibrated to meet the mean spread matching the error. Thus, various techniques have been developed to calibrate ensembles. that is, to remove the bias and to calibrate the spread. One of the most basic, but effective methods, is linear variance calibration [134,137]. In this method, one simply performs a linear regression analysis between the binned spread and the binned variance and fits a line with a slope and an intercept. The intercept is the bias, which is then subtracted from the mean. Then the spread is multiplied by the slope to produce a well calibrated ensemble. Other methods to accomplish calibration include quantile regression [138], Bayesian model averaging [139], and beyond.

3.6.4 Analog ensembles

Running multiple members of an ensemble can greatly increase the computational requirements of a study. An innovative method that accomplishes the same goal while relying on a single, high-quality model run is the analog ensemble (AnEn) method [140,141]. The AnEn technique begins with a sufficiently long (i.e., at least several months, but preferably a few years) database of historical forecasts and their verifying observations. A search is made of this database for the forecasts that best match the current forecast. Because those forecasts have matching observations, those observations become the pdf of the ensemble. It has been shown that the mean of this pdf is, on average, superior to the mean of a high quality ensemble, and that its spread better represents the uncertainty than that of a typical ensemble, even when calibrated [141]. The AnEn method has been successfully applied to wind-power forecasting [142,143] and wind-resource assessment [144].

3.7 Looking ahead

Mesoscale modeling is evolving along with advances in computing power and our need to more accurately model the boundary layer. Here we discuss a few of the directions where we expect changes in the coming decade and beyond.

3.7.1 Storm-scale prediction

Storm-scale predictions use a fine enough grid spacing to resolve individual storms. The motivation for this refinement is to explicitly resolve the effects of the storms instead of parameterizing them. Nowadays, storm-scale simulations typically have a horizontal grid spacing of a few kilometers (e.g., [145]). Simulations using this grid spacing are known as "convection permitting" because parameterizing the effects of deep convective clouds is not necessary. The benefit of the fine grid spacing should improve the performance of the forecasts, and in particular, the simulation of individual storms.

3.7.2 Scale-aware models

The parameterizations included in NWP models to account for physical processes not explicitly resolved by the models should modulate their effects as a function of the grid spacing (i.e., be "scale-aware"). This is a consequence of the sensitivity of the simulated weather to the model grid spacing. For example, the effects of deep convective clouds need to be parameterized at coarse grid spacing (e.g., $\geq$10 km), but progressively refined grids start to explicitly account for some convective effects and, if the horizontal grid is fine enough, explicitly represent these kinds of clouds. Hence, a scale-aware deep-convective parameterization (e.g., [146]) should account for this change in the model performance by gradually reducing its contribution for progressively refined grids. Other types of parameterizations present similar challenges. Developing scale-aware parameterizations is challenging because many internal parameters in the schemes may exhibit a dependence on the grid spacing. The importance of having scale-aware physics is becoming increasingly relevant given the tendency to perform multi-scale simulations.

3.7.3 Blended global/mesoscale models

Next-generation NWP models, such as the Model for Prediction Across Scales (MPAS; [147]—see grid in Figure 3.11) or the Finite Volume Cubed-Sphere Dynamical Core [98], are global models but are designed with a grid discretization that allows for variable-resolution meshes in different parts of the globe. These types of models allow the grid mesh spacing to be gradually refined near a region of interest, thus eliminating the problem of artificial errors being introduced at nested domain boundaries. Eliminating these sources of error is an attractive feature of these next-generation NWP models. As computational power continues to increase and these models continue to become more efficient, variable-resolution global NWP models will likely become the default choice of model, even for regional applications.

3.7.4 Seasonal to subseasonal prediction

Needs for longer term forecasts on the subseasonal to seasonal scales are emerging. Electric utilities need to plan for long-term maintenance, planners must prepare for fossil fuel purchases in advance, and traders are looking further into the future. To meet these needs as well as those of other industries, the meteorological community

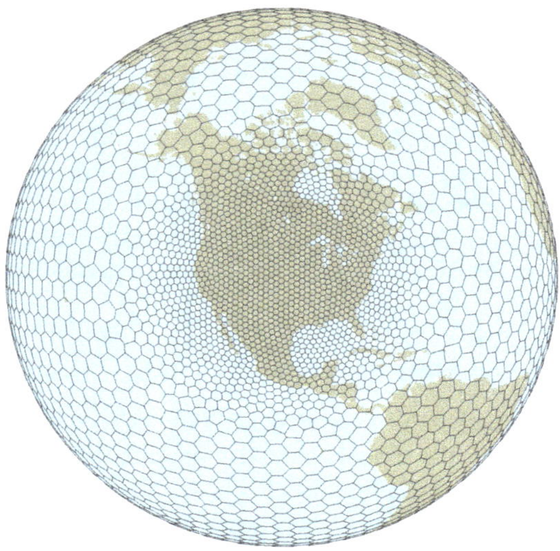

Figure 3.11 *Example of a Voronoi mesh used for the MPAS model, which allows for both quasi-uniform discretization over the sphere as well as local refinement (here over the United States). Figure courtesy of the MPAS development team*

is currently working to advance the state-of-the-science of forecasting at these scales. It is clear that the very short-term forecast is an initial value problem. In contrast, climate forecasting is a boundary value problem. These subseasonal to seasonal-scale forecasts are a blend of both problems, and depend critically on knowledge of the changing state of the ocean, land cover, stratospheric state, and other important forcings. National centers are now running ensemble systems for these forecast scales and service providers are smartly postprocessing the information to optimize and interpret the solutions for users' needs. The meteorology community is actively researching ways to improve forecasting at these time scales, and major advances are expected in the coming years.

3.7.5 Regime-dependent corrections

We saw that statistical postprocessing and ensemble calibration are important aspects of providing a best practices solution to a forecasting problem. These methods can be enhanced by recognizing that weather occurs in distinct regimes and that forecast models perform differently for the various regimes. Thus, if the regimes can be clustered, calibration methods can be configured specific to each regime if there are sufficient historical data. Regime-dependent forecasting and ensemble calibration have shown promise [148] including for renewable energy problems [149,150]. We expect to see this area of research evolve, especially as postprocessing advances toward using the full advantages of a gridded approach.

3.8 Summary and conclusions

This chapter on meteorological modeling has introduced the basic forces driving atmospheric flow, the general circulation, secondary circulation modifications to the flow, and atmospheric energetics. It then posed the atmospheric equations of motion and how they are discretized for NWP models. We have also described the various methods for initialization with best estimates of the current state using DA, including presenting the advantages and disadvantages of the major methods currently in use. We also discussed boundary conditions and physics parameterizations, as well as the use of NWP models for wind-energy applications. Because the atmosphere is a chaotic dynamical system, one must recognize and quantify the uncertainty in the simulations, through both parametric and structural uncertainty analyses. Both the forecast itself and its uncertainty should be calibrated for best results, and several methods to do so have been mentioned.

Atmospheric flow is much more than inflow to a wind plant simulation. It becomes the basis for a dynamic, fluid understanding of the forcing flow, and thus, the energy transfer that becomes the source for harvesting the wind energy. This energy source is not a constant continuous value. Instead, it varies greatly over the course of a day, whenever a front passes, in the lee of topography, near air–water boundaries, when the canopy changes, and when the upper boundary of the stratosphere or the lower boundary of ocean or land surface changes. All of these differences modify the flow substantially, and it is critical to capture this nonstationarity and inhomogeneity to effectively model wind power output. Thus, methods are being developed to couple the larger atmospheric flow to the wind plant and turbine scales to better capture these important characteristics [151–158]. As we move toward newer challenges of modeling wind plants in the ocean environment and in complex terrain, it will become increasingly more important to better model the atmospheric flow and the energy transfers through the scales to provide the information critical to harvesting wind power.

References

[1] Anthes, R., A. J. J. Cahir, A. B. Fraser, and H. A. Panofsky, 1981: *The Atmosphere, Third Ed.* Bell & Howell Co., Columbus, OH, 531 pp.

[2] Holton, J. R., 1972: *An Introduction to Dynamic Meteorology, Third Ed.* Academic Press, London, UK, 507 pp.

[3] Bjerknes, J., 1919: On the structure of moving cyclones. *Mon. Weather Rev.*, **47**, 95–99, doi:10.1175/1520-0493(1919)47<95:OTSOMC>2.0.CO;2.

[4] Stull, R. B., 1988: *An Introduction to Boundary Layer Meteorology*. Kluwer Academic Publishers, Boston, MA, 666 pp.

[5] Vanderwende, B., J. K. Lundquist, M. E. Rhodes, G. S. Takle, and S. I. Purdy, 2015: Observing and simulating the summertime low-level jet in central Iowa. *Mon. Weather Rev.*, **143**, 2319–2336.

[6] Skamarock, W. C., S.-H. Park, J. B. Klemp, and C. Snyder, 2014: Atmospheric kinetic energy spectra from global high-resolution nonhydrostatic simulations, *J. Atmos. Sci.*, **71**, 4369–4381, doi: 10.1175/JAS-D-14-0114.1.

[7] Haupt, S. E., P. Jimenez, J. A. Lee, and B. Kosovic, 2017a: Principles of meteorology and numerical weather prediction, in Kariniotakis, G., ed. *Renewable Energy Forecasting: From Models to Applications*. Elsevier, London, UK, pp. 3–28, doi:10.1016/B978-0-08-100504-0.00001-9.

[8] Lorenz, E. N., 1963: Deterministic nonperiodic flow. *J. Atmos. Sci.*, **20**, 130–141, doi:10.1175/1520-0469(1963)020<0130:DNF>2.0.CO;2.

[9] Lorenz, E. N., 1969: The predictability of a flow which possesses many scales of motion. *Tellus*, **21**, 289–307, doi:10.1111/j.2153-3490.1969.tb00444.x.

[10] Hoskins, B., 2013: The potential for skill across the range of seamless weather-climate prediction problem: A stimulus for our science. *Q. J. R. Meteorol. Soc.*, **139**, 573–584, doi:10.1002/qj.1991.

[11] Buizza, R., and M. Leutbecher, 2015: The forecast skill horizon. *Q. J. R. Meteorol. Soc.*, **141**, 3366–3382, doi:10.1002/qj.2619.

[12] Emanuel, K., and F. Zhang, 2016: On the predictability and error sources of tropical cyclone intensity forecasts. *J. Atmos. Sci.*, **73**, 3739–3747, doi:10.1175/JAS-D-16-0100.1.

[13] Bjerknes, V., 1904: Das Problem der Wettervorhersage, betrachtet von Standpunkt der Mechanik und Physik. *Meteorol. Z.*, **21**, 1–7.

[14] Richardson, L. F., 1922: *Weather Prediction by Numerical Process*. Cambridge: Cambridge University Press, 236 pp.

[15] Lynch, P., 2006: *The Emergence of Numerical Weather Prediction*. New York: Cambridge University Press, 292 pp.

[16] Lynch, P., 2008: The origins of computer weather prediction and climate modeling. *J. Comput. Phys.*, **227**, 3431–3444, doi:10.1016/j.jcp.2007.02.034.

[17] Burks, A. W., H. H. Goldstine, and J. von Neumann, 1946: Preliminary discussion of the logical design of an electronic computing instrument. Report to U.S. Army Ordnance Department, 54 pp. [Available online at https://library.ias.edu/files/Prelim_Disc_Logical_Design.pdf.]

[18] UCAR/NCAR AtmosNews, 2019: Turning the Tables on Chaos: Is the Atmosphere More Predictable Than We Assume? President's Corner by Richard A. Anthes, [Available online at http://www2.ucar.edu/atmosnews/opinion/turning-tables-chaos-atmosphere-more-predictable-we-assume-0, last accessed 3/19/19.]

[19] Pielke, R. A., 1984: *Mesoscale Meteorological Modeling*. New York: Academic Press, 612 pp.

[20] Dudhia, J., 1993: A nonhydrostatic version of the Penn State–NCAR Mesoscale Model: Validation tests and simulation of an Atlantic cyclone and cold front. *Mon. Weather Rev.*, **121**, 1493–1513, doi:10.1175/1520-0493(1993)121<1493:ANVOTP>2.0.CO;2.

[21] Stensrud, D. J., 2007: *Parameterization Schemes Keys to Understanding Numerical Weather Prediction Models*. New York: Cambridge University Press. 478 pp.

[22] Monin, A. S., and A. M. Obukhov, 1954: Basic laws of turbulent mixing in the surface layer of the atmosphere. *Tr. Akad. Nauk SSSR Geophiz. Inst.*, **24**, 163–187, [English translation available online at http://mcnaughty.com/keith/papers/Monin_and_Obukhov_1954.pdf.]

[23] Khain, A. P., K. D. Beheng, A. Heymsfield, *et al.*, 2015: Representation of microphysical processes in cloud-resolving models: Spectral (bin) microphysics versus bulk parameterization. *Rev. Geophys.*, **53**, 247–322, doi:10.1002/2014RG000468.

[24] Powers, J. G., J. B. Klemp, W. C. Skamarock, *et al.*, 2017: The weather research and forecasting model: Overview, system efforts, and future directions. *Bull. Am. Meteorol. Soc.*, **98**, 1717–1737, doi:10.1175/BAMS-D-15-00308.1.

[25] Skamarock, W. C., J. B. Klemp, J. Dudhia, *et al.*, 2008: A description of the Advanced Research WRF Version 3. *NCAR Tech. Note NCAR/TN-475+STR*, 113 pp., doi:10.5065/D68S4MVH.

[26] Warner, T. T., 2011: *Numerical Weather and Climate Prediction*. Cambridge University Press, New York, NY, 526 pp., ISBN 978-0-521-51389-0.

[27] Talagrand, O., 1997: Assimilation of observations, an introduction. *J. Meteorol. Soc. Jpn.*, **75**, 191–209, doi:10.2151/jmsj1965.75.1B_191.

[28] Whitaker, J., T. M. Hamill, X. Wei, Y. Song, and Z. Toth, 2008: Ensemble data assimilation with the NCEP Global Forecast System. *Mon. Weather Rev.*, **436**, 463–482, doi:10.1175/2007MWR2018.1.

[29] Hoke, J. E., and R. A. Anthes, 1976: The initialization of numerical models by a dynamic-initialization technique. *Mon. Weather Rev.*, **104**, 1551–1556, doi:10.1175/1520-0493(1976)104<1551:TIONMB>2.0.CO;2.

[30] Stauffer, D. R., and N. L. Seaman, 1990: Use of four-dimensional data assimilation in a limited-area mesoscale model. Part I: Experiments with synoptic-scale data. *Mon. Weather Rev.*, **118**, 1250–1277, doi:10.1175/1520-0493(1990)118<1250:UOFDDA>2.0.CO;2.

[31] Stauffer, D. R., and N. L. Seaman, 1994: Multiscale four-dimensional data assimilation. *J. Appl. Meteorol.*, **33**, 416–434, doi:10.1175/1520-0450(1994)033<0416:MFDDA>2.0.CO;2.

[32] Liu, Y., T. T. Warner, J. F. Bowers, *et al.*, 2008: The operational mesogamma-scale analysis and forecast system of the U.S. Army Test and Evaluation Command. Part I: Overview of the modeling system, the forecast products, and how the products are used. *J. Appl. Meteorol. Climatol.*, **47**, 1077–1092, doi:10.1175/2007JAMC1653.1.

[33] Deng, A., D. R. Stauffer, J. Dudhia, T. L. Otte, and G. K. Hunter, 2007: Update on analysis nudging FDDA in WRF-ARW. *8th WRF Users' Workshop*, Boulder, CO, National Center for Atmospheric Research, 4.8. [Available online at http://www2.mmm.ucar.edu/wrf/users/workshops/WS2007/abstracts/4-8_Deng.pdf.]

[34] Deng, A., D. Stauffer, B. Gaudet, *et al.*, 2009: Update on WRF-ARW end-to-end multi-scale FDDA system. *10th WRF Users' Workshop*, Boulder, CO, 1.9, 14 pp. [Available online at http://www2.mmm.ucar.edu/wrf/users/workshops/WS2009/WorkshopPapers.php.]

[35] Rogers, R. E., A. Deng, D. R. Stauffer, *et al.*, 2012: Application of the Weather Research and Forecasting model for air quality modeling in the San Francisco Bay area. *J. Appl. Meteorol. Climatol.*, **52**, 1953–1973, doi:10.1175/JAMC-D-12-0280.1.

[36] Otte, T. L., C. G. Nolte, M. J. Otte, and J. H. Bowden, 2012: Does nudging squelch the extremes in regional climate modeling. *J. Clim.*, **25**, 7046–7066, doi:10.1175/JCLI-D-12-00048.1.

[37] Schlatter, T. W., 2000: Variational assimilation of meteorological observations in the lower atmosphere: A tutorial on how it works. *J. Atmos. Sol. Ter. Phys.*, **62**, 1057–1070, doi:10.1016/S1364-6826(00)00096-1.

[38] Talagrand, O., and P. Courtier, 1987: Variational assimilation of meteorological observations with the adjoint vorticity equation. I: Theory. *Q. J. R. Meteorol. Soc.*, **113**, 1311–1328, doi:10.1002/qj.49711347812.

[39] Kalnay, E., 2003: *Atmospheric Modeling, Data Assimilation, and Predictability*. Cambridge University Press, New York, NY, 341 pp.

[40] Klinker, E., F. Rabier, G. Kelly, and J.-F. Mahfouf, 2000: The ECMWF operational implementation of four-dimensional variational assimilation. III: Experimental results and diagnostics with operational configuration. *Q. J. R. Meteorol. Soc.*, **126**, 1191–1215, doi:10.1002/qj.49712656417.

[41] Kalman, R. E., 1960: A new approach to linear filtering and prediction problems. *J. Basic Eng.*, **82**, 35–45, doi:10.1115/1.3662552.

[42] Kalman, R. E., and R. S. Bucy, 1961: New results in linear filtering and prediction theory. *J. Basic Eng.*, **83**, 95–108, doi:10.1115/1.3658902.

[43] Evensen, G., 1994: Sequential data assimilation with a nonlinear quasigeostrophic model using Monte Carlo methods to forecast error statistics. *J. Geophys. Res. Oceans*, **99**, 10143–10162, doi:10.1029/94JC00572.

[44] Anderson, J. L., 2001: An ensemble adjustment Kalman filter for data assimilation. *Mon. Weather Rev.*, **129**, 2884–2903, doi:10.1175/1520-0493(2001)129<2884;AWAKFF>2.0.CO;2.

[45] Bishop, C. H., B. J. Etherton, and S. J. Majumdar, 2001: Adaptive sampling with the ensemble transform Kalman filter. Part I: Theoretical aspects. *Mon. Weather Rev.*, **129**, 420–436, doi:10.1175/1520-0493(2001)1290420:< ASWTET>2.0.CO;2.

[46] Whitaker, J. S., and T. M. Hamill, 2002: Ensemble data assimilation without perturbed observations. *Mon. Weather Rev.*, **140**, 3078–3089, doi:10.1175/1520-0493(2002)130<1913:EDAWPO>2.0.CO;2; Corrigendum, **134**, 1722, doi:10.1175/MWR-D-11-00276.1.

[47] Yang, S.-C., M. Corazza, A. Carrassi, E. Kalnay, and T. Miyoshi, 2006: Comparison of local ensemble transform Kalman filter, 3DVAR, and 4DVAR in a quasigeostrophic model. *Mon. Weather Rev.*, **137**, 693–709, doi:10.1175/2009MWR2396.1.

[48] Anderson, J., T. Hoar, K. Raeder, *et al.*, 2009: The data assimilation research testbed: A community facility. *Bull. Am. Meteorol. Soc.*, **90**, 1283–1296, doi:10.1175/2009BAMS2618.1.

[49] Schwartz, C. S., G. S. Romine, R. A. Sobash, K. R. Fossell, and M. L. Weisman, 2015: NCAR's experimental real-time convection-allowing ensemble prediction system. *Weather Forecast.*, **30**, 1645–1654, doi:10.1175/WAF-D-15-0103.1.

[50] Houtekamer, P. L., and F. Zhang, 2016: Review of the ensemble Kalman filter for atmospheric data assimilation. *Mon. Weather Rev.*, **144**, 4489–4532, doi:10.1175/MWR-D-15-0440.1.

[51] Hacker, J. P., and W. M. Angevine, 2013: Ensemble data assimilation to characterize surface-layer errors in numerical weather prediction models. *Mon. Weather Rev.*, **141**, 1804–1821, doi:10.1175/MWR-D-12-00280.1.

[52] Bannister, R. N., 2017: A review of operational methods of variational and ensemble-variational data assimilation. *Q. J. R. Meteorol. Soc.*, **143**, 607–633, doi:10.1002/qj.2982.

[53] Epstein, E. S., 1969: Stochastic dynamic prediction. *Tellus*, **21**, 739–759, doi:10.1111/j.2153-3490.1969.tb00483.x.

[54] Gaspari, G., and S. E. Cohn, 1999: Construction of correlation functions in two and three dimensions. *Q. J. R. Meteorol. Soc.*, **125**, 723–757, doi:10.1002/qj.49712555417.

[55] Anderson, J. L., and S. L. Anderson, 1999: A Monte Carlo implementation of the nonlinear filtering problem to produce ensemble assimilations and forecasts. *Mon. Weather Rev.*, **127**, 2741–2758, doi:10.1175/1520-0493(1999)127<2741:AMCIOT>2.0.CO;2.

[56] Hamill, T. M., J. S. Whitaker, and C. Snyder, 2001: Distance-dependent filtering of background error covariance estimates in an ensemble Kalman filter. *Mon. Weather Rev.*, **129**, 2776–2790, doi:10.1175/1520-0493(2001)129<2776:DDFOBE>2.0.CO;2.

[57] Lei, L., D. R. Stauffer, and A. Deng, 2012: A hybrid nudging-ensemble Kalman filter approach to data assimilation in WRF/DART. *Q. J. R. Meteorol. Soc.*, **138**, 2066–2078, doi:10.1002/qj.1939.

[58] Schraff, C., H. Reich, A. Rhodin, *et al.*, 2016: Kilometre-scale ensemble data assimilation for the COSMO model (KENDA). *Q. J. R. Meteorol. Soc.*, **142**, 1453–1472, doi:10.1002/qj.2748.

[59] Hunt, B. R., E. J. Kostelich, and I. Szunyogh, 2007: Efficient data assimilation for spatiotemporal chaos: A local ensemble transform Kalman filter. *Physica D*, **230**, 112–126, doi:10.1016/j.physd.2006.11.008.

[60] Wang, X., D. M. Barker, C. Snyder, and T. M. Hamill, 2008: A hybrid ETKF-3D-Var data assimilation scheme for the WRF model. Part II: Real observation experiments. *Mon. Weather Rev.*, **136**, 5132–5147, doi:10.1175/2008MWR2445.1.

[61] Zhang, F., M. Zhang, and J. Poterjoy, 2013: E3DVar: Coupling an ensemble Kalman filter with three-dimensional variational data assimilation in a limited-area weather prediction model and comparison to E4DVar. *Mon. Weather Rev.*, **141**, 900–917, doi:10.1175/MWR-D-12-00075.1.

[62] Pan, Y., K. Zhu, M. Xue, *et al.*, 2014: A GSI-based coupled EnSRF-En3DVar hybrid data assimilation system for the operational Rapid Refresh

model: Tests at a reduced resolution. *Mon. Weather Rev.*, **142**, 3756–3780, doi:10.1175/MWR-D-13-00242.1.

[63] Lorenc, A. C., N. E. Bowler, A. M. Clayton, S. R. Pring, and D. Fairbairn, 2015: Comparison of hybrid 4DEnVar and hybrid-4DVar data assimilation methods for global NWP. *Mon. Weather Rev.*, **143**, 212–229, doi:10.1175/MWR-D-14-00195.1.

[64] Fraser, J., 2012: APS1 Upgrade of the ACCESS-G Numerical Weather Prediction System. NMOC Operations Bull. 93, Bureau of Meteorology, 29 pp. [Available online at http://www.bom.gov.au/australia/charts/bulletins/apob93.pdf.]

[65] Puri, K., G. Dietachmayer, P. Steinle, *et al.*, 2013: Implementation of the initial ACCESS numerical weather prediction system. *Aust. Meteorol. Oceanogr. J.*, **63**, 265–284. [Available online at http://www.bom.gov.au/jshess/docs/2013/puri.pdf (accessed 8 Aug 2019)]

[66] Tripathi, O. P., M. Baldwin, A. Charlton-Perez, *et al.*, 2016: Examining the predictability of the stratospheric sudden warming of January 2013 using multiple NWP systems. *Mon. Weather Rev.*, **144**, 1935–1960, doi:10.1175/MWR-D-15-0010.1.

[67] Rhodin, A., 2015: Multi scale VarEnKF localisation using wavelets. *International Symposium on Data Assimilation*, Kobe, Japan, RIKEN, Feb 2015. [Available online at http://www.data-assimilation.riken.jp/isda2015/program/pdf/8-4.pdf.]

[68] Buehner, M., R. McTaggart-Cowan, A. Beaulne, *et al.*, 2015: Implementation of deterministic weather forecasting systems based on ensemble-variational data assimilation at Environment Canada. Part I: The global system. *Mon. Weather Rev.*, **143**, 2532–2559, doi:10.1175/MWR-D-14-00354.1.

[69] Caron, J.-F., T. Milewski, M. Buehner, *et al.*, 2015: Implementation of deterministic weather forecasting systems based on ensemble-variational data assimilation at Environment Canada. Part II: The regional system. *Mon. Weather Rev.*, **143**, 2560–2580, doi:10.1175/MWR-D-14-00353.1.

[70] Bonavita, M., E. V. Hólm, L. Isaksen, and M. Fisher, 2016: The evolution of the ECMWF hybrid data assimilation system. *Q. J. R. Meteorol. Soc.*, **142**, 287–303, doi:10.1002/qj.2652.

[71] Gustafsson, N., J. Bojarova, and O. Vigne, 2014: A hybrid variational ensemble data assimilation for the HIgh Resolution Limited Area Model (HIRLAM). *Nonlinear Processes Geophys.*, **21**, 303–323, doi:10.5194/npg-21-303-2014.

[72] Gustafsson, N., and J. Bojarova, 2014: Four-dimensional ensemble variational (4D-En-Var) data assimilation for the HIgh Resolution Limited Area Model (HIRLAM). *Nonlinear Processes Geophys.*, **21**, 745–762, doi:10.5194/npg-21-745-2014.

[73] Japan Meteorological Agency, 2016: Joint WMO technical progress report on the global data processing and forecasting system and numerical weather prediction research activities for 2016. *JMA Tech. Note*. [Available online at www.jma.go.jp/jma/jma-eng/jma-center/nwp/report/2016_Japan.pdf.]

[74] Berre, L., O. Pannekoucke, G. Desroziers, S. E. Stefănescu, B. Chapnik, and L. Raynaud, 2007: A variational assimilation ensemble and the spatial filtering of its error covariances: Increase of sample size by local spatial averaging. *Proc. ECMWF Workshop on Flow-Dependent Aspects of Data Assimilation*, Reading, UK, ECMWF. 11–13 Jun 2007. [Available online at https://www.ecmwf.int/en/elibrary/8172-variational-assimilation-ensemble-and-spatial-filtering-its-error-covariances.]

[75] Berre, L., H. Varella, G. Desroziers, 2015: Modelling of flow-dependent ensemble-based background-error correlations using a wavelet formulation in 4D-Var at Météo-France. *Q. J. R. Meteorol. Soc.*, **141**, 2803–2812, doi:10.1002/qj.2565.

[76] Raynaud, L., L. Berre, and G. Desroziers, 2011: An extended specification of flow-dependent background-error variances in the Météo-France global 4D-Var system. *Q. J. R. Meteorol. Soc.*, **137**, 607–619, doi:10.1002/qj.795.

[77] Raynaud, L., L. Berre, and G. Desroziers, 2012: Accounting for model error in the Météo-France ensemble data assimilation system. *Q. J. R. Meteorol. Soc.*, **138**, 249–262, doi:10.1002/qj.906.

[78] Wang, X., and T. Lei, 2014: GSI-based four-dimensional ensemble-variational (4DEnsVar) data assimilation: Formulation and single-resolution experiments with real data for NCEP global forecast system. *Mon. Weather Rev.*, **142**, 3303–3325, doi:10.1175/MWR-D-13-00303.1.

[79] Kleist, D. T., and K. Ide, 2015a: An OSSE-based evaluation of hybrid variational-ensemble data assimilation for the NCEP GFS. Part I: System description and 3D-hybrid results. *Mon. Weather Rev.*, **143**, 433–451, doi:10.1175/MWR-D-13-00351.1.

[80] Kleist, D. T., and K. Ide, 2015b: An OSSE-based evaluation of hybrid variational-ensemble data assimilation for the NCEP GFS. Part II: 4DEnVar and hybrid variants. *Mon. Weather Rev.*, **143**, 452–470, doi:10.1175/MWR-D-13-00350.1.

[81] Wang, X., D. M. Barker, C. Snyder, and T. M. Hamill, 2008: A hybrid ETKF-3D-Var data assimilation scheme for the WRF model. Part I: Observing system simulation experiment. *Mon. Weather Rev.*, **136**, 5116–5131, doi:10.1175/2008MWR2444.1.

[82] Liu, C., Q. Xiao, and B. Wang, 2009: An ensemble-based four-dimensional variational data assimilation scheme. Part II: Observing system simulation experiments with the Advanced Research WRF (ARW). *Mon. Weather Rev.*, **137**, 1687–1704, doi:10.1175/2008MWR2699.1.

[83] Wang, X., 2010: Incorporating ensemble covariance in the Gridpoint Statistical Interpolation variational minimization: A mathematical framework. *Mon. Weather Rev.*, **138**, 2990–2995, doi:10.1175/2010MWR3245.1.

[84] Schwartz, C. S., Z. Liu, H.-C. Lin, and J. D. Cetola, 2014: Assimilating aerosol observations with a "hybrid" variational-ensemble data assimilation system. *J. Geophys. Res. Atmos.*, **119**, 4043–4069, doi:10.1002/2013JD020937.

[85] Bishop, C. H., D. Hodyss, P. Steinle, *et al.*, 2011: Efficient ensemble covariance localization in variational data assimilation. *Mon. Weather Rev.*, **139**, 573–580, doi:10.1175/2010MWR3405.1.

[86] Kuhl, D. D., T. E. Rosmond, C. H. Bishop, J. McLay, and N. L. Baker, 2013: Comparison of hybrid ensemble/4DVar and 4DVar within the NAVDAS-AR data assimilation framework. *Mon. Weather Rev.*, **143**, 452–470, doi:10.1175/MWR-D-12-00182.1.

[87] Zhao, C., X. Liu, Y. Qian, J. Yoon, and Z. Hou, 2013: A sensitivity study of radiative fluxes at the top of atmosphere to cloud-microphysics and aerosol parameters in the Community Atmosphere Model CAM5. *Atmos. Chem. Phys.*, **13**, 10969–10987, doi:10.5194/acp-13-10969-2013.

[88] Zhang, M., and F. Zhang, 2012: E4DVar: Coupling an ensemble Kalman filter with four-dimensional variational data assimilation in a limited-area weather prediction model. *Mon. Weather Rev.*, **140**, 587–600, doi:10.1175/MWR-D-11-00023.1.

[89] Clayton, A. M., A. C. Lorenc, and D. M. Barker, 2012: Operational implementation of a hybrid ensemble/4D-Var global data assimilation system at the Met Office. *Q. J. R. Meteorol. Soc.*, **139**, 1445–1461, doi:10.1002/qj.2054.

[90] Benjamin, S. G., S. S. Wygandt, J. M. Brown, *et al.*, 2016: A North American hourly assimilation and model forecast cycle: The Rapid Refresh. *Mon. Weather Rev.*, **144**, 1669–1694, doi:10.1175/MWR-D-15-0242.1.

[91] NCEP, 2003: The GFS Atmospheric Model. NCEP Office Note 442, 14 pp. [Available online at http://www.emc.ncep.noaa.gov/officenotes/newernotes/on442.pdf.]

[92] Warner, T. T., R. A. Peterson, and R. E. Treadon, 1997: A tutorial on lateral boundary conditions as a basic and potentially serious limitation to regional numerical weather prediction. *Bull. Am. Meteorol. Soc.*, **78**, 2599–2617, doi:10.1175/1520-0477(1997)078<2599:ATOLBC>2.0.CO;2.

[93] Obukhov, A. M., 1946: Turbulentnost v temperaturnoj – neodnorodnoj atmosfere (Turbulence in an Atmosphere with a Non-uniform Temperature). *Tr. Inst. Theor. Geofiz.*, 1, 95–115, AN SSSR.

[94] Foken, T., 2006: 50 Years of the Monin–Obukhov similarity theory. *Boundary Layer Meteorol.*, **119**, 431–447, doi:10.1007/s10546-006-9048-6.

[95] Dyer, A. J., and B. B. Hicks, 1970: Flux-gradient relationships in the constant flux layer. *Q. J. R. Meteorol. Soc.*, 96, 715–721.

[96] Dyer, A. J., 1974: A review of flux-profile relationships. *Boundary Layer Meteorol.*, **7**, 363–372, doi:10.1007/BF00240838.

[97] Garrett, J. R., 1994: *The Atmospheric Boundary Layer*. New York: Cambridge University Press, 336 pp.

[98] Chen, X, N. Andronova, B. van Leer, *et al.*, 2013: A control-volume model of the compressible Euler equations with a vertical Lagrangian coordinate. *Mon. Weather Rev.*, **141**, 2526–2544, doi:10.1175/MWR-D-12-00129.1.

[99] Kuznetsova, A. M., G. A. Baydakov, V. V. Papko, *et al.*, 2016: Field and numerical study of the wind-wave regime on the Gorky Reservoir. *Geogr., Environ., Sustainability*, **9**, 19–37, doi:10.15356/2071-9388_02v09_2016_02.

[100] Chen, S. S., W. Zhao, M. A. Donelan, and H. L. Tolman, 2013: Directional wind-wave coupling in fully coupled atmosphere-wave-ocean models: Results from CBLAST-Hurricane. *J. Atmos. Sci.*, **70**, 3198–3215, doi:10.1175/JAS-D-12-0157.1.

[101] Sullivan, P. P., J. C. McWilliams, and E. G. Patton, 2014: Large-eddy simulation of marine atmospheric boundary layers above a spectrum of moving waves. *J. Atmos. Sci.*, **71**, 4001–4027, doi:10.1175/JAS-D-14-0095.1.

[102] Gochis, D. J., W. Yu, and D. N. Yates, 2015: The WRF-Hydro model technical description and user's guide, version 3.0. *NCAR Technical Document*, 120 pp. [Available online at https://www.ral.ucar.edu/projects/wrf_hydro.]

[103] Maxwell, R. M., J. K. Lundquist, J. D. Mirocha, S. G. Smith, C. S. Woodward, and A. F. B. Thompson, 2011: Development of a coupled groundwater-atmosphere model. *Mon. Weather Rev.*, **139**, 96–116, doi:10.1175/2010MWR3392.1.

[104] Clifton, A., B.-M. Hodge, C. Draxl, J. Badger, and A. Habte, 2017: Wind and solar resource data sets. *WIREs Energy Environ.*, **7**, e276, doi:10.1002/wene.276.

[105] Draxl, C., A. Clifton, B. M. Hodge, and J. McCaa, 2015: The wind integration national dataset (WIND) toolkit. *Appl. Energy*, **151**, 355–366, doi:10.1016/j.apenergy.2015.03.121.

[106] James, E. P., S. G. Benjamin, and M. Marquis, 2017: A unified high-resolution wind and solar dataset from a rapidly updating numerical weather prediction model. *Renewable Energy*, **102**, 390–405, doi:10.1016/j.renene.2016.10.059.

[107] James, E. P., S. G. Benjamin, M. Marquis, 2018: Offshore wind speed estimates from a high-resolution rapidly updating numerical weather prediction model forecast dataset. *Wind Energy*, **21**, 264–284, doi:10.1002/we.2161.

[108] Lee, J. A., P. Doubrawa, L. Xue, A. J. Newman, C. Draxl, and G. Scott, 2019: Wind resource assessment for Alaska's offshore regions: Validation of a 14-year high-resolution WRF data set. *Energies*, **12**, 2780, https://doi.org/10.3390/en12142780.

[109] Stauffer, D. R., N. L. Seaman, and F. S. Binkowski, 1991: Use of four-dimensional data assimilation in a limited-area mesoscale model. Part II: Effects of data assimilation within the planetary boundary layer. *Mon. Weather Rev.*, **119**, 734–754, doi:10.1175/1520-0493(1991)119<0734:UOFDDA>2.0.CO;2.

[110] Miguez-Macho, G., G. L. Stenchikov, and A. Robock, 2004: Spectral nudging to eliminate the effects of domain position and geometry in regional climate model simulations. *J. Geophys. Res.*, **109**, D13104, doi:10.1029/2003JD004495.

[111] Miguez-Macho, G., G. L. Stenchikov, and A. Robock, 2005: Regional climate simulations over North America: Interaction of local processes with improved large-scale flow. *J. Clim.*, **18**, 1277–1246, doi:10.1175/JCLI3369.1.

[112] Fitch, A. C., J. B. Olson, J. K. Lundquist, *et al.*, 2012: Local and mesoscale impacts of wind farms as parameterized in a mesoscale NWP model. *Mon. Weather Rev.*, **140**, 3017–3038, doi:10.1175/MWR-D-11-00352.1.

[113] Volker, P. J. H., J. Badger, A. N. Hahmann, and S. Ott, 2015: The explicit wake parameterization v1.0: A wind farm parameterization in the mesoscale model WRF. *Geosci. Model Dev.*, 8, 3715–3731, doi:10.5194/gmd-8-3715-2015.

[114] Jiménez, P. A., J. Navarro, A. M. Palomares, and J. Dudhia, 2015: Mesoscale modeling of offshore wind turbine wakes at the wind farm resolving scale: A composite-based analysis with the WRF model over Horns Rev. *Wind Energy*, **18**, 559–566, doi:10.1002/we.1708.

[115] Hidalgo, A., J. Conte, P. A. Jiménez, J. Navarro, L. Prieto, and J. F. González-Rouco, 2014: *Impacts of a Cluster of Wind Farms on Wind Resource Availability and Wind Power Production Over Complex Terrain*. EWEA, Barcelona.

[116] Glahn, H. R., and D. A. Lowry, 1972: The use of model output statistics (MOS) in objective weather forecasting. *J. Appl. Meteor.*, **11**, 1203–1211, doi:10.1175/1520-0450(1972)011<1203:TUOMOS>2.0.CO;2.

[117] Woodcock, F., and C. Engel, 2005: Operational consensus forecasts. *Weather Forecast.*, **20**, 101–111, doi:10.1175/WAF-831.1.

[118] Gerding, S., and B. Myers, 2003: Adaptive data fusion of meteorological forecast modules. *3rd Conference on Artificial Intelligence Applications*, Long Beach, CA, 83rd Amer. Meteor. Soc., Annual Meeting, 9–13 Feb 2003.

[119] Myers, W., G. Wiener, S. Linden, and S. E. Haupt, 2011: A consensus forecasting approach for improved turbine hub height wind speed predictions. *Proc. WindPower 2011*, Anaheim, CA, 24 May 2011.

[120] Mahoney, W. P., K. Parks, G. Wiener, *et al.*, 2012: A wind power forecasting system to optimize grid integration. *IEEE Trans. Sustainable Energy*, **3**, 670–682, doi:10.1109/TSTE.2012.2201758.

[121] Haupt, S. E., W. P. Mahoney, and K. Parks, 2014: Wind power forecasting, in Troccoli, A., L. Dubus, and S. E. Haupt, eds. *Weather Matters for Energy*. Springer, 528 pp., ISBN: 978-1-4614-9220-7.

[122] Wilks, D. S., 2011: *Statistical Methods in the Atmospheric Sciences*, 3rd Ed. New York Academic Press, 704 p., ISBN:9780123850225.

[123] Hersbach, H., 2000: Decomposition of the continuous ranked probability score for ensemble prediction systems. *Weather Forecast.*, **15**, 559–570, doi:10.1175/1520-0434(2000)015<0559:DOTCRP>2.0.CO;2.

[124] Gneiting, T., and A. E. Raftery, 2007: Strictly proper scoring rules, prediction, and estimation. *J. Am. Stat. Assoc.*, **102**, 359–378, doi:10.1198/016214506 000001437.

[125] Matheson, J. E., and R. L. Winkler, 1976: Scoring rules for continuous probability distributions. *Manage. Sci.*, **22**, 1087–1096, doi:10.1287/mnsc. 22.10.1087.

[126] Ebert, E. E., and J. L. McBride, 2000: Verification of precipitation in weather systems: Determination of systematic errors. *J. Hydrol.*, **239**, 179–202, doi:10.1016/S0022-1694(00)00343-7.

[127] Davis, C., B. Brown, and R. Bullock, 2006a: Object-based verification of precipitation forecasts. Part I: Methodology and application to mesoscale rain areas. *Mon. Weather Rev.*, **134**, 1772–1784, doi:10.1175/MWR3145.1.

[128] Davis, C., B. Brown, and R. Bullock, 2006b: Object-based verification of precipitation forecasts. Part II: Application to convective rain systems. *Mon. Weather Rev.*, **134**, 1785–1795, doi:10.1175/MWR3146.1.

[129] Brown, B. G., R. Bullock, J. Halley Gotway, *et al.*, 2007: Application of the MODE object-based verification tool for the evaluation of model precipitation fields. *22nd Conf. on Weather Analysis and Forecasting/18th Conf. on Numerical Weather Prediction*, Amer. Meteor. Soc., Park City, UT, 10A.2. [Available online at https://ams.confex.com/ams/22WAF18NWP/techprogram/paper_124856.htm.]

[130] Developmental Testbed Center, 2019: Model Evaluation Tools (MET) User's Page. [Available online at https://dtcenter.org/met/users, last accessed 3/19/19.]

[131] Yang, B., Y. Qian, L. K. Berg, *et al.*, 2017: Sensitivity of turbine-height wind speeds to parameters in planetary boundary-layer and surface-layer schemes in the weather research and forecasting model. *Boundary Layer Meteorol.*, **162**, 117–142, doi:10.1007/s10546-016-0185-2.

[132] Zhao, Q., F. Zhang, T. Holt, C. H. Bishop, and Q. Xu, 2013: Development of a mesoscale ensemble data assimilation system at the Naval Research Laboratory. *Weather Forecast.*, **28**, 1322–1336, doi:10.1175/WAF-D-13-00015.1.

[133] Nakanishi, M., and H. Niino, 2006: An improved Mellor–Yamada Level-3 Model: Its numerical stability and application to a regional prediction of advection fog. *Boundary Layer Meteorol.*, **119**, 397–407, doi:10.1007/s10546-005-9030-8.

[134] Kolczynski, W. C., D. R. Stauffer, S. E. Haupt, N. S. Altman, and A. Deng, 2011: Investigation of ensemble variance as a measure of true forecast variance. *Mon. Weather Rev.*, **139**, 3954–3963, doi:10.1175/MWR-D-10-05081.1.

[135] Lee, J. A., W. C. Kolcynski, T. C. McCandless, and S. E. Haupt, 2012: An objective methodology for configuring and down-selecting an NWP ensemble for low-level wind prediction. *Mon. Weather Rev.*, **140**, 2270–2286, doi:10.1175/MWR-D-11-00065.1.

[136] Lee, J. A., S. E. Haupt, and G. S. Young, 2016: Down-selecting numerical weather prediction multi-physics ensembles with hierarchical cluster analysis. *J. Climatol. Weather Forecast.*, **4**, 156, doi:10.4172/2332-2594.1000156.

[137] Kolczynski, W. C., D. R. Stauffer, S. E. Haupt, and A. Deng, 2009: Ensemble variance calibration for representing meteorological uncertainty for atmospheric transport and dispersion modeling. *J. Appl. Meteorol. Climatol.*, **48**, 2001–2021, doi:10.1175/2009JAMC2059.1.

[138] Hopson, T., J. Hacker, Y. Liu, *et al.*, 2010: Quantile regression as a means of calibrating and verifying a mesoscale NWP ensemble. *Prob. Fcst Symp.*, Atlanta, GA, American Meteorological Society, 17–23 Jan 2010.

[139] Raftery, A. E., T. Gneiting, F. Balabdaoui, and M. Polakowski, 2005: Using Bayesian model averaging to calibrate forecast ensembles. *Mon. Weather Rev.*, **133**, 1155–1174, doi:10.1175/MWR2906.1.

[140] Delle Monache, L., T. Nipen, Y. Liu, G. Roux, and R. Stull, 2011: Kalman filter and analog schemes to postprocess numerical weather predictions. *Mon. Weather Rev.*, **139**, 3554–3570, doi:10.1175/2011MWR3653.1.

[141] Delle Monache, L., F. A. Eckel, D. L. Rife, B. Nagarajan, and K. Searight, 2013: Probabilistic weather prediction with an analog ensemble. *Mon. Weather Rev.*, **141**, 3498–3516, doi:10.1175/MWR-D-12-00281.1.

[142] Alessandrini, S., L. Delle Monache, S. Sperati, and J. N. Nissen, 2015: A novel application of an analog ensemble for short-term wind power forecasting. *Renewable Energy*, **76**, 768–781, doi:10.1016/j.renene.2014.11.061.

[143] Junk, C., L. Delle Monache, S. Alessandrini, G. Cervone, and L. von Bremen, 2015: Predictor-weighting strategies for probabilistic wind power forecasting with an analog ensemble. *Meteorol. Z.*, **24**, 361–379, doi:10.1127/metz/2015/0659.

[144] Vanvyve, E., L. Delle Monache, A. J. Monaghan, and J. O. Pinto, 2015: Wind resource estimates with an analog ensemble approach. *Renewable Energy*, **74**, 761–773, doi:10.1016/j.renene.2014.08.060.

[145] Yussouf, N., J. S. Kain, A. J. Clark, 2016: Short-term probabilistic forecasts of the 31 May 2013 Oklahoma tornado and flash flood event using a continuous-update-cycle storm-scale ensemble system. *Weather Forecast.*, **31**, 957–983, doi:10.1175/WAF-D-15-0160.1.

[146] Grell, G., and S. R. Freitas, 2014: A scale and aerosol aware stochastic convective parameterization for weather and air quality modeling. *Atmos. Chem. Phys.*, **14**, 5233–5250, doi:10.5194/acp-14-5233-2014.

[147] Skamarock, W. C., J. B. Klemp, M. G. Duda, L. D. Fowler, S.-H. Park, and T. D. Ringler, 2012: A multiscale nonhydrostatic atmospheric model using centroidal Voronoi tesselations and C-grid staggering. *Mon. Weather Rev.*, **140**, 3090–3105, doi:10.1175/MWR-D-11-00215.1.

[148] Greybush, S. J., S. E. Haupt, and G. S. Young, 2008: The regime dependence of optimally weighted ensemble model consensus forecasts of surface temperature. *Weather Forecast.*, **23**, 1146–1161, doi:10.1175/2008WAF2007078.1.

[149] McCandless, T. C., S. E. Haupt, and G. S. Young, 2016a: A regime-dependent artificial neural network technique for short-range solar irradiance forecasting. *Renewable Energy*, **89**, 351–359, doi:10.1016/j.renene.2015.12.030.

[150] McCandless, T. C., G. S. Young, S. E. Haupt, and L. M. Hinkelman, 2016b: Regime-dependent short-range solar irradiance forecasting. *J. Appl. Meteorol. Climatol.*, **55**, 1599–1613, doi:10.1175/JAMC-D-15-0354.1.

[151] Haupt, S. E., A. Anderson, L. Berg, *et al.*, 2015: First Year Report of the A2e Mesoscale to Microscale Coupling Project, Pacific Northwest Laboratory Report PNNL-25108, 124 pp.

[152] Mirocha, J. D., M. J. Churchfield, D. Muñoz-Esparza, *et al.*, 2018: Large-eddy simulation sensitivities to variations of configuration and forcing parameters in canonical boundary layer flows for wind energy applications. *Wind Energy Sci.*, **3**, 589–613, doi:10.5194/wes-3-589-2018.

[153] Rodrigo, J. S., M. Churchfield, and B. Kosovic, 2017: A methodology for the design and testing of atmospheric boundary layer models for wind applications, *Wind Energy Sci.*, **2**, 35–54, doi:10.5194/wes-2-35-2017.

[154] Pedersen, J. G., M. Kelly, S.-E. Gryning, and B. Brummer, 2013: The effect of unsteady and baroclinic forcing on predicted wind profiles in large eddy simulations: Two case studies of the daytime atmospheric boundary layer. *Meteorol. Z.*, **22**, 661–674, doi:10.1127/0941-2948/2013/0477.

[155] Jayaraman, B., J. Brasseur, S. E. Haupt, and J. Lee, 2016: Deviations from equilibrium in daytime atmospheric boundary layer turbulence arising from nonstationary mesoscale Forcing. *69th Annual Meeting of the APS Division of Fluid Dynamics*, 61 (20), Portland, Oregon, 22 Nov 2016.

[156] Muñoz-Esparza, D., J. K. Lundquist, J. A. Sauer, B. Kosović, and R. R. Linn, 2017: Coupled mesoscale-LES modeling of a diurnal cycle during the CWEX-13 field campaign: From weather to boundary-layer eddies. *J. Adv. Model. Earth Syst.*, **9**, 1572–1594, doi:10.1002/2017MS000960.

[157] Haupt, S. E., A. Anderson, R. Kotamarthi, *et al.*, 2017b: Second Year Report of the Atmosphere to Electrons Mesoscale to Microscale Coupling Project: Nonstationary Modeling Techniques and Assessment, Pacific Northwest Laboratory Report PNNL-26267, 156 pp.

[158] Rodrigo, J. S., M. Churchfield, and B. Kosovic, 2016: A wind energy benchmark for ABL modelling of a diurnal cycle with a nocturnal low-level jet: GABLS3 revisited. *J. Phys. Conf. Ser.*, 753, The Science of Making Torque from Wind, doi:10.1088/1742-6596/753/032024.

Chapter 4

Mesoscale to microscale coupling for high-fidelity wind plant simulation

Jeffrey D. Mirocha[1]

4.1 Introduction

Much of the discussion surrounding the development of the next generation of wind turbine and wind plant simulation capabilities is focused on increased resolution, more realistic representations of turbine components and controls, and improvements in scalability and efficiency to support ever more sophisticated and computationally expensive simulations. While such considerations are inarguably important, an often underemphasized factor contributing to the relevance of any such simulations to real-world applications is the representation of the complex and highly variable atmospheric flow fields in which real wind turbines operate. While the primary atmospheric drivers of wind power generation are the mean velocity and air density, which define the kinetic energy available for conversion to electricity, a multitude of highly variable characteristics of the atmospheric flow field from which energy is ultimately extracted also influence, sometimes strongly, many aspects of turbine and plant performance, impacting both power production and reliability. Spatial variabilities of wind speed and direction, particularly with height, the state of atmospheric turbulence, the presence of atmospheric waves, precipitation, icing, interactions with complex surfaces, as well as temporal variability related to weather, wave, and turbulence phenomena, and a host of other mesoscale and submesoscale atmospheric processes, are often limiting factors in energy extraction, strongly influence inter-turbine interactions (wake formation, recovery, and the efficacy of wake control), and are major drivers of stress loading and component fatigue.

The absence of adequate representation of the weather- and environment-dependent flow parameters determining critical aspects of the in-situ turbine and plant operating environment is manifested in the significant, widely reported plant underperformance and reliability issues plaguing the industry to date (Shaw *et al.*, 2008; [1,2]). The success of the next generation of high-fidelity wind plant simulation capabilities to address these underperformance and reliability issues will depend not only on more sophisticated wind plant physics and engineering modules but also on the

[1]Lawrence Livermore National Laboratory, Livermore, CA, USA

inclusion of realistic environmental drivers of the atmospheric flow state influencing turbine and plant performance and reliability.

While the need to improve the realism of the atmospheric flow state in wind plant simulation tools is gradually gaining acceptance, development of techniques to enable this capability has been slow, due to significant theoretical and practical issues involved in bridging the broad range of scales of atmospheric motions influencing wind turbine and plant operations. Inherent differences in how relevant atmospheric processes, especially turbulence, are modeled at different scales, as well as the different types of computational models used to simulate weather versus wind plant dynamics, must be bridged to enable seamless simulation of the atmospheric flow state from weather to wind plant scales, as required to capture the natural sources of flow state variability impacting turbine operation.

4.1.1 Overview of atmospheric simulation at meso and microscales

Wind turbines operate within the atmospheric boundary layer (ABL), the layer of the atmosphere adjacent to the surface, over which nearly continuous but highly variable turbulence motions efficiently distribute momentum and other constituents throughout its depth (see e.g., [3]). These turbulence motions are critically important to wind energy production for many reasons. A primary effect of turbulence is to slow the flow approaching the surface, via the vertical distribution of drag originating at the surface, resulting in a shear layer over which the wind speed (generally) increases with distance above. Turbulence also results in a rotation of the mean wind direction with height due to the Coriolis effect, which can result in significant turning of the wind vector with height throughout the midlatitudes, where many wind plants are located. In addition to the effects of turbulence on mean wind speed and direction, the correlated spatiotemporal fluctuations comprising the turbulence field impact the extraction of energy from the flow, while also producing stresses on turbine components. Further, characteristics of the turbulence field vary widely in time and space, due to changes in forcing brought about by myriad environmental and weather-related processes. While turbulence is critically important to numerous atmospheric and engineering applications, it remains, at a fundamental level, an incompletely understood aspect of fluid flow.

Despite the absence of a fundamental theoretical understanding of turbulence, all atmospheric simulation techniques must account for the ABL and the turbulence within it in some fashion in order to reproduce realistic flow characteristics. While turbulence can in principle be simulated to a very high degree of accuracy via direct numerical simulation (DNS) of the governing equations, practical limitations imposed by the very small mesh sizes and model time steps required to explicitly capture the full spectrum of turbulence scales, including those responsible for viscous dissipation, preclude DNS from widespread use as a general turbulence simulation technique. Standard simulation techniques for turbulent flows in most geophysical and engineering applications instead rely upon parameterizations to account for either the entire turbulence spectrum, or some portion thereof that is not explicitly resolved, in order for the simulation to provide realistic results while running quickly enough to be practical.

Atmospheric simulation techniques are often categorized both according to how much of the classical three-dimensional turbulence energy spectrum is explicitly resolved versus parameterized in a subgrid-scale (SGS) of subfilter-scale (SFS) model, and the type of formulation that the SGS or SFS model takes. While SGS and SFS are often used interchangeably, hereinafter we will typically refer to the unresolved scales as SFS, since the filtering operation separating resolved and unresolved (SFS) scales does not in general depend only upon the grid but also upon the numerical discretization scheme, the SFS turbulence parameterization itself, and any explicit filter that may be also employed.

While all numerical simulation methods for fluid flows fall under the umbrella of computational fluid dynamics (CFD), this terminology typically denotes simulations using sufficiently high resolution to capture small-scale turbulence motions. In geophysical applications, simulation of the much larger scales of weather ultimately forcing the flow conditions giving rise to the turbulence field is commonly denoted as numerical weather prediction (NWP). Due to the large spatial scales (continental to global) that must be spanned to capture the many environmental and atmospheric dynamics defining the weather and its changes over time, NWP simulations, of necessity, utilize mesh resolutions far too coarse to explicitly capture any motions associated with the classical three-dimensional turbulence energy cascade, such that are the domain of traditional CFD. As such, CFD and NWP models use very different formulations for the effects of unresolved scales of turbulence, which for NWP includes the entire spectrum of classical three-dimensional turbulence motions, while for CFD includes only the scales smaller than the model filter. With Δh representing a characteristic horizontal mesh spacing, NWP typically uses $\Delta h \gtrsim 10\,\mathrm{km}$ (historically much coarser), with traditional CFD typically using $\Delta h \lesssim 100\,\mathrm{m}$. Mesoscale denotes a range of scales between traditional NWP and CFD, roughly $10 \gtrsim \Delta h \gtrsim 1\,\mathrm{km}$. Mesoscale simulations are generally applied to only a subset of the Earth, such as a continent or region within, due to their use of higher resolution than traditional NWP, and hence much greater computational cost. However, despite their finer resolution, mesoscale mesh spacings are generally too coarse to resolve motions associated with the classical turbulence energy cascade and hence likewise rely on parameterizations similar to those used by NWP models to capture the effects of the unresolved turbulence field on the resolved-scale flow.

Due to the combination of different physical constraints and target applications, large-scale NWP and fine-scale CFD techniques have each been developed somewhat independently to high degrees of sophistication over many decades, with well-characterized techniques for turbulence closure at each of their respective scale ranges. However, the range of scales between traditional NWP and CFD, especially those at the finer end of mesoscale spectrum, represents an area of numerical simulation relatively unexplored until recently, following the advent of affordable high-performance computing (HPC). Cheaper computation has enabled the use of much finer mesh spacings in NWP and mesoscale simulations, the latter of which are now capable of capturing a portion of the classical three-dimensional turbulence spectrum, scales of motion for which such simulation paradigms and subgrid process models were not designed to work within. Concurrently, traditional CFD domains are being expanded to cover spatial extents approaching subsets of mesoscale domains, scales for which weather and environment-dependent process play a role in the evolution of the flow field, but for which physical process parameterizations of those phenomena are absent. Simulation within this new range of scales presents several practical and theoretical challenges are now being actively pursued to support many emerging applications.

Wind energy is among the preeminent applications straddling the interface between the disparate disciplines of mesoscale simulation and CFD. The scale of modern wind plants, spanning tens of kilometers in each direction, inhabits the range of scales traditionally simulated using mesoscale techniques. However, the need to explicitly resolve turbulent motions governing turbine–airflow interactions requires approaches germane to traditional CFD. Recognition of the importance of changing atmospheric and environmental inflow on wind plant performance requires the incorporation of scales of variability associated with traditional NWP. The development of new approaches and capabilities required to enable the seamless integration of weather and wind plant simulation, comprising widely disparate scale ranges, physical processes, and divergent simulation techniques and paradigms, is the focus of this chapter.

In the following two sections, several of the key issues, historical approaches, and paths currently underway in pursuit of a seamless multiscale simulation capability are described. The presentation begins with a discussion of microscale ABL simulation using the large-eddy simulation (LES) technique, beginning with basic setup, forcing, and historical idealized applications. Subsequently, extension of basic LES techniques to more complex and variable settings, via incorporation of higher fidelity weather and environmental information, is explored. Finally, progress and remaining research challenges toward enabling seamless simulation across the full spectrum of atmospheric motions spanning the mesoscale and microscale, as required for applicability to arbitrary meteorological and environmental conditions, are discussed.

4.2 Large-eddy simulation of the atmospheric boundary layer

The present state-of-the-science methodology for turbulence-resolving simulation of the ABL flow defining the wind plant operational environment is the LES technique [4–6]. LES represents a computationally affordable alternative to DNS for the ABL that is capable of capturing the primary scales of atmospheric turbulence motions responsible for most of the transport, stress, and variance within the flow, while avoiding the high computational costs of simulating the very fine scales of motion related chiefly to viscous dissipation. While the Reynold's averaged Navier–Stokes (RANS) approach retains applicability as more computationally efficient alternative when only lower frequency turbulence variability is required, LES enables the capture of the high-frequency variability associated with transient, evolving turbulence features, including extremes, as required to investigate the impacts of turbulence motions on energy extraction and stress loading, for example.

Both the RANS and LES flow equations are derived via application of a low-pass filter, to the governing Navier–Stokes equations (presented in Chapter 3 of this volume).

$$\tilde{\varphi}(x) = \int_{-a}^{a} G(x - y)\varphi(y)dy, \tag{4.1}$$

where G is a generic filter that removes scales of motion smaller than the filter width from a flow field variable φ. The resolved component $\tilde{\varphi}$, which contains only scales of motion larger than the filter width, is the field that is prognosed in LES.

Application of (4.1) to the governing flow equations yields a new term in the filtered flow equations,

$$\tau_{ij} = (\widetilde{u_i u_j} - \tilde{u}_i \tilde{u}_j), \tag{4.2}$$

which represents the impacts of scales removed by the filter on the resolved component of the solution. Here, subscripts identify the zonal (1), meridional (2), and vertical (3) directions, with corresponding velocity components commonly denoted u, v, and w, respectively. The term τ_{ij} expresses the difference between the products of the filtered variables, $\tilde{u}_i$, which are retained within the filtered LES equations, and therefore resolved within the simulation, and the filtered products of the full variables, u_i, which are not prognosed, since those include a component that has been removed by the filter. As such, a SFS stress parameterization must be used to account for the effects of those scales of motion removed by the filter, on the resolved-scale flow. As turbulence motions are ultimately dissipated by the very small scales of motion removed by the filter, a primary role of the SFS parameterization is to remove the proper amount of turbulence kinetic energy (TKE) from the larger resolved scales within the simulated flow. Further discussion of common strategies for the modeling of τ_{ij} in LES of the ABL is provided in Section 4.5.1.

When the filter function is defined as the well-known Reynold's average (e.g., [3]), the familiar RANS equations are recovered, for which the average of a product containing only one fluctuating quantity vanishes, leaving the familiar Reynolds stresses. However, general filters, such as those used in LES, do not possess that property, resulting in the retention of terms that modulate exchanges between resolved and SFS quantities (e.g., Ciofalo [7,8]). These interactions provide the LES technique with its superior time dependence relative to RANS techniques, making it the preferred technique for simulations requiring the capture of turbulence transients, variability, and extrema.

The price of the enhanced fidelity of LES relative to RANS is increased computational demand, due to the necessity of smaller time steps for the same mesh spacing, as required to maintain stable numerical integration in the presence of greater velocity extrema. However, with computational demands much closer to RANS than to the much more expensive DNS, LES has now become feasible using mainstream affordable HPC infrastructure and is therefore likely to remain the preeminent computational framework for wind plant applications requiring high-fidelity turbulence simulation for the foreseeable future.

The affordability of LES relative to DNS results from the explicit capture of only the scales of the turbulence spectrum chiefly responsible for the production of TKE (defined in Section 4.2.2) and its input into the downscale energy cascade. The impact of smaller scales, those modulating further downscale transport to the scales responsible for dissipation, is captured by the SFS model. Since the grid spacing required to capture the smallest scales also requires the smallest time steps, a significant cost saving is recovered by parameterizing the effects of energy transport and dissipation occurring at those scales, rather than resolving those explicitly, as in DNS.

Beyond computational affordability, the LES approach is well grounded on the physical basis that it is the largest scales within the turbulence spectrum which are most important to capture accurately, as those largest scales govern the primary energy production, transport processes, and stresses occurring within a turbulent flow. The largest scales are also those that are most strongly impacted by unique characteristics of the physical environment, including geometry and forcing. The smaller scales, via the mechanics of the downscale cascade process, become more isotropic and homogeneous, losing the imprint of the largest scales, and thereby submitting to general SFS parameterization strategies that account primarily for the dissipation of TKE (e.g., Nieuwstadt *et al.*, 1993).

While the largest scales of motion comprising the turbulence spectrum must be well captured in order for LES to be successful, the distribution of eddy sizes and other physical attributes of the turbulence field can vary considerably depending upon the environment and forcing. For this reason, the selection of key features of the computational setup, including domain size and mesh spacing, as well as specification of relevant forcing and boundary conditions, is critical to achieving high-fidelity LES of the ABL. Hereafter, we will describe several of these considerations in the context of wind energy applications and provide some common rules of thumb.

4.2.1 ABL LES setup

4.2.1.1 Forcing

The minimal set of forcing parameters for LES of ABL flows of interest to wind energy (those with nonzero mean velocity) include fluxes of momentum and sensible heat at the surface, which form the bottom boundary condition, and the horizontal pressure gradient, which forces the horizontal flow. Interactions between the mean forcing and surface thermodynamic stability and roughness determine many aspects of the state of the ABL and the turbulence field. Additional factors, such as large-scale advections and other physical processes, are not generally incorporated into LES. However, due to their potential importance in ABL flow features of relevance to wind plant simulation, methods to incorporate such physical forcing into LES are discussed in Section 4.4.

Since LES of the ABL cannot utilize sufficient resolution near the surface to permit a no-slip boundary condition (as in DNS or wall-resolved LES used in other applications), such simulations instead rely upon a wall model, which imposes the surface boundary conditions in terms of the fluxes of momentum (τ_{i3}^s) and, for non-neutral cases, sensible heat (H_S), as

$$\tau_{i3}^s = -C_D U(z_1) u_i(z_1), \tag{4.3}$$

and

$$H_S = -C_H U(z_1)[\theta_S - \theta(z_1)]. \tag{4.4}$$

where C_D and C_H are exchange coefficients for momentum and heat, $U = \sqrt{u^2 + v^2}$ is the scalar wind speed, $\theta = T(p_0/p)^{R/c_p}$ is the potential temperature, with T the temperature, p is the pressure, $p_0 = 1 \times 10^5$ Pa a reference pressure, R is the gas constant for dry air, c_p is the specific heat of dry air at constant pressure, θ_S is the surface value of the potential temperature, and z_1 is the lowest height above the surface at which velocity and potential temperature are computed. The exchange coefficients $C_\alpha = \kappa^2\left[\ln\left((z_1/z_{0,\alpha}) - \psi_\alpha(z/L)\right)\right]^{-2}$ in (4.3) and (4.4) are most commonly obtained from the Monin–Obukhov similarity theory (MOST; [9]), in terms of appropriate roughness lengths, $z_{0,\alpha}$ and stability functions $\psi_\alpha(z/L)$, for quantity α. Here $L = \left[-u_*^3\theta_{v0}\right]/[\kappa g H_S]$ is the Obukhov length, with $u_* = \left[\left(\tau_{13}^s\right)^2 + \left(\tau_{23}^s\right)^2\right]^{1/4}$, $\theta_{v0} = 300$ K a reference value of the virtual potential temperature, $\theta_v = \theta\,(1 + 0.61 q_v)$, where q_v is the water vapor mixing ratio, $\kappa = 0.4$ is the von Kármán constant, and g is the gravitational acceleration. In the limit of neutral conditions, for which $L = \infty$, $\psi\,(z/L) = 0$. Due to the interdependence of C_α, u_*, and H_S, their values are typically determined iteratively [10].

While τ_{i3}^s values can be obtained directly from the LES-resolved velocities (and an appropriate z_0 value), H_S values, as specified in (4.4), also require a surface (potential) temperature, θ_S. As LES models do not generally contain a parameterization for θ_S, LES is often forced instead by specifying H_S, with values obtained either from observations or from a larger scale atmospheric simulation which uses land surface models forced by the full suite of relevant physical processes such as radiative exchange, moist processes, and other surface physics. Alternatively, θ_S can also be obtained from observations or from a larger scale atmospheric model, allowing the simulation to compute corresponding H_S values (via (4.4)). Advantages of specifying θ_S rather than H_S are 2-fold; one is that the corresponding H_S values within the LES will reflect instantaneous fluctuations in the LES near-surface temperature and velocity fields, potentially increasing the fidelity of the simulations; another is the avoidance of an unphysical solution branch when H_S is used during moderately to strongly stable conditions [11]. When using θ_S rather than H_S, it is also important to initialize the atmospheric temperature field within the LES appropriately. Use of H_S does not require a matching atmospheric temperature field.

Above the surface, the mean flow in LES of the ABL is typically forced by a horizontal pressure gradient, commonly applied as a geostrophic wind, U_g, which represents an idealized force balance between the horizontal pressure gradient force and the Coriolis acceleration, in the absence of friction (e.g., [12]). The zonal and meridional components of U_g are given by $u_g = -f^{-1}\partial p/\partial y$ and $v_g = f^{-1}\partial p/\partial x$, where $f = 2\Omega\sin(\phi)$ is the Coriolis parameter, with Ω the Earth's angular rotation rate, and ϕ the latitude. Since geostrophic balance occurs only in the absence of friction, it never strictly holds within the ABL, where turbulence transports surface friction vertically, resulting in both a slowing and a rotation of the mean wind vector approaching the surface.

Despite the inapplicability of geostrophy within the ABL, it is often a reasonable approximation above, as evidenced by wind vectors typically coinciding with surfaces of constant pressure on synoptic weather maps (e.g., [12]). Due to the small magnitude of the Coriolis acceleration relative to other physical and dynamical forces, geostrophic balance is achieved over long timescales and represents integrated forcing. As such, the geostrophic wind does not represent ageostrophic pressure variability associated with smaller scale meteorological features, such as frontal passages, land and sea breezes, or thunderstorm outflows, for example.

During quiescent meteorological conditions, u_g and v_g are commonly estimated from the observed wind components occurring above the ABL top, H. However, it should be recognized that other physical processes beyond the mean horizontal pressure gradient associated with the geostrophic wind contribute to the mean flow, complicating any such estimation. As such, estimating u_g and v_g in the above-mentioned manner becomes increasingly tenuous under increasingly meteorological complexity.

The horizontal pressure force driving the LES flow must be specified at all heights within the computational domain. However, u_g and v_g cannot be accurately estimated within the ABL. The most common approach, therefore, is to specify constant values, obtained from above the ABL, throughout the domain. While the practice of using uniform geostrophic winds within the ABL is typical, baroclinicity, the variability of the geostrophic wind with height, does commonly occur and may represent an important forcing for ABL flow during certain conditions. However, as an estimation of baroclinicity is potentially even more difficult than the geostrophic wind components themselves, due to the multitude of potential sources of wind speed and direction variability with height, it is commonly ignored. Moreover, misrepresentation of other sources of wind variability as baroclinicity and incorporation as such into the geostrophic winds used to force LES will result in erroneous accelerations with unknown consequences.

Recognizing the difficulties of estimating geostrophic winds from observed winds, a common if not widely disclosed practice, is to slightly alter the values obtained from observations until observed wind speed and direction profiles within the ABL are approximately recovered. While this approach can be thought of as an inverse estimation method for the unknown geostrophic wind values, including baroclinicity, other sources of model error are implicitly included in geostrophic wind values deduced in this manner. However, if the goal of the simulation is to generate a turbulent flow consistent with observed mean wind speed and direction distributions, such an approach may be useful. Estimation of geostrophic wind components and their height variability from mesoscale model output, rather than direct observations, is discussed further in Section 4.4.1.

During convective conditions, the impacts of ignoring height variability, or using a slightly incorrect value of the geostrophic wind, would likely only negligibly impact the turbulence field, except perhaps some characteristics of convective morphology (e.g., spatial scales of convective rolls), as forcing during convective conditions is strongly governed by surface fluxes of heat and resulting buoyancy. However, during weakly convective, neutral and especially stable conditions, baroclinicity, and other processes may play a more prominent role in development of important ABL characteristics, including low-level jets (LLJs), as discussed further in Section 4.3.4.

4.2.1.2 Mesh spacing

Once the forcing for LES is specified, another key consideration is the choice of an appropriate mesh spacing. As Δh in LES is typically larger than the vertical mesh spacing Δz, Δh can be used to represent an effective scale for the smallest feature that can, in principle, be represented upon a mesh with that spacing. For an LES to satisfy the assumptions upon which the technique is formally predicated, Δh must be sufficiently fine that the effective filter scale (which is generally somewhat larger than Δh due to numerical discretization errors, as described in Section 4.2.2) be located within the inertial subrange of the classical three-dimensional turbulence energy cascade. This filter placement enables explicit resolution of the largest scales within the turbulence cascade, characterized by the integral scale, $L = \int_0^\infty \rho(r)\,dr$. Here, $\rho(r)$ defines the spatial autocorrelation function as a function of distance r, representing a characteristic distance over which a variable becomes uncorrelated with its previous position, or hence is no longer a part of the same turbulent eddy, giving an effective length scale for the dominant structures within a turbulent flow. For LES to function properly, Δh must be fine enough to capture not only L but also a portion of the downscale cascade process, which connects the largest resolved scale to dissipation which is handled within the SFS parameterization.

The wide range of naturally occurring variability in ABL structure and forcing results in a correspondingly broad range of characteristic length scales for ABL turbulence, and hence mesh spacings were required to capture the energetically important scales of the turbulence spectrum for a given flow. During strongly convective conditions, a relatively coarse mesh spacing of 100 m (with somewhat higher resolution in the vertical, approaching the surface) may be adequate to capture the primary scales of turbulence energy production and dynamics. Conversely, during stable conditions, a mesh spacing of just a few meters may be required.

For wind plant simulations designed to capture airflow turbine interactions on the chord scale or smaller, a mesh spacing capable of resolving those interactions would likely be sufficiently fine to capture L in most meteorological situations of relevance to wind energy (i.e., those with a nontrivial mean flow). However, as many wind plant CFD solvers use adaptive meshes to reduce computational overhead by enhancing resolution only near features of interest, care must be taken to provide adequate resolution also within the background flow in order to capture relevant turbulence scales sufficiently, rather than focusing mesh refinement only near the turbine components. Moreover, in mesoscale-to-microscale (MMC) frameworks, for which the wind plant simulation domain relies upon turbulent inflow provided by a coarser-resolution domain, care must be taken to ensure adequate mesh spacing on the coarser domain providing the turbulent inflow to the microscale domain in which those turbine-airflow interactions are resolved.

4.2.1.3 Turbulence generation

Once the forcing is specified, and an adequate mesh spacing is chosen, based on the expectation for L given the physical forcing occurring during a simulation period, the issue of generating the resolvable scales (those representable upon the chosen mesh) of the classical turbulence spectrum within the flow must be addressed. While interactions of the flow with turbine components can lead to the rapid generation of turbulence downstream, the background turbulence in the upstream flow encountering a turbine will strongly influence those interactions, as well as the wakes produced. However, when the forcing of background turbulence is weak, generating ABL turbulence can be tricky. A few of the more popular contemporary approaches applicable to LES of the ABL will be briefly described later, while more thorough discussions of many different approaches and multitude of issues involved in turbulence generation can be found in, e.g., Lund *et al.* [13], Moin and Mahesh [14], Keating *et al.* [15], and Mehta *et al.* [16].

The standard approach to generating the classical turbulence spectrum in LES of the ABL is to initialize a flow field with small perturbations to the velocity or temperature fields. These perturbations introduce heterogeneities that act to seed turbulence formation as the flow evolves in time. The evolution of the turbulence spectrum thereafter is largely determined by interactions with the underlying surface, which generate fluxes of heat and momentum (as well as scalars such as water vapor), or thermal instabilities aloft. The simulation is advanced from this perturbed initial state for a sufficient amount of time for the full spectrum of resolvable turbulence scales to develop and approach quasi-equilibrium with the specified forcing. Quasi-equilibrium can be established by examining statistics of the turbulent flow, as discussed in Section 4.2.2.

Turbulence morphology and its timescale of generation from initial perturbations vary considerably depending upon the nature of the forcing. Turbulence in the ABL is strongly modulated by the interaction of thermal instability (buoyancy) with differences in wind speed and direction speed with height (shear and veer, respectively). While shear always acts to generate turbulence motions, buoyancy can act to either augment or suppress turbulence, depending on its sign. Buoyant instability is typically controlled by the flux of sensible heat into the atmosphere from the surface, which heats or cools the air immediately above, leading to changes in the local air density that either drives or suppresses turbulence motions. Cloud-top cooling or advection of colder air above a warmer surface can also generate buoyant production of turbulence through similar mechanisms. In either case, the buoyancy-induced motions will rapidly develop a cascade of smaller scale motions via vorticity dynamics, quickly generating the turbulence spectrum. In the same manner that buoyancy augments turbulence development, adverse buoyancy acts to suppress it by limiting vertical motions. Surface cooling, such as often happens around sunset and overnight, extracts energy from the air just above the surface, thereby creating a shallow layer of colder, denser air that is resistant to the vertical excursions associated with turbulence. Advection of warmer air over a colder surface also acts to suppress turbulent overturning via the same mechanism.

The flow of air over a surface also generates turbulence via shear instability, due to the change in mean wind speed with height resulting from surface drag slowing the flow approaching the surface. Shear generation will increase in magnitude with an increase of either wind speed or surface roughness, each of which enhances near-surface shear. While the shear production mechanism alone will generate the full spectrum of ABL turbulence eventually, the timescale to approach equilibrium is longer than when buoyant generation dominates and can be very long (several hours) when the mean flow is weak or when buoyant instability acts to suppress turbulence.

Heterogeneity in the terrain and underlying surface cover can likewise augment turbulence generation in predominantly shear-driven flows via deflection of the flow around and over obstacles, which introduces additional spatial gradients that enhance the rate of shear production. However, in the absence of buoyant production, the rate of turbulence equilibration can still be quite slow unless the terrain is steep and well resolved within a simulation. While deflections induced by terrain can lead to meandering motions that superficially appear turbulent, spectra (see Section 4.2.2) can reveal if the expected inertial scales associated with the classical three-dimensional turbulence spectrum are present.

While convective instability and mean flow interactions with the surface will readily generate turbulence via buoyant and shear production (provided an adequate mesh resolution is chosen), stable conditions, characterized by negative values of H_S, consistent with $\theta_S < \theta(z_1)$, are more challenging to simulate, and increasingly so with increasing stability. Turbulence is likely to be much weaker and to develop much more slowly in stable than in neutral or convective conditions, with the potential to become globally intermittent under certain forcing scenarios (see Section 4.3.4). As such, very fine mesh spacings may be necessary to capture turbulent motions in the stable ABL.

In addition to requiring smaller mesh spacings, the absence of convective forcing can also significantly increase the fetch required for turbulence to approach equilibrium. Since the fetch consumes computational resources outside the area of interest, acceleration of turbulence development is of critical concern in LES, especially in wind energy applications, for which nontrivial mean flow rates result in significant advection of the developing flow within the computational domain.

While additional methods to accelerate turbulence development will be discussed in Section 4.4.2.1, the oldest and most commonly employed practice involves the use of periodic lateral boundary conditions (LBCs). Periodic LBCs allow flow exiting the computational domain to reenter from the opposite boundary or boundaries. As such, periodic LBCs create an effectively infinite fetch upon which ABL turbulence can develop for as long as required to achieve equilibration.

Figure 4.1 shows an instantaneous snapshot of wind speed at 100 m above the surface, near the top of the surface layer, often taken to be $0.1H$, from a representative LES of the ABL. The setup consists of neutral surface forcing ($H_S = 0$), flat terrain with a constant surface roughness length of $z_{0,m} = 0.1$ m, a constant, uniform

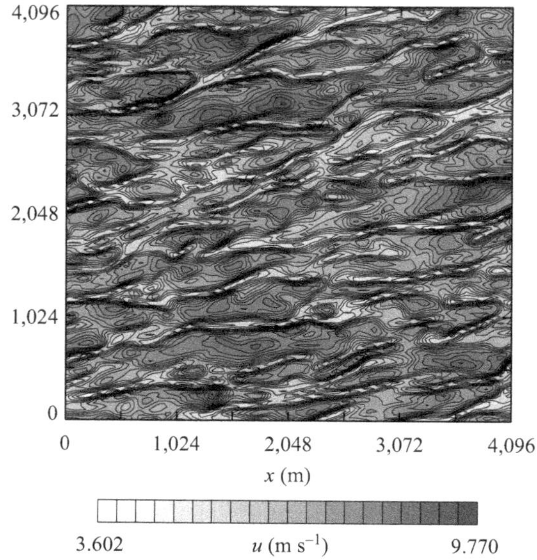

Figure 4.1 *Plan (x − y) view at 100 m above the surface of instantaneous zonal velocity component (u) from an LES of the neutral ABL of over flat, rough terrain (z_0 = 0.1 m) forced by a geostrophic wind of 10 m s^{-1} in the x-direction, and using periodic lateral boundary conditions*

geostrophic wind of $u_g = 10\,\text{m}\,\text{s}^{-1}$, from the west ($v_g = 0$), and periodic LBCs, as is typical of many such studies (e.g., [17,18]; Chow *et al.*, 2005). The domain size is 4,096 m in x and y by 1,500 m in z, using horizontal and vertical grid spacings of $\Delta h = 16$ m and $\Delta z = 4$ m (with a stretching of 5% per model grid point applied in the vertical direction). A weak temperature inversion was applied for $z > H = 1{,}000$ m to limit the vertical propagation of turbulence, with a Rayleigh damping layer applied within the inversion layer to relax the u and v profiles toward their geostrophic values and reduce the amplitude of waves which could reflect from the model top and contaminate the solution.

Figure 4.1 reveals several attributes of such canonical LES setups, including a broad range of variability in wind speed, a range of resolved length scales, periodicity of the solution, and a slight anticlockwise rotation of the flow, due to the Coriolis acceleration.

4.2.2 LES assessment

While the flow depicted in Figure 4.1 appears turbulent, it is not possible to ascertain the fidelity of the turbulence field from an instantaneous snapshot, rather its statistical behavior must instead be examined. Given the broad range of meteorological and surface factors potentially influencing ABL and turbulence dynamics, it is important to assess if a particular LES setup is satisfying the assumptions on which the technique is predicated (i e., that L is adequately captured with the chosen Δh) for a given set of forcing conditions. The metrics most commonly used to assess LES performance include mean values of velocity, TKE, turbulent stresses (τ_{ij}), and spectra of flow quantity $\alpha(F_\alpha)$, with α representing a velocity or scalar field.

The resolved components of TKE and τ_{ij}, which are of primary importance in evaluating the adequacy of Δh to capture L (as opposed to the SFS components, which are useful primarily to assess the SFS model itself), can be straightforwardly computed using the eddy covariance technique as $TKE = 1/2\left(\overline{\acute{u}_i^2}\right)$ and $\tau_{ij} = \overline{\acute{u}_i\acute{u}_j}$, where overbars represent an average over time or space, and $\acute{a}$ represents an instantaneous departure from that average, $\acute{a} = a - \bar{a}$ (see e.g., [3]).

An example of assessing the validity of LES setups using different horizontal mesh spacings Δh is provided in Figure 4.2, which shows profiles of the streamwise vertical stress $\tau_{13} = \overline{\acute{u}\acute{w}}$. Here, horizontal plane averages were used to compute the instantaneous quantities (e.g., $\acute{u}_i(x,y,z) = \bar{u}_i(z) - u_i(x,y,z)$) at each height, as well as the product $\acute{u}\acute{w}$. Due to the significant spatiotemporal variability of τ_{13} (and similar quantities) in turbulent flows, averaging in time or space may be required to reveal their statistical behavior. The profiles shown in Figure 4.2 are 4-h averages, computed from plane-average vertical profiles computed at 1-min intervals.

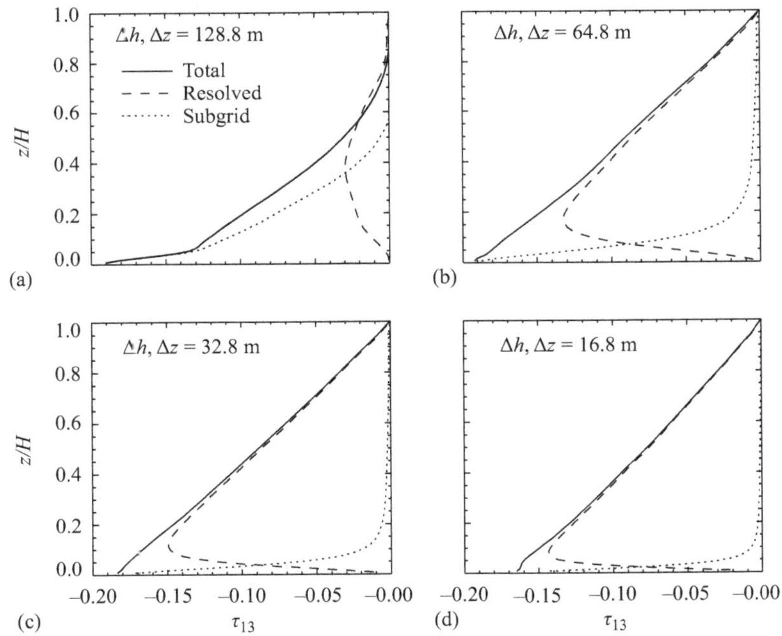

Figure 4.2 Vertical profiles of streamwise vertical stress component $\tau_{13} = \overline{u'w'}$ obtained from idealized LES as described in Figure 4.1, here showing sensitivity to model horizontal resolution Δh with values of (a) 128, (b) 64, (c) 32, and (d) 16 m, as the vertical mesh spacing of 8 m is held constant

Each panel in Figure 4.2 shows the total, resolved, and SFS components of τ_{13}, with the different panels showing how each component varies with changes of Δh. The simulation using the coarsest resolution (Figure 4.2(a)) contains very little resolved stress, indicating that the mesh resolution is too coarse to capture L and the key turbulence dynamics upon which the LES technique is predicated, for this set of forcing conditions. In this case, the SFS model is producing nearly all of the stress throughout the ABL, which is inconsistent with the role for which it is designed, which is to supply proper forcing for resolved-scale turbulent motions. Refining the mesh by a factor of two (Figure 4.2(b)) results in the effective capture of L and a sufficient portion of resolved turbulence dynamics throughout most of the ABL, with the SFS model serving its proper role.

Further increases of resolution (Figure 4.2(c) and (d)) only minimally impact the partitioning between resolved and SFS components far above the surface; however, they do allow a greater proportion of the stress field to be resolved nearer to the surface. In all cases, the resolved stresses eventually vanish approaching the surface, due to the characteristic behavior of eddy size to scale with height above the surface in the near-surface region (e.g., Tennekes and Lumley, 1972), a process which effectively reduces the eddy size below the filter width at some height above the surface, irrespective of the chosen mesh resolution. Hence, while LES is designed to explicitly resolve almost all of the motions producing stresses throughout most of the ABL, the SFS model provides nearly all of the stresses over the first few computational grid points above the surface (with a wall model such as described in Section 4.2.1 providing the stresses at the surface).

A commonly observed feature of LES exhibited by Figure 4.2 is the slight difference in magnitude of the stresses across the three simulations explicitly resolving the majority of the stress field (Figure 4.2(a)–(c)). These differences are due in part not only to the higher resolution setups resolving a greater proportion of the stress, especially near the surface, but also to the effects of the grid aspect ratio, $\alpha = \Delta h/\Delta z$. Multiple studies (e.g., [19–21]) have demonstrated the impact of α on LES profiles of stresses and other quantities arising due to the interplay between the energy dissipation provided by the SFS model and the dissipative effects of various numerical solution methods, each of which varies with changes to model grid spacing in the vertical and horizontal directions. The sensitivity of LES results to the multiple configuration choices available, including α, Δh, the numerical solution procedure, and the SFS model (to be discussed in subsequent sections), leads the experienced LES practitioner to the interpretation that any particular LES solution represents one member of a probability density function for the "true" solution, which is unknowable beyond a certain envelope. Checking characteristics of the resolved energetics (stress, TKE, and spectra) can verify the proper functioning of LES, but not the ultimate accuracy of the solution.

Spectra of flow parameters, such as velocity, can also be used to assess the state of resolved turbulence within a simulation. Spectra, which can be straightforwardly computed using Fourier transforms of timeseries or spatial transects (e.g., [3]), indicate the relative amounts of variability occurring within the given time series or transect contributed by each frequency f captured within the span of the data sample and its grid spacing or time resolution. For idealized, homogeneous LES with constant forcing, spectra should contain a maximum of energy near the frequency corresponding to the integral scale of turbulence, representing variability associated with structures of size L, followed at higher frequencies by a region of inertial cascade scales, decreasing with a slope of $f^{-5/3}$ [22]. Other slopes may appear over portions of the spectra, such as the f^{-1} decrease associated with turbulence production scales at frequencies between the integral scale and the inertial cascade (e.g., [23]). At yet higher frequencies, spectra from LES will show a rapid attenuation of power at frequencies approaching the filter cutoff, due to the removal of higher frequency information by some combinations of the grid spacing and numerical discretization effects, or an explicit filter, if used. While pseudo-spectral solvers (which use spectral and finite numerical methods in the horizontal and vertical directions, respectively) can resolve energy very close to the grid Nyquist frequency $f_N = \pi/(\Delta h)$, less exact solvers, such as finite difference and finite volume schemes, begin to attenuate energy at lower frequencies, with significant attenuation of scales as large as $5 - 7\Delta h$ (see e.g., [24]). Often spectra are displayed in compensated form, as the product of the power and the frequency (modifying the slope of energy decay in the production and inertial ranges to f^0 and $f^{-2/3}$, respectively), which may more clearly indicate the low-frequency spectral peak and also possess the property of the area under the spectra being proportional to the TKE (e.g., [3,25]).

Figure 4.3 shows compensated spectra for the vertical (w, Figure 4.3(a)) and streamwise (u, Figure 4.3(b)) components of velocity, from the same simulations as depicted in Figures 4.1 and 4.2. As with the resolved stress profiles, spectra show smaller magnitudes distributed over a narrower range of values for the most coarsely resolved simulation, indicating that the expected turbulence cascade is not being adequately captured. All higher resolution simulations produce similar magnitudes of lower frequency power, indicating capture of integral scale motions. Further increases of resolution show a greater range of inertial scales being resolved, with dissipation occurring at higher frequencies. Another characteristic feature of LES

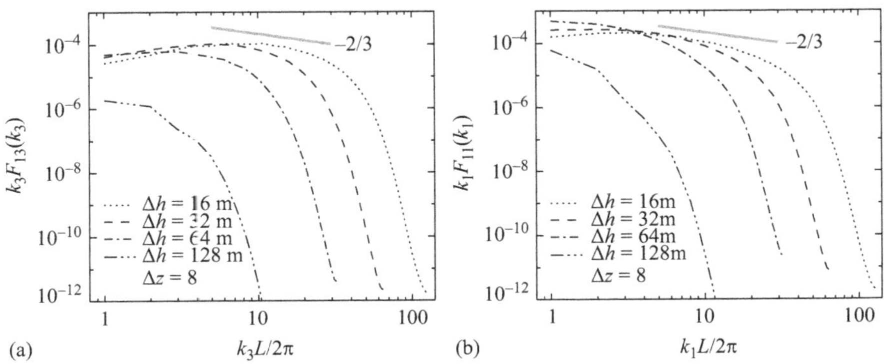

Figure 4.3 Spectra of the (a) vertical and (b) zonal velocity components, taken in the x-direction, from idealized LES as shown in Figures 4.1 and 4.2

of idealized flows over flat, uniform terrain for which mechanical production dominates the TKE budget (i.e., stable to weakly convective conditions) is the tendency of increasing resolution to shift the spectral peak to higher frequencies, reducing the power associated with the lowest frequencies. The greater spectral power at lower frequencies often observed in coarsely resolved LES occurs as the result of the tendency of coarser mesh spacing to amplify the correlations of the elongated streaks aligned with the streamwise flow direction that often appears in such idealized LES setups (see e.g., Figure 4.1). Increases of resolution tend to decrease the coherence of the largest resolved structures by capturing a broader range of energetic eddy sizes, thereby attenuating the streaks. This breakup of the elongated streamwise streaks is manifested in the greater reduction of low-frequency spectral power for the u relative to the w-velocity component, as u is approximately aligned in the streamwise direction for this particular setup (see Figure 4.1).

Another metric commonly used to assess the performance of LES in steady, homogeneous setups is the agreement of U with its expected MOST counterpart, the familiar logarithmic distribution, or "log law" within the surface layer. Figure 4.4 shows profiles of time-averaged U using different horizontal mesh resolutions, again showing the sensitivity of the solution to α. The failure of LES to reproduce a logarithmic distribution of wind speed, even with the log law used to specify the surface boundary conditions (Equation (4.3)), is a frequently cited issue in LES that is discussed further in Section 4.5.3.

While Δh defines one necessary component of the computational mesh, the size of the domain must also be selected appropriately to ensure capture of the largest scales impacting the turbulence field. For typical forcing conditions, L is commonly taken as similar to H. In such cases, domain sizes of at least $3H$ are recommended

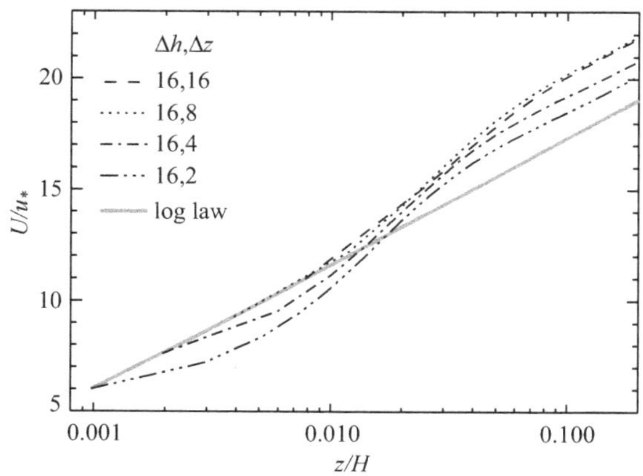

Figure 4.4 Plane-averaged horizontal wind speed distributions relative to the expected logarithmic distribution (log law), from idealized LES as shown in Figures 4.1 and 4.2

over the area of interest, in order to fully resolve a few of the largest features, and also to prevent any artifacts associated with a turbulence development fetch or periodicity of the domain, if periodic LBCs are used, from contaminating the solution. Due to the much smaller characteristic values of L occurring during stable conditions, the stably stratified ABL can typically be simulated with much smaller domains than those required for convective conditions. However, estimating H during stable conditions can be problematic, since H may not be well defined or may vary widely over time. Further, the stable ABL is often influenced by waves and other mesoscale and submesoscale processes with much larger characteristic length scales than L or H, which may require very large domains to capture (see e.g., Section 4.3.4).

Many historical LES studies of the ABL have been conducted using idealized setups, similar to those depicted in Figure 4.1, featuring horizontally homogeneous and steady forcing conditions, as well as use of periodic LBCs. Such setups permit computation of the above-described assessment metrics in either time or space, with spatial averaging and computation of spectra permissible over homogeneous directions. For setups with no homogeneous directions, these quantities must instead be computed in time. A rule of thumb is that the quantities be computed over sufficient spatial or temporal extents to capture a few of the largest structures associated with the classical turbulence spectrum, but not so large so as to include variability associated with larger scale wave, mesoscale, or submesoscale motions. Typically, spatial extents of a few kilometers, or temporal periods of 10–20 min, are utilized as approximate ranges to capture most of the turbulence signal with a minimum of contamination from larger scales. In many idealized LES setups, such as shown in Figure 4.1, mesoscale variability is implicitly filtered from the solution. While evidence exists for a "spectral gap" in the continuum of energetic motions in long-term atmospheric timeseries data (e.g., [3]), the atmosphere often exhibits a smoother distribution of energy over timescales of hours to days, requiring care in the selection of an appropriate timescale for computation of spectra and other turbulence quantities (e.g., [26]).

4.2.3 *Unsteady conditions*

While the previous discussion has considered forcing conditions that are steady in time, periodic LBCs are also applicable to certain types of time-varying forcing that still satisfy periodicity across the extent of the computational domain. One example of meteorological variability relevant to wind energy applications that may be introduced into LES using periodic LBCs is the typical diurnal cycle of the ABL. Diurnal variability arises due to changes in the surface energy balance, typically resulting in warming of the surface during the daylight hours, which initiates surface-based convection, followed by a cooling during the evening and overnight, which stabilizes the ABL.

The cycle of surface warming and cooling forces the ABL through changes of sensible and latent heat exchanged between the ABL and the surface (see Section 4.3.1.1), the former representing a direct conductive exchange of thermal energy, the latter involving exchanges of latent energy due to phase changes of water. Positive values of sensible and latent energy represent a transfer of energy from the surface to the air. The presence of water upon the surface or within the soil or plant matter reduces the magnitude of sensible heating via the uptake of energy that would otherwise heat the ground and air above it, to instead evaporate water, or melt snow or ice. The presence of moisture at the surface therefore generally results in a reduction of heating and buoyant TKE production, producing shallower ABL depths and smaller TKE values. Since water vapor is lighter than dry air, latent fluxes do contribute to buoyant TKE production; however, as sensible heating typically dominates ABL and TKE dynamics, latent heating is usually ignored.

Typical diurnal variability within the ABL may be straightforwardly simulated within LES simply by specifying time-varying values of H_S. The subsequent warming and cooling of the air above the surface, coupled with turbulent transport, will readily generate characteristic diurnal evolution of ABL growth and decay. The same diurnal variability can be achieved by instead varying the surface temperature, provided that a parameterization for H_S in terms of surface-to-air temperature gradient is implemented. Due to the simplicity of implementation, and elimination of the reliance on a parameterization to convert temperature differences to fluxes, H_S is more commonly used. However, in moderately to strongly stable conditions, θ_S should instead be used to avoid an unphysical solution branch that may be encountered when using large negative H_S values within the MOST surface boundary condition framework [11]. Since strong stability requires weak mean winds, such conditions are not typically relevant to wind energy simulations.

Another meteorological change that can be incorporated into periodic LES is variability of the large-scale pressure gradient force, expressed through the geostrophic wind (see Section 4.3.1.1), which drives the evolution of the mean horizontal winds. The combination of variability of wind speed and H_S can produce a significant range of ABL variability. One type of variability arising from the interaction of mean winds and surface fluxes is the large-scale morphology of turbulence structure, which can be characterized by the dimensionless ratio of the height of the boundary layer to the Obukhov stability parameter L (see Section 4.3.1.1). Smaller values of H/L, either due to low wind speeds, which reduce u_*, or to greater convective instability, due to larger H_S values, lead to predominantly cellular convective morphology. A sufficient increase of the mean wind speed, or a corresponding decrease H_S, will result instead in elongated convective roll structures, with more coherent roll morphologies for larger values of H/L. During diurnal cycles with a relatively constant large-scale geostrophic forcing, roll structures are likely to be observed during the morning and evening hours, with more cellular structures during the peak of surface heating.

Similar changes of surface-based instability to those occurring over land during the diurnal cycle can also occur over bodies of water due to large-scale advections of air with different thermal properties. Incorporation of the effects of large-scale advections, which can capture turbulence morphology and other characteristics, into periodic LES, is discussed in Section 4.4.1.

4.2.4 Stable conditions

The stable portion of a typical diurnal cycle, following the transition of H_S from positive to negative values, usually occurring during late afternoon, is of particular relevance to wind plant simulation. While strongly stable conditions, which only happen in the absence of a significant mean wind speed, are not generally relevant to wind energy applications, weakly to moderately stable conditions, for which a sufficient level of mechanical generation is maintained by the mean flow to prevent the strong thermal stratification that leads to strong stability, are highly relevant. In addition to being the most common conditions occurring on Earth's surface, weakly to moderately stable conditions often produce optimal conditions for wind power production.

The enhanced wind energy potential often occurring during stable conditions results from several different physical processes. One such process is the reduction of strength and vertical extent of turbulent mixing of lower momentum air from near the surface, where friction retards the flow, allowing faster wind speeds at typical turbine rotor heights.

Another process that occurs during stable conditions in the presence of sloping terrain is cold air drainage, characterized by faster moving air adjacent to the surface that results from the higher density of air cooled by the underlying surface being accelerated downhill by gravity. The sinking of the colder air results in warmer air being drawn downward from aloft, creating a circulation pattern that can persist for several hours in the absence of other stronger forcings. Drainage flows are often shallow and relatively weak; however, in certain locations and under the right conditions, it can contribute to the wind energy resource.

A third important physical process also driven by thermal contrasts is the land–sea breeze cycle. This cycle is driven by density changes resulting from the different rates of change of air temperature occurring over land versus large bodies of water in response to the solar cycle. Such breezes can be quite strong and extend many kilometers beyond the land/water interface, making it an important source for both land-based and offshore wind farms located near the shoreline. While the basic physics of the land–sea breeze is well understood, accurate prediction is complicated by the role of synoptic meteorology, which can augment or oppose the diurnal forcing, as well the impact of surface parameters, such as soil moisture and vegetation on land, and water temperature and sea state offshore, each of which can strongly influence the fluxes of momentum and heat, impacting ABL structure and evolution of low-altitude winds.

During strongly forced simulations (convective or wind-driven), the assessment metrics described in Section 4.3.2 should provide reliable guidance on the adequacy of a given LES setup to ensure proper functioning of the technique. However, during stable conditions, assessment may require more critical analysis. For example, during calm, clear conditions, strong stabilization may occur, leading to periods of very weak turbulence. If resolved turbulence motions disappear during LES of stable conditions and do not redevelop within a few hours (which may occur following LLJ formation, for example, as described in the next subsection), it is likely that mesh spacing needs to be reduced.

4.2.4.1 Nocturnal low-level jets

Perhaps the preeminent physical process influencing wind energy production during stable conditions is the LLJ. LLJs, defined in various ways but most generally as wind speed maxima occurring within the lowest few hundred meters above the surface (e.g., [27]), frequently produce wind speeds that significantly exceed geostrophic values. LLJs have many generation mechanisms depending on their location and time of development (e.g., [28]; Whiteman *et al.*, 1967; [29]). A commonly occurring LLJ mechanism impacting several major wind resources worldwide, including the Great Plains region of the United States, is the nocturnal LLJ (NLLJ). NLLJs typically form in the evening, following the waning of convectively generated turbulence motions within the ABL, which reduces the effective drag force, resulting in a temporary acceleration of flow aloft in the formerly more well-mixed middle and upper portions of the ABL (e.g., [30,31]). NLLJs typically persist for several hours, exhibiting an inertial oscillation characterized by variations in magnitude over several hours, depending upon the latitude (e.g., [32]).

While the enhanced wind speeds generated by LLJs increase the energy available for conversion to electricity, the accompanying increase of vertical shear can also amplify stress loading on turbine components. In addition, the potential for wave activity, the stable background environment during NLLJs, can lead to various wave phenomena that can also impart damaging stress loads to turbine components [33]. Kelvin-Helmholtz instabilities, for example, often occur in the presence of sheared flow within a statically stable background environment ($\partial\theta/\partial z > 0$; see e.g., [34]), conditions typical during NLLJs (e.g., [35]). The resulting wave activity can amplify and eventually break, resulting in strong turbulence generation within the region of the breaking wave. The resulting turbulence may then mix heat and momentum sufficiently to reduce both the mean shear and thermal stratification below the criteria for further wave generation. However, as the localized wave-generated turbulence decays, the background conditions giving rise to the initial instability may again generate sufficient shear and thermal stratification to reestablish the instability, resulting in subsequent wave breaking. This type of turbulence variability, referred to as global intermittency, to distinguish it from the inherent intermittency characterizing normal turbulent flows (e.g., [36]) can be particularly important for wind energy, due to impacts on both the mean flow speed and potentially harmful stress loading events.

NLLJs can be readily generated within an LES by incorporating Coriolis accelerations and specifying negative H_S values, or $\theta_S < \theta(z_1)$. However, additional atmospheric factors beyond surface energy exchange can play significant roles in defining important NLLJ characteristics, including their strength, duration, height above the surface, and time variability. These additional atmospheric factors, including baroclinicity, advections, and deformation frontogenesis (e.g., [32,37]), are generally not included in LES. However, due to their importance, the method to incorporate these additional atmospheric effects into LES models used for wind plant simulation is an active area of research, with approaches under development described in Section 4.3.1.

While LES of the stable ABL can be quite computationally expense due to the small mesh sizes and associated model time steps required to capture the small turbulence structures, including the NLLJ increases this burden considerably due to the simultaneous requirement of large enough computational domain sizes to capture relevant NLLJ forcing mechanisms, such as the previous afternoon's convective ABL, and any wave motions that may develop within the stable ABL (of order a few kilometers). The extreme computational demand of NLLJs and associated wave phenomena has resulted in few studies appearing in the literature to date (e.g., [38]). A potential path forward for the investigation of such computationally intensive phenomena is to utilize adaptive meshing to concentrate the fine-scale meshing only where it is required to capture the small length scales associated with the turbulence, while using a coarser mesh to represent the larger scale forcing mechanisms. A discussion of different types of mesh refinement and potential issues with ABL flow simulation is provided in Section 4.3.2.

4.2.4.2 Lateral boundary conditions

While the existence of wind turbine impacts on the downstream flow would seem to preclude the utility of periodic LBCs in wind plant simulation, the complications and increased computational expenses associate with alternative methods for specifying LBCs (as described in Section 4.3.2) motivate the use of periodic LBCs whenever possible. Numerous applications can benefit from the efficiency and simplicity of periodic LBCs. Idealized LES using periodic LBCs can be used to investigate a range of relevant flow physics impacting plant design and operation, providing useful information without the need to resolve wind turbine impacts within the LES (e.g., [39]). Idealized infinite wind plant studies with repeatable spatial arrangements of turbines may be readily simulated using periodic LES (e.g., [40,41]). For simulations requiring both unadulterated inflow and representations of turbine/airflow interactions (e.g., the generation of wakes), periodic LES can be utilized as an "auxiliary" simulation domain, within which a background turbulent ABL flow field can be efficiently generated for simultaneous or later use within the simulation resolving turbine/airflow interactions. Both synchronous, wherein both the auxiliary and wind plant simulation domains run concurrently (e.g., [42,43]) and asynchronous approaches, for which the periodic LES is run as a "precursor" simulation, with the flow field saved for later use during the wind plant simulation (e.g., [44,45]), have shown promise as a wind plant-simulation technique incorporating the utility of periodic LBCs.

Both synchronous and asynchronous simulation frameworks have specific advantages and disadvantages. An advantage of synchronous coupling is that both the resolution and the forcing conditions for the auxiliary simulation can be matched to the wind plant domain for the specific application. Asynchronous coupling would require the existence of a precomputed flow field matching the forcing conditions, at the required resolution If the LES of interest utilized a finer resolution than the archived auxiliary result, a fetch within the wind plant LES would be required for the generation of smaller scales from the coarser-sale inflow, increasing the computational cost.

A disadvantage of the synchronous approach is the additional computational overhead required at run time to run two simulations instead of one. While the asynchronous approach removes the requirement of conducting parallel simulations, disadvantages include the upfront cost of running a sufficiently broad suite of simulations to match the extant forcing conditions associated with the wind plant LES, as well as the overhead associated with archival and retrieval of those fields. Further research is required to quantify how closely the forcing for the archived flow must match that of the extant situation to be useable, and the extent of associated fetch requirements for downscaling an archived flow at coarser resolution to a finer mesh within the wind plant LES domain.

While periodic LBCs, either in ABL simulations without turbine interactions or as auxiliary simulations to augment wind plant LES, efficiently and robustly generate turbulent ABL flows (provided sufficient mesh spacing), their applicability remains restricted to a relatively limited range of environmental, terrain, and forcing conditions. Methods to both extend the applicability of periodic LBCs to a wider range of unsteady and heterogeneous scenarios that can still be considered quasiperiodic, and to replace periodic LBCs with other data sources for highly heterogeneous conditions, are under development, as described in the following section.

4.3 Enabling multiscale simulation

The basics of ABL LES described in Section 4.2 focused primarily on capturing the range of energetically important scales of atmospheric motion associated with the classical three-dimension turbulence spectrum, using idealized setups featuring steady and homogeneous surface characteristics and forcing. This range of scales however represents only a portion of the spectrum of atmospheric motions typically present in realistic ABL flows, which evolve in structure and dynamics from a variety of sources of variability in forcing associated with a broad range of atmospheric and environmental drivers. Discussions of drainage flows, land and sea breeze cycles, and the NNLJ are just a few examples of the continuum of forcings relevant to ABL flow and turbulence dynamics.

This section describes methods to incorporate several of these important sources of atmospheric and environmental variability into ABL LES, as required to expand applicability of the LES technique to the unsteady and heterogeneous conditions defining typical wind farm operating conditions. We begin by describing several methods to expand the utility of LES using periodic LBCs, which are still useful in many applications. Methods to replace periodic LBCs in highly heterogeneous settings, for which the use of periodic LBCs can no longer be justified, with inflow from observations or, more commonly, mesoscale simulations, are also described, along with a number of associated issues involved in downscaling of mesoscale flows to LES scales.

4.3.1 Methods to extend the applicability of periodic LES

Historically, the impacts of larger scale atmospheric processes impacting ABL evolution have not been incorporated into periodic LES in part due to difficulties inherent in their measurement but also due to the potential to violate restrictions imposed by the use of periodic LBCs. However, under certain conditions, some larger scale atmospheric process can be regarded as sufficiently horizontally homogeneous over the spatial scale of a wind plant simulation domain that their effects can be incorporated into idealized LES with minimal violation of the assumption of periodicity.

One manner to incorporate large-scale meteorological forcing terms into periodic LES is to simply add time-varying but horizontally homogeneous source terms to the governing equations for momentum and any scalars (e.g., temperature) prognosed within the periodic LES domain. The source term, often referred to as a tendency, reflects the contribution of a specific process to the time rate of change of the associated quantity. As most mesoscale weather models collect the individual tendencies from each relevant physical process (e.g., advection, turbulence, radiation, latent processes, turbulence fluxes) within each time step of the integration cycle, each individual term contributing to the evolution of the field variable may be archived during a mesoscale simulation and subsequently made available for incorporation into an offline LES.

While the incorporation of additional source terms into offline LES is rather trivial, the difficulty is obtaining accurate estimates of these parameters. As described previously, estimates of the geostrophic winds and large-scale advective tendencies can, in principle, be determined from observations; however, reliable observation-based determination requires higher spatiotemporal sampling than is typically available from routine atmospheric measurements. Even in well-instrumented field campaigns occurring in quasi-idealized settings and flow conditions, determination of advections, as with the geostrophic wind values, can be difficult and often involves some assumptions (e.g., Mirocha *et al.*, 2004; [46,47]).

Given the difficulties involved in estimating LES forcing parameters from observations, a potentially more tractable method is to use a mesoscale atmospheric model, for which the values of all relevant atmospheric flow and forcing parameters are available. Sanz Rodrigo *et al.* [48] outline a methodology for the estimation of geostrophic wind speed variability and large-scale advections from a mesoscale model, and their incorporation into a periodic LES. While their approach shows promise, determination of appropriate advections and geostrophic winds from mesoscale model output is not always straightforward. First, as with real data, the simulated wind field above the ABL may be influenced by myriad factors beyond the large-scale pressure distribution, complicating determination of appropriate geostrophic wind values. Second, while the horizontal pressure gradients defining the geostrophic winds can be computed at each location on the model grid, local pressure variability may contain smaller scale or transient components that, due to the long timescale of Coriolis acceleration, are not consistent with the large-scale geostrophic balance. Incorporating such information into the geostrophic forcing applied within an LES will result in erroneous accelerations (although the impacts of these erroneous accelerations are not known). In addition, the advective tendencies and geostrophic winds deduced from mesoscale simulations may contain considerable high-frequency variability associated with other processes, especially when higher resolutions are employed ($\Delta h \lesssim 5$–10 km). Such variability must be filtered to obtain the slowly varying, larger scale component of the forcing. While averaging in space or time effectively filters the high-frequency variability, little guidance presently exists regarding optimal filtering strategies for general conditions.

While the ageostrophic pressure fluctuations removed by the filtering operation could potentially be applied as a separate, non-geostrophic acceleration within the LES, such an approach would require careful execution. For example, a fine enough mesoscale mesh may resolve scales of pressure variability that can also be resolved within the LES domain, such as those caused by terrain or convective structures. Similar arguments hold for the advections, which may likewise contain contributions from local-scale features and processes. Mesoscale tendencies incorporating forcing associated with structures and processes also resolved within the LES would result in duplicate representation therein (double counting). Additionally, tendencies computed from high resolution mesoscale simulations may not strictly satisfy periodicity, due to the potential capture of fine-scale structure that does not "average out" to a mean value over the footprint of the LES domain. As such, averaging of the mesoscale tendencies in time or space may be required. While mesoscale models can potentially provide the time history of all relevant atmospheric parameters required to parse the forcing into different spatiotemporal scales and processes, techniques to accomplish this parsing and generate appropriate forcing terms for offline periodic LES under general conditions do not yet exist.

While promising approaches to extend the applicability of the relatively economical approach of using periodic LBCs to a wider range of wind-energy-relevant meteorological conditions are under development, periodic LBCs still impose restrictions on not only the type of meteorological complexity that can be considered but also on characteristics of the underlying and surrounding surface conditions, which also must be either horizontally homogeneous or periodic. While various methods have been developed to extend the use of periodic LBCs to increasingly nonperiodic and heterogeneous conditions (see e.g., review in [16]), such approaches can be quite involved and rely upon approximations that become tenuous in highly nonuniform settings. For many meteorological conditions and environmental settings, the use of periodic LBCs becomes untenable and other methods must instead be utilized.

4.3.2 Coupling LES to mesoscale model output at lateral boundaries

A general multiscale wind plant LES capability applicable to arbitrary locations and conditions requires an alternative to the use of periodic LBCs. Measurements could, in principle, be used to specify the LBCs of a microscale LES domain; however as described previously, measurements do not contain the full suite of required parameters at both the spatiotemporal resolution and the spatial coverage to specify all of the required boundary information. In addition, errors associated with measurements and their extrapolation to a given mesh may result in an imbalanced initial state that may spawn spurious transients. As such, mesoscale simulations, which contain capabilities to assimilate data, spin up a balanced atmospheric state, and dynamically downscale to microscale mesh spacings, can provide a superior and more general means of specifying LBCs for LES.

Many atmospheric simulation frameworks support various dynamical downscaling capabilities enabling generation of lateral inflow conditions for LES at a range of spatial scales, including those resolving the three-dimensional turbulence field. Due to the prohibitive computational costs of simultaneously resolving both large- and fine-scale motions over domains of sufficient size (continental to global) to capture both the largest scales of forcing and the smallest scales of turbulence, dynamic downscaling approaches utilize variable mesh spacing, allowing a relatively coarse background resolution to capture the largest scales of forcing, while subsets of the domain can be refined locally to capture smaller scale features, including turbulence.

Two classes of dynamic downscaling approaches are available within the most widely used publicly available atmospheric research simulation codes. The most common approach, used within the Advanced Regional Prediction System (ARPS; [49]), the Coupled Ocean/Atmosphere Mesoscale Prediction System ([50]), the Consortium for Small Scale Modeling [51], the Meso-NH model [52], the Regional Atmospheric Modeling System ([53]), and the Weather Research and Forecasting (WRF; [54]) model, is grid nesting, whereby a rectangular subset of a "parent" domain is instantiated as a separate "child" domain, as shown in Figure 4.5(a). Each nest shares the same model top with the parent domain; however, it may utilize higher resolution data at the surface. The mesh within the nested domain is typically refined in each of the horizontal directions, as shown in Figure 4.5(b), using an integer mesh refinement ratio, typically between 2 and 10, although larger ratios have been used. Vertical mesh refinement is less common but is gaining popularity following the introduction of a vertical nesting capability for concurrent nested domain simulations in the WRF model [55].

While all of the domains are coupled dynamically in a nested simulation, the different domains may use different physical process parameterizations appropriate to the resolution of the given domain. Initial conditions within the nested domain may either be generated from the same dataset used to initialize the parent domain or may be interpolated from the bounding domain solution. LBCs are interpolated down to the nest resolution from the bounding domain solution. Nesting can be either "one-way," for which the nest solution does not provide any forcing for the solution on the parent domain, or "two-way," for which information from the nested domain is aggregated to provide forcing for the parent-domain solution within the region overlapping the nest.

Nesting has achieved widespread popularity within many mesoscale applications, from regional weather and climate prediction to resource assessment for wind energy. More recently, the grid nesting approach has been pursued down to LES scales, with various degrees of success. The growing number of studies using coupled mesoscale-to-LES simulations is gradually leading to improved characterization of the approach, identification of best practices, and development of techniques to improve the methodology (e.g., Muñoz-Esparza *et al.*, 2018; Rai *et al.*, 2017a,b; [55,56]).

Several newer multiscale atmospheric simulation systems are providing more flexible mesh refinement approaches than traditional grid nesting. The Non-hydrostatic Unified Model of the Atmosphere ([57]), for example, supports mesh

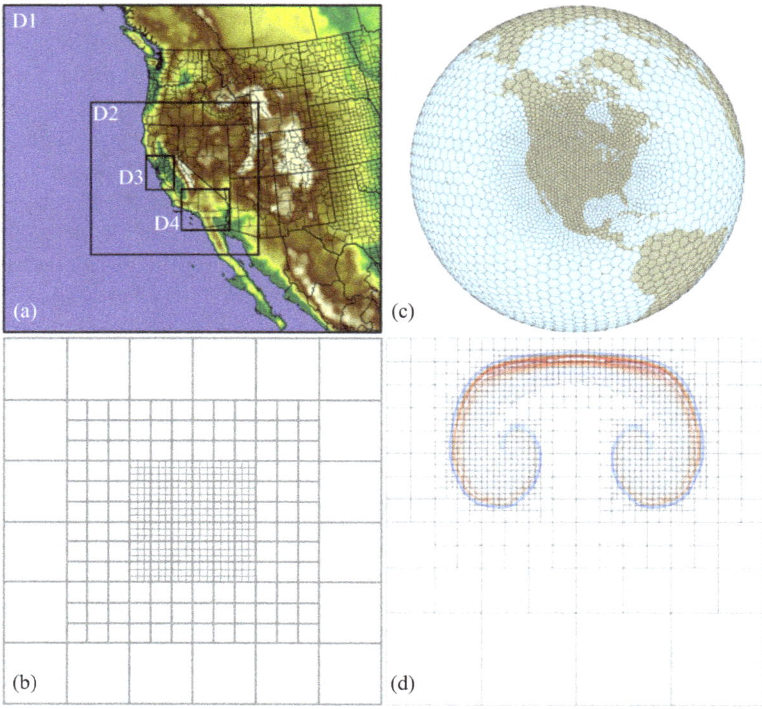

Figure 4.5 Examples of mesoscale atmospheric model mesh refinement strategies, including grid nesting (e.g., WRF; (a and b)), Voronoi tesselations (e.g., MPAS; (c)) and adaptive (e.g., NUMA; (d))

refinement over individual grid cells, rather than rectangular blocks of grid cells, enabling locally enhanced resolution as required to capture fine-scale structures as those move through the domain (see Figure 4.5(d)). Such an approach could vastly improve the efficiency of wind plant LES, where the characteristic turbulence length scale can vary widely in time and space due to localized turbine–airflow interactions. The Model for Prediction Across Scales (MPAS; [58]) and Energy Earth Exascale Model (E3SM; [59]) instead utilize a smoothly refining mesh consisting of Voronoi and associated Delaunay tessellations (see Figure 4.5(c)).

While the smoothly refining meshes used by MPAS and E3SM are in principle extensible to LES scales, the impacts of their more complicated grid cell geometries to turbulence resolving flows have not yet been reported upon within the literature. Further, while smoothly refining mesh capabilities may improve upon block rectangular grid nesting when downscaling within the mesoscale realm, it is unclear how well the approach will work at interfacing mesoscale and LES scales, or under further refinement within the LES regime. While abrupt mesh refinement leads to certain well-known issues, including the introduction of certain anomalies into the flow, and the need for a fetch upon which turbulence may equilibrate to the finer mesh (see discussion in Section 4.4.2.1), some research (e.g., [60,61]) suggests that abrupt mesh refinement may generate smaller errors overall than gradual reductions in the mesh, which, by maintaining larger eddy viscosities, delay the onset of smaller scales motions, thereby increasing the distance required for turbulence to develop and equilibrate. Further research is needed to quantify the benefits and drawbacks of different mesh refinement approaches in both MMC and downscaling within LES.

Given the improved flexibility afforded by the current and expanding realm of multiscale atmospheric simulation models available, mesoscale simulations have and will continue to become an increasingly popular method for providing LBCs for microscale wind plant LES. LES may either be embedded within the same code used for the dynamic downscaling process or may be conducted within a separate solver. In either case, two significant issues stand in the way of enabling this approach for general conditions, one being the generation of turbulence within the LES domain receiving mesoscale LBCs, while the other is the modeling of SFS turbulence effects within a range of scales just beyond the coarsest LES domain, known as the terra incognita (described in Section 4.4.2.2).

4.3.2.1 Turbulence generation: mesoscale to LES

Any mesh spacing coarser than the integral turbulence scale L will filter out essential scales of motion associated with the classical three-dimensional turbulence cascade. For an LES domain receiving boundary conditions from a mesoscale simulation using $\Delta h > L$, those missing scales of motion must develop within the turbulence-resolving domain in order to recover accurate flow and turbulence information. The physical processes governing the development of turbulence from mesoscale inflow are similar to those discussed in the context of periodic LES (Section 4.3.5), with the crucial distinction that unlike domains using periodic LBCs, which allow developing turbulence to exit and reenter the domain as many times as is necessary to approach equilibrium, LES using mesoscale inflow must achieve rapid turbulence equilibration within a subset of the simulation domain upwind from the area of interest, since flow exiting the domain does not reenter.

As with the periodic LES, sufficiently strong convective forcing will generate the full spectrum of resolvable turbulence motions over relatively very short fetch. However, as wind energy applications feature flows with significant mean velocities, the distance traversed as the spectrum develops can be significant, especially in the absence of convective forcing.

To address the computational overhead associated with the sometimes very long fetches required for turbulence to develop, various methods have been developed to hasten turbulence development within microscale LES domains not using periodic LBCs. Such methods generally fall into one of the following three broad categories: (i) superposition of a spectrum of correlated turbulence motions onto mean flow profiles at the LES inflow plane(s), (ii) application of stochastic perturbations to one or more flow field variables near the inflow plane(s), and (iii) use of turbulent flow generated from a precursor simulation, such as an offline LES. Each of these methods has specific benefits and drawbacks and is likely to be optimal for different applications.

Methods that impart the entire spectrum of turbulence at the inflow, either as a superposition of turbulence onto a mean flow component or utilizing turbulence information generated from a separate simulation, can, in principle, achieve the shortest fetch within the LES of interest, provided that the turbulence information contained within the inflow extends to the effective resolution of the simulation. However, if the specified turbulence information is coarser than that resolvable within the LES, some fetch will be required for equilibration to the finer mesh. This fetch can be significant, due to the long timescale required for energetics of well-resolved scales to establish a new balance with dissipation following a mesh refinement. As described more fully in Section 4.5.1, many popular SFS stress models parameterize the SFS stress as the product of a grid-dependent length-scale and resolved-scale gradient within the flow. As flow crosses the mesh refinement interface, the immediate reduction of the length-scale results in an immediate reduction of the SFS stresses, thereby reducing the dissipation of resolved scale structures, which are inherited from the coarser mesh, and therefore do not contain gradients of sufficient magnitude to balance the reduced length scale. Once the smaller scales resolvable upon the finer mesh form, the greater spatial gradients contributed by those smaller scale structures deeper within the energy cascade balance the reduced length scale, leading to a new balance between production and dissipation. However, the temporary disconnect allows the energetics of well-resolved features to increase anomalously, resulting in temporary overshoots of resolved stress and TKE. Once the full range of resolvable inertial scales develops, additional fetch is required for those overshoots to diminish to correct values (see e.g., [62]). While the fetch associated with any grid-scale mismatch can be eliminated by generating turbulent inflow data at whatever scale is used within the LES of interest, generating inflow data at very fine resolution contributes significant additional computational overhead.

When using an auxiliary simulation computed synchronously (see Section 4.2.5), both the mesh spacing and forcing of the auxiliary simulation can be specified to exactly match the LES of interest, resulting in very modest fetch requirements. However, when using a precomputed turbulence database, an additional consideration is the effect of any mismatch between the forcing used to generate the database and that within the LES of interest, as any such mismatch will likewise necessitate a fetch upon which inflow turbulence must equilibrate to the local forcing. Guidance on fetch requirements as a function of the type and magnitude of mismatch, as well as impacts on various quantities of interest, does not presently exist. Such information is required to optimally populate the database with the smallest number of simulations to adequately span the range of required inflow conditions, to select the appropriate inflow file to match extant flow and forcing conditions during a simulation, including switching between databases during time-varying forcing, and to ensure a sufficient fetch to produce an equilibrated turbulence field at a location of interest.

The high computational cost of auxiliary simulations can be offset somewhat by instead using a synthetic turbulence generator to provide spectrally correlated turbulence information consistent with a mean flow profile; however, the computation of three-dimensional correlated turbulence information is quite involved (e.g., [63]), and while simpler and more efficient methods, for example, using two-dimensional planes (e.g., [64]) have been developed, such methods require a-priori knowledge of several flow parameters, including H, and profiles of mean velocity and variances. These required flow parameters can likewise be obtained from auxiliary simulations; however, such an approach limits cost savings relative to simply using the auxiliary simulation results directly.

While precursor simulations can be useful for many applications, a significant caveat is that any such simulations will also require forcing and specification of their LBCs. Often the LES producing statistics for synthetic methods themselves use idealized setups, including homogeneous forcing and periodic LBCs (e.g., [65,66]). Moreover, synthetic turbulence methods have been likewise developed thus far only for neutral conditions over flat terrain ([64]; Veers, 1998; [67]). While the use of auxiliary and synthetic methods may comprise an optimal workflow for many applications, their efficacy for real-world setups featuring high degrees of heterogeneity and time variability is not well known at present and requires further examination.

Given the complexity, computational overhead, and restrictions of the above-described methods, approaches based on superposition of small amplitude perturbations onto the resolved inflow variables have shown promise as a potentially simpler, more efficient, and more general method to accelerate turbulence development at the mesoscale–LES interface. While perturbation approaches have shown mixed performance in various historical applications (e.g., [68,69]), recent work has demonstrated that perturbing flow field variables within a specific range of spatiotemporal scales and magnitudes can significantly accelerate the development of turbulence upon a mesoscale inflow field under a range of flow conditions. The efficacy of more recent perturbation approaches results from the careful selection of perturbations characteristic that optimally triggers the development of turbulence via amplification of nonlinear modes inherent in the governing flow physics. Triggering the natural development of turbulence under the influence of forcing conditions germane to a particular flow scenario can potentially provide more rapid turbulence equilibration than specification of a correlated turbulence field that does not match the extant flow conditions. Further, use of relatively simple geometric and amplitude characteristics for the perturbations makes the approach simple to implement and computationally inexpensive.

Several perturbation methods have recently been examined using LES domains nested within mesoscale simulations, with both velocity and temperature being perturbed over different spatiotemporal scales, with varying results. Mirocha *et al.* [70] examined the addition of small amplitude ($\cong e$m s^{-1}) horizontally sinusoidal perturbations with a period of approximately 2 km to the u and v momentum equations in three different flow configurations, neutral and convective flow over flat terrain, and neutral flow over sinusoidally varying terrain. The perturbations, applied as a forcing term to the momentum equations, were shown to significantly accelerate turbulence generation in neutral cases, while slightly improving the rate of turbulence development in convective cases as well, relative to unperturbed simulations. This turbulence generation was still too slow to provide good spectral agreement with an offline simulation using periodic LBCs (taken to approximate the expected solution) within the nested LES domain upon which the inflow perturbations were applied. However, nesting a second, finer LES within the first LES domain produced good agreement with the expected solution therein (although perturbations were not applied on the second LES domain since the inflow to this domain from the bounding LES contained resolved turbulence already).

Reasons for the relatively slow development of turbulence in Mirocha *et al.* [70] included excessively large spatial scales of the perturbations and impartation of relatively small local gradients resulting both from the large spatial scales of the perturbations, their sinusoidal amplitude variability, and the relatively slow timescale of the application of the forcing. All of these factors resulted in less than optimal development of correlated turbulence structures at the proper scales to rapidly initiate the turbulence cascade. Muñoz-Esparza *et al.* [71] and Muñoz-Esparza *et al.* [65] addressed these shortcomings and significantly improved the technique by developing the stochastic cell perturbation method (SCPM; Figure 4.6). The SCPM consists of rectangular patches of stochastically generated perturbations applied instantaneously to the potential temperature field. The motivation for perturbing potential temperature rather than velocity is that the motions resulting from buoyant accelerations acting through the governing flow equations result in correlated velocity and temperature structures, leading to more rapid generation of a correlated turbulence spectrum. Perturbing velocity directly may lead to correlations that are not consistent with the governing flow equations and hence are dissipated rather than seeding turbulence production. Use of rectangular patches of a fixed size, with stochastic values for the

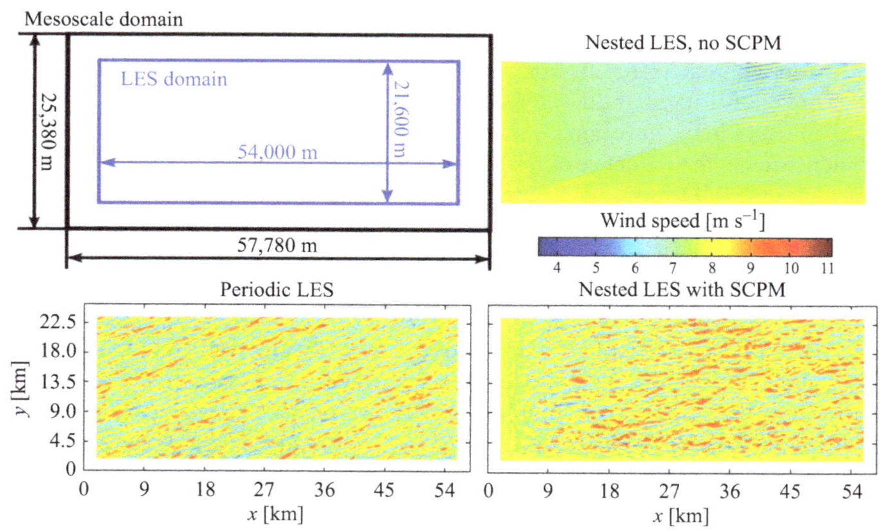

Figure 4.6 The stochastic cell perturbation method (SCPM) applied to LES nested within a mesoscale simulation, showing the acceleration of turbulence and improved agreement with the expected solution, obtained from a periodic LES, relative to an unperturbed simulation

amplitudes on each patch, generates features over a distribution of sizes, centered about the nominal patch size, due to the random nature of perturbation values on adjacent patches. Instantaneous application of the perturbations in discrete patches with no blending between the patches likewise generates a distribution of gradients across various patch boundaries.

Horizontal dimensions of $8\Delta h$ were found to be optimal sizes for the perturbation cells, as smaller cells were too small to initiate resolved buoyant motions sufficient to rapidly generate the turbulence spectrum (due to dissipation resulting from the numerical solution procedure) while larger cells generated large, persistent flow structures that required a significant time or distance to attenuate. Amplitude sensitivities indicated that perturbations much smaller than 1K were insufficient to rapidly trigger buoyant correlations, while those much larger than 1K produced excessive convective activity that again required a long timescale to attenuate to an appropriate turbulence spectrum. The amplitude of the perturbations required to achieve the optimal stimulation of turbulence was related to the inflow velocity via the perturbation Eckert number, $Ec = U_g^2/(C_p\acute{\theta})$, where $\acute{\theta}$ is the magnitude of the perturbation. Determination that optimal results are achieved with $Ec = 0.2$ provides the perturbation amplitude in terms of the geostrophic wind speed.

It was also discovered that applying three adjacent strips of cell perturbations along each inflow boundary, and refreshing the perturbation values over the advective timescale of the flow speed near the surface, yielded the best performance. Fewer strips of perturbations did not as rapidly introduce three-dimensional correlations into the developing turbulence field, while a larger number of strips interfered with the developing buoyancy-induced perturbations beginning to form further within the domain. Refreshing the patches more frequently than the advective timescale of the near-surface flow effectively generated smaller patches with larger amplitudes by superposing new patches onto old patches that had not yet been broken up by turbulence motions, while refreshing more slowly resulted in gaps between the strips of patches. Since flow speed generally increases with height, use of the near-surface advective timescale to refresh the perturbation may yield gaps between the strips at greater heights. The technique may benefit from use of height-dependent perturbation refresh timescale in cases with significant mean velocity shear.

Muñoz-Esparza *et al.* [65] compared the above-described SCPM to the synthetic method of Xie and Castro [64] in the simulation of neutral flow over flat terrain, for which the SCPM performed slightly better relative to a stand-alone periodic LES run with the same geostrophic and surface forcing, taken to be the true solution. Subsequently, the SCPM was applied during a diurnal cycle, where it produced turbulence statistics that agreed much better with observed turbulence than those from a corresponding unperturbed simulation [72]. Subsequently, new scaling relationships for the perturbation magnitudes during stable and convective conditions were developed to extend the efficacy of the approach to more general ABL flow conditions [73].

More recent work [74] examining the application of perturbations to the velocity components, as in Mirocha *et al.* [70], but applied stochastically over similar patches to those of Muñoz-Esparza *et al.* [65], has shown further improvement in the rate of equilibration under certain flow conditions. Optimization of the SCPM to specific forcing conditions, including height and stability dependence of the perturbation magnitude and timescale, variations of cell geometries, and combinations of temperature and velocity components, shows significant promise as an economical and easy-to-implement turbulence inflow generation approach applicable to the full range of meteorological and environmental conditions impacting wind plants.

4.3.2.2 The terra incognita

An additional challenge to that of efficient turbulence generation from mesoscale inflow into an LES domain is the generation of suitable mesoscale flow information to drive wind plant simulations. While any NWP or mesoscale simulation can provide inflow for a nested LES domain, the fidelity requirements of wind energy applications place unique demands on inflow characteristics not traditionally emphasized in atmospheric model assessment. For this reason, dynamic downscaling methods, such as described earlier (Section 4.3.1), are often employed in an effort to increase the fidelity of the near-surface flow field. However, a potential drawback common to all dynamic downscaling approaches is the representation of a portion of the computational domain with mesh resolutions that are finer than those for which their physical process parameterizations, often inherited from NWP models, were designed ($\Delta h \gtrsim 10\,\text{km}$), but not fine enough to use an LES SFS model ($\Delta h \lesssim 100\,\text{m}$). The range of scales between traditional NWP and LES, often called the "terra incognita" or "gray zone," presents significant challenges, due to both the structural forms of most multi-resolution atmospheric models, as well as the absence of appropriate SFS closure techniques for turbulence and other physical processes for use at scales between traditional NWP and LES [75].

The construction of both NWP models' dynamical cores and the physical process parameterizations that provide thermodynamic forcing is based on several assumptions that are valid when applied at the traditional coarse NWP grid spacings used to capture the synoptic-scale weather features spanning continental to planetary scales. These coarse grid spacings enable simplifications based upon the assumption of horizontal homogeneity and the appropriateness of ensemble mean relationships to parameterize the impacts of physical processes occurring within the large areas spanned by large grid cells. These assumptions are not well justified at higher resolutions that use smaller grid cells and resolve greater horizontal variability. Further, the structural form of NWP models to parameterize physical processes, such as turbulence transport, in the vertical direction only, is predicated upon the assumption that grid spacing is sufficiently coarse in the horizontal directions to justify the exclusion of explicit horizontal communication. The resulting vertical distributions of model fields prescribed by the one-dimensional parameterizations comprising the model physics package thereby influence the three-dimensional flow field primarily through the advections and pressure gradient accelerations contained within the model's dynamical core. A small background horizontal diffusion, parameterized independently from the vertical diffusion, is often included to stabilize the numerical solution method.

Errors associated with the terra incognita are encountered when utilizing traditional NWP parameterizations at scales too fine for the above-described assumptions upon which those schemes are based to hold. In such cases, the model physics and structural forms each contribute to an incorrect representation of horizontal transport processes and feedback mechanisms. The incorrect representation of horizontal transport and variability in high-resolution mesoscale simulation can be revealed by conducting LES with the same forcing as the mesoscale simulation, and computing vertical profiles from the LES, averaged horizontally over the same footprint as the mesoscale grid cell. The LES results will reveal increasing variability of the mean profile characteristics as the footprint size is reduced, while a mesoscale parameterization, in the absence of explicit knowledge of the footprint size, will mix each vertical column of grid cells toward the same ensemble mean value, appropriate for a large NWP-scale footprint.

A second impact of incorrect horizontal transport and feedback mechanisms within the terra incognita is the generation spurious convective roll structures during convective conditions with a moderate to strong mean flow (see [76]). The anomalous roll characteristics can be similarly confirmed via comparison with LES using the same forcing as the mesoscale simulation.

A third potential error inherent in terra-incognita simulation is the "double counting" of turbulence transport that occurs when the grid spacing becomes fine enough for some of the transport to be partially resolved, resulting in the resolved transport being superimposed upon the parameterized transport, the latter of which is assumed to account for the entire spectrum of turbulent motions.

Given the nature of NWP model physics parameterizations and structures, further increases of resolution within the terra-incognita may exacerbate issues associated with modeling within the terra-incognita by pushing the modeling framework further from its design envelope. One approach to ameliorate potential artifacts resulting from modeling within the terra-incognita is to use a large mesh refinement ratio for the dynamic downscaling, effectively leapfrogging the terra incognita scale range. However, other deleterious issues can be encountered using such an approach, including the discontinuous representation of atmospheric and terrain features, and boundaries between different surfaces, for example, that may introduce other artifacts with unknown impacts.

Despite these potential errors arising from using resolutions within the terra incognita, other aspects of flow simulations important to wind plant simulation may be substantially improved with increased resolution due to enhanced representation of terrain features, land-cover characteristics, and smaller scale mesoscale meteorological processes. Teasing apart the benefits of increased resolution relative to the drawbacks of violating the assumptions of the NWP framework within the terra incognita is difficult and is also likely to be case specific. While little is presently known about the quantitative impacts of resolving terra-incognita scales within the wind energy context, recent examinations of mesoscale-to-LES downscaling on quantities of interest to wind energy by Rai *et al.* [77] have corroborated the findings of Ching *et al.* [76] on the deleterious impact of fine mesh spacing on the representation of convective roll structures, demonstrating that using $\Delta h < H$ produces spurious artifacts.

Given the numerous potential benefits of increased resolution relative to poorly understood and difficult to quantify drawbacks, simulation within the terra incognita is increasing in popularity. To address the issues of turbulence parameterization within the terra incognita, fundamentally new scale aware and three-dimensional PBL schemes will be required, with several methods under development. Approaches have ranged from the use empirical relationships to modify the parameterized transport [78,79] or the turbulence length scale [80] to account for the proportion of transport that is increasingly resolved under mesh refinement, to reformulation of the SFS fluxes to either smoothly blend with a three-dimensional turbulence closure as the mesh spacing is decreased [81,82] or to be fully three-dimensional by construction [83]. Further development of such parameterizations is likely to lead to significant improvements in multiscale simulation for many ABL applications, including wind energy.

4.3.2.3 Terrain-following coordinates

Many mesoscale models, such as WRF, utilize a terrain-following vertical coordinate which conforms to the terrain at the surface while flattening with height above. Use of such a coordinate simplifies the treatment of boundary conditions at the surface, where the terrain and the bottom coordinate coincide. Use of a terrain-following coordinate requires a coordinate transformation within the numerical solution procedure, introducing the computation of terms arising from the Jacobian transformation matrix. These additional terms can introduce errors due to numerical approximations in their computation, such as in the calculation of horizontal derivatives on a non-horizontal terrain following grid. Such numerical errors often become more problematic with increasing resolution, as terrain slopes become larger as more fine-scale features are resolved. Alternatives to terrain-following coordinates are an important consideration in many wind energy applications for which enhanced resolution of fine-scale terrain features is desired.

Various approaches exist to ameliorate numerical errors associated with terrain-following vertical coordinates. The simplest approach is to smooth the terrain to reduce the steepest slopes. An obvious drawback of such an approach is the removal of features of interest, which likewise compromises the solution accuracy. Some numerical formulations (such as those used by the WRF model) contain stabilization parameters (i.e., damping) that can be increased over steep terrain; however, their efficacy is limited, while their impact on solution accuracy at fine scales is not well understood.

Yet another option is to couple the mesoscale model to a separate fine-scale solver with a gridding methodology better suited to steep terrain. However, as slope impacts on numerical accuracy become considerable even over the modest slopes encountered in routine mesoscale simulations, addressing only the fine-scale portion of a simulation does nothing to mitigate the impacts of numerical errors on the inflow.

The removal of numerical errors due to terrain following vertical coordinates without compromising other aspects of simulation fidelity requires modification of the numerical solution method itself. One such approach is to use higher order numerical stencils in the calculation of horizontal derivatives. Implementation of modified or alternative stencils into the COSMO model, for example, has demonstrated significant reductions in numerical errors up to slopes of 60 degrees in idealized setups [84].

While alternative finite-difference stencils have shown improved fidelity for moderately steep slopes, very steeper slopes will require an alternate gridding method. One set of approaches that can be implemented directly into existing mesoscale models is cut-cell or immersed boundary methods (IBMs). These methods replace the need to map the computational grid around objects of interest with the ability to instead use a nonconforming (i.e., Cartesian or uniform) grid which intersects such features. The impact of such "immersed" features on the solution is incorporated either as additional cell faces or as forces in the governing equations that approximate a desired boundary condition.

While IBMs have enjoyed widespread success in many engineering applications [85–88], atmospheric applications [89–92] present unique challenges due to the relative coarseness of the mesh resolutions used in atmospheric relative to engineering applications. Since IBMs apply fictitious forces that approximate effects of the boundary, rather than exact boundary conditions, errors associated with the method, which increase with decreasing resolution, can become considerable at coarse resolutions. Further examination is required to determine the tradeoff between reduced numerical errors associated with the elimination of grid distortion versus errors contributed by the inexact boundary treatment of IBM methods.

While improved numerical stencils may adequately address numerical errors over moderate slopes while avoiding errors contributed by IBMs at coarse resolutions, a benefit of IBMs is the ability to resolve arbitrary slopes, such as canyons, escarpments, and urban structures. The IBM implemented into the WRF model has been applied to capture the physics of slope-induced gravity flows over steep terrain ([93]; Figure 4.7) and the dispersion of contaminants through an urban area with resolved buildings [94,95]. One drawback of the IBM within the WRF model, which uses a

Figure 4.7 Simulated downslope flow (blue shading) following movement of the shadow front (dark gray shading) during the evening over Granite Mountain, Utah. Simulations were conducted using the WRF model with the immersed boundary method to represent the steep terrain

pressure-based vertical coordinate, is accounting for the time evolution of the height of vertical coordinate surfaces, due to the impact of thermodynamic forcing on the pressure field. The computational overhead required to recompute projections that depend on the vertical distance between model grid points and the immersed boundary at each model time step are considerable; however, the algorithm could be streamlined to recompute the heights based on an acceptable error threshold.

While additional research is required to determine the optimal method of mitigating numerical errors for a given application, based on the resolved slopes, model resolution, and other considerations, improved numerical stencils and IBMs show considerable promise to improve the fidelity of high-resolution flow simulation for wind energy applications.

4.3.3 Online versus offline coupled simulations

The issues associated with downscaling from mesoscale to LES described in this section apply whether the downscaling is done within a single atmosphere simulation framework that supports mesh refinement from mesoscale to LES internally or whether separate solvers are used for the mesoscale and the microscale simulations. Given the capabilities of many atmospheric simulation frameworks to refine the mesh from mesoscale to LES scales internally, thereby capturing all relevant scales of motion and forcing for the microscale domain, one could envision implementing the wind plant modules (e.g., turbine and load parameterizations) directly into the atmospheric solver. Such an approach would eliminate the complexity of using a separate microscale solver, including the transfer of various boundary condition information. However, a key drawback of existing atmospheric codes for wind plant simulation is their use of mesh refinement capabilities that are not well suited for high-resolution wind plant simulation needs. As described in Section 4.3.2, most of the popular publicly available multiscale atmospheric simulation codes utilize either a block rectangular grid nesting approach or smoothly refining polygonal tessellations (MPAS, ESM3). These mesh-refinement strategies are suitable for refining local geographical areas within a large-scale atmospheric simulation but may not be optimal for refining multiple portions of microscale subdomains as required to resolve flow features resulting from interactions with multiple turbines or turbine components, whose orientations may be changing in time, thereby requiring frequent updating of the mesh refinement region. While actuator disk and line wind turbine parameterizations have been implemented into the WRF atmospheric code [96–100], those implementations are intended for the investigation of aggregated turbine impacts, such as wake effects. More flexible meshing strategies capable of locally refining the mesh around turbine components or flow structures of interest are required for higher fidelity turbine-resolving applications.

Since many microscale code bases possess flexible and efficient gridding algorithms better suited for high-fidelity wind plant simulation needs, an argument could be made for expanding the domains of such code bases to include mesoscale atmospheric simulation as well and perform any downscaling to LES within. However, at present no such microscale code bases with mesoscale simulation capabilities exist. One factor impeding the construction of a microscale code base capable of also conducting mesoscale simulations is the broad suite of atmospheric and surface physics parameterizations required to properly simulate the mesoscale flow. A second consideration is the specification of initial and boundary condition information from large-scale data sources, such as NWP models and atmospheric analysis and reanalysis products. Many of the widely used atmospheric simulation models contain a suite of preprocessing procedures to interface seamlessly with many such datasets, including procedures to reduce spurious high-frequency transients associated with imbalanced initialization fields (e.g., [101]). Further, some models also contain methods to assimilate data, either from measurements or from the forecast or analysis product used to supply the LBCs, into the simulation interior, to relax specific parameter values at specified locations during the simulation (e.g., [102]).

Implementation of the abovementioned capabilities into a microscale wind plant simulation environment with a flexible yet complex gridding capability poses such a significant undertaking that it is unlikely to materialize in the near future. As such, offline coupling using separate solvers for the atmospheric and wind plant components is likely to be the only option for multiscale wind plant simulation requiring high-fidelity representations of turbine–airflow interactions for the foreseeable future. Given the requirement of using separate solvers for the mesoscale and microscale portions of the simulation, several choices must be made regarding how to optimally couple the two simulation frameworks, depending upon the problem of interest. One such choice is at which spatial scale (mesh spacing and domain size) to switch from the atmospheric solver to the wind plant simulator. While the more efficient gridding and computational scaling achievable within the new generation of microscale solvers suggest optimization of computational efficiency when coupling at the coarsest possible grid spacings, the absence of physical process parameterizations that could potentially influence turbulence generation and other important flow characteristics within the microscale code base suggests that any such efficiency gains may be offset by compromised solution accuracy. With respect to turbulence generation, mesoscale atmospheric and surface physical processes act primarily on the largest scales of the turbulence spectrum, as those scales are more sensitive to the particulars of extant forcing conditions than the inertial scales, which are primarily generated from the downscale energy cascade. Therefore, downscaling to LES within the atmospheric solver before exchanging flow information with the microscale domain is likely to yield more physically accurate and realistic turbulent flows for the wind plant environment. Further downscaling within the wind plant simulator can be pursued to capture finer-scale turbine–airflow interactions (with the caveats described in Section 4.3.2.1).

In addition to coupling the solutions at the microscale lateral inflow boundaries, forcing terms representing the impacts of physical processes not parameterized within the microscale domain, can be obtained from the mesoscale simulation and applied internally, as described in Section 4.3.1, but in this case without any restrictions imposed by the use of periodic LBCs; the atmospheric forcing terms need not be horizontally uniform or periodic and could instead incorporate natural mesoscale heterogeneity within the footprint of the microscale simulation domain.

While incorporation of internal mesoscale forcing within the microscale solver could potentially lead to improved representation of ABL characteristic, including the evolving turbulence spectrum, one caveat associated with the incorporation of forcing related to features with fine enough scales to be well resolved within the microscale solver is a likelihood of phase errors between the forcing and evolving features within the microscale domain. An example would be the application of time-varying surface fluxes obtained from a land surface model running within a mesoscale simulation. As energy fluxes reflect the local, instantaneous values of surface and flow field parameters, including clouds, soil moisture, wind speed and surface-to-air temperature and moisture differences, using flux values resolved within the mesoscale simulation within a separate microscale solver, within which features of the resolved-scale flow (and surface parameters, if applicable), will inevitably evolve to a different distribution of instantaneous features and hence inconsistencies between the applied forcing and the resolved-scale flow field. The impact of such phase errors on relevant quantities of interest is not known and merits investigation.

An additional issue requiring further examination for offline simulation is the impact of different formulations of the governing equations on the coupling of flow information at the domain boundaries, particularly the lateral and top boundaries. A primary difference between many mesoscale and microscale simulation codes is their treatment of compressibility effects, with mesoscale codes requiring density variability due to the approximately exponential decrease of density with height in the atmosphere, resulting in large vertical variations of density over depths typically simulated in such models. Many microscale solvers, which have historically been applied to much shallower depths, are formulated to enforce incompressibility, resulting in uniform density throughout the computational domain, including with height. This mismatch in the treatment of density (and other fields) can result in spurious acceleration when mesoscale atmospheric data are used as boundary conditions for incompressible solvers over domains of depths greater than a few kilometers. Currently, little guidance exists on methods to modify the mesoscale inflow prior to use as inflow to an incompressible offline simulation model, or to modify the offline solver's incompressible equation set to accommodate compressibility effects. Issues related to offline coupling of solvers with different numerical formulations are discussed in greater detail in Chapter 6 of this volume.

4.3.3.1 Top and bottom boundary conditions

As described in the preceding section, microscale simulation domains typically span much shallower depths than those required for mesoscale atmospheric simulation. Microscale wind plant simulation domains typically extend only to heights necessary to capture ABL dynamics directly impacting turbine performance and ABL energetics, $L_z \cong 1.5H_{\max}$, where $H_{\max}$ is the maximum value of H during a simulation. The additional depth beyond $H_{\max}$ provides for a layer above the ABL over which to apply physically appropriate boundary conditions. One commonly used upper boundary condition used in ABL simulation is a capping inversion, an increase of the mean potential temperature profile $\bar{\theta}$ above H, with values of $\partial\bar{\theta}/\partial z \cong 0.01\,\mathrm{K\,m^{-1}}$ typical. The adverse buoyancy force imposed by the increase of

$\bar{\theta}$ limits the vertical excursions of turbulence motions in a natural manner. In addition to the capping inversion, numerical damping is likewise commonly imposed to limit the unphysical amplification of wave activity arising from interactions with the model top. One popular approach is Rayleigh damping [103], which relaxes flow components toward specified values as a decreasing function of distance from the model top. Boundary conditions at the model top typically include free slip for the horizontal velocities and vanishing of vertical velocity and fluxes of scalars and momentum.

In simple meteorological situations, the impacts of other potentially relevant forcing terms at the microscale model top, including cloud, radiative, or synoptic-scale weather processes, can be ignored with little consequence. For more meteorologically complex situations, upper boundary conditions could, in principle, be expanded to represent relevant additional forcing factors; however, such approaches have not yet been developed sufficiently to have appeared in the literature. Again, the internal application of tendencies of such processes, obtained from a mesoscale simulation, as described in Section 4.4.1, can be explored with a view toward improving the upper boundary condition in offline microscale simulations.

While the basic implementation of the bottom boundary condition common to many atmospheric LES codes is described in Section 4.3.1.1, for offline simulations, values of some of the required surface parameters within the microscale domain (temperature, moisture, or fluxes thereof) can be obtained directly from a full-physics atmospheric simulation over the area of interest. The use of surface data obtained in this manner could potentially provide more accurate parameter values within the microscale simulation than use of a simpler parameterization internal to the microscale solver. However, the potential for phase errors, as discussed above, must be considered. Additional discussion of relevant issues related to the typical surface boundary condition implementation into LES for ABL applications is provided in Section 4.5.3.

4.4 Additional challenges facing high-fidelity multiscale simulation

4.4.1 LES SFS models

While several books and review papers describe numerous turbulence SFS modeling approaches in great detail and under a wide spectrum of applications (e.g., [104–108]), this section will provide a brief overview of issues and approaches from the atmospheric perspective, targeting the generation of high-fidelity ABL flow and turbulence information for wind plant applications.

As described in Section 4.1.1, DNSs are simply too computationally expensive to be useful as a general ABL flow simulation framework, leaving the LES technique as an intermediate fidelity approach between DNS and less time-accurate but computationally cheaper RANS methods. However, while LES is growing in use and acceptance, a handful of theoretical and practical limitations are highlighted herein both to constrain expectations, especially in challenging settings, and also inspire improvements in current capabilities.

The major theoretical shortcoming of the LES technique is the absence of an accepted universal theory of turbulence to definitively address how to properly model the SFS stresses. In the absence of a formal theoretical grounding, SFS models have instead been developed from a series of assumptions and empirical relationships. However, the lack of a firm theoretical grounding for SFS closure is generally thought to be of only marginal consequence in most applications, due to the explicit capture of the most energetically important scales of motion. With the scales of motion responsible for most of the transport and energy production well resolved, the SFS model has only to account for the effects of smaller scales, which, having undergone some downscale cascade, are thought to be sufficiently isotropic and homogeneous to submit to a generalized parameterization which accounts primarily for the proper mean energy dissipation. Given the above, the impact of the SFS model in LES is expected to diminish with increasing resolution, allowing an increasingly broad range of the inertial subrange to be explicitly resolved, further reducing reliance on the SFS model and any associated errors.

While the proliferation of affordable HPC infrastructure is providing the ability to perform LES at ever-increasing resolutions, the SFS model will always play a significant role where the scales of motion influencing turbulence transport and energy production are forced to approach the grid scale of the simulation. This occurs in the presence of stable thermal stratification, where adverse buoyancy suppresses turbulence motions in the vertical direction, and also in regions where flow impinges upon a surface or an obstacle within the flow, such as a terrain feature or a wind turbine. In the ABL, the surface boundary conditions play a key role in defining aspects of the flow and turbulence field which impact wind turbine performance, with many studies showing that errors near the surface can propagate upward and impact the entire ABL flow [109110]. Given the unique demands of wind plant LES, the role of the SFS model is likely to be magnified relative to other applications. Herein, we will briefly describe several of the more common approaches to LES SFS modeling used in ABL flow applications of relevance to wind energy.

A key requirement of any LES SFS model is to dissipate the proper amount of TKE from the resolved-scale turbulence within the flow, since the scales of motion responsible for viscous dissipation are not resolved. The oldest and most popular approaches to parameterizing the SFS stresses involve the use of an eddy-viscosity coefficient, which, by analogy with molecular diffusion, functions to transfer energy from resolved to SFS scales, in proportion to the rate of strain of the resolved-scale flow. This relationship can be expressed as

$$\tau_{ij} - \frac{1}{3}\delta_{ij}\tau_{kk} = -2\vartheta_t\widetilde{S}_{ij}, \tag{4.5}$$

where τ_{ij} is the desired SFS stress (strictly, the deviatoric portion thereof, which remains after the hydrostatic stresses, those defining the pressure field, are subtracted off), $\widetilde{S}_{ij} = 1/2\left(\left(\partial\widetilde{u}_i/\partial x_j\right) + \left(\partial\widetilde{u}_j/\partial x_i\right)\right)$ is the rate of strain tensor, where $x_j, j = 1,2,3$ denote the Cartesian coordinate directions, the tilde represents the resolved portion of the flow field (as opposed to the SFS stresses, the effects of which are being parameterized) and ϑ_t is the aforementioned eddy viscosity coefficient.

The most basic formulation for ϑ_t is to make it proportional to the product of a length scale and a velocity scale [6], with the velocity scale in turn formulated as the product of the same length scale and the magnitude of the strain rate tensor, which defines a characteristic scale of velocity changes occurring over that length scale. These assumptions yield the classic constant-coefficient Smagorinsky model [111],

$$\vartheta_t = (c_s \Delta)^2 |\tilde{S}|. \tag{4.6}$$

where c_S is the model constant, taken to be time-and space invariant, and Δ is a length scale, related to the size of the computational mesh, typically as $\Delta = (\Delta x \Delta y \Delta z)^{1/3}$.

While the constant-coefficient Smagorinsky formulation is simple, easy to implement, and remains widely used due to its simplicity and long history, several aspects of its formulation limit its accuracy. One issue is that no universal value of the parameter c_S exists for general flow conditions and applications. While $c_S = 0.16$ was determined by Lilly [6] to provide the correct rate of energy dissipation under the assumptions of isotropy and equilibrium within the energy cascade, such assumptions are violated in the nonequilibrium, heterogeneous, and anisotropic conditions characterizing general ABL flow. To account for the mismatch between the equilibrium c_S value and nonequilibrium flow conditions, improved results have been demonstrated in various configurations using a range of values $0.1 \lesssim c_s \lesssim 0.25$; however, no single value is superior over general flow conditions and LES configurations.

Another drawback of the constant-coefficient Smagorinsky formulation is excessive dissipation in the near-wall region for wall-bounded flows when using a constant c_s value (e.g., [112]). Various damping functions have been developed to reduce the excessive dissipation near surfaces (e.g., [113–115]); however, as with the use of different c_S values, a general wall-damping approach applicable to a broad range of configurations has likewise not been demonstrated.

To address the inability of the Smagorinsky closure to account for the nonequilibrium flow conditions often encountered in the ABL, ϑ_t can instead be formulated in terms of SGS TKE, e, as

$$\vartheta_t = c_{e,\Delta} \Delta \sqrt{e} \tag{4.7}$$

Deardorff [116]. The well-known 1.5-order TKE closure, for example, solves one additional time-dependent equation for e, following a 1.5-order closure (using the hierarchy of [117]; see e.g., [3]). While the incorporation of the additional transport equation adds complexity and introduces more empirical coefficients, the 1.5-order TKE model represents a significant improvement in many ABL flow applications and has therefore enjoyed widespread use.

Despite the improvements afforded by the 1.5-order TKE closure, it too suffers from fundamental limitations inherent to all linear eddy-viscosity models. First, all such closures are absolutely dissipative, hence do not account for backscatter, the upscale transfer of energy that effectively reduces the mean dissipation rate. Linear eddy-viscosity models also exhibit low correlations between modeled and actual SFS stresses in a priori tests, those comparing SFS stresses with those obtained from data or DNS results (e.g. [118,119]). Third, linear eddy-viscosity models stipulate perfect alignment between the eigenvectors of the SFS stress and fluid strain tensors, while significant misalignment has been observed (e.g., [120,121]). Further, eddy viscosity models are isotropic by definition and therefore unable to account for anisotropies such that occur near stable layers and solid surfaces, for example (e.g. [109]).

While LES using the above-described Smagorinsky and 1.5-order TKE SFS models can provide good performance in a-posteriori tests (comparing LES results with observations) in idealized ABL simulations, more challenging applications featuring significant heterogeneity and anisotropies, such as encountered under more general meteorological and environmental conditions, expose the limitations of such approaches.

For more complicated ABL flow conditions, the limitations of simple linear eddy viscosity SFS models can be ameliorated or avoided entirely by the use of more sophisticated SFS modeling approaches. One popular method that addresses one of the fundamental limitations of the constant-coefficient Smagorinsky model is to compute the model coefficient (c_S) dynamically, based on local, instantaneous flow conditions. Such dynamic formulations allow the local, instantaneous dissipation to adjust to the local, instantaneous resolved stress field, and can even approximate backscatter via locally negative coefficient values. Use of variable coefficient values leads to improved agreement between local production and dissipation, providing superior performance in many respects, including enhanced resolution of smaller scale motions (within the range of resolvable scales imposed by the grid and model numerical scheme), more robust turbulence representation during stable conditions, and a reduction of the characteristic excessive dissipation occurring near solid boundaries [122].

The improved performance of dynamic models comes at the expense of more complicated implementations and increased computational overhead, much of which results from the use of a filtering procedure to project the model solution onto a coarsened mesh, as required to obtain estimates of the resolved model stress field.

The first and simplest dynamic SFS model, the dynamic Smagorinsky model (DSM; [122]), uses one coarsened mesh level, defined by $\alpha\Delta$ (where $\alpha = 2$), to compute the resolved-scale stress field at $\alpha\Delta$. This information is utilized within the Smagorinsky closure framework to obtain a value of c_S, assuming its value does not change across scales. This scale-invariant version of the DSM has been shown to work well provided that the filter scale is much smaller than the integral scale of the resolved turbulence field [123]; however, near solid surfaces and during stable conditions, the DSM can underpredict the SFS stress field [112,124]

These shortcomings of the scale-invariant DSM can be ameliorated by introducing scale dependence [112], achieved by incorporating a second coarser filtering level $\alpha^2\Delta$, which, combined with the $\alpha\Delta$ projection, provides an estimate of how the coefficient varies across the two filter scales, thereby enabling projection of an appropriate coefficient value for the SFS scale.

One potential issue with dynamic SFS models is that very small or negative local coefficient values can result in numerical instabilities, requiring stabilization. One common method to stabilize dynamic procedures is to average the coefficient values over homogeneous flow directions (planes or lines with similar flow, forcing and surface conditions). A more general approach for heterogeneous conditions is to average the coefficients along Lagrangian pathlines [125,126]. Local filtering of the coefficient values using a different dynamic SFS modeling approach has also been applied by Chow *et al.* (2005). Addition stabilization of dynamic models can be provided by preventing the coefficient values from falling below a specified value (e.g., "clipping").

While the above-described dynamic SFS models have been shown to improve important aspects of ABL simulation relative to constant-coefficient models, they are linear eddy viscosity models and, as such, impose an incorrect alignment between eigenvectors of the stress- and strain-rate tensors. Alternative SFS model formulations either ameliorate or avoid this deficiency. One such approach is to formulate an SFS model that is nonlinear at the tensor level. The Nonlinear Backscatter and Anisotropy (NBA) model of Kosović [127], for example, augments a linear eddy viscosity model with addition terms incorporating interactions between strain and rotation tensors. The model explicitly accounts for backscatter, improves eigenvector (mis)alignment, and also captures normal stress effects that are likewise not correctly treated by linear eddy viscosity approaches.

The NBA model has shown improved performance relative to constant-coefficient linear eddy-viscosity models in several ABL setups [21,127,128]; however, dynamic SFS models have been shown to perform somewhat better than the NBA approach in some low-order flow statistics, such as improved agreement with the expected logarithmic velocity distribution near the surface, and an ability to resolve a greater range of smaller scale motions [129], as well as in turbulence transition and equilibration under mesh refinement [70,130]. One benefit of the NBA approach relative to dynamic SFS models is a much simpler implementation and reduced execution time, due to the absence of any filtering requirements. Further, the superior performance obtained using the dynamic models in the previously cited studies was dependent upon the incorporation and tuning of an associated near-wall stress model, as discussed later, none of which is required for the NBA model.

While the NBA model improves several aspects of model performance, including addressing the incorrect eigenvector alignment inherent in linear eddy viscosity approaches, alternative formulations have been proposed that eliminate the use of an eddy viscosity entirely. One such approach is the scale similarity model which, as the name suggests, assumes a similarity in form between the SFS velocities and the smallest scale motions explicitly resolved within the flow, then uses those SFS velocities to compute the model stresses directly. The simplest scale similarity model uses the actual resolved velocity field as a proxy for the SFS scales [118], while other methods use higher order approximations of the SFS velocities, obtained via successive inversion of a Taylor series expansion (e.g., [131,132]) or an iterative deconvolution process (e.g., Chow *et al.*, 2005; [133,134]).

Relative to eddy-viscosity formulations, scale-similarity models reduce dissipation, capture backscatter more physically than via negative eddy viscosity coefficient values, do not posit incorrect alignment between the eigenvectors of the stress and strain, and properly capture normal stress effects. However, as the stresses from only a portion of the entire range of scales filtered from the flow (up to the grid Nyquist frequency, f_N) can be represented in this manner (e.g., [135]), scale similarity models have been shown to provide too little dissipation overall. To provide the required dissipation, scale similarity models are often combined with an eddy viscosity approach into a mixed model, with the eddy viscosity model providing the additional dissipation from scales beneath the mesh resolution, as required to ensure the correct overall TKE dissipation. Bardina *et al.* [118] combined the scale-similarity model with the standard Smagorinsky closure, while Zäng *et al.* (1993) combined it with a scale-invariant DSM, each showing improved SFS model dynamics while providing correct overall dissipation rates. While some deleterious impacts of the eddy viscosity concept are reintroduced in the mixed model framework, the effects are thought to be small, relative to the improvements afforded by the scale similarity component.

An insightful interpretation of and justification for the mixed-model approach is provided by Carati *et al.* [136], who differentiate between two distinct contributions to the SFS stress field occurring in LES using finite volume and finite difference solvers. The first class of stresses, arising from scales beneath f_N, and hence too small to be resolved on the computational mesh, is designated as traditional SGS stresses. A second class of SFS stresses arises from scales of motion large enough to be captured on the model grid but that have been attenuated due to errors in the discrete numerical differencing operators used to solve the flow equations, which attenuate energy at scales $f_D > f_N$ The scales of motion occurring between f_D and f_N are in principle resolvable upon the mesh and can be approximately reconstructed using the higher order methods described earlier. Such explicit filtering and reconstruction approaches improve SFS model dynamics in two ways, first by reconstructing the stresses arising from missing scales of that would otherwise be resolved upon the grid (using perfect numerical methods) and also by properties of the reconstruction process that reduce the deleterious effects of numerical errors overall (e.g. [137,138]).

Chow *et al.* (2005) used explicit filtering and reconstruction in combination with the dynamic eddy-viscosity SGS model of Wong and Lilly [139] to produce the dynamic reconstruction model (DRM). The DRM has been shown to provide superior performance to traditional linear eddy viscosity SFS models in several ABL flow studies, from canonical neutral flow over flat terrain (Chow *et al.*, 2005; [129,130]) to complex terrain and stable conditions [38,140,141]. While the DRM using the lowest-order of velocity reconstruction (level-0) has shown improved results relative to the 1.5 TKE model in simulations over complex terrain, higher orders of reconstruction triggered instabilities when using the Wong and Lilly SGS model [142]. These instabilities were remedied by replacing the Wong and Lilly SGS stress model with the 1.5-order TKE SGS model, the combination of which showed improved results relative to use of the standard 1.5-order TKE SGS model alone. The combination of explicit filtering and reconstruction with other SGS models presents a potentially promising approach; however care must be taken when blending models to ensure that the correct overall dissipation and other criteria are maintained

Kirkil *et al.* [129], Mirocha *et al.* [130], and Mirocha *et al.* [70] compared the DRM model of Chow *et al.* (2005), the LASD model of Bou-Zeid *et al.* [125], the

NBA model of Kosović [127] as well as the standard linear eddy viscosity Smagorinsky and 1.5-order TKE models in a few idealized, canonical boundary layer flow simulations using the WRF model. These comparisons showed a general progression of improvements using the least-to-most complicated SFS models, in terms of near-surface velocity profiles, turbulence statistics, and equilibration of mean flow and turbulence characteristics under mesh refinement, both from coarse to fine LES, and from mesoscale to LES. While limited in scope, these studies, coupled with other studies in the literature, suggest that more sophisticated SGS closures may confer significant advantages for wind plant simulations in the ABL and merit further investigation in that context.

One additional consideration when using dynamic SFS models implemented into finite difference or finite volume solvers is the potential need for enhancement of the SFS stress field near solid boundaries. Such approaches have been shown to predict too little SFS stress near the surface, relative to implementations into pseudo-spectral solvers, due to the enhanced attenuation of resolved scale motions arising from the use of less precise numerical methods. This issue of underpredicted near-surface stresses while using the DRM implemented into the ARPS atmospheric model [49] was addressed in Chow *et al.* (2005) by incorporating the near-wall stress model of Brown *et al.* (1990) to augment the near-surface stress field. This model uses concepts from the canopy model literature to formulate an augmentation to the near-surface SGS stress field as

$$\tau_{i3} = -\int_0^{H_{nw}} C_{nw} a(z) \left| \bar{\tilde{u}} \right| \bar{\tilde{u}}_i dz. \tag{4.8}$$

where $i = 1, 2$, C_{nw} is a scaling factor, H_{nw} is the height above the surface to which the near-wall stress is applied, and $a(z)$ is the shape function, which is a decreasing function of height. This approach was shown to recover good agreement with the expected logarithmic velocity distribution near the surface when combined with the DRM in the ARPS model (Chow *et al.*, 2005). The Brown *et al.* (1990) approach was also successfully combined with the DRM and LASD models implemented into the WRF model, which utilizes similar model discretization and numerical methods as ARPS. Slightly different shape functions $a(z)$ and coefficient values C_{nw} were utilized in the WRF model, and between the DRM and LASD stress models (see [129]), due to slight differences in the numerical solver and SFS model formulations.

A new modulated gradient dynamic SFS model was recently developed that exhibits good agreement with the expected logarithmic profile near the surface in a finite difference solver without the requirement of a separate near wall stress model; however, the model has thus far only been validated in an idealized, neutral, flat plate setup [143]. Further validation of the approach in more complicated setups is likely to afford many insights into how to better parameterize SFS stresses near the surface in finite difference and finite volume solvers.

While the above-described approaches are among the most commonly encountered in the literature on ABL simulation, several other potential approaches exist, including methods based on solving a set of transport equations for the SFS stresses (e.g., [144]), methods based on probability density functions (e.g., [145]), and implicit LES, for which no explicit SFS model is employed at all, with the filter defined by the grid and numerical methods [146]. These approaches all avoid the use of an eddy viscosity and associated errors entirely, however, can be quite complicated to implement and have not been well validated in ABL simulations.

To address above-discussed difficulties encountered using traditional LES SFS models near solid boundaries, hybrid RANS/LES techniques (e.g., [147]), including the familiar detached eddy simulations [148], have also been developed. These approaches utilize a RANS formulation near the wall, or in other locations within the computation domains amenable to a reduced order representation, while using LES in areas requiring higher fidelity. Such approaches, while popular in aeronautical applications, have likewise not yet been widely validated in ABL simulation.

The unique demands of wind plant LES, including flow interactions with complicated, heterogeneous, and time-varying surface and meteorological forcings, in addition to turbine components, each of which possess unique physics occurring across a wide range of scales, require further validation of SFS models within this setting. An important consideration is to examine the reproduction of parameters of specific interest to wind plant engineering, such as those contributing to energy extraction and fatigue loading. Traditional ABL validation studies have focused primarily on lower order flow statistics, often presented as averages (see e.g., Section 4.2.6). Wind plant optimization additionally requires the accurate capture, and characterization of the range of instantaneous, three-dimensional structures and extreme events, a tall order given that a dearth of appropriate validation datasets. While remote-sensing technologies capable of measuring ever finer and wider scales of motion are undergoing continuous improvement, another possible path forward is to develop methods to use data from the operating turbines themselves as sensors. Stress loading and other machine performance data represent responses to instantaneous structures occurring within the ABL flow. As such, implementation of turbine power and load modules within LES for comparison with operating turbines may provide a novel validation pathway for wind plant LES and SFS parameterization.

4.4.2 Flow transition at coarse-to-fine LES refinement

The boundary between mesoscale and LES is one interface within a multiscale simulation where the turbulence field requires augmentation, such as through the application of turbulence information or perturbations, as described in Section 4.3.2.1. However, under subsequent mesh refinement of one LES to finer-scale LES, another turbulence transition is encountered, resulting from a lag between the reduction of the grid spacing and the development of finer scale turbulence motions resolvable on the finer mesh. As described in Section 4.3.2.1, the length scale of this transition can be extensive. Figure 4.8 shows the turbulence transition an equilibration process under mesh refinement from a coarse LES (left panels) to a finer LES nested within (middle panels), with the expected solution for the nested domains obtained from stand-alone LES domains with the same configurations and forcing as the nested domains, but using periodic LBCs. Results using four different SFS parameterizations are shown,

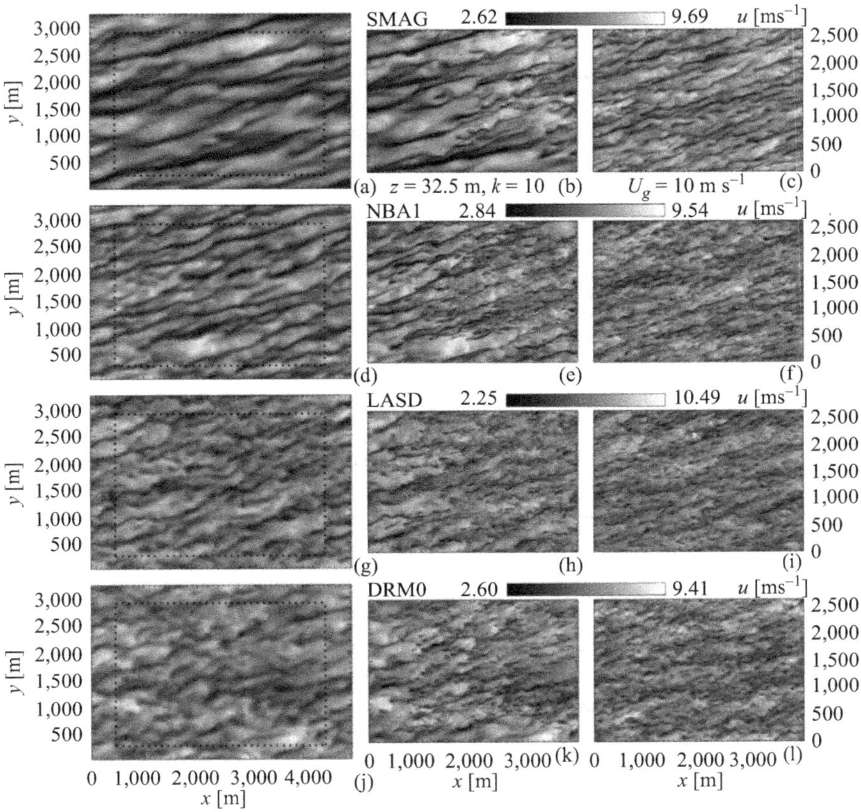

Figure 4.8 Instantaneous horizontal cross sections of the u-velocity component at 100 m above the surface from idealized LES of geostrophically forced flow over flat rough terrain under neutral stability conditions, using four different SFS parameterizations, here showing the impact of the SFS stress model on the equilibration of turbulence subsequent to refinement of the horizontal mesh spacing. Left panels (a,d,g,j) show the outer coarse LES ($\Delta h = 30$ m); middle panels (b,e,h,k) show solution on the refined mesh ($\Delta h = 10$ m), driven by inflow from the coarse mesh; right panels (c,f,i,l) show LES using the same domain size and grid spacing as the nested domain, but using periodic LBCs

including the Smagorinsky model (SMAG, a,b,c), the NBA model formulated without SFS TKE (NBA1; d,e,f), the DRM with level-zero velocity reconstruction, (DRM0; g,h,i), and the LASD (j,k,l) model. While use of more sophisticated SFS models can reduce the equilibration distance substantially, a significant distance is still required for both the formation of smaller scales representable upon the finer mesh and the reduction of TKE anomalies.

Research is underway to examine if the SCPM or some variant thereof can be used to accelerate the transition of turbulence under mesh refinement. Proper perturbations may attenuate the largest resolved scales sufficiently to ameliorate excessive TKE generation, while also accelerating the formation of inertial scales, which may more rapidly reconnect production scales with dissipation.

4.4.3 Bottom boundary condition

As mentioned in Section 4.1, the common implementation of the MOST to parameterize surface energy exchange in LES may contribute errors to simulated ABL flow and turbulence characteristics. One potential source of error results from the application of MOST to conditions that do not satisfy the assumptions of homogeneity and steadiness on which the theory is based. While corrections have been formulated to extend the basic theory, which is based on neutral flow within the shear-generated turbulence regime, to stable and unstable conditions, featuring buoyancy and anisotropy, deficiencies of MOST within non-neutral regimes are likewise well documented (e.g., [149,150]). A second source of error when using MOST in LES is due to its implementation, which typically applies the MOST relationships to each grid cell adjacent to the surface within the LES domain. Such an implementation poses a mismatch between the theory, which applies to the mean velocity, and the instantaneous fluctuations captured by the evolving LES flow field (e.g., [151]). A combination of these factors (and others, including errors contributed by the SFS parameterizations and model numerical solution procedures) results in a commonly observed failure of ABL simulations using MOST to reproduce corresponding flow characteristics, such as expected vertical distributions of wind speed and other quantities, within the surface layer, even when applied in highly idealized simulations closely conforming to the assumptions of the theory. Moreover, mainly due to the lack of a preferable alternative, MOST is also used in unsteady and heterogeneous applications even further removed from the assumptions upon which the theory is based.

As errors in the parameterization of surface exchange inevitably propagate into the surface layer and ABL above, improved parameterizations of surface exchange, suitable to heterogeneous and unsteady conditions, and appropriate for application to the instantaneous velocity fluctuations resolved within LES, are desirable. As such, many attempts have been made to improve the fidelity of near-surface flow characteristics in a variety of conditions, including the previously discussed use of damping functions and improved SFS models. Despite the utility of these alternative procedures for specific applications, no generalized framework for improved surface-layer flow simulation in arbitrary environments has been developed and thus remains an active area of research.

One approach that improves the fidelity of near-surface flow in applications involving tall vegetated canopies is the use of explicit plant canopy parameterizations. Such parameterizations utilize drag terms directly impacting the momentum equations within the plant canopy of height h_c, e.g.,

$$F_i = -(C_d + C_{sf})aVu_i, \tag{4.9}$$

where C_D and C_{sf} are coefficients corresponding to form drag and skin friction contributed by the blockage of flow and viscous drag effects of flow over the surface areas of plant matter, respectively, a is the frontal area of drag elements per unit volume, and V is the scalar wind speed. Depending on the functional form of $a(z)$, (4.9) can capture both the nonalgorithmic increase of mean wind speed observed in vegetated canopies, as well as the large vertical shear values, and corresponding enhancement of shear-generated turbulence production, immediately above the canopy. Modifications to other physical processes within the canopy, including SFS TKE, radiative transfer, and gas exchange physics, may also be incorporated in more sophisticated models to better capture energy exchanges required for various applications (see review by [152]).

Figure 4.9 shows vertical distributions of wind speed (Figure 4.10(a)) and components of the SFS TKE (Figure 4.9(b)) from a simulation of geostrophically forced flow over a uniform plant canopy using the explicit canopy model of Shaw and Patton [153] implemented into the WRF model [93]. Note the addition of a "wake scale" SFS TKE, which augments the traditional 1.5-order SFS TKE, the latter of which is based on the grid scale of the simulation only, and assumes a downscale energy cascade in the absence of plant material, which alters SFS TKE production and dissipation.

As described in Section 4.4.1, canopy models have also been applied to LES over surfaces characterized by small roughness lengths, in efforts to improve agreement with expected logarithmic velocity distribution in neutral flow conditions (e.g., Chow *et al.*, 2005; [129]). Recently, Arthur *et al.* [93] developed a pseudocanopy model (PCM) based on a hybrid of the Shaw and Patton [153] and Brown *et al.* (1990) canopy models. The PCM is intended to improve agreement with the expected law-of-the-wall behavior in LES when using simple constant-coefficient eddy viscosity LES parameterizations, as an alternative to having to implement more complicated and expensive approaches.

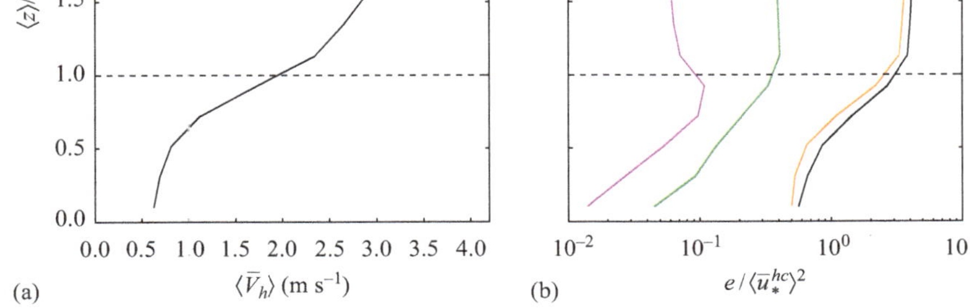

Figure 4.9 Plane averaged horizontal wind speed $\langle\overline{V_h}\rangle$ (a) and TKE (b), including the total e_T, resolved e_R, grid-scale SGS $\langle\overline{e_{SGS}}\rangle$, and wake scale SGS $\langle\overline{e_W}\rangle$ components, from LES over a uniform canopy of height $h_c = 20$ m, using the explicit canopy parameterization of Shaw and Patton [153] implemented into the WRF model

The PCM applies drag terms to the momentum equations over a specified depth; however, the drag term is not based on characteristics of a plant canopy but rather upon the surface C_D value obtained from MOST (see Section 4.2.1). This value of C_D is then distributed as a decreasing function of height, up to h_c, rather than being used at the surface to compute a surface stress. The motivation to distribute C_D across the first few model grid cells arises from the recognition that errors in both

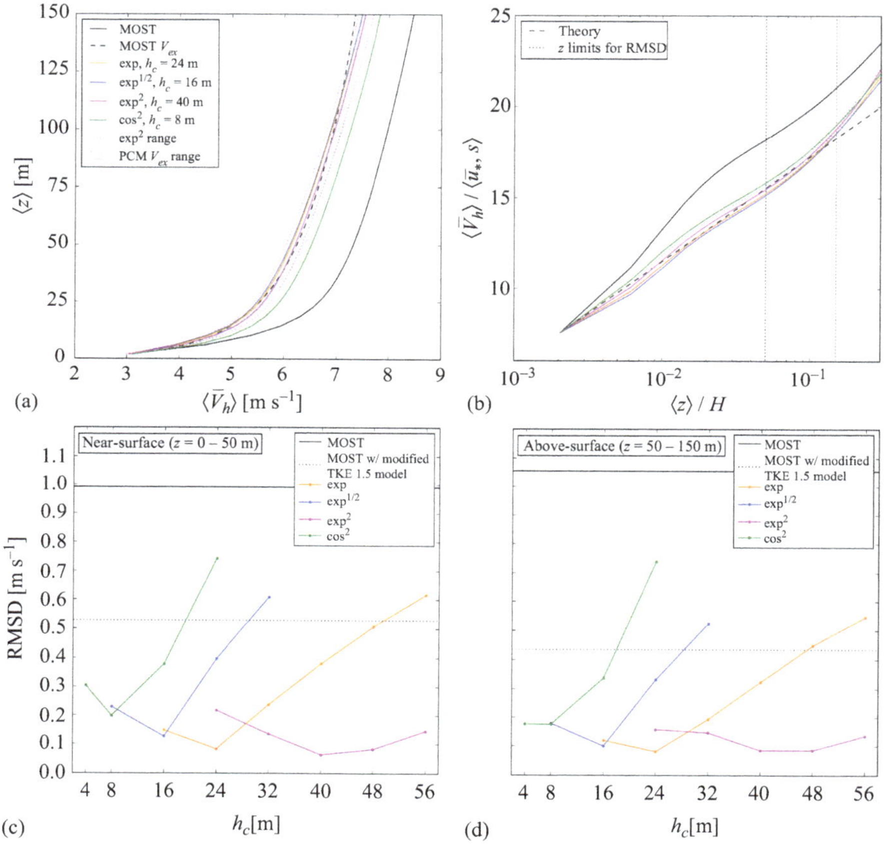

Figure 4.10 Wind speed profiles in the lower ABL as functions of height (a) and as departures from the theoretical log-law behavior (b), using the 1.5 TKE SFS model. Dashed black lines show expected behavior; solid black lines show results using the standard MOST framework; colored lines show results using concepts from explicit plant canopy parameterizations. Root mean square differences between the simulations and the expected solution, integrated over heights of 0–50 m (c) and 50–100 m (d) using the standard MOST approach (solid black lines), application of near-wall damping (black dotted lines), and different canopy shape functions and heights (colored lines) replacing the standard MOST approach

the numerical discretization used by the model and in the SFS stress parameterization each play a role in the failure of many LES to capture the expected MOST behavior despite its application as a surface stress boundary condition. Applying the MOST-derived surface C_D value as a distributed drag, rather than as a stress (as is done in Brown *et al.*, 1990), allows the drag force to influence flow speed directly, rather than as accelerations arising from the divergence of the stresses, which may implicitly incorporate errors from other sources.

Figure 4.10 shows an example of application of the PCM to improve capture of the expected logarithmic layer in the surface layer in a canonical idealized geostrophically forced LES over a flat, rough boundary, with $z_0 = 0.1$. The solid lines in Figure 4.10(a) and (b) show plane-averaged horizontal wind speed profiles, using both standard MOST (black) and four different PCM shape functions (colors). The dashed lines show the expected logarithmic profile, $u(z) = (u_*/\kappa)\ln(z/z_0)$, based upon the surface u_* value obtained using (4.3). Similarly, the gray shading represents the range of expected profiles using the surface u_* values obtained from simulations using the PCM with the other four shape functions, each using the h_c value minimizing its RMSD value shown in Figure 4.10(d). Here, RMSD is the root mean square difference between the simulated mean wind speed profile, and the expected distribution using the surface u_* value, over a particular depth, either from the surface to 50 m (Figure 4.10(c)) or between 50 and 150 m (Figure 4.10(d)), the latter representing a characteristic wind turbine rotor swept area. The red shading represents the variability of expected profiles using the PCM with the height function that decays as the square of the exponential, $\exp^2$, which shows the least sensitivity to changes of h_c.

While the approaches of Arthur *et al.* [93] and Brown *et al.* (1990) represent largely ad-hoc solutions that do not directly address fundamental sources of error, they nevertheless provide practical improvements in fidelity, while also contributing insights into how surface layer physics in LES may be improved.

4.4.4 Data assimilation

Due to the existence of many potential sources of error in the simulation of ABL flow (as detailed in preceding sections), any given simulation may fail to capture a desired characteristic of the flow field, such as a wind speed or direction profile, for example, to a sufficient degree of accuracy for a given application. In cases for which the reproduction of specific flow characteristics is desired, such as in forensic studies of unique events, for example, various methods have been developed to nudge parameters of the model solution toward specified values during a simulation.

Target parameter values may be incorporated into the simulation supply chain in a variety of ways. The most common method is through the data set used to initialize and force the model. Initialization and forcing data used for limited area simulations are typically obtained from global NWP model forecasts or atmospheric analysis products, the latter of which are constrained during their evolution by the observed state of the atmosphere over time. Each of these types of global atmospheric simulation products themselves relies upon a variety of observational data to form their initial and boundary conditions, and any additional data used to constrain the simulation. These initial, boundary and forcing data, in turn, rely upon the projection of measurements taken at different times and locations around the Earth, to specific times and locations comprising the model integration domain. Uncertainties in both the parameter values obtained from the wide variety of measurement platforms utilized, each with unique accuracies and precisions, combined with errors related to the projection of those values, result not in an exact snapshot of the state of the system but instead a rather coarsely gridded estimate of atmospheric and surface states, within an envelope of uncertainty. Subsequent simulation within a forecast or analysis model initialized and forced with those inexact representations of the atmosphere and surface states state introduces additional discrepancies between the simulation and the realized state of the system through internal sources of model error, such as inexact numerical methods, sparse surface cover and terrain representation, and imperfect physical process parameterizations, for example.

For global NWP forecasts, these sources of inaccuracy combine to cause the state of the system to diverge from the realized state, leading to the well-known decrease of forecast skill over timescales of a few to several days. Analysis products seek to improve upon forecasts for retrospective studies by incorporating observations of the realized state of the system throughout the simulation. However, the incorporation of observed data into the simulations does not yield perfect agreement with observed state of the system due to the previously discussed combination of internal model errors, which still influence the solution despite the constraints of the observations, and errors inherent in the measurement and projection of the observations to the model grid. The net result of these sources of error inherent in all global forecast and analysis products is that all limited area models initialized and forced with such data sets will inherit those errors at the outset of any simulation.

While measurements may be instead used to directly specify initial and boundary conditions for a microscale simulation, the discussion in Sections 4.2.1.1 and 4.3.1 motivates the use of data from gridded forecast or analysis products, despite the earlier discussed limitations, except in highly idealized situations.

Irrespective of the source of initial and boundary data, subsequent simulation within a limited area model will inevitably lead to a different representation of the atmospheric state than that represented by the measurements or the larger scale simulation data set. Given the finer resolution and often more sophisticated physical process parameterizations used by limited area models, the fidelity of features relevant to wind energy simulation is typically substantially improved relative to those contained within the larger scale simulation output. However the timing and location of finer-scale phenomena of interest, such as thunderstorm outflows, frontal passages, wave events, or other mesoscale and submesoscale processes, will invariably be shifted in space and time relative to their actual occurrence, due to the limited area simulation developing its own internally consistent dynamical state.

The ability of limited area simulations to capture desired features of atmospheric flows at high fidelity, albeit with associated phase errors, provides considerable value to many applications, such as process studies that do not require specific time or location accuracy. However, for cases requiring very close agreement with an observed quantity or feature, the evolution of the state within a limited area simulation can be steered toward the observed realization via the application of accelerations within the simulations domain that relax model parameters toward specified values. Such so-called nudging may take different forms to target specific applications and outcomes.

Two popular relaxation methods (each provided within the WRF model) include so-called "analysis" and "obs" nudging. Analysis nudging can be used to relax parameter values at every grid point within the simulation domain, or some subset thereof, toward the same gridded forecast or analysis dataset used to specify the initial and boundary conditions.

Analysis nudging can help to constrain the drift of the limited area simulation from the parent forecast or analysis product; however if the forecast or analysis product fails to capture the characteristic of interest, the limited area model is unlikely to do so either. In such cases obs nudging provides a method to locally relax parameter values toward a set of observations at a given set of locations within the simulation domain, with the strength of nudging a function of distance from the observation location. Obs nudging thereby can include local area measurements generally not included in the larger scale model initialization or forcing suite. As such, obs nudging can improve agreement between simulated and observed characteristics of a flow field. A caveat of obs nudging is the potential to perturb some parameter values, or a portion of the simulation domain, away from the consistent internal representation of the system state within the rest of the simulation domain. Another potential risk is that obs nudging may reduce the accuracy of fine-scale or high-frequency flow features due to relaxation toward observed values that are either too coarse to capture fine-scale or high-frequency fluctuations, or to observations that capture such features well, but represent a realized atmospheric state that is out of phase with that represented internally within the simulation. As the strength of nudging is typically formulated as a function of the difference between the simulated quantity and its specified value, such nudging will more strongly constrain larger magnitude excursions, such as turbulence fluctuations, potentially resulting in unphysical representations of high-frequency variability. For this reason, common practice is to either apply nudging only at resolutions too coarse to capture high-frequency variability, or to apply the nudging to locations surrounding the area of interest, such as on bounding domains in nested simulations. Such approaches can result in the selective influence on lower frequency flow characteristics, while allowing the higher frequency content, such as turbulence, to evolve naturally without any direct impacts of the nudging. Spectral nudging [154], which has been used in regional scale modeling to constrain only the largest scales of forcing while allowing finer-scale regional detail to evolve, represents an approach that may be extensible to turbulence resolving scales.

Another methodology to selectively constrain only the low-frequency component of wind speed and direction profiles directly within the turbulence resolving portion of a simulation domain is to nudge not the actual resolved velocity values but instead the height- and time-dependent geostrophic wind values used to force the flow. Due to the expected departure of the flow from geostrophy within the ABL, geostrophic wind profile characteristics within the ABL should not be expected to resemble those of the resolved-scale flow, but rather should result in the resolved-scale flow attaining the specified target values. An iterative framework could be used to determine the optimal geostrophic wind values best matching the target profiles for a given flow scenario. Alternatively, the methodology could be formulated to instead determine nudging terms to be applied within a simulation, removing the need for expensive iteration-based a-priori determinations. Heuristics for the determination of appropriate nudging parameters would be required due to both the lag time between changes to the geostrophic winds and the resulting flow field, and inexact knowledge of how ABL flow characteristics respond to changes of the geostrophic forcing under various conditions (see Section 4.2.1.1). Such heuristics can be developed from training simulations.

A caveat of these approaches is the incorporation of other sources of model error into the applied geostrophic wind values. Due to these other sources of error being model and configuration dependent, geostrophic profiles or nudging heuristics are likely not transferrable between different models and configurations. Moreover, the relatively slow action of the geostrophic forcing renders this approach appropriate only for slowly evolving flow conditions. Similar modifications to the resolved pressure field reflecting ageostrophic pressure accelerations could potentially be developed to incorporate higher frequency forcing, subject to the caveats discussed in Section 4.3.1. Despite these limitations, the geostrophic forging methodology could potentially provide a relatively straightforward and practical method for obtaining turbulence and other high-frequency flow information consistent with a given set of lower frequency flow characteristics, with utility to many wind energy applications.

References

[1] Bailey, B. H.: The financial impact of wind plant uncertainty, NAWEA Symposium, Boulder, CO, August 6, 2013.

[2] Clifton, A., A. Smith, and M. Fields, 2016: Wind Plant Preconstruction Energy Estimates: Current Practice and Opportunities, Technical Report, National Renewable Energy Laboratory, NREL/TP-5000-64735.

[3] Stull, R. B., 1988: *An Introduction to Boundary Layer Meteorology*, Kluwer Academic Publishers, Dordrecht, 666 pp.

[4] Deardorff, J. W., 1970: A numerical study of three-dimensional turbulent channel flow at large Reynolds numbers, *J. Fluid Mech.*, **41**, 453–480, doi:10.1017/S0022112070000691.

[5] Deardorff, J. W., 1974: Three-dimensional numerical study of the height and mean structure of a heated planetary boundary layer, *Bound.-Layer Meteorol.*, **7**(1), 81–106.

[6] Lilly, D. K., 1967: The representation of small-scale turbulence in numerical experiment, in Proc. IBM Scientific Computing Symposium on Environmental Sciences, IBM, White Plains, NY.

[7] Ciofalo, M., 1994: Large-Eddy Simulation: A Critical Survey of Models and Applications, in *Advances in Heat Transfer*, Vol. 25, Chapter 4, J.P. Hartnett and T.F. Irvine, Jr., eds., Academic Press, pp. 321–419.

[8] Leonard, A., 1974: Energy cascade in large-eddy simulations of turbulent fluid flows, *Adv. Geophys.*, **18**, 237–248, DOI: 10.1016/S0065-2687(08)60464-1.

[9] Monin, A. S., and A. M. Obukhov, 1954: Basic laws of turbulent mixing in the surface layer of the atmosphere, *Tr. -Akad. Nauk SSSR Geofiz. Inst.*, 24, 163–187, 1954. English translation by John Miller, 1959.

[10] Moeng, C.-H., 1984: A large-eddy-simulation model for the study of planetary boundary-layer turbulence, *J. Atmos. Sci.*, **41**(13), 2052–2062, https://doi.org/10.1175/1520-0469(1984)041<2052:ALESMF>2.0.CO;2.

[11] Basu, S., A. A. M. Holtslag, B. J. H. Van de Wiel, A. F. Moene, and G.-J. Steeneveld, 2008: An inconvenient "truth" about using sensible heat flux as a surface boundary condition in models under stably stratified regimes, *Acta Geophys.* **56**, 88–99, https://doi.org/10.2478/s11600-007-0038-y.

[12] Holton, J. R., 2012: *An Introduction to Dynamic Meteorology*, Fifth Edition. Elsevier Academic Press, London, 552 pp.

[13] Lund, T., X. Wu, and K. D. Squires, 1998: Generation of turbulent inflow data for spatially-developing boundary layer simulations, *J. Comput. Phys.*, **140**, 233–258.

[14] Moin, P., and K. Mahesh, 1998: Direct numerical simulation: a tool in turbulence research, *Annu. Rev. Fluid Mech.*, **30**, 539–578.

[15] Keating, A., U. Piomelli, E. Balaras, and H. Kaltenbach, 2004: A priori and a posteriori tests of inflow conditions for large-eddy simulation, *Phys. Fluids*, **16**(12), 4696–4712.

[16] Mehta, D., A. H. van Zuijlen, B. Koren, J. G. Holierhoek, and H. Bijl, 2014: Large eddy simulation of wind farm aerodynamics: a review, *J. Wind Eng. Ind. Aerodyn.*, **133**, 1–17, https://doi.org/10.1016/j.jweia.2014.07.002.

[17] Andrén, A., A. R. Brown, J. Graf, *et al.*, 1994: Large-eddy simulation of a neutrally stratified boundary layer: a comparison of four computer codes, *Q. J. R. Meteorolog. Soc.*, **120**, 1457–1484.

[18] Moeng, C., J. Dudhia, J. Klemp, and P. Sullivan, 2007: Examining the two-way grid nesting for large-eddy simulation of the PBL using the WRF model, *Mon. Weather Rev.*, **135**, 2295–2311, doi: 10.1175/MWR3406.1.

[19] Ercolani, G., C. Gorlé, C. Corbari, and M. Mancini, 2017: RAMS sensitivity to grid spacing and grid aspect ratio in large-eddy simulations of the dry neutral atmospheric boundary layer, *Comput. Fluids*, 146, 59–73.

[20] Brasseur, J. G., and T. Wie, 2010: Designing large-eddy simulation of the turbulent boundary layer to capture law-of-the-wall scaling, *Phys. Fluids*, **22**, 021303.

[21] Mirocha, J. D., J. K. Lundquist, and B. Kosović, 2010: Implementation of a nonlinear subfilter turbulence stress model for large-eddy simulation in the Advanced Research WRF Model, *Mon. Weather Rev.*, **138**, 4212–4228, https://doi.org/10.1175/2010MWR3286.1.

[22] Kolmogorov, A. N., 1941: The local structure of turbulence in incompressible viscous fluid for very large Reynolds numbers, *Proc. USSR Acad. Sci.*, **30**, 299–303 (in Russian).

[23] Katul, G. G., C. D. Geron, C-I. Hsieh, B. Vidakovic, and A. B. Guenthere, 1998: Active turbulence and scalar transport near the forest atmosphere interface, *J. Appl. Meteorol.*, **37**, 1533–1546, https://doi.org/10.1175/1520-0450(1998)037<1533:ATASTN>2.0.CO;2.

[24] Skamarock, W. C., 2004: Evaluating mesoscale NWP models using kinetic energy spectra, *Mon. Weather Rev.*, **132**, 3019–3032, https://doi.org/10.1175/MWR2830.1.

[25] Kaimal, J. C., and J. J. Finnigan, 1994: *Atmospheric Boundary Layer Flows*, Oxford University Press, New York, 289 pp.

[26] Vickers, D., and L. Mahrt, 2003: The cospectral gap and turbulent flux calculations, *J. Atmos. Ocean Tech.*, **20**, 660–672. https://doi.org/10.1175/1520-0426(2003)20<660:TCGATF>2.0.CO;2.

[27] Whiteman, C. D., X. Bian, and S. Zhong, 1997: Low-level jet climatology from enhanced rawinsonde observations at a site in the Southern Great Plains, *J. Appl. Meteorol.*, **36**, 1363–1376.

[28] Bonner W. D., 1968: Climatology of the low-level jet, *Mon. Weather Rev.*, **96**, 833–850.

[29] Song, J., K. Liao, R. L. Coulter, and B. M. Lesht, 2005: Climatology of the low-level jet at the southern great plains atmospheric boundary layer experiments site, *J. Appl. Meteorol.*, **44**, 1593–1606.

[30] Blackadar, A. K., 1957: Boundary layer wind maxima and their significance for the growth of nocturnal inversions, *Bull. Amer. Meteorol. Soc.*, **38**, 283–290.

[31] Buajitti, K., and Blackadar, A. K., 1957: Theoretical studies of diurnal wind structure variations in the planetary boundary layer, *Q. J. R. Meteorol. Soc.*, **83**, 486–500.

[32] Lundquist, J. K., 2003: Intermittent and elliptical inertial oscillations in the atmospheric boundary layer, *J. Atmos. Sci.*, **60**, 2661–2673.

[33] Kelley, N. D., B. J. Jonkman, and G. N. Scott, 2006: The Great Plains turbulence environment: its origins, impact and simulation, in National Renewable Energy Laboratory Conference, Paper No. NREL/CP-500-40176.

[34] Drazin, P. G., and W. H. Reid, *Hydrodynamic Instability*, Cambridge University Press, Cambridge, 1981, 525 pp.

[35] Blumen, W., R. Banta, S. P. Burns, *et al.* 2001: Turbulence statistics of a Kelvin-Helmholtz billow event observed in the nighttime boundary layer during the Cooperative Atmosphere-Surface Exchange Study field program, *Dyn. Atmos. Oceans*, **34**, 180–204.

[36] Mahrt, L., 1999: Stratified atmospheric boundary layers, *Bound.-Layer. Meteorol.*, **90**, 375–396, https://doi.org/10.1023/A:1001765727956.

[37] Blumen, W., 1997: A model of inertial oscillations with deformation frontogenesis, *J. Atmos. Sci.*, **54**, 2681–2692.

[38] Zhou, B., and F.K. Chow, 2014: Nested large-eddy simulations of the intermittently turbulent stable atmospheric boundary layer over real terrain, *J. Atmos. Sci.*, **71**(3), 1021–1039. DOI 10.1175/JAS-D-13-0168.1.

[39] Mirocha, J. D., M. J. Churchfield, D. Muñoz-Esparza, *et al.*, 2018: Large-eddy simulation sensitivities to variations of configuration and forcing parameters in canonical boundary layer flows for wind energy applications, *Wind Energy Sci.*, **3**, 589–613, https://doi.org/10.5194/wes-3-589-2018.

[40] Calaf, M., C. Meneveau, and J. Meyers, 2010: Large eddy simulation study of fully developed wind-turbine array boundary layers, *Phys. Fluids*, 22, 015110, https://doi.org/10.1063/1.3291077.

[41] Yang, X., S. Kang, and F. Sotiropoulos, 2012: Computational study and modeling of turbine spacing effects infinite aligned wind farms, *Phys. Fluids*, **24**(11), 115107, https://doi.org/10.1063/1.4767727.

[42] Bokharaei, V. S., P. Bauwererts, and J. Meyers, 2016: Wind-farm layout optimization using a hybrid Jensen-LES approach, *Wind Energy Sci.*, **1**, 311–325, https://doi.org/10.5194/wes-1-311-2016.

[43] Stevens, R. J. A. M., J. Graham, and C. Meneveau, 2014: A concurrent precursor inflow method for large eddy simulations and applications to finite length wind farms, *Renew. Energy*, **68**, 46–50.

[44] Gaudet, B. J., A. Deng, D. R. Stauffer, and A. Gassmann, 2012: Eddy seeding for improved WRF-LES simulations using realistic lateral boundary conditions, in: WRF Users' Workshop. National Center for Atmospheric Research, Boulder, CO.

[45] Thomas, T. G., and J. J. R. Williams, 1999: Generating a wind environment for large eddy simulation of bluff body flows, *J. Wind Eng. Ind. Aerodyn.*, **82**, 189–208, doi:10.1016/S0167-6105(99)00037-9.

[46] Bosveld, F. C., P. Baas, E. van Meijgaard, E. I. F. de Bruijn, G.-J. Steeneveld, and A. A. M. Holtslag, 2014: The third GABLS intercomparison case for evaluation studies of boundary-layer models. Part A: Case selection and set-up, *Bound.-Layer Meteorol.*, **152**(2), 133–156.

[47] Bosveld, F. C., P. Baas, G.-J. Steeneveld, *et al.*, 2014: The third GABLS intercomparison case for evaluation studies of boundary-layer models. Part B: Results and process understanding, *Bound.-Layer Meteorol.*, **152**(2), 157–187.

[48] Sanz Rodrigo, J., M. Churchfield, and B. Kosović, 2017: A methodology for the design and testing of atmospheric boundary layer models for wind energy applications, *Wind Energy Sci.*, **2**, 35–54, https://doi.org/10.5194/wes-2-35-2017.

[49] Xue, M., K. K. Droegemeier, and V. Wong, 2000: The Advanced Regional Prediction System (ARPS): a multi-scale nonhydrostatic atmospheric simulation and prediction model. Part I: Model dynamics and verification, *Meteorol. Atmos. Phys.*, **75**(3–4), 161–193.

[50] Hodur, R. M. 1997: The Naval Research Laboratory's Coupled Ocean/Atmosphere Mesoscale Prediction System (COAMPS), *Mon. Weather Rev.*, **142**, 3284–3302.

[51] Baldauf, M., A. Seifert, J. Förstner, D. Majewski, M. Raschendorfer, and T. Reinhardt, 2011: Operational convective-scale numerical weather

prediction with the COSMO model: description and sensitivities, *Mon. Weather Rev.*, **139**(12), 3887–3905.

[52] Lac., C., J.-P. Chaboureau, V. Masson, *et al.*, 2018: Overview of the Meso-NH model version 5.4 and its applications, Geosci. *Model Dev.*, **11**, 1929–1969, https://doi.org/10.5194/gmd-11-1929-2018.

[53] Pielke, R. A., W. R. Cotton, R. L. Walko, *et al.*, 1992: A comprehensive meteorological modeling system-RAMS, *Meteorol. Atmos. Phys.*, **49**, 69–91.

[54] Skamarock, W. C., J. B. Klemp, J. Dudhia, *et al.*, 2008: A Description of the Advanced Research WRF Version 3, Report No. NCAR/TN-4751STR, National Center for Atmospheric Research, Boulder, CO.

[55] Daniels, M. H., K. A. Lundquist, J. D. Mirocha, D. J. Wiersema, and F. K. Chow, 2016: A new vertical grid nesting capability in the Weather Research and Forecasting (WRF) Model, *Mon. Weather Rev.*, **144**, 3725–3747, http://dx.doi.org/10.1175/MWR-D-16-0049.1.

[56] Talbot, C., E. Bou-Zeid, and J. Smith, 2012: Nested mesoscale large-eddy simulations with WRF: performance in real test cases, *J. Hydrometeor.*, **13**(5), 1421–1441, doi:f10.1175/JHM-D-11-048.1g.

[57] Kelly, J. F., and F. X. Giraldo, 2012: Continuous and discontinuous Galerkin methods for a scalable three-dimensional nonhydrostatic atmospheric model: limited-area mode, *J. Comp. Phys.*, **231**, 7988–8008.

[58] Skamarock, W. C., J. B. Klemp, M. G. Duda, L. Fowler, S.-H. Park, and T. D. Ringler, 2012: A multi-scale nonhydrostatic atmospheric model using centroidal Voronoi Tessellations and C-grid staggering, *Mon. Weather Rev.*, **240**, 3090–3105, doi:10.1175/MWR-D-11-00215.1.

[59] Xie, S., W. Lin, P. J. Rasch, *et al.*, 2018: Understanding cloud and convective characteristics in version 1 of the E3SM atmosphere model, *J. Adv. Model Earth Syst.*, **10**, 2618–2644, doi:10.1029/2018MS001350.

[60] Piomelli, U., S. Kang, F. Ham, and G. Iaccarino, 2006: Effect of discontinuous filter width in large-eddy simulations of plane channel flow, in: Proc. 2006 Summer Program, Stanford, CA, Center for Turbulence Research, pp. 151–162.

[61] Vanella, M., U. Piomelli, and E. Balaras, 2008: Effect of grid discontinuities on large-eddy simulation statistics and flow fields, *J. Turbul.*, **9**, 1–23.

[62] Mirocha, J. D., and K. A. Lundquist, 2017: Impacts of vertical mesh refinement in nested large eddy simulations using the Weather Research and Forecasting model, *Mon. Weather Rev.*, **145**(8), 3025–3048, https://doi.org/10.1175/MWR-D-16-0347.1.

[63] Spalart, P. R., and A. Leonard, 1985: Direct numerical simulation of equilibrium turbulent boundary layers, in Proc. 5th Symp. On Turbulent Shear flows, Ithaca, NY.

[64] Xie, Z. T., and I. P. Castro, 2008: Efficient generation of inflow conditions for large eddy simulations of street-scale flows, *Flow Turbul. Combust.*, **81**, 449–470, https://doi.org/10.1007/s10494-008-9151-5.

[65] Muñoz-Esparza, D., B. Kosović, J. van Beek, and J. D. Mirocha, 2015: A stochastic perturbation method to generate inflow turbulence in large-eddy

simulation models: application to neutrally stratified atmospheric boundary layers, *Phys. Fluids*, **27**, 035102, doi: http://dx.doi.org/10.1063/1.4913572.

[66] Churchfield, M. J., and E. Qoun, 2018: Coupling mesoscale and microscale atmospheric dynamics for wind plant simulations in complex terrain, in: American Meteorological Society's 23rd Symposium on Boundary Layers and Turbulence, 11-15 June 2018, Oklahoma City, OK, USA.

[67] Ghate, A. S., and S. K. Lele, 2017: Subfilter-scale enrichment of planetary boundary layer large eddy simulation using discrete Fourier-Gabor modes, *J. Fluid Mech.*, **819**, 494–539, doi:10.1017/jfm.2017.187.

[68] Akselvoll, K., and P. Moin, 1993: Application of the Dynamic Localization Model to Large Eddy Simulation of Turbulent Flow Over a Backward-Facing Step, in *Engineering Applications to Large Eddy Simulation*, U. Piomelli and S. Ragab, eds., ASME, New York.

[69] Akselvoll, K., and P. Moin, 1995: Large Eddy Simulation of Turbulent Confined Coannular Jets and Turbulent Flow over a Backward Facing Step, Report TF-63, in: Thermosciences Div., Dept. Mech. Eng., Stanford University, Stanford, CA 94305.

[70] Mirocha, J. D., B. Kosović, and G. Kirkil, 2014: Resolved turbulence characteristics in large-eddy simulations nested within mesoscale simulations using the Weather Research and Forecasting model, *Mon. Weather Rev.*, **142**, 806–831, https://doi.org/10.1175/MWR-D-13-00064.1.

[71] Muñoz-Esparza, D., B. Kosović, J. D. Mirocha, and J. van Beek, 2014: Bridging the transition from mesoscales to microscale turbulence in atmospheric models, *Bound.-Layer Meteorol.*, **153**(3), 409–440, https://doi.org/10.1007/s10546-014-9956-9.

[72] Muñoz-Esparza, D., J. K. Lundquist, J. A. Sauer, B. Kosović, and R. R. Linn, 2017: Coupled mesoscale-LES modeling of a diurnal cycle during the CWEX-13 field campaign: from weather to boundary-layer eddies, *J. Adv. Model. Earth Syst.*, **9**, 1572–1594, doi:10.1002/2017MS000960.

[73] Muñoz-Esparza, D., and B. Kosović, 2018: Generation of inflow turbulence in large-eddy simulations of nonneutral atmospheric boundary layers with the cell perturbation method, *Mon. Weather Rev.*, **146**, 1889–1909, https://doi.org/10.1175/MWR-D-18-0077.1.

[74] Mazzaro, L. J., E. Koo, D. Muñoz-Esparza, J. K. Lundquist, and R. Linn, 2018: Force-based perturbations for turbulence generation in mesoscale-to-LES grid-nesting applications, in: American Meteorological Society's 23rd Symposium on Boundary Layers and Turbulence, 11–15 June 2018, Oklahoma City, OK, USA.

[75] Wyngaard, J. C., 2004: Toward numerical modeling in the "Terra Incognita," *J. Atmos. Sci.*, **61**, 1816–1826.

[76] Ching, J., R. Rotunno, M. LeMone, *et al.*, 2014: Convectively induced secondary circulations in fine-grid mesoscale numerical weather prediction models, *Mon. Weather Rev.*, **142**, 3284–3302, https://doi.org/10.1175/MWR-D-13-00318.1.

[77] Rai, R. K., L. K. Berg, B. Kosović, J. D. Mirocha, M. S. Pekour, and W. J. Shaw, 2017: Comparison of measured and numerically simulated turbulence statistics in a convective boundary layer over complex terrain, *Bound.-Layer Meteorol.*, **163**(1), 69–89, https://link.springer.com/article/10.1007/s10546-016-0217-y.

[78] Honnert, R., V. Masson, and F. Couvreux, 2011: A diagnostic for evaluating the representation of turbulence in atmospheric models at the kilometric scale, *J. Atmos. Sci.*, **68**, 3112–3131. https://doi.org/10.1175/JAS-D-11-061.1.

[79] Shin, H. H., and S.-Y. Hong, 2015: Representation of the subgrid-scale turbulent transport in convective boundary layers at gray-zone resolutions, *Mon. Weather Rev.*, **143**(1), 250–271.

[80] Ito, J., H. Niino, M. Nakanishi, and C.-H. Moeng, 2015: An extension of the Mellor–Yamada model to the terra incognita zone for dry convective mixed layers in the free convection regime, *Bound.-Layer Meteorol.*, **157**(1), 23–43.

[81] Boutle, I. A., J. E. J. Eyre, and A. P. Lock, 2014: Seamless stratocumulus simulation across the turbulent gray zone, *Mon. Weather Rev.,* **142**(4), 1655–1668.

[82] Efstathiou, G. A., R. J. Beare, S. Osborne, and A. P. Lock, 2016: Grey zone simulations of the morning convective boundary layer development, *J. Geophys. Res.*, **121**(9), 4769–4782.

[83] Jiménez, P., B. Kosović, A. Martilli, S.-E. Haupt, J. Olson, and J.-W. Bao: A three dimensional PBL parameterization to improve wind forecasting at sub-kilometer scales, in: International Conference on Future Technologies in Wind Energy, October 24–26, 2017, Boulder, CO, USA.

[84] Zängl, G., 2012: Extending the numerical stability limit of terrain-following coordinate models over steep slopes, *Mon. Weather Rev.*, **140**, 3722–3733. https://doi.org/10.1175/MWR-D-12-00049.1.

[85] Mittal, R., and G. Iaccarino, 2005: Immersed boundary methods, *Annu. Rev. Fluid Mech.*, **37**, 239–261.

[86] Peskin, C., 1972: Flow patterns around heart valves: a numerical method, *J. Comp. Phys.*, **10**, 252–271.

[87] Sotiropoulos, F., and X. Yang, 2014: Immersed boundary methods for simulating fluid–structure interaction, *Prog. Aerosp. Sci.*, **65**, 1–21. https://doi.org/10.1016/j.paerosci.2013.09.003.

[88] Iaccarino, G., and R. Verzicco, 2003: Immersed boundary technique for turbulent flow simulations, *Appl. Mech. Rev.*, **56**, 331–347.

[89] DeLeon, R., M. Sandusky, and I. Senocak, 2018: Simulations of turbulent flow over complex terrain using an immersed-boundary method, *Bound.-Layer Meteorol.*, **167**(3), 399–420. doi:10.1007/s10546-018-0336-8.

[90] Ma, Y., and H. Liu, 2017: Large eddy simulations of atmospheric flows over complex terrain using the immersed boundary method in the Weather Research and Forecasting Model, *Bound.-Layer Meteorol.*, **165**, 421–445.

[91] Steppeler, J., H. W. Bitzer, M. Minotte, and L. Bonaventura, 2002: Nonhydrostatic atmospheric modeling using a z-coordinate representation, *Mon. Weather Rev.*, **130**, 2143–2149.

[92] Tseng, Y., and J. Ferziger, 2003: A ghost-cell immersed boundary method for flow in complex geometry, *J. Comp. Phys.*, **192**, 593–623.

[93] Arthur, R. S., J. D. Mirocha, and K. A. Lundquist, 2018: Using a canopy model framework to improve large-eddy simulations of the atmospheric boundary layer in the Weather Research and Forecasting model, *Mon. Weather Rev.*, **147**(1), 31–52.

[94] Lundquist, K. A, F. K. Chow, and J. K. Lundquist, 2010: An immersed boundary method for the Weather Research and Forecasting Model, *Mon. Weather Rev.*, **138**, 796–817.

[95] Lundquist, K. A, F. K. Chow, and J. K. Lundquist, 2012: An immersed boundary method enabling large-eddy simulations of flow over complex terrain the WRF Model, *Mon. Weather Rev.*, **140**, 3936–3955.

[96] Aitken, M. L., B. Kosović, J. D. Mirocha, and J. K. Lundquist, 2014: Large eddy simulation of wind turbine wake dynamics in the stable boundary layer using the Weather Research and Forecasting model, *J. Renew. Sustain. Energy*, **6**, http://dx.doi.org/10.1063/1.4885111.

[97] Vanderwende, B., B. Kosović, J. K. Lundquist, and J. D. Mirocha, 2016: Simulating effects of a wind-turbine array ising LES and RANS, *J. Adv. Model. Earth Syst.*, **8**, 1376–1390.

[98] Marjanovic, N., J. D. Mirocha, F. K. Chow, B. Kosović, and J. K. Lundquist, 2017: Implementation of a generalized actuator line wind turbine parameterization in the Weather Research and Forecasting model, *J. Renew. Sustain. Energy*, **9**, 063308, https://doi.org/10.1063/1.4989443.

[99] Mirocha, J. D., B. Kosović, M. L. Aitken, and J. K. Lundquist, 2014: Implementation of a generalized actuator disk wind turbine model into the weather research and forecasting model for large-eddy simulation applications, *J. Renew. Sustain. Energy*, **6**, 013104, doi: http://dx.doi.org/10.1063/1.4861061.

[100] Mirocha, J. D., D. A. Rajewski, N. Marjanovic, *et al.*, 2015: Investigating wind turbine impacts on near-wake velocity and turbulence characteristics using profiling lidar data and large-eddy simulations with an actuator disk model, *J. Renew. Sustain. Energy*, 7, doi: http://dx.doi.org/10.1063/1.4928873.

[101] Peckham, S. E., T. G. Smirnova, S. G. Benjamin, J. M. Brown, and J. S. Kenyon, 2016: Implementation of a digital filter initialization in the WRF model and its application in the rapid refresh, *Mon. Weather Rev.*, **144**, 99–106, https://doi.org/10.1175/MWR-D-15-0219.1.

[102] Liu, Y., T. T. Warner, J. F. Bowers, *et al.*, 2008: The operational mesogamma-scale analysis and forecast system of the U.S. Army test and evaluation command. Part I: Overview of the modeling system, the forecast products, and how the products are used, *J. Appl. Meteorol. Clim.*, **47**, 1077–1092, https://doi.org/10.1175/2007JAMC1653.1.

[103] Klemp, J. B., J. Dudhia, and A. D. Hassiotis, 2008: An upper gravity-wave absorbing layer for NWP applications, *Mon. Weather Rev.*, **136**, 3988–4004, DOI:10.1175/2008MWR2596.1.

[104] Lesieur, M., and O. Métais, 1996: New trends in large-eddy simulations of turbulence, *Annu. Rev. Fluid Mech.*, **28**, 45–82, https://doi.org/10.1146/annurev.fl.28.010196.000401.

[105] Meneveau, C., and J. Katz, 2000: Scale-invariance and turbulence models for large-eddy simulation, *Annu. Rev. Fluid Mech.*, **32**, 1–32, https://doi.org/10.1146/annurev.fluid.32.1.1.

[106] Pope, S. B., 2000: *Turbulent Flows*, Cambridge University Press, Cambridge, 771 pp.

[107] Sagaut, P., 2006: *Large Eddy Simulation for Incompressible Flows*, Springer-Verlag, Berlin, Heidelberg, 558 pp.

[108] Rogallo, R. S., and P. Moin, 1985: Numerical simulations of turbulent flows, *Annu. Rev. Fluid Mech.*, **16**, 99–137, https://doi.org/10.1146/annurev.fl.16.010184.000531.

[109] Juneja, A., and J. G. Brasseur, 1999: Characteristics of subgrid-resolved-scale dynamics in anisotropic turbulence, with application to rough-wall boundary layers, *Phys. Fluids*, **11**, 3054, https://doi.org/10.1063/1.870164.

[110] Khanna, S., and J. G. Brasseur, 1998: Three-dimensional buoyancy- and shear-induced local structure of the atmospheric boundary layer, *J. Atmos. Sci.*, **55**, 710–743.

[111] Smagorinsky, J., 1963: General circulation experiments with the primitive equations, *Mon. Weather Rev.*, **91**, 99–152, https://doi.org/10.1175/1520-0493(1963)091<0099:GCEWTP>2.3.CO;2.

[112] Porté-Agel, F., C. Meneveau, and M. B. Parlange, 2000: A scale dependent dynamic model for large eddy simulation: application to a neutral atmospheric boundary layer, *J. Fluid Mech.*, 415, 261–284.

[113] Mason, P. J., and D. J. Thomson, 1992: Stochastic backscatter in large-eddy simulations of boundary layers, *J. Fluid Mech.*, **242**, 51–78.

[114] Piomelli, U., and E. Balaras, 2002: Wall-layer models for large-eddy simulations, *Annu. Rev. Fluid Mech.*, **34**, 349–374, https://doi.org/10.1146/annurev.fluid.34.082901.144919.

[115] Van Driest, E. R., 1956: On turbulent flow near a wall, *J. Aeronaut. Sci.*, **23**(11), 1007–1011, https://doi.org/10.2514/8.3713.

[116] Deardorff, J. W., 1980: Stratocumulus-capped mixed layers derived from a three-dimensional model, *Bound.-Layer Meteorol.*, **18**, 495–527.

[117] Mellor, G. L., and T. Yamada, 1974: A hierarchy of turbulence closure models for planetary boundary layers, *J. Atmos. Sci.*, **31**, 1791–1806.

[118] Bardina, J., J. H. Ferzinger, and W. C. Reynolds, 1983: Improved Turbulence Models Based on Large Eddy Simulation of Homogeneous, Incompressible, Turbulent Flows, Technical Report TF-19, Department of Mechanical Engineering, Stanford University, Stanford, CA.

[119] Liu, S., C. Meneveau, and J. Katz, 1994: On the properties of similarity subgrid-scale models as deduced from measurements in a turbulent jet, *J. Fluid Mech.*, **275**, 83–119.

[120] Bou-Zeid, E., C. Higgins, H. Huwald, and C. Meneveau, 2010: Field study of the dynamics and modelling of subgrid-scale turbulence in a

stable atmospheric surface layer over a glacier, *J. Fluid Mech.*, **665**, 480–515, doi:10.1017/S0022112010004015.

[121] Higgins, C. W., M. B. Parlange, and C. Meneveau, 2003: Alignment trends of velocity gradients and subgrid-scale fluxes in the turbulent atmospheric boundary layer, *Bound.-Layer Meteorol.*, **109**, 59–83.

[122] Germano, M., U. Piomelli, P. Moin, and W. H. Cabot, 1991: A dynamic subgrid-scale eddy viscosity model, *Phys. Fluids*, **3**, 1760–1765.

[123] Moin, P., K. Squires, W. Cabot, and S. Lee, 1991: A dynamic subgrid-scale model for compressible turbulence and scalar transport, *Phys. Fluids*, **3**A, 2746–2757.

[124] Kleissl, J., C. Meneveau, and M. B. Parlange, 2003: On the magnitude and variability of subgrid-scale eddy-diffusion coefficients in the atmospheric surface layer, *J. Atmos. Sci.*, **60**, 2372–2388.

[125] Bou-Zeid, E., C. Meneveau, and M. B. Parlange, 2005: A scale-dependent Lagrangian dynamic model for large eddy simulation of complex turbulent flows, *Phys. Fluids*, **17**, 025105.

[126] Meneveau, C., T. Lund, and W. Cabot, 1996: A Lagrangian dynamic subgrid-scale model of turbulence, *J. Fluid Mech.*, **319**, 353–385.

[127] Kosović, B., 1997: Subgrid-scale modeling for the large-eddy simulation of high-Reynolds-number boundary layers, *J. Fluid Mech.*, **336**, 151–182, DOI: https://doi.org/10.1017/S0022112096004697.

[128] Kosović, B., and J. A. Curry, 2000: A large eddy simulation study of a quasi-steady, stably stratified atmospheric boundary layer, *J. Atmos. Sci.*, **57**, 1052–1068.

[129] Kirkil G., J. D. Mirocha, F. K. Chow, and E. Bou-Zeid. 2012: Implementation and evaluation of dynamic subfilter-scale stress models for large-eddy simulation using WRF, *Mon. Weather Rev.*, **140**, 266–284.

[130] Mirocha, J. D., G. Kirkil, E. Bou-Zeid, F. K. Chow, and B. Kosović, 2013: Transition and equilibration of neutral atmospheric boundary layer flow in one-way nested large-eddy simulations using the Weather Research and Forecasting model, *Mon. Weather Rev.*, **141**, 918–940, http://dx.doi.org/10.1175/MWR-D-11-00263.1.

[131] Katopodes, F. K., R. L. Street, and J. H. Ferziger, 2000a: Subfilter-scale scalar transport for large-eddy simulation, in: Preprints, 14th Symp. on Boundary Layers and Turbulence, Aspen, CO, Amer. Meteor. Soc., pp. 472–475.

[132] Katopodes, F. K., R. L. Street, and J. H. Ferziger, 2000b: A Theory for the Subfilter-Scale Model in Large-Eddy Simulation, Tech. Rep. 2000-K1, Environmental Fluid Mechanics Laboratory, Stanford University, 19 pp.

[133] Stolz, S., and N. A. Adams, 1999: An approximate deconvolution procedure for large-eddy simulation, *Phys. Fluids*, **11**, 1699–1701.

[134] Stolz, S., and N. A. Adams, and L. Kleiser, 2001: An approximate deconvolution model for large-eddy simulation with application to incompressible wall-bounded flows, *Phys. Fluids*, **13**, 997–1015.

[135] Domaradzki, J. A., and N. A. Adams, 2002: Direct modelling of subgrid scales of turbulence in large eddy simulations, *J. Turbul.*, **3**, N24, 1–19, DOI:10.1088/1468-5248/3/1/024.

[136] Carati, D., G. S. Winckelmans, and H. Jeanmart, 2001: On the modeling of the subgrid-scale and filtered-scale stress tensors in large-eddy simulation, *J. Fluid Mech.*, **441**, 119–138.

[137] Chow, F. K., and P. Moin, 2003: A further study of numerical errors in large-eddy simulations, *J. Comput. Phys.*, **184**, 366–380.

[138] Lund, T. S., 1997: On the use of discrete filters for large eddy simulation, Annual Research Briefs, Center for Turbulence Research, Stanford University/NASA Ames, pp. 83–95.

[139] Wong, V. C., and D. K. Lilly, 1994: A comparison of two dynamic subgrid closure methods for turbulent thermal-convection, *Phys. Fluids*, **6**, 1016–1023.

[140] Zhou, B., and F. K. Chow, 2011: Large-eddy simulation of the stable boundary layer with explicit filtering and reconstruction turbulence modeling, *J. Atmos. Sci.*, **68**(9), 2142–2155. DOI:10.1175/2011JAS3693.1.

[141] Zhou, B., and F. K. Chow, 2012: Turbulence modeling for the stable atmospheric boundary layer and implications for wind energy, *Flow Turbul. Combust.*, **88**, 255–277. DOI 10.1007/s10494-011-9359-7.

[142] Chow, F. K., and R. L. Street, 2009: Evalution of turbulence closure models for large-eddy simulation over complex terrain: flow over Askervein Hill, *J. Appl. Meteorol. Climatol.*, **48**(5), 1050–1065. DOI: 10.1175/2008JAMC1862.1.

[143] Khani, S., and F. Porté-Agel, 2017: A modulated-gradient parametrization for the large eddy simulation of the atmospheric boundary layer using the Weather Research and Forecasting model, *Bound.-Layer Meteorol.*, **165**(3), 385–404.

[144] Findikakis, A. N., and R. L. Street, 1979: An algebraic model for subgrid-scale turbulence in stratified flows, *J. Atmos. Sci.*, **36**, 1934–1949.

[145] Gicquel, L. Y. M., and P. Givi, 2002: Velocity filtered density function for large eddy simulation of turbulent flows *Phys. Fluids*, **14**(3), 1196–1213

[146] Boris, J. P., F. F. Grinstein, E. S. Oran, and R. L. Kolbe, 1992: New insights into large eddy simulation, *Fluid Dyn. Res.*, **10**, 199–228.

[147] Chaouat, B., 2017: The state of the art of hybrid RANS/LES modeling for the simulation of turbulent flows, *Flow Turbul. Combust.*, **99**, 279–327. https://doi.org/10.1007/s10494-017-9828-8.

[148] Spalart, P. R., 2009: Detached-eddy simulation, *Annu. Rev. Fluid Mech.*, **41**, 181–202.

[149] Kaimal, J. C., J. C. Wyngaard, D. A. Haugen, *et al.*, 1976: Turbulence structure in the convective boundary layer, *J. Atmos. Sci.*, **33**, 2152–2169.

[150] Mahrt, L., 1998: Stratified atmospheric boundary layers and breakdown of models, *Theoret. Comput. Fluid Dyn.*, **11**, 263–279, https://doi.org/10.1007/s001620050093.

[151] Sullivan, P. P., J. C. McWilliams, and C.-H. Moeng, 1994: A subgrid-scale model for large-eddy simulation of planetary boundary-layer flows, *Bound.-Layer Meteorol.*, **71**(3), 247–276.

[152] Patton, E. G., and J. J. Finnigan, 2012: Chap. 24, Canopy turbulence, in: *Handbook of Environmental Fluid Dynamics*, Vol. 1, H. J. S. Fernando, ed., CRC Press, Boca Raton, FL, pp. 311–328.

[153] Shaw, R. H., and E. G. Patton, 2003: Canopy element influences on resolved-and subgrid-scale energy within a large-eddy simulation. *Agric. For. Meteorol.*, **115**, 5–17.

[154] von Storch, H., H. Langenberg, and F. Feser, 2000: A spectral nudging technique for dynamical downscaling purposes, *Mon. Weather Rev.*, **128**, 3664–3673.

[155] Arthur, R. S., K. A. Lundquist, J. D. Mirocha, and F. K. Chow, Topographic effects on radiation in the WRF model with the immersed boundary method: implementation, validation and application to complex terrain, *Mon. Weather Rev.*, https://doi.org/10.1175/MWR-D-18-0108.1.

[156] Brown, A. R., J. M. Hobson, and N. Wood, 2001: Large-eddy simulation of neutral turbulent flow over rough sinusoidal ridges, *Bound.-Layer Meteorol.*, **98**, 411–441.

[157] Chow, F. K., R. L. Street, M. Xue, and J. H. Ferziger, 2004: Explicit filtering and reconstruction turbulence modeling for large-eddy simulation of neutral boundary layer flow, *J. Atmos. Sci.*, **62**, 2058–2077.

[158] Mirocha, J. D., B. Kosović, and J. A. Curry, 2005: Vertical heat transfer in the lower atmosphere over the Arctic Ocean during clear-sky periods, *Bound.-Layer Meteorol.*, **117**, 37–71, http://dx.doi.org/10.1007/s10546-004-1130-3.

[159] Nieuwstadt, F. T. M., P. J. Mason, C.-H. Moeng, and U. Schumann, 1993: Large-eddy simulation of the convective boundary layer: a comparison of four computer codes, in *Turbulent Shear Flows,* Vol. 8, Durst *et al.*, eds., Springer-Verlag, Berlin, 431 pp.

[160] Rai, R. K., L. K. Berg, M. Pekour, *et al.*, 2017: Spatiotemporal variability of turbulence kinetic energy budgets in the convective boundary layer over both simple and complex terrain, *J. App. Meteorol. Clim.*, **56**(12), 3285–3302, https://doi.org/10.1175/JAMC-D-17-0124.1.

[161] Shaw, W. J., J. K. Lundquist, and S. J. Schreck, 2009: Workshop on research needs for wind resource characterization, *Bull. Amer. Meteorol. Soc.*, **90**, 535–538, http://dx.doi.org/10.1175/2008BAMS2729.1.

[162] Tennekes, H., and J. L. Lumley, 1974: *A First Course in Turbulence*, MIT Press, Cambridge, MA, 300 pp.

[163] Veers, P. S., 1988: Three-Dimensional Wind Simulation, SAND88-0152, Sandia National Laboratory, Albuquerque, NM.

Chapter 5

Atmospheric turbulence modelling, synthesis, and simulation

Jacob Berg[1] *and Mark Kelly*[1]

The starting point of wind turbine operation is the incoming wind. Wind turbines are positioned in the atmospheric boundary layer (ABL), the lower approximately 1 km of the atmosphere; here the wind tends to be dominated by turbulent structures generated through the transfer of momentum and heat with the Earth's surface, as well as interaction with the free atmosphere above governed by large-scale motion.

In this chapter, we will look at the turbulence affecting wind turbines from a turbulence-simulation point of view. This means that the focus will be on the properties of atmospheric turbulence which directly affect the performance and operation of wind turbines.

In the ABL, turbulence is produced by mean wind shear and enhanced or destructed by buoyancy effects. This results in profiles of the various turbulence quantities across wind turbine rotors. Examples include the mean wind speed itself; second order moments, like variances and stresses; and turning of the mean wind speed and even length scales of turbulence. The degree to which a wind turbine will be affected by the turbulence in the ABL depends on its size such as rotor diameter and hub height, its power generating properties such as thrust coefficient, as well as on the applied controller which ultimately decides the operation window of the turbine. Simulations of atmospheric turbulence can guide us in quantifying the effects.

5.1 Introduction

5.1.1 Notation and ensemble averaging

Following, e.g., [1], we denote any three-dimensional (3-d) variable which fluctuates in space and time as $\tilde{a}_i(\mathbf{x}, t)$, where $i = 1, 2, 3$. We will use Einstein notation throughout the chapter with implicit summing over repeated indices. We introduce Reynolds decomposition with ensemble averaging, i.e. we define the decomposition

$$\tilde{a}_i(\mathbf{x}, t) = A_i(\mathbf{x}, t) + a_i(\mathbf{x}, t), \tag{5.1}$$

[1] Department of Wind Energy, Technical University of Denmark, Roskilde, Denmark

where $A_i(\mathbf{x}, t)$ is the mean part and $a_i(\mathbf{x}, t)$ is the fluctuation part. Denoting the ensemble-average operation by $\langle\cdot\rangle$, we have

$$\langle\tilde{a}_i(\mathbf{x}, t)\rangle = \langle A_i(\mathbf{x}, t)\rangle + \langle a_i(\mathbf{x}, t)\rangle = A_i(\mathbf{x}, t); \tag{5.2}$$

i.e. the average of an average is the average, $\langle A_i\rangle = A_i$, and the average of a fluctuation is zero: $\langle a_i\rangle = 0$. Furthermore, we have the *distributive* and *commutative* properties. In the case of two variables, a_i and b_i, we have

$$\langle\tilde{a}\tilde{b}\rangle = \langle(A + a)(B + b)\rangle = \langle AB\rangle + \langle ab\rangle, \tag{5.3}$$

since the product of a mean and a fluctuation variable is zero. The covariance term, $\langle ab\rangle$, equals zero in the case when a and b are uncorrelated.

5.1.2 Defining the notion of turbulence simulations

Following the well-known text of Pope [2], we will use the term *turbulence simulation* when we refer to a time-dependent solution to the governing equations, wherein the velocity field is described by $\tilde{u}_i(\mathbf{x}, t)$. The term *turbulence model* then refers to the solutions of mean quantities such as ensemble averages, U_i and $\langle u_i u_j\rangle$.*

In applications solely focusing on wind-resource assessment (estimation of 10-min mean wind speed U and long-term statistics of U), it is often most practical to solve the Reynolds-averaged Navier–Stokes (RANS) equations, with simple closure to account for the mean effects of turbulence. Common examples of the latter are two-equation models such as k–ϵ and k–ω [3]. On the other hand, if we want to estimate the loads on wind turbines, we will need the time-resolved turbulent flow field – and hence a turbulence simulation approach is required. Such methods can roughly be divided into three categories:

- Turbulence synthesis via spectral/quasi-linearized modelling
- Large eddy simulation (LES)
- Direct numerical simulation (DNS).

The last item, DNS, solves the full Navier–Stokes (N–S) equations taking into account all scales from the smallest scale in the flow (the Kolmogorov length scale of dissipation, $\eta = \nu^{3/4}\epsilon^{-1/4}$) to the largest energy-containing scale in question (ℓ). Here, ν is the kinematic viscosity and ϵ is the dissipation rate of turbulent kinetic energy (TKE), which under Reynolds-number similarity equals the rate of production of kinetic energy and hence $\epsilon = u^3/\ell$ [1]. The extent of the inertial subrange depends on the Reynolds number, *Re*, which is the ratio between the inertial forces and the viscous forces in the flow:

$$Re = \frac{u\ell}{\nu}, \tag{5.4}$$

where u is the characteristic scale of turbulent velocity fluctuations in the flow and ν is the kinematic viscosity.

*Unless otherwise stated, we will use a right-handed coordinate system aligned with the mean wind.

To do DNS thus requires that we choose a numerical resolution Δ smaller or equal to the smallest scale in the flow, η. Setting $\Delta = \eta$, we need

$$N = \frac{\ell}{\eta} = \left(\frac{u\ell}{\nu}\right)^{3/4} = Re^{3/4} \tag{5.5}$$

in each of the three spatial directions, or $N^3 = Re^{9/4}$ points to resolve a cubic volume ℓ^3; the number N in (5.5) corresponds to the (ratio of) scales defining the inertial subrange. For a Reynolds number of $\sim 10^8$ (such as can be found in the ABL, where $\ell \sim 1{,}000$ m, $u \sim 1$ m/s and $\nu \sim 10^{-5}$ m/s^2), this would require close to 10^{18} grid points in the simulation, and hence computer resources far beyond what any existing machine is capable of today – or even far into the future. This leaves us with the first two bullet-points of the list of methods given earlier: LES, which is a spatially truncated version of DNS in which only the energy-containing and some inertial-range scales are resolved, and turbulence synthesis approaches based on spectral forms. Within the latter, we include the spectral-tensor turbulence model of Mann [4]; it is based on the rapid distortion theory (RDT), which is in turn a reduced N–S equation methodology [2,5].

5.2 Simulating turbulence for wind turbine applications

The effect of atmospheric turbulence on the operation of wind turbines has classically been divided in two categories:

1. the effect on the wind resource and power output of the wind turbines and
2. the effect on turbine loads.

These two categories fall under (a) wind-resource assessment and (b) site suitability and turbine design (e.g. IEC 61400-1, Appendix B [6]). When wind turbines had hub heights below 80 m, and were situated on relatively flat terrain, this distinction was somewhat meaningful. In many facets of the wind energy industry, until recently the atmosphere was treated as purely neutral (i.e. turbulence assumed to be produced by mean wind shear), ignoring the complications of buoyancy – and also the finite height of the ABL. Today, the picture is different. Offshore installation and wind turbine deployment in complex terrain has advanced. The latter development makes the division of wind turbine applications into resource- and site-assessment problematic: the speed-up effect of having compressed stream-lines above hilltops and ridges is accompanied by increased production of turbulence. In addition, larger rotor sizes and 'taller turbines' increase the importance of buoyancy, the capping inversion, and the Coriolis force as factors in characterizing the atmospheric flow around the turbines.

Before delving into the different types of turbulence modelling, we will first take a look at the turbulence in the ABL. Then we describe the dominant methodologies along with their strengths and limitations for the simulation of atmospheric flow including the turbulent structures important for wind turbines.

5.3 Turbulence in the atmospheric boundary layer

The starting point of turbulence in the atmosphere is the famous N–S equation, which expresses the 3-d vectorial momentum balance of fluids; i.e. the motion of liquids or gases and, in our case, thus the lower atmosphere [1,7]. In order to arrive at the final set equations for ABL flow, we include the Coriolis force due to the rotation of the Earth in the N–S. Furthermore we define a base state of constant density ρ_0, and associated potential (adiabatic [8]) temperature θ_0 and pressure p_0, around which we can linearize the pressure gradient (the only term in the N–S equation which includes density[†]). We ultimately end up with an equation describing the velocity $\tilde{u}_i \equiv \tilde{u}_i(\mathbf{x}, t)$, pressure fluctuation $\tilde{p}' \equiv \tilde{p}'(\mathbf{x}, t)$, and adiabatic temperature fluctuations $\tilde{\theta}' \equiv \tilde{\theta}'(\mathbf{x}, t)$, the latter two being departures from their base states $\{p_0, \theta_0\}$:

$$\frac{\partial \tilde{u}_i}{\partial t} + \tilde{u}_j \frac{\partial \tilde{u}_i}{\partial x_j} = -\frac{1}{\rho_0}\frac{\partial \tilde{p}'}{\partial x_i} + \nu \frac{\partial^2 \tilde{u}_i}{\partial x_j \partial x_j} + \frac{g}{\theta_0}\tilde{\theta}'\delta_{i3} - \varepsilon_{ijk}\Omega_j \tilde{u}_k. \tag{5.6}$$

Furthermore, we have the incompressibility condition

$$\frac{\partial \tilde{u}_i}{\partial x_i} = 0, \tag{5.7}$$

and a conservation equation for internal energy via the temperature ($\tilde{\theta} = \theta_0 + \tilde{\theta}'$):

$$\frac{\partial \tilde{\theta}}{\partial t} + u_i \frac{\partial \tilde{\theta}}{\partial x_i} = \alpha \frac{\partial^2 \tilde{\theta}}{\partial x_i \partial x_i} + R, \tag{5.8}$$

where α is the thermal diffusivity and R represents radiation processes.

Here we have assumed that the air is dry. But in reality, water vapour, liquid water, and the processes of phase change should be included through additional terms and scalar equations. Since these effects mainly become important further away from the surface compared to the height at which wind turbines operate, the assumption seems justified.

In the ABL, the wind is directly affected by the presence of the Earth's surface. This manifests itself as a drag force (friction) which is responsible for the exact magnitude and direction of the wind close to the surface: at the surface, the wind speed must be zero, and hence there exists a downward turbulent flux of momentum, $\langle uw \rangle < 0$, balanced by a positive vertical gradient in the mean wind, $dU/dz > 0$. This gradient is the main source of production of turbulence. It is also called (mean) shear production (SP) and constitutes the cornerstone of the conservation equation for TKE. With Reynolds averaging denoted by capital letters ($\tilde{u}_i = U_i + u_i, \tilde{p}' = P + p$, and $\tilde{\theta} = \Theta + \theta$) and angle brackets, the TKE E can be defined as

$$E \equiv \frac{1}{2}\langle u_i u_i \rangle. \tag{5.9}$$

[†]Assuming the Boussinesq approximation to convert density fluctuations to temperature fluctuations.

Taking the dot product of velocity and the momentum equation (5.6) leads to the TKE balance

$$\frac{\partial E}{\partial t} + U_j \frac{\partial E}{\partial x_j} = \underbrace{-\langle u_i u_j \rangle \frac{\partial U_i}{\partial x_j}}_{\text{SP}} + \underbrace{\frac{g}{\theta_0} \langle u_i \theta \rangle \delta_{i3}}_{\text{BP}} - \underbrace{\frac{\partial \langle u_j E \rangle}{\partial x_j}}_{\text{TT}} - \underbrace{\frac{1}{\rho_0} \frac{\partial \langle u_i p \rangle}{\partial x_i}}_{\text{PD}} - \varepsilon. \tag{5.10}$$

This includes terms representing the SP, buoyant production (BP) or destruction, turbulent transport (TT), pressure transport (PD), and the rate of viscous dissipation ε. The last is defined as

$$\epsilon = \nu \left\langle \frac{\partial u_i}{\partial x_j} \frac{\partial u_i}{\partial x_j} \right\rangle. \tag{5.11}$$

It is noted that no terms include the Coriolis force; its only role is to redistribute energy between the three velocity components (e.g. [1]). Together with appropriate boundary conditions, the TKE equation describes the turbulence in the ABL and, as such, describes all flow features which affect wind turbines – and thus those which we want to mimic with our numerical simulations.

5.3.1 Surface-layer scaling and Monin–Obukhov similarity theory

Monin–Obukhov (M–O) similarity theory (MOST) is in essence a *bottom-up* formulation that describes atmospheric turbulence in terms of turbulent surface fluxes, in homogeneous conditions. MOST disregards terrain effects such as canopies, roughness changes and internal boundary layer growth, as well as effects due to variations in terrain elevation such as hills.

Close to the surface, at a given height, there exists a (local) balance between SP and dissipation. Denoting the turbulent momentum flux at the surface by $u_\star^2 \equiv -\langle uw \rangle|_0$ and choosing characteristic velocity-fluctuation and length scales $u_\star$ and κz, we obtain the famous 'log-law'

$$U = \frac{u_\star}{\kappa} \ln\left(\frac{z}{z_0}\right), \tag{5.12}$$

which describes the mean wind close to the surface in homogeneous conditions. Here z_0 is the roughness length and is the height at which $U = 0$ according to (5.12), and $\kappa = 0.4$ is the von Karman constant.

The log-law above is the 'neutrally stratified' wind profile, and part of a broader class of profiles in MOST. In general, we can write the wind shear as

$$\frac{dU}{dz} = \frac{u_\star}{\kappa z} \varphi_m \left(\frac{z}{L}\right), \tag{5.13}$$

where we have now introduced a non-dimensional function $\phi_m(z/L)$, with the non-dimensional argument z/L, where the Obukhov length

$$L = -\frac{\theta_0}{\kappa g} \frac{u_\star^3}{\langle u_3 \theta \rangle} \tag{5.14}$$

is defined as the ratio between the SP of TKE and the BP (or destruction) term in the TKE equation [1,9]. The argument z/L thus determines the effective strength of the buoyancy effect at a given height z. Integrating (5.13), we can find the so-called M–O wind profile for non-neutral stratification,

$$U = \frac{u_\star}{\kappa}\left[\ln\left(\frac{z}{z_0}\right) - \Psi_m\left(\frac{z}{L}\right)\right], \tag{5.15}$$

using the non-dimensional 'stability correction' function

$$\Psi_m\left(\frac{z}{L}\right) = \int\left[1 - \phi_m\left(\frac{z}{L}\right)\right] d\ln z. \tag{5.16}$$

In situations of strong convection (unstable conditions), the warm surface heats the relatively colder atmosphere, i.e. $\langle u_3\theta\rangle > 0$ and $L < 0$, and turbulence is enhanced and generated by conversion of potential (thermal) energy to TKE. This generates large turbulent structures and a very tall ABL. The picture tends to be reversed at night, with strong cooling from the surface (stable conditions), where the strong stratification of the ABL prohibits the development of turbulence. The best way to get an idea of the stability on a given day is to look at the smoke from a chimney: nearly laminar horizontal plumes are signatures of stable stratification, while erratically growing plumes correspond to unstable stratification. In the case of $1/L = 0$, and hence $\phi_m(z/L) = 1$ and $\Psi_m(z/L) = 0$, we have neutral conditions in which the buoyancy has vanishing effect, then (5.15) reverts to the log-law (5.12). Neutral conditions are often found in conditions dominated by cloud cover in mid-latitudes.

The three types of boundary layers presented here – convective (unstable), stable, and neutral – are often denoted *the canonical boundary layers*. They correspond to solutions of the N–S and TKE equations that arise from including only the dominant terms (SP, BP or destruction, and viscous dissipation, [9]) in the TKE equation (5.10). Plots of flow fields for these three types of ABL, along with corresponding profiles of mean wind speed and turbulence properties, are presented in Figures 5.1–5.3.

Again MOST only applies close to the surface, where surface fluxes govern the flow. In the LES simulations used to produce the figures, to simulate actual ABL conditions, we have also included a temperature inversion defining the 'top' of the ABL with a stable layer (most visible as a non-turbulent layer at $z \gtrsim z_i$, see Section 5.3.2); this gives rise to so-called top-down effects. In reality, the daily variation in surface heat flux present in most places on the Earth makes a clear distinction of the three types somewhat difficult, since the solutions are influenced by a transient behaviour [11].

Analytical forms of the non-dimensional stability functions have been determined empirically from measurements. For example, Högström [12] proposed

$$\text{stable conditions:} \qquad \phi_m\left(\frac{z}{L}\right) = 1.0 + 4.8z/L \tag{5.17}$$

$$\text{unstable conditions:} \qquad \phi_m\left(\frac{z}{L}\right) = (1.0 - 19.3z/L)^{-1/4}, \tag{5.18}$$

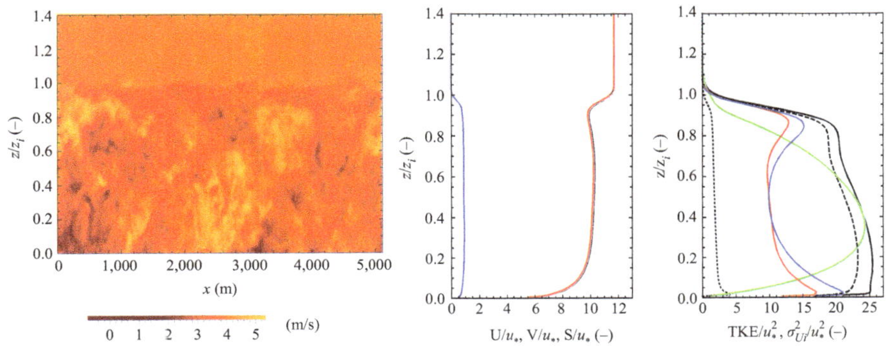

Figure 5.1 LES of a convective ABL. Left: instantaneous x–z planar view of the horizontal wind. Middle: profiles of non-dimensionalized mean wind components, U (red line), V (blue line), and mean horizontal wind speed S (black line). Right: profiles of non-dimensionalized variance, σ_u^2 (red line), σ_v^2 (blue line), σ_w^2 (green line), and non-dimensionalized turbulent kinetic energy, resolved TKE (dashed black line), e (dotted black line) and total TKE (solid black line). All heights are scaled with the height of the ABL, z_i. The simulation is done with the pseudo-spectral LES model documented in [10]

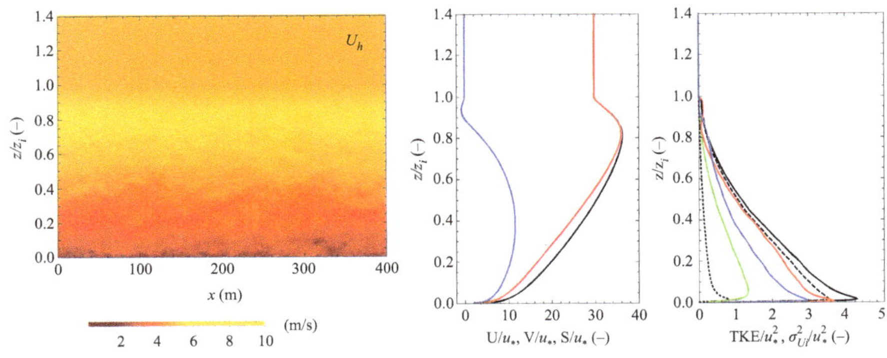

Figure 5.2 As in Figure 5.1, but for a stable ABL

with similar expressions by others [7,13,14] also finding common use.[‡] In addition to treatment of momentum (via the non-dimensional wind shear ϕ_m and wind profile correction function Ψ_m), MOST also applies to temperature and other scalars, with non-dimensional functions introduced in similar ways as presented here for U.

[‡]There are also physical arguments that in free convection $\phi_m \propto (z/L)^{-1/3}$, with corresponding forms for unstable conditions reflecting such, e.g. [15].

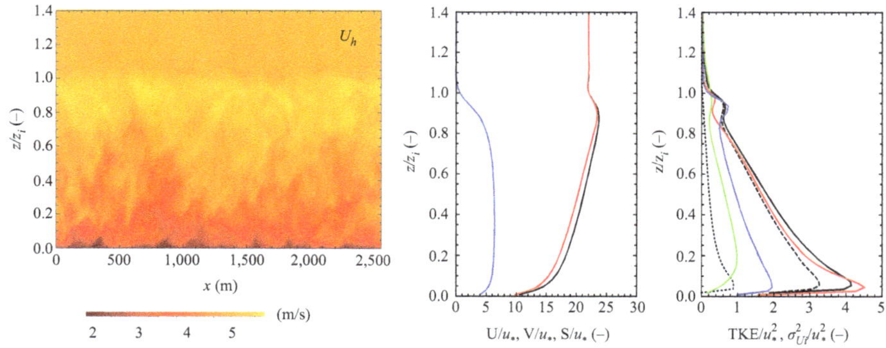

Figure 5.3 As in Figure 5.1, but for a conditionally neutral ABL (neutral, but with capping temperature inversion)

In the surface layer, nearly constant fluxes mean that the standard deviations of the wind velocity component fluctuations $\{\sigma_u, \sigma_v, \sigma_w\}$ can be assumed proportional to the surface friction velocity $u_\star$:

$$\sigma_u = Au_\star, \qquad \sigma_v \sim 0.75\sigma_u, \quad \text{and} \quad \sigma_w \sim 0.5\sigma_u, \tag{5.19}$$

with the constant $A \approx 2.2$–2.5 [14,16,17]. With the definition

$$I_u(z) \equiv \frac{\sigma_u}{U(z)} \tag{5.20}$$

for (streamwise) turbulence intensity, which in the IEC standard [6] plays a central role (more about this later), we arrive at

$$I_u(z) = \frac{\kappa A}{\ln(z/z_0)} \simeq \frac{1}{\ln(z/z_0)}. \tag{5.21}$$

In many wind-engineering applications and in the IEC standard, a power-law wind profile is often employed instead of (5.12). Over a limited vertical range (for example over the rotor area of a wind turbine) the power law

$$U(z) = U_{\text{ref}} \left(\frac{z}{z_{\text{ref}}} \right)^{\alpha} \tag{5.22}$$

is often a good approximation, with α not varying significantly with height [18]. To calculate the power-law exponent α, for wind measurements U_1 and U_2 at heights z_1 and z_2, respectively, one uses the form

$$\alpha = \frac{\ln(U_2/U_1)}{\ln(z_2/z_1)}; \tag{5.23}$$

this follows from seeing that $\alpha = d(\ln U)/d(\ln z)$ in (5.22). The two wind profile formulations (log-law and the power-law) can be related by equating the wind shear (dU/dz) evaluated at z_{ref} for the two expressions. This results in

$$\alpha_{\text{log-law}} = \frac{1}{\ln\left(z_{\text{ref}}/z_0\right)} \tag{5.24}$$

This is identical to the expression obtained for the turbulence intensity in (5.21), using $A = 2.5$.

For non-neutral conditions [9] considered the TKE rate equation non-dimensionalized by viscous dissipation rate; including only SP, viscous dissipation, and BP/destruction in stationary conditions, they derived a simple expression $\alpha(I_u, z/L)$. That is, ignoring turbulence transport, the relation between α and I_u can be modified to account for stability according the effects of buoyancy on TKE; casting the form of [9] in a general way consistent with M–O theory and the related work of [19], we have

$$\alpha = I_u\left(1 + \beta_k \frac{z}{L}\right). \tag{5.25}$$

This can be applied to, e.g., probabilistic loads calculations with shear, as done by [19]. In stable conditions, $L > 0$ and the wind shear increases, with $\beta_k \sim 5$. On the other hand, for unstable conditions $L < 0$ and the shear decreases; however, in unstable conditions, we note that one may generally not neglect turbulent transport (e.g. [20]).

5.3.2 Above the surface layer: typical wind turbine rotor heights

At the top of the ABL at $z \sim z_i$, the mean atmospheric lapse rate $d\Theta/dz$ becomes significantly positive, and hence the 'free' atmosphere is stably stratified. The lapse rate is somewhere between 1 and 10 K/km, hence turbulence ceases to exist. This stable layer on top of the ABL is responsible for the entrainment of warmer air (larger Θ) into the ABL, which as a consequence grows. That is, $dz_i/dt > 0$, making the whole ABL non-stationary even in the canonical convective, stable and neutral boundary layer cases. A stable layer is indeed present in all three canonical boundary layers depicted in Figures 5.1–5.3, where turbulence is absent. There has been a tendency in the wind energy community to ignore the top-down effect of stable stratification above the ABL, since the effect tends to impact high altitudes above the hub height of first-generation wind turbines. The modern development of large wind turbines, however, forces us to change our view – and necessitates introduction of a stable 'free' atmospheric layer. Various studies with LES have shed light on the relation between the free atmosphere and the operation of wind farms [21–24]. In analogy with *bottom-up* in MOST, the introduction of a free atmospheric lapse rate and entrainment can also be denoted *top-down* [1], and its effect on the wind profile can be significant [25].

At $z \sim z_i$ the wind is often close to being in geostrophic balance (particularly in the mid-latitudes), i.e. there exists a balance between the horizontal pressure gradient force and the Coriolis force; both forces are also present inside the ABL. The latter is responsible for the directional change of mean wind speed with height (also called wind veer). Just like the effect of the capping inversion, this effect has also been largely ignored in the wind energy community due to limited wind veer across small rotor diameters. Significant veer has been observed over modern turbine rotor extents and can affect wind turbine production and loads [26].

In order to capture even more realistic scenarios than what is contained within the three canonical boundary layers, the temporal variations in surface temperature, heat flux, and geostrophic velocity need to be included, as well as geostrophic shear and advection terms representing large-scale forcing [27,28].

5.4 Which characteristics of turbulence affect wind turbines?

From an engineering point of view, it is only important to resolve characteristics of atmospheric turbulence which affect the operation of wind turbines. In the turbulence community concerned with advanced fundamental research, the focus on departures from Kolmogorov scaling in high Reynolds number flows has led to insights into universal properties of turbulent flow. The intermittency observed is quantified by anomalous scaling of structure functions and explained by multifractal theories (e.g. [29,30]). To what extent do we need to take such findings into account, when simulating turbulence within the context of wind energy? Due to its complexity and demands on resources in regards to turbulence modelling, the answer has still not been settled. What seems much more relevant – for now at least – is the quantification of the sensitivity of wind turbines to the different physical variables contained in the IEC standard. A recent study [31] (denoted 'W2L' hereafter) examined precisely this topic, addressing the question

Which physical variables are responsible for turbine loading?

Following the IEC standard, atmospheric turbulence is expressed basically through a few measurable quantities: mean wind speed at hub height, U; vertical shear (exponent) of the mean wind, α; streamwise turbulence intensity, I_U; turbulence length scale, L; and a variable quantifying the anisotropy of turbulence, Γ (as we will see later, Γ is one of the external parameters of the Mann model). Furthermore, we have density of air, ρ; yaw angle, ϕ_h; flow inclination (yaw in the vertical direction), ϕ_v; and a parameter indicating the change of wind direction with height, the wind veer $\Delta\phi$. In total, there are nine external parameters which can in principle be obtained from observations around a wind turbine. There is therefore no explicit reference to buoyancy, one of two mechanisms through which turbulence is produced in the ABL. Instead the effects of stability (e.g. low shear/high turbulence intensity in convective conditions and high shear/low turbulence intensity in stable conditions) are represented via the variables listed above, which are used as inputs to load-calculation models.

By establishing practical bounds for and the interdependence of the nine parameters in non-complex terrain, a load database was created by conducting a very large number of simulations with the aeroelastic tool HAWC2 [32]; this was accomplished also simulating turbulence input fields via the Mann model (see Section 5.5.2.2), using the 10 MW DTU test turbine [33]. Site-specific lifetime fatigue loads were estimated in different IEC wind climates for different turbine component loads through various methods, including polynomial chaos expansion (PCE). Besides providing a fast way to estimate turbine loading, the methodology gives the sensitivity of the loads to the nine external parameters, through calculation of Sobol sensitivity indices. The Sobol index expresses the influence of input variance (the variance of for example mean wind speed at hub height) on the variance of the output (the variance of, for example, the tower base fore–aft moment). Results from the distributions of loads in the database are presented in Table 5.4. The largest Sobol indices – and hence the most important parameters – are the mean wind speed (U) and the variance of turbulent fluctuations (σ_u). The wind shear exponent α was also found to have some influence, especially on the blade-bending moments. Turbulence characteristics such as the length scale L and anisotropy factor Γ are somewhat less important in most of the cases; meanwhile, the wind veer $\Delta\phi$, air density ρ, yaw ϕ_h, and flow inclination ϕ_v had negligible effect on the fatigue loads. Similar results were found when looking at distributions of site specific loads. It should be remembered that the turbine used in this study is the 10 MW test turbine of DTU wind energy with a hub height of 119 m and rotor diameter of 178 m, and that the range of variables is limited to values within realistic operational settings in non-complex terrain: yaw and flow inclinations are, for example, limited to $\pm 10°$. One should therefore be cautious when extrapolating the conclusions to other turbines and to ranges in the variables covering complex flow situations, such as larger wind shears and flow inclinations.

In the W2L study, the turbulent flow fields were synthesized via the Mann spectral-tensor model, thus implicitly (as we will soon learn) carrying the assumptions of constant (effective) shear, vertical homogeneity, and Gaussian velocity component fluctuations. All these assumptions can be violated in various situations in the ABL, and so should be questioned. LES, on the other hand, does not rely on such assumptions but is instead dependent on the numerical set-up, discretization method, and mesh. Mesh size and resolution are known to particularly affect the moments of turbulence fluctuations such as the variance, which the W2L study identified as a crucial parameter for the fatigue loads.

Table 5.1 PCE-based Sobol sensitivity indices for high-fidelity database distribution of fatigue loads. Since the polynomial expansion from which the Sobol indices are calculated includes cross-terms between different variables, the sum of indices can exceed one. © 2018. Reproduced, with permissions, from [31]

Loads (moments)	**External variables**								
	U	σ_u	α	L	Γ	$\Delta\varphi$	$\bar{\varphi}_h$	$\bar{\varphi}_v$	ρ
Tower base fore–aft	0.42	0.65	0.01	0.03	0.02	0.01	0.00	0.00	0.01
Tower base side-side	0.62	0.42	0.05	0.04	0.04	0.02	0.02	0.02	0.02
Blade root flapwise	0.20	0.64	0.19	0.02	0.01	0.00	0.01	0.00	0.02
Blade root edgewise	0.22	0.54	0.25	0.05	0.03	0.01	0.01	0.03	0.01
Tower top yaw	0.14	0.85	0.00	0.02	0.01	0.00	0.00	0.00	0.01
Main shaft torsion	0.48	0.53	0.01	0.06	0.01	0.01	0.01	0.01	0.01

5.5 Synthetic turbulence and standard industrial approach

A number of 'simplified' models exist, which do not solve the fluid equations of motion *per se*; they instead solve either simplified versions of the N–S equation or involve empirical representation consistent with observations (and sometimes with basic fluid-dynamics theory). Such models facilitate generation of time series of turbulent velocity fields for input to wind turbine loads calculations. In this section, we focus on spectrally based models, because of their efficacy and widespread industrial use.

5.5.1 Statistical attempts

A basic starting point, which is basically unrelated to the equations of motion, is to use a recursive Markov process for the streamwise velocity component fluctuation $u(t)$:

$$u(t + \Delta t) = \gamma u(t) + \xi(t), \tag{5.26}$$

where ξ is random and is normally distributed for each t, having unit variance and zero mean. Since turbulence components in the ABL are usually nearly Gaussian (and non-Gaussianity does not significantly affect turbine loads [34]), the Gaussian assumption for ξ is tenable. The 'memory' (or correlation) parameter γ determines the variance of such an autoregressive process: $\sigma_u^2 = 1/(1 - \gamma^2)$. This simple methodology has been extended (e.g. [35]) for use with both turbulence and gusts, offering some insight within research, but is not in standard use.

5.5.2 Standard spectral models

The standard method used in wind energy (IEC 61400-1 Ed.3, Appendix B [6]) is to generate time series of a (statistically stationary)[§] random velocity vector field based on spectra of turbulence components. Two spectral models are prescribed in the IEC standard [6]:

1. An adaptation of the classic Kaimal *et al.* component spectra [36] plus an empirical coherence model extended from Davenport [37,38], essentially the 'Sandia method' of Veers [39].
2. The spectral-tensor model of Mann [4].

According to the study of [38], the empirical coherence formulation (5.37) tends to have problems at low frequencies and large separations; accordingly the IEC 61400-1 (editions 3 and 4 [6,40]) recommends the Mann [4] model. Below we give the necessary background for, and details of, these two approaches.

[§]In wind energy applications, stationarity is invoked in practice by assuming that the wind statistics (e.g. means and standard deviations) are constant per averaging period; this averaging period is typically 10 min, though one can also find and use, e.g., 30 min averages. However, one must take care that data analysed do not violate this assumption, or possibly enforce such an assumption by *de-trending* the turbulent velocity signal for each averaging period.

5.5.2.1 Kaimal spectra with exponential coherence model

As prescribed by the IEC 61400-1 standard, the Kaimal-based method consists of two parts: a prescription for velocity-component spectra and an empirical parameterization for the transverse coherence of streamwise velocity fluctuations (the normalized cross-spectrum of u at two different points in space, covered in detail below). To start, it is useful to define a 1-d spectrum that we typically can calculate from time series, e.g. that corresponding to streamwise ('longitudinal') velocity fluctuations $u(t)$ measured by an anemometer. The temporal spectrum is defined through the *autocovariance* (covariance of $u(t)$ with itself lagged by time τ), i.e.

$$R_{11}(\tau) \equiv \langle u_1(t)u_1(t+\tau)\rangle \tag{5.27}$$

which in $\{u, v, w\}$ notation we just write as $\langle u(t)u(t+\tau)\rangle$. The *power spectral density*, or 'spectrum' corresponding to $u(t)$ times itself, is a single-point statistic that can be defined as

$$S_{11}(\omega) = \frac{1}{2\pi}\int_{-\infty}^{\infty} R_{11}(\tau)e^{-i\omega\tau}d\tau = \frac{1}{\pi}\int_{0}^{\infty} R_{11}(\tau)e^{-i\omega\tau}d\tau \tag{5.28}$$

where the angular frequency is defined $\omega \equiv 2\pi f$ and the factor of 2 with lower limit change arises because $\rho_{uu}(-\tau) = \rho_{uu}(\tau)$. Conversely, the covariance can be expressed in terms of the spectrum,

$$R_{11}(\tau) = \int_{-\infty}^{\infty} S_{11}(\tau)e^{-i\omega\tau}d\omega, \tag{5.29}$$

so it is said that $R_{ij}(\tau)$ and $S_{ij}(f)$ form a 'Fourier transform pair'.¶

In practice, one typically calculates, e.g., $S_{11}(f)$ without calculating the autocovariance $R_{11}(\tau)$, by taking the (discrete) Fourier transform of the time series, often accomplished by a 'fast' Fourier transform (FFT) implementation. Then, the spectrum $S_{11}(f)$ is found via

$$S_{11}(f) = \frac{|\mathrm{FFT}\,\{u(t)\}|^2}{2\pi T}$$

where T is the record length (e.g. 600 s).‖

The forms for component spectra adopted by the IEC follow from the classic normalized 1-d power-spectra forms of Kaimal [36]; here, the non-dimensionalization

¶We note that for non-periodic functions, mathematically, one cannot exactly use a Fourier transform for a random process, but instead of (5.28) one can use a Fourier–Stieltjes integral, as described in, e.g., Ch. 15 of [1] or Appendix E cf [2]. However, the calculations used are essentially the same in practice.

‖Let the reader beware that some FFT routines give one-sided transforms corresponding to discrete frequencies of $f = \{1/T, 2/T, 3/T \ldots\}$ and the right-hand side of (5.28), while others give 'two-sided' output corresponding to negative and then positive frequencies; different FFT routines may also have a factor other than $1/\pi$ or $1/2\pi$ in their definition. One can also see from (5.29) that since $R_{11}(0) = \langle u^2\rangle = \sigma_u^2$, the one-sided spectrum (from 0 to ∞) integrates to half the variance, $\sigma_u^2/2$; this is often a good 'sanity check' for the calculated spectra.

is per $\langle u_k^2 \rangle / f$. The IEC implementation uses a one-sided spectral form $S_k(f)$ corresponding to $u_k(t)$, defined such that

$$\int_0^\infty S_k(f) df = \langle u_k^2 \rangle, \tag{5.30}$$

i.e. with a factor of 2 compared to (5.29) above. The spectra are expressed as

$$\frac{f\, S_k(f)}{\langle u_k^2 \rangle} = \frac{4fL_k/V_{\text{hub}}}{(1+6fL_k/V_{\text{hub}})^{5/3}}. \tag{5.31}$$

The IEC prescribes

$$\{\sigma_1, \sigma_2, \sigma_3\} = \{1, 0.8, 0.5\}\sigma_1, \tag{5.32}$$

i.e. very similar to (5.19). Furthermore, it sets the length scale

$$L_k = \{8.1, 2.7, 0.66\}\Lambda_1, \tag{5.33}$$

where $\Lambda_1 = \min\,[0.7z, 42\,\text{m}]$.

Following the streamwise component example (5.28), the spatial correlation in the cross-wind directions of $u(t)$ is relatable as a function of frequency by the *coherence*. The coherence is defined through the two-point cross-covariance, which for this instance can be written as

$$R_{11}(\tau, \Delta y, \Delta z) \equiv \langle u_1(t, y, z) u_1(t+\tau, y+\Delta y, z+\Delta z) \rangle.$$

Just as the (Fourier) integral of the single-point autocovariance $R_{11}(\tau)$ gives the spectrum $S_{11}(f)$ according to (5.28), the two-point cross-covariance $R_{11}(\tau, \Delta y, \Delta z)$ analogously gives a (two-point) statistic known as the cross-spectrum:

$$\chi_{11}(\omega, \Delta y, \Delta z) = \frac{1}{2\pi} \int_{-\infty}^{\infty} R_{11}(\tau, \Delta y, \Delta z) e^{-i\omega\tau} d\tau. \tag{5.34}$$

The coherence is simply the two-point cross-spectrum normalized by the corresponding single-point autospectra, which in the case of $u_1 u_1$ is simply

$$\text{coh}_{11}(\omega, \Delta y, \Delta z) = \frac{|\chi_{11}(\omega, \Delta y, \Delta z)|^2}{[S_{11}(\omega)]^2}. \tag{5.35}$$

This streamwise component case can be generalized from $\{i, j\} = \{1, 1\}$ to any $\{i, j\}$; the coherence can be expressed generally then as

$$\text{coh}_{ij}(\omega, \Delta y, \Delta z) = \frac{|\chi_{ij}(\omega, \Delta y, \Delta z)|^2}{S_i(\omega) S_j(\omega)}. \tag{5.36}$$

The IEC 61400-1 gives a prescription for coherence of streamwise fluctuations as

$$\mathrm{coh}_{11}(f, r) = e^{-12\sqrt{(fr/V_{\mathrm{hub}})^2+(0.12r/L_1)^2}}, \tag{5.37}$$

where $r = \sqrt{(\Delta y)^2 + (\Delta z)^2}$ is the distance perpendicular to the mean wind direction, and the length scale $L_1 = 8.1\Lambda_1$ is the same as used in the dimensionless spectrum (5.31).

5.5.2.2 Mann model: rapid-distortion theory with eddy lifetime

The Mann spectral-tensor model [4] uses RDT [2,5,40], building upon it with a sub-model for eddy-life time. In contrast to the empirical forms of (5.31) and (5.37), RDT is a solution of the quasi-linearized N–S equation. Also in contrast to the empirical forms for the streamwise component, the Mann-model solves for all three velocity components in 3-d Fourier space, implicitly including 3-d coherences between all components; this is done for constant shear dU/dz over flat terrain. The latter assumption is in contrast to the $1/z$ dependence of shear in the surface layer (5.13); but used as an effective rotor-average (and considering the reduced z-dependence of dU/dz generally found above the surface layer** where modern turbine rotors are often located), this assumption has been quite successful. The advantage of assuming constant shear, as we will experience shortly, is the utilization of Fourier components in all three directions.

Spectral tensor

To start, we now consider velocity components in 3-d Fourier space. The Mann-model assumes homogeneous turbulence, meaning that a spectral velocity tensor can be defined through the 3-d Fourier transform of the velocity covariance R_{ij}. The 3-d covariance tensor

$$R_{ij}(\boldsymbol{r}) = \langle u_i(\boldsymbol{x})u_j(\boldsymbol{x}+\boldsymbol{r})\rangle \tag{5.38}$$

involves a separation $\boldsymbol{r} = \{r_1, r_2, r_3\}$ between two points in space and does not depend on $\boldsymbol{x}$ assuming homogeneous turbulence. $R_{ij}(\boldsymbol{r})$ can be seen as the generic 3-d version of the 1-d (in time) autocovariance $\langle u_1(t)u_1(t+\tau)\rangle$ given in (5.27); the latter simply had a time-lag τ instead of spatial separation $\boldsymbol{r}$ and considered u_1 times itself instead of any u_iu_j. Now the general 3-d (spatial) version of the 1-d spectrum (5.28) can be written

$$\Phi_{ij}(\boldsymbol{k}) = \frac{1}{(2\pi)^3}\iiint R_{ij}(\boldsymbol{r})e^{-i\boldsymbol{k}\cdot\boldsymbol{r}}d\boldsymbol{r}, \tag{5.39}$$

where $\boldsymbol{k} = (k_1, k_2, k_3)$ is the 3-d wavenumber vector. Thus, $R_{ij}(\boldsymbol{r})$ and $\Phi_{ij}(\boldsymbol{k})$ form a Fourier transform pair, so basically the spectral tensor Φ_{ij} is the 3-d Fourier transform of the ensemble-mean of u_i times u_j. For example, $\Phi_{11}(\boldsymbol{k})$ is the 3-d version of the power spectrum of streamwise velocity component. The power spectrum of $u(t)$ in

**This is above the surface layer, but with the exception of rotors approaching, the entrainment zone at the top of the ABL.

the mean wind (streamwise) direction can be obtained from (5.39), similar to how it was obtained in (5.28):

$$S_{11}(k_1) = \frac{1}{2\pi}\int_{\infty}^{\infty} R_{11}(\{r_1, 0, 0\})e^{-ik_1 r_1}dr_1; \tag{5.40}$$

for the more general case of $S_{ij}(k_1)$, one replaces R_{11} with R_{ij}.

To connect or compare with typical measurements taken over time, $R_{11}(r_1)$ and $S_{11}(k_1)$ in (5.40) must be converted into $R_{11}(\tau)$ and $S_{11}(\omega)$ of (5.28);[††] to do this one may invoke Taylor's 'frozen turbulence' hypothesis

$$x = Ut, \quad k_1 = \omega/U = 2\pi f/U. \tag{5.41}$$

However, as the Mann-model solves for $\Phi_{ij}(\boldsymbol{k})$, we must relate it to the 1-d spectra that we typically calculate from measurements. This follows from considering that the cross-spectrum in three dimensions can be written as [42]

$$\chi_{ij}(k_1, \Delta y, \Delta z) = \iint \Phi_{ij}(\boldsymbol{k})\mathrm{e}^{\mathrm{i}(k_2\Delta y + k_3\Delta z)}dk_2 dk_3, \tag{5.42}$$

which for zero separation ($\Delta y = \Delta z = 0$) becomes a single-point statistic, the 1-d spectrum

$$S_{ij}(k_1) = \chi_{ij}(k_1, 0, 0) = \iint \Phi_{ij}(\boldsymbol{k})dk_2 dk_3. \tag{5.43}$$

Mann-model implementations do the integral in (5.43) numerically, giving single-point spectra as output.

Rapid-distortion theory

The N–S equation (5.6) for the full velocity (with Reynolds decomposition $\widetilde{u_i} \equiv U_i + u_i$) can be simplified by linearizing it (retaining terms to first order in u_i) and neglecting the viscous term in the high Reynolds-number limit:

$$\frac{Du_i}{Dt} = -u_j\frac{\partial U_i}{\partial x_j} - \frac{1}{\rho}\frac{\partial p}{\partial x_i}. \tag{5.44}$$

Here u_i is the instantaneous ith velocity component, p is the pressure, ρ is the density, and g is the acceleration due to gravity; we neglect the Coriolis force, following [4]. Taking the divergence of (5.6) and using the Reynolds decomposition $\widetilde{p} = P + p$ leads to the Poisson equation

$$\frac{1}{\rho}\nabla^2 p = -2\frac{\partial U_j}{\partial x_i}\frac{\partial u_i}{\partial x_j} - \frac{\partial^2}{\partial x_i \partial x_j}(u_i u_j - \langle u_i u_j\rangle). \tag{5.45}$$

The *rapid-distortion limit* assumes that the time scale implied by the shear (scaling as the reciprocal of $\partial U_i/\partial x_j$) is much shorter than the turbulence time scale ($\sim u_i u_i/\varepsilon$).

[††]Now S_{11} can be obtained via (5.40) by measuring wind speeds with LIDAR aimed upwind (nearly simultaneously at different r_1), but in practice, we are still most often dealing with time series.

In this limit, then the right-most term becomes negligible, making (5.45) linear in p; meanwhile, (5.44) is linear in u_i.

Because the rapid-distortion equations are linear in the fluctuations u_i and p, any given Fourier mode (of either) evolves independently. Thus, we will Fourier transform the above equations to solve per wavenumber. But the total derivative and independence of Fourier modes imply that **k** itself is a function of time, with the advective part of (D/Dt) implying that $dk_k/dt = -k_j \partial U_j/\partial x_k$ [4]; this allows the spatial derivative portion to be written in terms of $\mathbf{k} = k_k$. Fourier transforming and combining the transformed (5.44) and (5.45) to eliminate p, then if we consider that in the atmosphere $\partial U_j/\partial x_j = \partial U/\partial z$ (flow over relatively uniform ground), one arrives at the rapid-distortion equation to be solved:

$$\frac{D}{Dt} dZ_i\left(\boldsymbol{k}(t)\right) = \left(2\frac{k_i k_1}{k^2} - \delta_{i1}\right)\left(\frac{dU}{dz}\right) dZ_3\left(\boldsymbol{k}(t)\right), \tag{5.46}$$

where the Fourier–Stieltjes integral[‡‡] form [1,2,43]

$$u_i(\boldsymbol{x}) = \int \mathrm{e}^{\mathrm{i}\boldsymbol{k}\cdot\boldsymbol{x}} \mathrm{d}Z_i(\boldsymbol{k}) \tag{5.47}$$

is employed to write (in effect) the equation for evolution of the Fourier modes dZ_i. The wavenumber evolution equation $dk_3/dt = -k_1 \partial U/\partial z$ gives

$$k_k(t) = \{k_1, k_2, k_{3,0} - k_1 t \partial U/\partial z\}.$$

Along with $k_k(t)$, (5.46) is solved using the initial conditions $dZ_i(k_i(0))$. The spectral tensor and Fourier–Stieltjes 'modes' are related by

$$\Phi_{ij}(\boldsymbol{k}) = \frac{\langle dZ_i^*(\boldsymbol{k}) dZ_j(\boldsymbol{k})\rangle}{dk_1 dk_2 dk_3}, \tag{5.48}$$

where $()^*$ denotes complex conjugate. The initial $dZ_i(k_k(0))$ is also defined via (5.48) based on an initial isotropic spectral tensor

$$\Phi_{ij}(k_k(0)) = \frac{E(k)}{4\pi k^2}\left(\delta_{ij} - \frac{k_i k_j}{k^2}\right) \tag{5.49}$$

where the isotropic energy spectrum follows the von Kármán [44] form

$$E(k) = \alpha\varepsilon^{2/3} L^{5/3} \frac{(kL)^4}{(1 + (kL)^2)^{17/6}}. \tag{5.50}$$

Here L is basically the length scale corresponding to the isotropic spectral peak, $\alpha = 1.7$ is the universal Kolmogorov constant, and ε is the rate of (viscous) dissipation of TKE. L and ε are two of the three parameters of the Mann model, with the latter often being replaced simply by $\alpha\varepsilon^{2/3}$ because $E(k)$ is proportional to it. The solution of the Fourier rapid-distortion equation (5.46) must now be found at some time, t, 'after' the initial isotropic condition, for each wavenumber.

‡‡The integration in (5.47) is over all the wavenumber space [43].

Eddy-lifetime and solution

In order to find a distortion time per eddy size (k^{-1}), Mann [4] constructed a velocity scale $v_{\text{eddy}} \sim \sqrt{\int_k^\infty E(k)dk}$ for this size based on the von Kármán energy spectrum (5.50), and subsequently a time scale $\tau_M(k) \sim k^{-1}/v_{\text{eddy}}$, giving

$$\tau_M(k) = \frac{\Gamma}{dU/dz} \frac{(kL)^{-2/3}}{\sqrt{{}_2F_1\left(1/3, 17/6; 4/3.(-1/(kL)^2)\right)}}. \tag{5.51}$$

Since ${}_2F_1\{1/3, 17/6; 4/3, (-kL)^{-2}\} \simeq [1 + 3.07(kL)^{-2}]^{-1/3}$ as shown in [17],[§§] we see expression (5.51) means that smaller and smaller eddies (higher k) are modelled to be distorted for smaller durations. The constant Γ is known as the eddy-lifetime parameter or anisotropy parameter and is the third constant of the Mann-model; it dictates the relative duration of rapid distortion at all scales – and thus how *anisotropic* the turbulence becomes. A value of $\Gamma = 0$ means the output spectrum is simply the isotropic initial spectrum.

With the eddy-lifetime $\tau_M(k)$ (5.51) then Mann [4] gives closed form solutions for dZ_i and subsequently $\Phi_{ij}(\mathbf{k}(\tau))$ as a function of wavenumber, in terms of the initial isotropic spectrum (with parameters $\alpha\varepsilon^{2/3}$ and L), multiplied by a function of $\tau dU/dz$; i.e. each component spectrum looks like $E(k)$ of (5.50) multiplied by a function of $\Gamma(kL)^{-2/3}[1 + 3.07(kL)^{-2}]^{1/3}$. From the Φ_{ij}, one can then obtain 1-d frequency spectra $S_{ij}(f)$ via (5.41) and (5.43), as explained above. Due to the length of the solution's expression, we refer the reader to [4] for details.

Basically, the solution is as such: an initial isotropic von Kármán spectrum $E(k)$ with peak length scale L is distorted per wavenumber due to the shear dU/dz, where higher values of Γ correspond to longer (more) distortion. As shown in Figure 4 of [4], the streamwise spectral peak scale grows with Γ (from $\sim 2L$ to $5.5L$ over the range Γ from 2–4), while the smaller peak scale of $S_{33}(k_1)$ depends more weakly on Γ (shrinking from $\sim 0.6L$ to $0.5L$ over the same range); thus, the anisotropy parameter Γ acts to separate the peaks of $S_{11}(f)$ and $S_{33}(f)$, primarily by 'stretching' the streamwise peak to larger scales.

Standard IEC use of the spectral tensor model

The IEC 61400-1 Ed.3 standard [6] prescribes spectral-tensor model parameters based largely on the analysis of [4], setting $\Gamma = 3.9$, while setting the other two parameters based on the height and turbine class. The isotropic (undistorted, initial) turbulence length scale L is prescribed to be

$$L = 0.7\lambda_1 = 0.7 \min[0.7z, 42\,\text{m}]$$

[§§]From (5.51) with the hypergeometric function [45], $\tau_M \propto k^p$, where $p \to -1$ for $kL \ll 1$ and $p \to -2/3$ in the inertial range where $kL \gg 1$. This is in part due to the weak dependence on kL in Gauss' hypergeometric function ${}_2F_1$ [45].

where λ_1 is the same scale used for the coherence prescription.The other parameter is given indirectly by an expression equivalent to

$$\alpha\varepsilon^{2/3} = 0.44L^{-2/3}\sigma_1^2$$

where the IEC 61400-1 Ed.3 [6] defined

$$\sigma_1 \equiv I_{\mathrm{ref}}\,(0.75V_{\mathrm{hub}} + 5.6\,\mathrm{m/s})$$

based on the turbulence class reference intensity I_{ref} and hub-height wind speed.

Site-specific use; measurements and Mann-model input parameters

Just as in the first tests of the model [4,46], it is possible to obtain the model parameters via 3-d sonic anemometer measurements. Because the model output scales with $\alpha\varepsilon^{2/3}L^{5/3}$ (via the initial isotropic spectrum to be distorted), all possible output spectra can be calculated simply for a range of Γ and k_1 (or k_1L). Then this output can be fit to measure spectra from 3-d sonic anemometers, interpolating between different Γ-based results per the other two parameters [4,47]. Examples of typical spectra measured the above 2/3 hub-height is shown in Figure 5.4, along with the chi-squared fit of Mann-model result based on S_{11}, S_{22}, S_{33}, and S_{13} following the fitting method of [4]. One can see from the figure that in the inertial range, the spectral-tensor model replicates the universal '5/3-law' and '4/3-law' of Kolmogorov: $S_{11}(k_1)|_{k_1L\gg1} \to (9/55)\alpha^{2/3}\varepsilon k_1^{-5/3}$ where $S_{33}(k_1)|_{k_1L\gg1} = S_{22}(k_1)|_{k_1L\gg1} = (4/3)S_{11}(k_1)|_{k_1L\gg1}$.

Even before 2005, it was observed that the parameters prescribed by the IEC 61400-1 Ed.3 [6] can vary from those found by fitting spectra [49]. More recently, Kelly [17] found for a number of sites and flow regimes (simple, coastal, and forested) that L can often be smaller than the IEC-prescribed values (often due to stably stratified conditions), as well as occasionally being much larger (due to, e.g., unstable conditions), with L possessing a non-symmetric distribution over the long

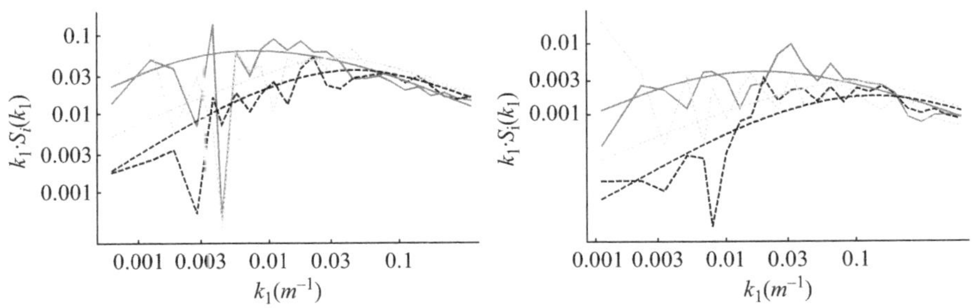

Figure 5.4 Observed spectra (jagged lines) with Mann-model fit (smooth curves) from z = 80 m at Høvsøre on the west coast of Denmark [48]. Left: slightly unstable case from offshore, Γ = 2.5 and L = 41 m; right: stable case over land, Γ = 2.9, L = 14 m. Solid/Grey is u-component, dashed/black is w-component, and dotted/light-grey is v-component

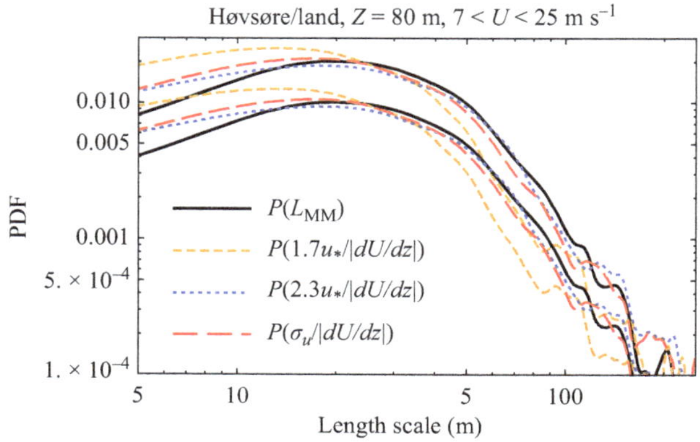

Figure 5.5 Probability density function of turbulent length scale from 1 year of observations at Høvsore from the homogeneous eastern land sectors, from [17]. Solid (black): L from fits to spectra; dotted (blue): 'mixing-length' formulation with revised constant; dashed (yellow): Peña [50] form; long-dashed (red): form (5.52)

term. A distribution of L from 1 year of 10-min spectra, corresponding to Figure 5.4, demonstrates this and is shown in Figure 5.5. Further, [17] also showed that the turbulence length scale L can be obtained without sonic anemometers, using measured σ_u and shear:

$$L \approx \frac{\sigma_{u,\text{obs}}}{|dU/dz|}. \tag{5.52}$$

The efficacy of this expression for obtaining the distribution $P(L)$, towards probabilistic loads calculations, is also shown in Figure 5.5. Since turbine fatigue loads are affected by stability through stability's effect on U, σ_u, L, and dU/dz (e.g. [51]), then the estimate (5.52) for L facilitates (probabilistic) loads calculations without the need for 3-d sonic anemometers or stability measurements. However, while $\alpha\varepsilon^{2/3}$ can then be found via σ_u^2, there is no simple method yet to find Γ. Contrary to the IEC-prescribed value of 3.9, Γ typically varies between 2 and 4, most commonly falling around 3; a neutral asymptotic value of 3.1 can be inferred from [17], consistent with the stability distributions found in [52]. This is further demonstrated by Figure 5.6, which shows the probability density of Γ from 1 year of spectra 80 m above homogeneous land. Γ is affected by stability (e.g. [51]), but not in a simply described way [53].

5.5.3 Extensions of the spectral-tensor model

de Mare and Mann [54] constructed a model for non-stationary (variable) eddy-lifetime for application towards, e.g., meandering wakes. Chougule *et al.* [53,55] modified the spectral tensor model by including buoyancy in the vertical velocity

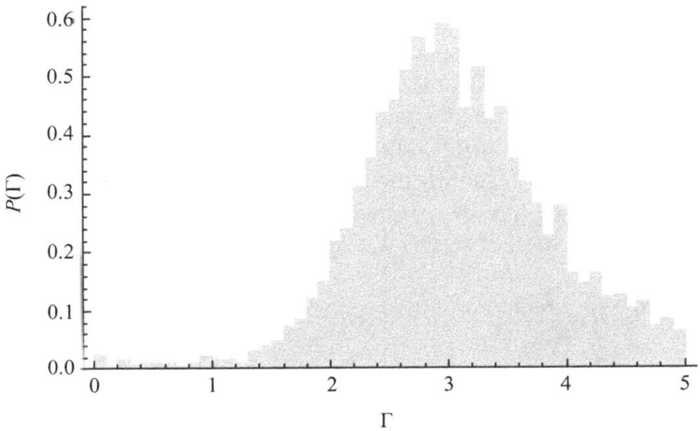

Figure 5.6 Probability density function of anisotropy parameter Γ, from the same observations as in Figure 5.5

equation. Segalini and Arnqvist [56] also modified the Mann-model for stably stratified conditions, additionally modifying the eddy lifetime for the effect of stratification. However, models including extension for atmospheric stability are not yet quite usable for more unstable (convective) conditions, and obtaining their input parameters from direct (sonic anemometer heat flux) measurements is not trivial [57,58]. Use of the original three-parameter model with site-specific L and dU/dz may currently be more tenable as it includes the effects of stability (as described above); a comparison of both methods [17] and [58] has yet to be made. Ongoing and remaining work in this area includes translation of standard measurements to obtain site-specific Γ, translation of measurements to buoyant-spectral tensor model parameters without spectra, modification for the effect of the Coriolis force, and adaptation for convective (more unstable) simulation.

5.5.3.1 Turbulence synthesis

For practical application, the spectral forms and coherence – whether explicitly set as in Veers' method [39] or implicit as in the Mann-model [4] – must be used to generate time series of turbulent wind for inputs to load simulations and calculations.

Synthesis from Veers/Sandia-type models

The details of simulation for prescribed Kaimal spectra and cross-wind coherence of u can be found in [39], with extensions of this (e.g. v- and w-components) found in [59,60].

The basic premise starts with using single-point spectra such as (5.31) to populate the diagonals of an $N \times N$ matrix, B_{mn}. Each index (from 1 to N) corresponds to a point in the 2-d plane being simulated perpendicular to the mean wind; note that the m, n *do not* correspond to dimensions as in the previous notation, but rather here to numerical indices referring to different locations in the plane. Then using a formulation for coherence such as (5.37) is used to calculate the off-diagonals of the B-matrix by multiplying the coherence corresponding to the distance between the points $(y,z)_m$ and $(y,z)_n$ by $\sqrt{S_1(f)|_m S_1(f)|_n}$ (for the u case). Random phases are assigned for the diagonals $B_{m=n}$ and zero phase assigned for $B_{m\neq n}$ via phase matrix $X_{mn} = \delta_{mn} e^{i\theta_{\mathrm{rand}}}$. Fourier coefficients of the speed for each of the N points are calculated by solving $H_{pq}(f)H^*_{qp}(f) = B_{pq}$ for $H(f)$, noting that H_{pq} summed over p or q corresponds to the Fourier coefficient for point q or p, respectively. The Fourier coefficients are found by the matrix multiplication of H_{pq} by X_{pq} by the unit vector 1:

$$V_p(f) = \sum_{q=1}^{p} H_{pq}(f) X_{qq}(f). \tag{5.53}$$

For each p, then V_p is found using inverse FFT, thus generating N correlated time series by linear combinations of N independent stochastic (e.g. Gaussian) processes. Further details can be found in [39].

Synthesis from Mann spectral-tensor model

Given the spectral tensor (or component spectrum or spectra) of turbulent velocity fluctuations, a general method for Fourier simulation of a so-called turbulence box – i.e. time series of one or more turbulence velocity components in two or three dimensions – is given by [46]. Basically, the Fourier–Stieltjes integral (5.47) is approximated by a discrete Fourier series within a box of dimensions $\Lambda_i = \{\Lambda_1, \Lambda_2, \Lambda_3\}$,

$$u_i(\mathbf{x}) = \sum_k e^{i\mathbf{k}\cdot\mathbf{x}} C_{ij}(\mathbf{k}) n_j(\mathbf{k}), \tag{5.54}$$

where the C_{ij} are coefficients found from $\Phi_{ij}(\mathbf{k})$ and n_j are independent Gaussian-distributed (stochastic) complex variables with unit variance. The discrete $\mathbf{k}$ have components $k_i = [-N_i/2, -N_i/2+1 \ldots N_i/2]2\pi/\Lambda_i$; the sum is done over all wave vectors. The Fourier coefficients are found by solving

$$C^*_{ik}(\mathbf{k}) C_{jk}(\mathbf{k}) = \frac{(2\pi)^3 \Phi_{ij}(\mathbf{k})}{(\Lambda_1 \Lambda_2 \Lambda_3)}, \tag{5.55}$$

i.e.

$$C_{ij}(\mathbf{k}) = \frac{(2\pi)^{3/2}}{(\Lambda_1 \Lambda_2 \Lambda_3)^{1/2}} A_{ij}(\mathbf{k}) \qquad \text{with} \quad A^*_{ik} A_{jk} = \Phi_{ij}. \tag{5.56}$$

Note for simulations where the length Λ_1 is not much bigger than L, that the above form must be adapted to account for the finite extent of the 'turbulence box' size, as shown in [46]; the latter work suggests this is the case if Λ_1 is not at least $3L$, or $\{\Lambda_2, \Lambda_3\}$ are not at least $8L$. Using a random-number generator to simulate the Gaussian variable n_j per each (3-d) wavenumber, (5.54) is calculated (inverse FFT) to get $u_i(\mathbf{x})$.

As pointed out by [46], the resultant field $u_i(\mathbf{x})$ is both incompressible and implicitly includes coherent phase information based on the linearized equations of motion. The latter, with dU/dz, causes, e.g., tilted elongated streak patterns; this is shown in Figure 5.7. Further, this model does not arbitrarily set the phase for two-point spectra

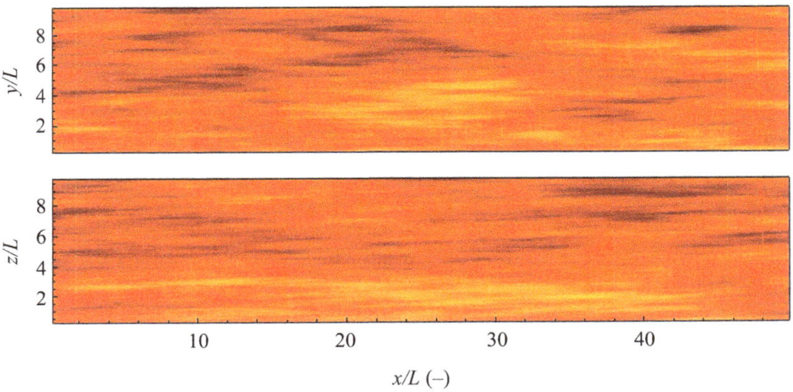

Figure 5.7 *Streamwise velocity component (u) from Mann-model simulation with $L = 30\,m$ and $\Gamma = 3$, side-view (top) and top-view (bottom). Fluctuation amplitude ranges from $-3\sigma_u$ (dark red) to $+3\sigma_u$ (yellow)*

but rather provides it as part of the solution; this allows, e.g., coherent drifting lateral motions as noted by [61]. However, the Mann-model does not treat the Coriolis force and gives no information about $\langle uv \rangle$.

5.6 Large eddy simulation

LES goes back to the late 1960s and was first developed in order to study the ABL [62–64].

The notion of LES of being too complicated, too expensive, and hence giving too inaccurate solutions is with the next generation of very large turbines being challenged. With hub heights above 100 m and blade tips reaching above 200 m, the assumptions made in, for example, the Mann model as described in Section 5.5.2.2 may become invalid. All the terms in the TKE equation (5.10) need to be included, this statement especially holds for the buoyancy production/destruction term, which only vanishes very close to the surface.

An aspect about large wind turbines which also favours the use of LES is that of scale. Larger turbines mean smaller operational impact from smaller unresolved turbulent scales, and the concept of only resolving the *larger energy containing eddies* is thus justified in wind turbine applications as, for example, fatigue load calculations. In addition, big hub height of the turbine moves the turbine away from the layer which again is hard to resolve due to vanishing turbulent scales.

5.6.1 The fundamentals

The governing equations are formed by applying a spatial filter, G, to the N–S equation, i.e. performing a convolution according to

$$\mathbf{u}^r(\mathbf{x},t) = \int\int\int_{-\infty}^{\infty} G(\mathbf{r},\mathbf{x})\tilde{\mathbf{u}}(\mathbf{x}-\mathbf{r},t)d\mathbf{r}, \tag{5.57}$$

so that the velocity field is split into [1]

$$\tilde{u}_i = u_i^r + (\tilde{u}_i - u_i^r) = u_i^r + u_i^s. \tag{5.58}$$

The explicit formulation of the filter, G, depends on the discretization scheme and numerical methodology used in the LES approach. G is non-zero on an interval $\{-\Delta/2; \Delta/2\}$, where the filter size Δ is linked to the size of the numerical mesh of the simulation [65]. Recasting the equations in spectral (Fourier) space, the filter operation can explicitly define the boundary between resolved ($k < \Delta_k$) and unresolved ($k > \Delta_k$) scales, where $\Delta_k = 2\pi/\Delta$. The advantage of utilizing a spectral cut-off filter in LES is counterbalanced by the implications of using periodic boundary conditions. Since this is not feasible in the vertical direction due to the surface, pseudo-spectral models are used, in which the vertical direction is treated in physical space.

Applying the convolution (5.57) to the governing equation (5.6), we end up with the LES momentum equation

$$\frac{\partial u_i^r}{\partial t} + u_j^r \frac{\partial u_i^r}{\partial x_j} = -\frac{1}{\rho_0}\frac{\partial p^r}{\partial x_i} + \frac{1}{\rho_0}\frac{\partial \tau_{ij}}{\partial x_i} - \frac{g}{\theta_0}\theta^r\delta_{i3} - \varepsilon_{ijk}\Omega_j u_k^r. \tag{5.59}$$

Here we have introduced the subfilter stress

$$\tau_{ij} = \rho_0[u_i^r u_j^r - (\tilde{u}_i\tilde{u}_j)^r] \tag{5.60}$$

due to spatial filtering, which includes contributions from scales smaller than the filter size Δ; this may be seen by breaking τ_{ij} up into more terms, including cross products between resolved and unresolved variables [66]. For this reason, τ_{ij} is also named the sub-grid-scale (SGS) stress, and it is a term that needs to be modelled – and thus the Achilles heel of LES is presented. It is the LES equivalent to the Reynolds stresses in RANS modelling, where the averaging procedure is performed as ensemble and time-averaging, and hence all statistics are stationary (constant in time). In order to compare the relative importance between the Reynolds stresses in LES and RANS, we perform an ensemble average of (5.59), as introduced in Section 5.1.1, and compare the Reynolds stress terms:

$$\underbrace{\langle u_i^r u_j^r \rangle - \frac{\langle \tau_{ij} \rangle}{\rho}}_{\text{LES}} = \underbrace{\langle u_i u_j \rangle}_{\text{RANS}}. \tag{5.61}$$

From (5.61), we can see that parameterization or *modelling* of the Reynolds stresses plays a more significant role in RANS compared to LES; in the latter, part of the stress $\langle u_i u_j \rangle$ is directly resolved by the flow in the simulation. If we choose Δ well into the inertial range, we furthermore have that

$$\langle u_i^r u_j^r \rangle > \frac{\langle \tau_{ij} \rangle}{\rho}. \tag{5.62}$$

Following a Kolmogorov scaling argument [67], one can show that

$$\langle u_i u_j \rangle \sim \ell^{7/3}, \tag{5.63}$$

and hence the majority of Reynolds stresses is carried by the larger scales resolved by LES. This is in principle the reason why LES works at all. Since the viscous dissipation is associated with very small scale much smaller than the filter size, Δ, this term vanishes and has thus been left out of (5.59).

In a similar fashion, spatially filtered equations for conservation of mass and for scalar fields (such as for example temperature) can be constructed:

$$\frac{\partial u_i^r}{\partial x_i} = 0, \tag{5.64}$$

and

$$\frac{\partial \theta^r}{\partial t} + u_j^r \frac{\partial \theta^r}{\partial x_j} = -\frac{\partial \tau_{\theta j}}{\partial x_j} + S, \tag{5.65}$$

where S represents various source terms and

$$\tau_{\theta i} = u_i^r \theta^r - (\tilde{u}_i \theta^r)^r \tag{5.66}$$

is the scalar SGS flux of θ.

5.6.2 SGS models

There exist a large number of SGS models, i.e. formulations of the τ_{ij}, and related scalar SGS terms like $\tau_{\theta i}$. A general assumption is to require the filter size Δ to satisfy $\Delta \ll \ell$, i.e. fall within the inertial subrange, where ℓ is some length scale characterizing the dominant energy-containing scales of the turbulence [66]. There are several families (generations) of basic SGS closure, which we outlined below.

5.6.2.1 Smagorinsky models: first-order and O(1.5) closure

Of all SGS models, the classic Smagorinsky model [62,68] is the most widespread. Like most SGS models which follow from it, the basis is a (filtered) flux-gradient relation [69]

$$\tau_{ij} = -2K_m S_{ij}^r, \tag{5.67}$$

where the strain rate tensor of resolved velocities is

$$S_{ij}^r = \frac{1}{2}\left(\frac{\partial u_i^r}{\partial x_j} + \frac{\partial u_j^r}{\partial x_i}\right), \tag{5.68}$$

and K_m is chosen proportional to the product of filter scale Δ and characteristic subgrid velocity scale $(S^r \Delta)$:

$$K_m = (C_s \Delta)^2 S^r. \tag{5.69}$$

Here C_s is the Smagorinsky constant and S^r is the magnitude of the strain-rate tensor S_{ij}^r, so that $S^r = (2S_{ij}^r S_{ij}^r)^{1/2}$ [2]. In homogeneous surface layers $K_m = \kappa z u_\star$, and (5.67) is consistent with the log-law profile (5.12). The value of C_s is typically chosen between 0.15 and 0.3 for atmospheric applications.

A popular extension of the above is the Deardorff model in which

$$K_m = C_K l \sqrt{e}, \tag{5.70}$$

where C_K is a constant and

$$l = \min\left[0.76\frac{e^{1/2}}{N}, \Delta\right] \tag{5.71}$$

is a stability-corrected length scale [70]. Here $N = \sqrt{(g/\theta_0)d\theta/dz}$ is the Brunt-Väisälä frequency, g is the acceleration due to gravity, e is the SGS TKE. A prognostic transport equation, de/dt, is solved in order to obtain e. This equation takes on the same form as (5.10) with $E \rightarrow e$ and $u_i \rightarrow u_i^r$. The dissipation rate is parameterized via

$$\mathscr{E} = C_e \frac{e^{3/2}}{l}, \tag{5.72}$$

with the constant dissipation constant, $C_e = 0.19 + 0.74l/\Delta$. This choice of C_e guarantees a correct spectral slope in the inertial subrange [71]. Finally it can be shown that the three constants are related through

$$C_e = C_K^3 C_S^{-4}. \tag{5.73}$$

In strictly neutral stratification in which $l \equiv \Delta$, $C_K = 0.1$ when $C_S = 0.18$.

5.6.2.2 Advanced Smagorinsky-type closures

A key problem with the Smagorinsky formulation is that it uses a flux-gradient relation with scalar (isotropic) effective viscosity, where the energy flows from large scales to small scales (forward cascade) at all times [69]. Small-scale turbulence, on the other hand, often possesses significant anisotropy (due in the ABL to, e.g., the surface or stability), and TKE can be transferred from small scales to large scales ('backscatter') [30]. A variety of SGS models have been developed to address these two consequences of Smagorinsky-based SGS models. Kosović [72] derived a constitutive rate equation for subgrid stress which explicitly included backscatter, while [73] introduced stochastic backscatter into the SGS model. To deal with the effect of near-surface anisotropy, [74] added a mean-shear term to the flux-gradient relation (5.60), such that the subgrid stress converges to the RANS limit as the wall is approached, and the Deardorff–Smagorinsky form (5.60) is recovered farther from the surface.

Alternatively, 'Dynamical Smagorinsky' SGS models estimate C_s based on the Germano identity [75], i.e. applying the double-filter method [65,76] and thus allowing C_s to vary during a simulation. A number of methods can be used to do so, with C_s (or C_K) potentially varying in one or more dimensions as well as in time [77].

A yet more advanced, 'third generation' of Smagorinsky-style closures has become popular due to its robustness and flexibility. Known as path-averaged or Lagrangian-scale-dependent SGS models, these involve updating of C_K as a function of time and space, taking into account the immediate upstream (advective) influence on $C_K(x, y, z, t)$. We refer the reader to, e.g., [78,79] for further description.

5.6.2.3 Higher order SGS closures

Following [80], numerous closures solving for rate equations of stress and higher order products of velocity components have been derived. However, they tend to be too expensive in terms of computing resources and can be numerically unstable. Stress-budget models were made for poorly resolved LES [81] (and waves [82]), with [81] implemented and tested by [83], while giving some promising results, these also have limitations [84] and have not been widely adopted due to computational cost.

5.6.2.4 Boundary conditions

Close to the ground (or large surfaces), the scales of motion get smaller (recalling $l \sim \kappa z$ in the surface layer), and hence resolution and the dependence of LES on the SGS model becomes increasingly important. The most common approach is to enforce MOST behaviour at the lower boundary. This means applying a surface stress, $\tau_{i3} = \rho \langle u_i' u_3' \rangle$ where the momentum flux $\langle u_i' u_3' \rangle$ is obtained from the corresponding MOST similarity function, Ψ_m:

$$\tau_{i3} = -\rho \left(\frac{\kappa}{\ln(z_1/z_0) - \Psi_m(z_1/L)} \right)^2 U_i(z_1)^2. \tag{5.74}$$

Here z_1 is the lowest resolved height above ground, normally identified as half the height of the first cell. Strictly speaking, (5.74) is only valid for $z_1 \gg z_0$. Since the Obukhov length, L, is a function of the momentum flux itself, an iteration procedure is needed when simulating non-neutral stratified flow for which $\Psi_m(z/L) \neq 0$. Boundary conditions for scalars (c) are set up in a similar way using similarity functions (Ψ_c).

5.6.3 *Numerical approach*

The governing equations of LES are solved with a method of discretization. For atmospheric turbulence, most models apply either a pseudo-spectral approach ([10,24,76,78,85–88], among others) or a finite volume approach (e.g. [89–91]). In the pseudo-spectral approach, the horizontal directions are solved in spectral space by the utilization of FFT, hence applying periodic boundary conditions. In the vertical direction, finite differencing is applied. For generic studies of homogeneous ABLs, this method has been very successful, but even in non-homogeneous conditions where the coordinates are transformed to follow the local surface shape have proved valuable [34,92,93]. The main advantage of the pseudo-spectral models is the explicit filtering, i.e. filtering in spectral Fourier space. The main disadvantage is the periodic boundary conditions in the horizontal directions (although for purely homogeneous flow studies this might actually be an advantage due to faster generation of turbulence due to recycling, and therefore ultimately less computer time needed). Finite volume models, on the other hand, filter in physical space, which only implicitly allows decomposition of variables into resolved and SGS parts by scale [2]. A benefit of finite volume models is the lack of periodic boundary conditions: especially in non-homogeneous conditions, like simulations over variable terrain and roughness, periodic boundaries can potentially add unwanted constraints to the LES domain set-up and consequently affect the solutions.

A primary ansatz of LES is to assume the resolved fluxes are much larger than the corresponding SGS fluxes, by positioning the filter scale Δ within the inertial subrange [69], i.e. spatial resolution is vital if correct scaling of the resolved turbulence fields should be obtained. Recent studies have therefore focused on the effect of increasing the resolution, e.g. with the use of Smagorinsky-type SGS models [10,86].

In Figure 5.8, we show some examples of the impact of resolution on profiles of turbulence statistics. We use the pseudo-spectral LES model documented in [10,86] with the Deardorff SGS model with stability corrected length scale (5.70) and (5.71). For a conditional neutral flow, we look at four different mesh sizes; 128^3, 256^3, 512^3, and $1,024^3$, on a domain of $2,560\,\text{m} \times 2,560\,\text{m} \times 1,792\,\text{m}$. First we look at the ratio of SGS to resolved Reynolds stress, R, i.e. we explore the inequality in (5.62). We observe a notable increase in R with resolution as expected. It was shown in [94] that a value R 1 in the first cell is a necessary but not sufficient condition in order to obtain true surface layer scaling (log-law in neutral conditions). The horizontal variances, σ_u and σ_v (the simulation includes the Coriolis force) are especially at the surface dependent on the mesh, but the dependence is visible in the whole profile. Finally, we also show a surrogate for the turbulence intensity, $\sqrt{\text{TKE}}/S$. S is the horizontal wind speed. Besides, when very close to the surface, the profiles seem to collapse, which can be advantageous in LES wind energy application context.

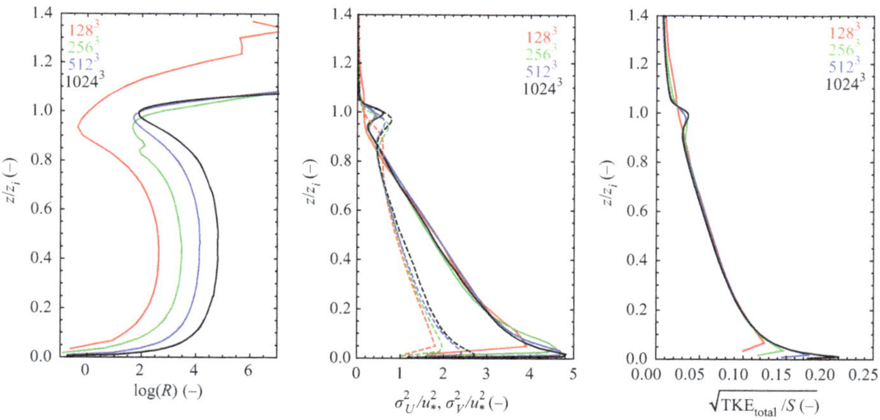

Figure 5.8 LES of inversion-capped (conditionally) neutral ABL with four different mesh sizes: 128^3 (red lines), 256^3 (green lines), 512^3 (blue lines), and $1,024^3$ (black lines). Domain size is 2,560 m × 2,560 m × 1,792 m. Profiles are calculated at equivalent times after start-up, $20\,z_i/u_\star$. Left: ratio of SGS to resolved fluxes, R. Middle: horizontal variances, σ_u (solid lines) and σ_v (dashed lines). Right: surrogate turbulence intensity, $\sqrt{TKE}/S$. The LES runs have been performed with the pseudo-spectral model documented in [10,86]

5.7 Final remarks

In this chapter, we have presented different methodologies used to simulate atmospheric turbulence as it appears around wind turbines in the ABL. The area is vast, so we have chosen to present the most commonly used methodologies currently used in industry and academic research. We have presented a practical and conceptual treatment of the area and have accordingly omitted some details. It is therefore our hope that readers desiring a more comprehensive understanding will find the reference list to be a useful supplement to the content of this chapter.

The absence of wind turbines in the material presented in this chapter necessitates review of the chapters in this volume concerned with wind farm flow, in order to provide a more complete picture of the influence of turbulent flow on the operation of wind turbines. Further, siting of wind turbines in non-homogeneous conditions demands additional attention, as does coupling to larger mesoscale motions.

References

[1] Wyngaard JC. Turbulence in the Atmosphere. New York, NY: Cambridge University Press; 2010.

[2] Pope SB. Turbulent Flows. Cambridge: Cambridge University Press, 770 pp; 2000.

[3] Wilcox DC. Turbulence Modeling for CFD. La Cañada, CA, USA: DCW Industries; 1994.

[4] Mann J. The spatial structure of neutral atmospheric surface-layer turbulence. Journal of Fluid Mechanics. 1994;273:141–168.

[5] Savill AM. Recent developments in rapid-distortion theory. Annual Review of Fluid Mechanics. 1987;19:531–575.

[6] International Electrotechnical Commission. IEC 61400-1 Ed.3 Wind Turbines – Part 1: Design Requirements; 2005.

[7] Kaimal JC, Finnigan JJ. Atmospheric Boundary Layer Flows. New York, NY: Oxford University Press; 1994.

[8] Bohren CF, Albrecht BA. Atmospheric Thermodynamics. New York, NY, USA: Oxford University Press; 1998.

[9] Kelly M, Larsen G, Dimitrov NK, *et al.* Probabilistic meteorological characterization for turbine loads. Journal of Physics: Conference Series. 2014;524(1):012076.

[10] Sullivan PP, Patton EG. The effect of mesh resolution on convective boundary layer statistics and structures generated by large-eddy simulation. Journal of the Atmospheric Sciences. 2011;68:2395–2415.

[11] Kumar V, Kleissl J, Meneveau C, *et al.* Large eddy simulation of a diurnal cycle of the atmospheric boundary layer: Atmospheric stability and scaling issues. Water Resources Research. 2006;42(6):W06D09.

[12] Högström U. Non-dimensional wind and temperature profiles in the atmospheric surface layer. Boundary-Layer Meteorology. 1988;42:263–270.

[13] Businger JA, Wyngaard JC, Izumi Y, *et al.* Flux-profile relationships in the atmospheric surface layer. Journal of the Atmospheric Sciences. 1971;28: 181–189.

[14] Panofsky HA, Dutton JA. Atmospheric Turbulence. New York, NY: Wiley; 1984.

[15] Carl DM, Tarbell TC, Panofsky HA. Profiles of wind and temperature from towers over homogeneous terrain. Journal of the Atmospheric Sciences. 1973;30(30):788–794.

[16] Caughey S, Wyngaard JC, Kaimal J. Turbulence in the evolving stable boundary-layer. Journal of the Atmospheric Sciences. 1979;36(6):1041–1052.

[17] Kelly M. From standard wind measurements to spectral characterization: turbulence length scale and distribution. Wind Energy Science Discussions. 2018; DOI 10.5194/wes-2018-14. Available from: https://www.wind-energ-sci-discuss.net/wes-2018-14/.

[18] Kelly M. Uncertainty in Vertical Extrapolation of Wind Statistics: Shear-Exponent and WAsP/EWA Methods. Roskilde, Denmark: Wind Energy Dept., Risø Lab/Campus, Danish Tech. Univ. (DTU); 2016. DTU Wind Energy E-0121(EN).

[19] Dimitrov NK, Natarajan A, Kelly M. Model of wind shear conditional on turbulence and its impact on wind turbine loads. Wind Energy. 2015;18(11): 1917–1931.

[20] Pedersen J, Gryning SE, Kelly M. On the structure and adjustment of inversion-capped neutral atmospheric boundary-layer flows: Large-eddy simulation study. Boundary-Layer Meteorology. 2014;153:43–62.

[21] Abkar M, Porté-Agel F. The effect of free-atmosphere stratification on boundary-layer flow and power output from very large wind farms. Energies. 2013;6(5):2338–2361.

[22] Abkar M, Porté-Agel F. Mean and turbulent kinetic energy budgets inside and above very large wind farms under conventionally-neutral condition. Renewable Energy. 2014;70:142–152.

[23] Allaerts D, Meyers J. Large eddy simulation of a large wind-turbine array in a conventionally neutral atmospheric boundary layer. Physics of Fluids. 2015;27:065108.

[24] Allaerts D, Meyers J. Boundary-layer development and gravity waves in conventionally neutral wind farms. Journal of Fluid Mechanics. 2017;814:95–130.

[25] Kelly M, Cersosimo RA, Berg J. A universal wind profile for the inversion-capped neutral atmospheric boundary layer. Quarterly Journal of the Royal Meteorological Society. 2019;145(720):982–992.

[26] Natarajan A, Dimitrov N, Madsen P, *et al.* Demonstration of a Basis for Tall Wind Turbine Design, EUDP Project Final Report. Roskilde, Denmark: Wind Energy Dept., Risø Lab/Campus, Danish Tech. Univ. (DTU); 2016. DTU Wind Energy E-0108(EN). EUDP 11-II 64011-0352.

[27] Rodrigo JS, Allaerts D, Avila M, *et al.* Results of the GABLS3 diurnal-cycle benchmark for wind energy applications. Journal of Physics: Conference Series. 2017;854(1):012037. Available from: http://stacks.iop.org/1742-6596/854/i=1/a=012037.

[28] Sanz Rodrigo J, Churchfield M, Kosovic B. A methodology for the design and testing of atmospheric boundary layer models for wind energy applications. Wind Energy Science. 2017;2(1):35–54. Available from: https://www.wind-energ-sci.net/2/35/2017/.

[29] Bershadskii A. Some generalization of self-similarity. Europhysics letters. 1997;39(6):587–592.

[30] Frisch U. Turbulence – The Legacy of A. N. Kolmogorov. Cambridge: Cambridge University Press; 1995.

[31] Dimitrov N, Kelly M, Vignaroli A, *et al.* From wind to loads: wind turbine site-specific load estimation using databases with high-fidelity load simulations. Wind Energy Science Discussions. 2018;2018:1–39. Available from: https://www.wind-energ-sci-discuss.net/wes-2018-18/.

[32] Larsen TJ, Hansen MA. How to HAWC2, the User's Manual. DTU Wind Energy; 2012.

[33] Bak C, Zahle F, Bitsche R, *et al.* Description of the DTU 10 MW Reference Wind Turbine. Technical University of Denmark, Department of Wind Energy; 2013. I-0092.

[34] Berg J, Troldborg N, Sørensen NN, *et al.* Large-eddy simulation of turbine wake in complex terrain. Journal of Physics: Conference Series. 2017;854(1):012003.

[35] Peinke J, Barth S, Bottcher F, *et al.* Turbulence, a challenging problem for wind energy. Physica A – Statistical Mechanics and Its Applications. 2004; 338(1–2):187–193.

[36] Kaimal JC, Wyngaard JC, Izumi Y, *et al.* Spectral characteristics of surface-layer turbulence. Quarterly Journal of the Royal Meteorological Society. 1972;98:563–589.

[37] Davenport AG. The spectrum of horizontal gustiness near the ground in high winds. Quarterly Journal of the Royal Meteorological Society. 1961;87: 194–211.

[38] Saranyasoontorn K, Manuel L, Veers PS. A comparison of standard coherence models for inflow turbulence with estimates from field measurements. Journal of Solar Energy Engineering-Transactions of the ASME. 2004;126(4): 1069–1082.

[39] Veers PS. Three-Dimensional Wind Simulation. Sandia National Laboratories; 1988. SAND88-0152.

[40] Townsend AA. The Structure of Turbulent Shear Flow. Cambridge: Cambridge University Press; 1976.

[41] International Electrotechnical Commission TC88 (ed.). IEC 61400-1 Ed.4.0 Wind Energy Generation Systems – Part 1: Design Requirements. Geneva, 2019.

[42] Lumley JL. Stochastic Tools in Turbulence. New York, NY, USA: Academic Press, Inc. (Dover Publications 2007), 194 pp; 1970.

[43] Batchelor GK. The Theory of Homogeneous Turbulence. Cambridge: Cambridge University Press; 1953.

[44] von Kármán T. Progress in the statistical theory of turbulence. Proceedings of the National Academy of Sciences. 1948;34(11):530–539.

[45] Abramowitz M, Stegun IA. Handbook of Mathematical Functions with Formulas, Graphs, and Mathematical Tables; 9th printing. New York, NY, USA: Dover; 1972.
[46] Mann J. Wind Field Simulation. Probabilistic Engineering Mechanics. 1998;13(4):269–282.
[47] Dimitrov NK, Natarajan A, Mann J. Effects of normal and extreme turbulence spectral parameters on wind turbine loads. Renewable Energy. 2017;101: 1180–1193.
[48] Pena Diaz A, Floors R, Sathe A, *et al.* Ten years of boundary-layer and wind-power meteorology at Høvsøre, Denmark. Boundary-Layer Meteorology. 2016;158:1–26.
[49] Mann J. The spectral velocity tensor in moderately complex terrain. Journal of Wind Engineering and Industrial Aerodynamics. 2000;88:153–169.
[50] Peña Diaz A, Gryning SE, Mann J. On the length-scale of the wind profile. Royal Meteorological Society Quarterly Journal. 2010;136(653): 2119–2131.
[51] Sathe A, Mann J, Barlas TK, *et al.* Influence of atmospheric stability on wind turbine loads. Wind Energy. 2013;16:1013–1032.
[52] Kelly M, Gryning SE. Long-term mean wind profiles based on similarity theory. Boundary-Layer Meteorology. 2010;136(3):377–390.
[53] Chougule A, Mann J, Kelly M. Vertical cross-spectral phases in atmospheric flow. Journal of Physics: Conference Series. 2014;555(1):012017.
[54] de Mare MT, Mann J. On the space-time structure of sheared turbulence. Boundary-Layer Meteorology. 2016;160(3):453–474.
[55] Chougule A, Mann J, Kelly M. Influence of Atmospheric Stability on the Spatial Structure of Turbulence. The Danish Technical University (DTU); 2013.
[56] Segalini A, Arnqvist J. A spectral model for stably stratified turbulence. Journal of Fluid Mechanics. 2015;781:330–352.
[57] Chougule AS, Mann J, Kelly M, *et al.* Modeling atmospheric turbulence via rapid distortion theory: Spectral tensor of velocity and buoyancy. Journal of the Atmospheric Sciences. 2017;74(4):949–974.
[58] Chougule A, Mann J, Kelly M, *et al.* Simplification and validation of a spectral-tensor model for turbulence including atmospheric stability. Boundary-Layer Meteorology. 2018;167:371–397.
[59] Kelley ND. Full Vector (3-D) Inflow Simulation in Natural and Wind Farm Environments Using An Expanded Versions of the SNLWIND (Veers) Turbine Code. Golden, CO, USA: National Renewable Energy Laboratory; 1992. NREL/TP-442-525.
[60] Kelley ND, Jonkman BJ. Overview of the TurbSim Stochastic Inflow Turbulence Simulator. Golden, CO, USA: National Renewable Energy Laboratory; 2007. NREL/TP-500-41137.
[61] Chougule A, Mann J, Kelly M. Vertical cross-spectral phases in neutral atmospheric flow. Journal of Turbulence. 2012;13(1):1–13.

[62] Lilly DK. The representation of small-scale turbulence in numerical simulation experiments. In: Proc. IBM Scientific Computing Symp. on Environmental Sciences. Yorktown Heights, NY: IBM Thomas J. Watson Research Center; 1967. p. 195–210. Form No. 320-1951.

[63] Deardorff JW. A numerical study of three-dimensional turbulent channel flow at large Reynolds numbers. Journal of Fluid Mechanics. 1970;41(1970): 453–480.

[64] Deardorff JW. Numerical investigation of neutral and unstable planetary boundary layers. Journal of the Atmospheric Sciences. 1972;29(1):91–115.

[65] Pope SB. Ten questions concerning the large-eddy simulation of turbulent flows. New Journal of Physics. 2004;6(1):35:1–24.

[66] Wyngaard JC. Toward numerical modeling in the "Terra Incognita". Journal of the Atmospheric Sciences. 2004;61:1816–1826.

[67] Wyngaard JC, Cote OR. Cospectral similarity in the atmospheric surface layer. Quarterly Journal of the Royal Meteorological Society. 1972;98(417): 590–603.

[68] Smagorinsky J. General circulation experiments with the primitive equations. I. The basic experiment. Monthly Weather Review. 1963;91:99–164.

[69] Wyngaard JC. Toward numerical modeling in the 'Terra Incognita'. Journal of the Atmospheric Sciences. 2004;61:1816–1826.

[70] Moeng CH. A large-eddy simulation model for the study of planetary boundary-layer turbulence. Journal of the Atmospheric Sciences. 1984;41:2052–2062.

[71] Moeng CH, Wyngaard JC. Spectral analysis of large-eddy simulations of the convective boundary layer. Journal of the Atmospheric Sciences. 1988;45(23):3573–3587.

[72] Kosović B. Subgrid-scale modelling for the large-eddy simulation of high-Reynolds-number boundary layers. Journal of Fluid Mechanics. 1997;336:151–182.

[73] Mason PJ, Thomson DJ. Stochastic backscatter in large-eddy simulations of boundary layers. Journal of Fluid Mechanics. 1992;242:51–78.

[74] Sullivan PP, McWilliams JC, Moeng CH. A subgrid-scale model for large-eddy simulation of planetary boundary-layer flows. Boundary-Layer Meteorology. 1994;71:247–276.

[75] Germano M, Piomelli U, Moin P, *et al.* A dynamic subgrid-scale eddy viscosity model. Physics of Fluids A: Fluid Dynamics. 1991;3:1760–1765.

[76] Porte-Agel F, Meneveau C, Parlange MB. A scale-dependent dynamic model for large-eddy simulation: Application to a neutral atmospheric boundary layer. Journal of Fluid Mechanics. 2000;415:261–284.

[77] Anderson WC, Basu S, Letchford CW. Comparison of dynamic subgrid-scale models for simulations of neutrally buoyant shear-driven atmospheric boundary layer flows. Environmental Fluid Mechanics. 2007;7:195–215.

[78] Bou-Zeid E, Meneveau C, Parlange M. A scale-dependent Lagrangian dynamic model for large eddy simulation of complex turbulent flows. Physics of Fluids. 2005;17:W01409.

[79] Stoll R, Porté-Agel F. Dynamic subgrid-scale models for momentum and scalar fluxes in large-eddy simulations of neutrally stratified atmospheric boundary layers over heterogeneous terrain. Water Resources Research. 2006;42.

[80] Zeman O, Jensen NO. Modification of turbulence characteristics in flow over hills. Quarterly Journal of the Royal Meteorological Society. 1987;113:55–80.

[81] Hatlee SC, Wyngaard JC. Improved subfilter-scale models from the HATS field data. Journal of the Atmospheric Sciences. 2007;64(5):1694–1705.

[82] Kelly M, Wyngaard JC, Sullivan PP. Application of a subfilter-scale flux model over the ocean using OHATS field data. Journal of the Atmospheric Sciences. 2009;66(10):3217–3225.

[83] Ramachandran S, Wyngaard JC. Subfilter-scale modelling using transport equations: Large-eddy simulation of the moderately convective atmospheric boundary layer. Boundary-Layer Meteorology. 2011;139(1):1–35.

[84] Ramachandran S, Wyngaard JC. An a priori study of the subfilter-scale conservation equations for the stably stratified atmospheric surface layer. Quarterly Journal of the Royal Meteorological Society. 2011;140(681):1304–1314.

[85] Moeng CH. A large-eddy-simulation model for the study of planetary boundary-layer turbulence. Journal of the Atmospheric Sciences. 1984;41(13):2052–2062.

[86] Sullivan PP, Weil JC, Patton EG, *et al.* Turbulent winds and temperature fronts in large-eddy simulations of the stable atmospheric boundary layer. Journal of the Atmospheric Sciences. 2016;73(4):1815–1840.

[87] Stevens RJAM, Wilczek M, Meneveau C. Large-eddy simulation study of the logarithmic law for second- and higher-order moments in turbulent wall-bounded flow. Journal of Fluid Mechanics. 2014;757:888–907.

[88] Bou-Zeid E, Vercauteren N, Parlange MB, *et al.* Scale dependence of subgrid-scale model coefficients: An a priori study. Physics of Fluids. 2008;20:115106.

[89] Churchfield MJ, Vijayakumar G, Brasseur JG, *et al.* Wind energy-related atmospheric boundary layer large-eddy simulation using OpenFOAM. In: 19th Symposium on Boundary Layers and Turbulence, Keystone, CO; 2010.

[90] Churchfield MJ, Lee S, Michalakes J, *et al.* A numerical study of the effects of atmospheric and wake turbulence on wind turbine dynamics. Journal of Turbulence. 2012;13:N14.

[91] Bechmann A, Sørensen NN. Hybrid RANS/LES method for wind flow over complex terrain. Wind Energy. 2010;13(1):36–50.

[92] Sullivan PP, McWilliams JC, Patton EG. Large eddy simulation of marine atmospheric boundary layers above a spectrum of moving waves. Journal of the Atmospheric Sciences. 2014;71:4001–4027.

[93] Shamsoddin S, Porté-Agel F. Large-eddy simulation of atmospheric boundary-layer flow through a wind farm sited on topography. Boundary-Layer Meteorology. 2017;163:1–17.

[94] Brasseur JG, Wei T. Designing large eddy simulation of the turbulent boundary layer to capture law-of-the-wall scaling. Physics of Fluids. 2010;22(2):1–21.

Chapter 6
Modeling and simulation of wind-farm flows

Matthew J. Churchfield[1] *and Patrick J. Moriarty*[1]

6.1 Introduction

The subject of this chapter is the modeling and simulation of the flow through full wind farms, also known as wind plants, which are collections of many wind turbines within a region working together to supply power to the grid like any other type of power plant. This topic is interesting and exciting because of the increasing variety of and uses for wind plant flow modeling and simulation. For example, Figure 6.1 shows the computed flow field from a high-fidelity simulation of the Lillgrund offshore wind farm using computational fluid dynamics (CFD) with large-eddy simulation (LES) turbulence modeling and actuator line rotor aerodynamics models. There are many ways to model this type of flow, ranging from very simple models of wind turbine wakes and their interactions that can run in a matter of seconds on a laptop to high-fidelity, turbulence-resolving LES of the atmospheric flow and wakes that require days of use of thousands of cores on a high-performance computing cluster. The uses of wind plant flow simulations are equally varied. Wind plant layout designers run hundreds or thousands of simulations over all possible wind directions and speeds using simple models to predict a future wind plant's annual energy production. Researchers use LESs of the wind plant to study phenomena such as wind turbine wake interactions with each other and with the atmospheric boundary layer (ABL). As computational resources become more readily available, it is even becoming common for turbine manufacturers to use CFD of some form to model wind plants in complex terrain to help in the wind turbine siting and selection process. We are beginning to see the expensive, high-fidelity tools being used to create data and knowledge that improves or enables the creation of new cheaper, lower fidelity tools. We are also beginning to see the low- and high-fidelity tools used together to perform uncertainty quantification of energy estimates and optimization of wind plant layouts. As new computing technologies mature, such as graphics processing unit (GPU)-based computing, the future of full wind plant flow modeling is exciting.

In simulating the wind plant flow field, there are two main components of the flow that must be considered: the inflow wind and wind turbine wakes. There are

[1]National Renewable Energy Laboratory, Golden, CO, USA

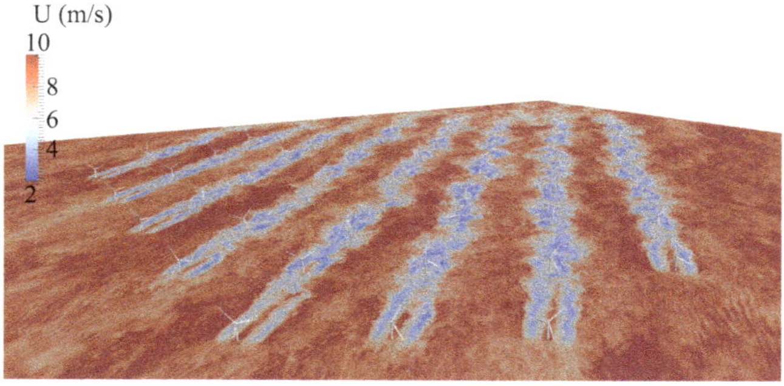

Figure 6.1 A horizontal plane at wind turbine hub height showing contours of instantaneous velocity from a fluid flow simulation of the Lillgrund offshore wind farm. The wind is aligned with a main row direction, and wind turbine wakes are clearly visible in blue shades

also two main methods employed to model these wind plant flow features: those that employ some form of CFD, which is relatively computationally expensive, and those that do not and are much more computationally inexpensive.

The inflow wind is the "fuel" to the wind plant, so it is a critical component of the wind plant flow modeling process. The wind contains the kinetic energy that wind turbines convert to electrical energy. The wind also contains shear, veer, and turbulence that create damaging fatigue loads applied to the wind turbines. Electricity generation and component damage that requires maintenance are driving factors in the cost of energy. The inflow wind is a challenge to model for many reasons, including the fact that it has a complex time series with fluctuations spanning a range of timescales. There are long timescales that follow weather patterns with timescales of days or a week. Diurnal variations cause daily fluctuations. Atmospheric turbulence causes variations with short timescales ranging from seconds to minutes. Additionally, the inflow wind has a vertical structure that is affected by atmospheric stability, surface roughness, and terrain. Within and surrounding a wind farm, the winds may vary horizontally because of changes in roughness, terrain, obstacles, and large-scale turbulent structures, as well as wind turbine wakes.

The wind turbine wake is an area of reduced wind speed and increased turbulence behind an operating turbine created by the extraction of momentum from the wind, so it is also a critical wind plant flow modeling component. As turbines are placed in wind farms, the interactions between turbines and wakes result in overall power losses relative to stand-alone turbines; a typical wind farm may experience a 10% energy loss as a result of wakes over an annual period. Wake models are critical for predicting energy losses from interactions between turbines within the wind-farm environment and the surrounding atmosphere. Perhaps as important as the energy lost, the uncertainty in wake loss estimates using industry standard models is currently in the range of 20%–50% [1]. The impact of wakes is most influenced by the local wind speed, turbine operating behavior, and background atmospheric turbulence characteristics.

When we refer to CFD-based methods, we mean methods that numerically solve some form of the governing equations of fluid flow requiring significant computational cost. It makes most sense to view the working fluid—air—from a continuum basis, so the equations that govern wind plant flows are the Navier–Stokes equations. This cost can come from the need to use multiple compute cores, long compute times, or a combination of both. By non-CFD-based methods, on the other hand, we mean methods that are computationally much less expensive than CFD-based methods. A non-CFD-based method may solve a simplified version of the Navier–Stokes equations, or may be kinematic and based on empirical relationships derived from observation of the behavior of inflow winds, wakes, and their interactions. Wind plant flow modeling should be thought of as modular and flexible.

There is no reason that CFD-based and non-CFD-based methods cannot be mixed. For example, one can use non-CFD generated inflow winds as inflow boundary conditions for a CFD-based wind plant simulation. The CFD solver will rapidly take these inflow boundary conditions and transform them to something that satisfies the governing equations as the flow advects downstream toward the turbines. Or conversely, one may use CFD to generate a realistic wind field that is then advected through a non-CFD-based wind plant simulation using Taylor's frozen turbulence hypothesis with superposition of non-CFD-generated wind turbine wakes. Explanations of a series of potential methods are explained in the following sections.

The material is arranged as follows: in Section 6.2, we provide motivation by discussing the question of "why simulate the wind plant flows?"; in Section 6.3, we present a broad discussion of wind plant flow simulation approaches, both CFD- and non-CFD-based, and how inflow winds and wakes are generated and put together; in Section 6.4, we discuss validation of wind plant flow simulation; and in Section 6.5, we discuss future developments.

6.2 Why simulate the flow through wind plants?

There are many reasons to simulate the flow through wind plants. We will discuss a number of important applications of wind plant flow simulation, but this list is not exhaustive. As wind energy becomes more popular, the list of reasons grows. Depending on a wind plant modelers use case, they may employ different models depending on their application. Simple models may be sufficient for predicting power output of offshore wind farms with regular layouts, whereas the most high-fidelity models may be needed for determining wake impacts on turbine site suitability in complex terrain. As computing power accelerates, the complexity of models may follow suit; however, at the start of the design process or within multidisciplinary optimization, the simplest models will continue to be of use. The number of turbines modeled in a wind plant may also be determined by the fidelity of the model and the availability of computing resources. There is also often a trade-off between speed and accuracy. And users must make a choice between computational turnaround time and acceptable levels of uncertainty. As model complexity increases, more details about the atmospheric environment, turbine design, and operation are also required.

This will simultaneously increase the amount of higher fidelity observations required to initialize, drive, and validate these models.

6.2.1 Improved physical understanding

Wind plant flows are truly complex and attaining a solid understanding of the flow physics is a challenge. The background flow, and the fuel to the wind plant, is the wind within the ABL. The Earth's ABL is turbulent and is influenced by the local or even regional topography and the regional-scale weather. Over land, the ABL undergoes a diurnal cycle in which the thermal stratification changes from stable at night to unstable during the day and back again the next night, which highly influences the structure and strength of the turbulence and the mean vertical structure of the wind speed. The Earth's rotation and curvature introduce Coriolis effects, which can cause wind direction change with height and horizontally over wind plant length scales. The wind plant is embedded in this highly complex boundary layer, and each turbine produces wakes that interact with the ABL, with other wakes, and with other turbines. The flow disturbance caused by the wind turbines can cause gravity wave propagation if the conditions are stably stratified. Large wind plants form something similar to a boundary layer embedded within the ABL that can persist for kilometers and interact with other wind plants. The range of scales of interest starts from 1 m at the smallest turbine wake scale to tens of kilometers or more between operating wind plants. With this range of flow complexity and all of the interrelated flow features, there are many unanswered questions about wind plant flows, and wind plant flow simulation tools are helping to answer these questions.

An example of an area in which more physical understanding is needed is the effect of atmospheric stability on wakes. The topic of atmospheric stability is well covered by Stull [2]. The momentum deficit of a wind turbine wake decays with downstream distance, mainly as a result of turbulent transport of momentum. Wakes also meander as they advect downstream, which is in part caused by the larger than rotor scale turbulent structures within the ABL. The intensity and structure of atmospheric turbulence is highly dependent on the thermal stratification of the ABL. For example, at night, the land surface radiates heat and often becomes cool relative to the air above it. It will, thus, become a heat sink causing the air near the ground to be cooler and denser than the air above. This is stable stratification, which tends to damp turbulence. The turbulence intensity under stable stratification is often low, and the size of the turbulent structures is small relative to other conditions. During daytime, the sun radiates heat to the land surface making it warm relative to the air above. Heat is then conducted from the ground into the air causing warmer, less dense air beneath cooler air above. This is unstable stratification in which turbulence levels are increased because of buoyant motions, and the size of the turbulent structures are increased relative to other conditions. It is clearly understood that the wake momentum deficit decay and meandering behavior are sensitive to turbulence intensity and structure, but reduced-order models for wakes still need much improvement. This is especially true in the area of turbulence modeling, reflecting our lack of detailed knowledge of these physical processes. On the other hand, high-fidelity LES can capture the dependence of wake deficit decay and meandering on different types of inflow turbulence. The work of Debnath [3], among others, provides good examples of this. However, the amount of information generated is so large and complex that the task of distilling it into a simple, robust, widely applicable, physics-based turbulence model remains unachieved, leading researchers to look toward new turbulence modeling ideas and even data-driven turbulence models that use artificial intelligence, e.g., King *et al.* [4].

A critical example of a wind plant flow-induced feature that is not fully understood is the effect of wakes on wind plant power and wind turbine mechanical loads. Within a wind plant, wakes from upstream turbines advect downstream and interact with downstream turbines. Those wakes may merge with wakes from other turbines, a process that has not been well quantified. Many rows downstream into the wind plant, wind turbines are often subject to inflows containing multiple merged wakes. Without a solid understanding of wake merging, accurate predictions of wind plant power output under these conditions are difficult. Furthermore, wakes are complex because they are generated by a rotor that imparts torque and hence rotation on the wake flow. For some distance downstream, the blade root and tip vortices remain at the wake edge and core. The wake shear layer generates its own turbulence. The bulk wake deficit also meanders laterally and vertically as it advects downstream. From the perspective of a turbine subject to this wake, the low frequency part of the turbulence spectrum and the coherence of the turbulence are modified in the wake relative to the inflow wind. The fact that the wake meanders means that the wake is constantly sweeping across the face of a waked downstream rotor, an effect that on its own is turbulence-like. Some attempts have been made at quantifying the spectra and coherence of turbulence in wakes generated with LES [5], but more work must be done to fully understand wake turbulence and the process of modeling it in simpler wake models.

A last example of a fundamental wind plant flow physics question that has not yet been fully addressed is the generation of gravity waves by the wind plant itself and the effect of those gravity waves on the wind plant. Gravity waves are vertical oscillations in the atmosphere created by disturbances such as mountains or even wind farms. Some work in using wind plant flow simulation to better understand gravity wave effects has been performed by Allaerts and Meyers [6]. Through LES, they find that wind plants can, indeed, generate gravity waves that can propagate upstream and affect the performance of the wind plant.

There are countless other examples and clearly a lot of work that must be done to better understand wind plant flow physics. Experimental campaigns will certainly be part of this effort, but the sheer size of wind plants and the cost and limitations of the current generation of measurement devices, such as scanning lidar, necessitate the use of high-fidelity, physics-based, wind-plant-simulation tools used in conjunction. LES is currently the simulation tool with the most promise in providing accurate, high spatiotemporal resolution wind plant flow data under a variety of conditions for the purpose of better understanding this complex flow. LES though requires heavy computational resources over days or weeks to simulate minutes to hours of a wind plant flow. A single LES run of a wind plant can easily yield terabytes of data. A major challenge is how to distill all of the data into meaningful information that can then be used to improve the computationally efficient, reduced-order wind plant flow models that are used to design, optimize, and even operate wind farms. The future may see further integration of observations with simulations for a range of models and applications from fundamental research to wind plant operations in real time.

6.2.2 Design

One of the main uses of wind plant flow modeling is to design the wind plant for maximal energy production while minimizing reliability impacts. Simpler, more computationally efficient wind plant flow simulation tools, such as WAsP [7], Wind-Farmer [8], and OpenWind [9], are typically used for this purpose. Given probability distributions of a potential wind-plant site's wind conditions and a map of the topography and surface conditions, these tools can be efficiently used to compute power over all wind direction sectors and wind speed and turbulence bins to give an estimate of annual energy production. Such tools usually employ linear flow models to predict the undisturbed wind resource. Many tools, however, are beginning to employ more complicated Reynolds-averaged Navier–Stokes (RANS) CFD flow models that have better performance in complex terrain conditions. New research [10] into layout optimization has emphasized the importance of accounting for uncertainty within wind resource observations and modeling tools that can greatly impact the optimal layout.

These tools also contain more simplistic wake models, such as the commonly used Jensen model [11,12], which is the subject of further discussion in Section 6.3.1.2. The wind speed range of importance for power losses is between turbine start-up wind speed and above rated wind speed (often 15 m/s [13]), where all of the turbines within the wind farm will be operating at or above the rated power.

In addition to estimated flow fields and wake interaction, the optimization of wind-farm layouts must also consider constraints, such as land area, soil, noise, flicker, road construction, and cable lengths. Thus, the final design may not always be driven by performance alone.

Site suitability predictions for turbines are also important for wind-farm designers, although they are typically done by the turbine manufacturers, who have access to turbine design details, and not the wind site developers. To determine the loads encountered by each turbine in the plant, designers rely on estimates of atmospheric and wake-induced turbulence intensity and extreme events, which are not well quantified within wind-farm environments, to meet requirements designated by the International Electrotechnical Commission (IEC) design standard [14]. The wind and turbulence conditions are specified by the standard, but designers can refine them using observations and additional simulations. For each specific site, the controller of individual turbines is adjusted to reduce loads based on the local environment, and optimal turbine placement ensures minimal power loss to meet structural loading requirements. Unlike power losses, the range of wind speeds at which the wake impacts structural loading is unknown and a topic for ongoing research.

6.2.3 Wind-farm control

Considering the wind-farm layout is fixed after construction, recent research work has focused on controlling wind-farm wake interactions in real time to increase power and reliability. More accurate wind-farm simulation tools will enable control designers to create and tune more advanced wind-farm control schemes. In initial stimulation studies, researchers reduced front row performance to allow more energy to pass into downstream rows, resulting in a net wind-farm power increase. However, large power increases from simulation were not fully realized in the field because of an oversimplification of the unsteady flow field and turbine model. More recently, researchers have used the idea of wake steering to shift wakes away from downstream turbines or modify the overall wind-farm flow field [15]. This flow modification has produced even greater potential power gains, by augmenting the flow inside the farm. Field testing of wake steering has shown promise [16], but more research is required. The largest potential benefit from wake steering may come from titled rotors, which can direct the wake toward the ground and entrain higher speed flow from above the wind farm. Large tilt angles would require downwind turbine designs, which are not currently in widespread use.

6.2.4 Special cases of interest and forensic analysis

Often wind-farm owners will have turbines within their farms that have underperformance or reliability issues, which require further study and development of mitigation strategies. This is common in wind farms built in complex terrain and designed using simple wind plant modeling tools that are unable to capture the complex terrain ABL interactions. In this case, the plant owner will often deploy additional instrumentation and also enlist higher fidelity models to better understand the flow at a specific problem turbine site. Once the problem behavior is better understood, solutions can range from simple tuning of individual turbine control systems to, in extreme cases, turbine relocation.

6.2.5 Design of experiments

In the research community, the combination of simulation tools and experimental design has improved the experimental execution and data quality by allowing researchers to run experiments virtually before testing (e.g., [17]). Typically, these design studies employ high-fidelity modeling tools that mimic the instruments operating in realistic wind-farm atmospheric environments. Such simulations provide experimentalists detailed information on instrumentation performance and also an opportunity to optimize sensor placement and operation to maximize their value prior to deployment. Within industry, wind-farm developers use lower order simulation tools to determine optimal locations of meteorological tower deployments during the preconstruction measurement phase that will provide them with the optimal resource information prior to building their wind farms. Given the synergistic benefits, further blending of simulating and observations will continue for both research and industry applications.

6.3 Simulation approaches

In this section, we give an overview of different wind plant flow modeling approaches. We first discuss approaches using non-CFD-based methods, and then we detail CFD-based methods. In both discussions, we describe how the two main flow components, the inflow winds and wind turbine wakes, are generated. We then talk about how these flow components are combined to model the entire wind plant flow. In this section, we build upon the models described in other chapters and also the previous wind-plant model overview publications [18–23].

6.3.1 Noncomputational-fluid-dynamics-based approaches

Non-CFD approaches are used primarily because of their lower computational cost, many of which are built on the simplest models originally developed in the 1970s [24]. The non-CFD approaches often rely on multiple models of different physical processes that are superimposed to create a complete wind plant environment. The superposition begins with a model of the inflow wind, on top of which the wake impacts are added from individual turbines and, at the largest scales, wind-farm effects.

6.3.1.1 Inflow wind generation

Because of its complexity, it is convenient to consider the inflow wind as a mean and a fluctuating part. The mean is usually derived with time averaging, and the length of time averaging is important, especially because the vertical structure of the inflow winds changes throughout the day. Typical averaging times range from minutes to an hour. With short averaging durations like these, the atmospheric turbulence is filtered away, but the diurnal variation of the vertical structure of the wind is preserved. In this way, the inflow winds are decomposed mathematically as

$$\mathbf{u} = \bar{u} + \mathbf{u}' \tag{6.1}$$

where $\mathbf{u}$ is the instantaneous wind vector, the overbar denotes the mean component, and the prime denotes the fluctuating component.

Mean inflow wind

We begin with a brief discussion of models of the mean wind. For many applications, the mean part of the inflow wind is all that is needed. For example, in performing a wind resource assessment, the mean wind profile can be applied to the rotor swept area to compute a mean power. By gathering information about the inflow wind speed and direction probability distribution and the distribution of different atmospheric stability states over long times, and using it with an appropriate wake model, the annual energy production of a wind farm can be estimated.

Over relatively flat terrain with homogeneous surface roughness and heating, the most popular approach to modeling the mean wind profile is through the use of Monin–Obukhov similarity theory, which can be used to express the mean wind profile as a function of height and atmospheric stability

$$\bar{u} = \frac{u_*}{\kappa}\left[\ln\left(\frac{z}{z_0}\right) - \psi\left(\frac{z}{L}\right)\right] \tag{6.2}$$

where $\bar{u}$ is the horizontal component of the wind vector (note that this profile does not separate the two horizontal Cartesian components, u_1 and u_2, so wind direction variation with height is not captured), u_* is the friction velocity, κ is the von Kármán constant, z is height, z_0 is the surface roughness, ψ is an atmospheric-stability dependent function, and L is the Obukhov length, which is a measure of atmospheric stability. The form of the atmospheric stability-dependent function is discussed by various researchers, including Businger [25], Dyer [26], and Paulson [27].

Another model for the mean wind profile is a simple power law

$$\bar{u} = \bar{u}(z_{ref}) \left(\frac{z}{z_{ref}} \right)^{\alpha} \tag{6.3}$$

where z_{ref} is some reference height and α is the shear exponent with typical values between 0.1 and 0.4. Smaller shear exponents produce flatter profiles resembling those of the unstably stratified ABL, whereas larger values produce higher shear profiles resembling those of the stably stratified ABL.

The models for the mean wind profiles given in (6.2) and (6.3) are appropriate for flat or very gentle terrain with surface roughness homogeneity, and they deviate from true mean wind profiles at some point approaching the top of the ABL. Luckily, this point is often well above wind turbines. Additional models are required for deviation from these canonical conditions. Models for the effect of topography, changes in surface roughness, and localized wakes of blunt structures such as buildings are discussed in detail in [7].

For example, various models exist that account for the effect of local topography on the mean wind profile. Often these models use potential flow theory to compute local accelerations and decelerations of the flow over complex terrain. The WAsP theory manual [7] discusses one such model in detail. That model solves for the potential flow solution on a cylindrical grid centered upon the site of interest (the axis of the cylinder extends vertically, and the cylinder is terrain conforming) using Fourier–Bessel series. Additionally, a model for the flow perturbation in the turbulent region of the boundary layer, where potential flow theory does not hold, is given. Other good references are Walmsley *et al.* [28] and Troen and De Baas [29].

The WAsP manual [7] also describes a method for computing the effect of surface roughness changes, such as at the land–sea interface. The idea is that an internal boundary layer begins to form at the roughness change location and extends downstream. At some point far from the roughness change, the wind profile will come to a new equilibrium, but upstream of that equilibrium region, the logarithmic profile of (6.2) does not hold. A three-part logarithmic profile is given in [7].

It also may be important to incorporate larger scale flow features, such as from mesoscale phenomena, which impact the spatial distribution of winds across a wind farm such that the idea of a single inflow model is less valid. This topic is discussed in detail in Chapter 4 of this book.

Fluctuating inflow wind

There are also applications in which only the mean inflow wind profile is not enough. For example, in estimating fatigue loads on turbines within a wind farm, the fluctuating part of the inflow wind must also be represented. In loads analysis, the time history of inflow turbulence spanning the entire rotor disk is necessary. Real turbulence has definite structure with spatial and temporal correlation, so modeling the turbulent part of the inflow wind with simple noise is not a viable option. More sophisticated models are required. Examples of commonly used models for the fluctuating component of the flow for wind energy applications are the models of Veers [30] and Mann [31]. Figure 6.2 shows a single instance of synthetically generated turbulence from the Veers methods in a plane normal to the flow. This turbulence would then be superimposed upon a mean inflow profile.

Because the Veers and Mann models require the generation of spectral tensors or cross-correlations at each point in the flow cross section, which are frequency- or wave-number-dependent, and then the inverse Fourier transform of each of these at every cross-section point, they become computationally expensive for large numbers of field points. This is usually not a problem when generating turbulence to span a single turbine rotor for loads calculations. However, there may be times when one wants to synthetically generate turbulence for a much larger region, such as the inflow to an entire wind farm. In that case, the Veers and Mann methods are only feasible if the spanwise and vertical resolution is quite coarse compared to the single turbine

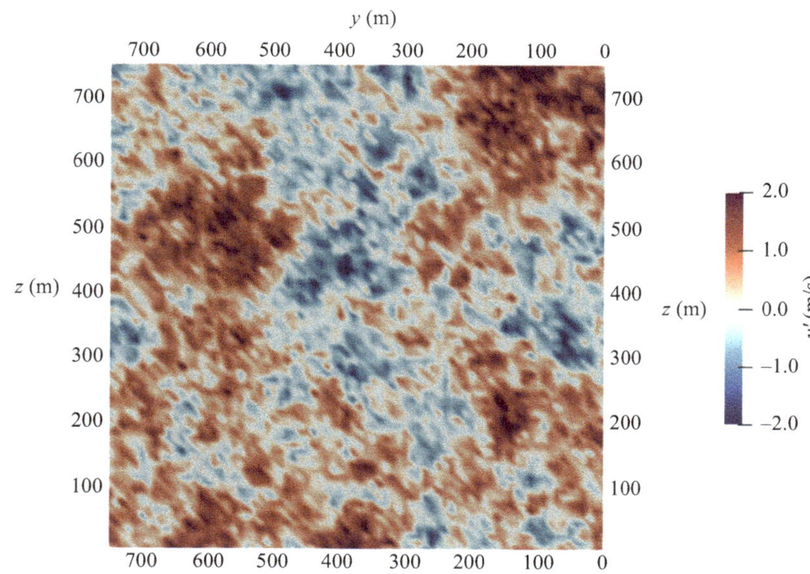

Figure 6.2 A single plane of synthetically generated turbulence using the Veers model [30]. The turbulence has 11.5% turbulence intensity

case. If the smaller scales are important, there are methods to fill in these more isotropic scales. Such "enrichment" methods include that of Ghate and Lele [32], and the fractal interpolation of Sim *et al.* [33].

6.3.1.2 Wake modeling

As with most simulation tool development, wake models began as simple steady-state models that gradually increased in complexity and fidelity through the addition of more physical processes and unsteadiness. The full range of fidelities continues to be used today for various purposes. Figure 6.3 shows a range of fidelity for wake models developed at the National Renewable Energy Laboratory (NREL), where the blue color represents the low speed flow of the wake, and the red is higher speed free-stream flow. In this figure, the rotor plane is on the left-hand side and the flow is from left to right. The models range from the relatively simple FLOw Redirection and Induction in Steady State (FLORIS) model [34], based on the models by Bastankhah and Porté-Agel [35,36], shown on the top of the figure to highly resolved LES at the bottom of the figure using NREL's Simulator fOr Wind Farm Applications (SOWFA) code [37]. The middle figure is that of NREL's FAST.Farm [38] based on the Technical University of Denmark's (DTU's) dynamic wake meandering (DWM) model [39]. Note the different details resolved by each wake model and evidence of unsteadiness, or wake meandering, from the wake deviating from the centerline of the domain. As expected, the high resolution LES model reproduces fine scale turbulent wake structures, while the lowest fidelity model primarily captures the steady mean wake shape. We will provide some details on the models behind Figure 6.3(a) and (b) in this section, whereas the model for Figure 6.3(c) is explained in Section 6.3.2.

Wakes and their subsequent models can be categorized into different scales; from the near wake of a single turbine with rotating coherent vortices to the turbulent meandering far wake of a single turbine, to the merged unsteady interactions of multiple wakes in wind farms, and to the large-scale wake of the overall wind farm. Simulated examples of these are shown in Figure 6.4. All such wakes result from a single source: the energy extraction from turbines within farms.

For individual turbines, the wake can be separated into the near and far wake. Many wind turbine wake models are used to estimate velocity deficits in the far wake, which is typically assumed to begin 2–3 diameters downstream of an operational turbine. Because most farms have turbine spacing distances of at least 5 diameters in the dominant wind directions and the near wake is physically complicated to model, the near wake is often neglected for simplicity.

The near wake is important for individual turbine aerodynamics and serves as an initial condition for the far wake. The near wake is sensitive to the blade aerodynamics and small-scale turbulence interactions, whereas the far wake is most sensitive to the overall rotor thrust and larger atmospheric turbulent length scales. Wind turbine near wake modeling and its impact on wind turbine aerodynamics is discussed in further detail in Volume 2, Chapter 1 of this book. In this chapter, we focus on the behavior and modeling of the far wake, which is most important for wind-farm energy estimation as well as turbine-to-turbine and interplant wake interactions. The majority of models described in this chapter are invalid and/or inaccurate in the near wake.

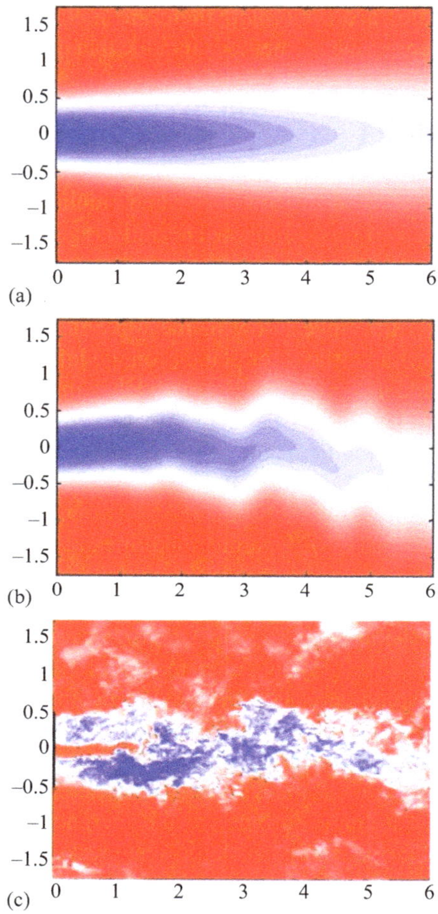

Figure 6.3 Examples of wake model fidelity: flow from left to right and turbine disk on left-hand side of each graphic: (a) a quasi-steady-state model from NREL's FLORIS model, (b) unsteady dynamic wake meandering model, and (c) large eddy simulation model of wake

The transition between near and far wake is a turbulent area where the coherent vortices from the blade tips and root break down and the wake resembles that of a more traditional turbulent bluff body type wake shape. Sørensen [40] developed a formal estimate where the near wake ends based on the number of turbine blades, N_b, tip-speed ratio, λ, and turbulence intensity, Ti,

$$\left(\frac{l}{R}\right)_{nearwake} = -\left(\frac{6.22}{N_b \lambda C_T} + 3\right) \log\left(\frac{Ti}{3}\right) \tag{6.4}$$

where l is the length of the near wake region downstream of the turbine and R is the turbine radius.

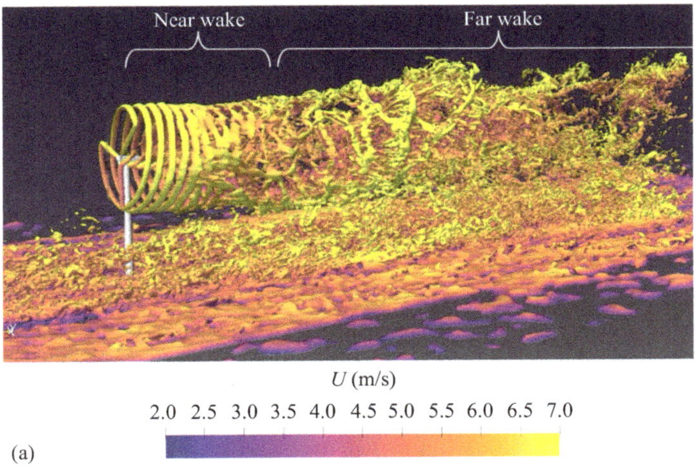

(a)

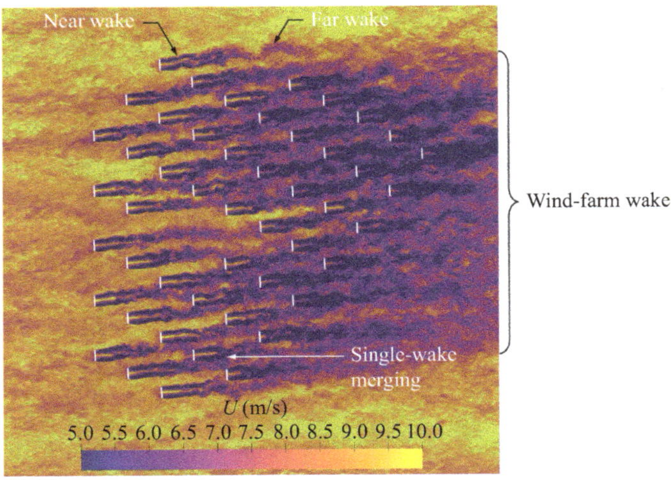

(b)

Figure 6.4 *An overview of wind turbine wakes: (a) the detailed structure of a single turbine wake shown with instantaneous isosurfaces of vorticity magnitude colored by velocity magnitude and (b) the complex interactions of many wakes within a full wind farm shown with a hub-height horizontal contour plane of instantaneous horizontal velocity as viewed from above (the white lines represent the rotor disk). Both images were created from data generated by computational fluid dynamics wind-farm modeling*

The wakes of the tower and nacelle are also often neglected. However, some research [41] has suggested that instabilities from the nacelle wake and blade root vortices, which are within the inner core of the overall turbine wake, interact with blade tip vortices, influencing near wake behavior. Therefore, under some situations, it may be advantageous to model the tower and nacelle wakes.

Self-similarity

In the far wake, the idea of self-similar velocity profiles [42] is useful for understanding the fundamental evolution of the wake and serves as a simple estimate of wake behavior across different inflow and turbine operating conditions. Many non-CFD based wake models use self-similarity as a foundational assumption. A commonly used set of self-similar equations is based on infinite Reynolds number solutions of the Navier–Stokes equations for circular wakes. In these equations, the wake velocity profile is only a function of the downstream distance from the rotor and Gaussian in shape [43,44]. Assuming laminar inflow, it can be shown that

$$\Delta u_{\max} \propto x^{-2/3} \tag{6.5}$$

$$b \propto x^{1/3} \tag{6.6}$$

where x is the downstream distance from the rotor, $\Delta u_{\max}$ is the maximum wake deficit velocity, and b is the mean wake width. According to Crespo and Hernandez [45], self-similarity can be extended to wake turbulence intensity, Ti as well.

$$Ti \propto x^{-1/3} \tag{6.7}$$

Recently, Wosnik and Dufresne [46,47] also introduced similarity solutions for the wake rotational velocity.

$$W_s \propto x^{-1} \tag{6.8}$$

where W_s is the rotational swirl velocity.

These similarity solutions assume that the flow is dominated by the turbulence in the wake and the solutions are invalid in the presence of an ABL with shear. In practice, some [48] have suggested that because of the atmospheric turbulence, the actual scaling law for the wake velocity deficit is closer to $x^{-1/2}$. Although these equations are useful for estimates, more complex relationships are typically used. Beyond downstream distance, the most significant variables that influence wake behavior also include rotor diameter, thrust, and turbulence intensity.

Individual turbine wake models

The most basic wake models still in use today are those of individual turbine wakes that are steady state. These models are superimposed on the atmospheric inflow and can be combined to create interactions of multiple wakes within the farm. They can also be coupled with larger wind plant effects to model impacts at the larger scale.

The simplest single turbine wake model developed by Froude is derived from a one-dimensional balance of mass, momentum, and energy for the fluid in a stream tube passing through a porous pressure disk. The theory was applied to wind energy and propeller systems by Betz, Lanchester, and Joukowsky [49]. The theory states that the wake velocity deficit in the far field is twice the induced velocity at the rotor disk, where the induced velocity is defined as the velocity reduction relative to the free-stream value. The induced velocity is often represented as its ratio to the free-stream velocity and the variable a. According to this classical theory, the induced velocity ratio is only a function of the rotor thrust coefficient, where $C_T = 4a(1 - a)$ [50].

At optimal conditions for power generation, the Betz limit states the induced velocity ratio, $a = 1/3$, and that the wake velocity is one-third of the free-stream velocity. Ideally, turbines in the field will be operating close to these conditions below rated wind speed. Above the rated wind speed, wake velocities will be higher because of less efficient operation to maintain constant power.

A popular wake model still in use today is the Jensen wake model [11,12], which is an extension of Betz one-dimensional stream tube theory and assumes a two-dimensional axisymmetric wake. The model uses a top-hat-shaped velocity profile across the wake in the lateral direction. The model also introduces wake dissipation that allows the wake velocity deficit to gradually recover with distance from the rotor plane. And it includes a simple model for wake expansion, where the wake width expands linearly with distance downstream. The wake velocity is expressed as a function of downstream distance.

$$\frac{U_{wake}}{U_\infty} = 1 - \frac{2a}{(1 + 2\alpha x/D)^2} \tag{6.9}$$

where U_{wake} is the constant velocity across the wake, U_∞ is the free-stream velocity, x is the distance downstream of the rotor, D is the rotor diameter, α is the wake decay coefficient, and a is the induction factor, which can be expressed in terms of the thrust coefficient, C_T

$$a = \frac{1 - \sqrt{1 - C_T}}{2} \tag{6.10}$$

The wake width, b, in the Jensen model expands as a simple linear relationship with downstream distance.

$$b = D + 2\alpha x \tag{6.11}$$

The wake decay coefficient is a critical input for this model and is assumed to be proportional to the observed turbulence intensity. As an influential input parameter, this constant is often tuned to existing wind-farm operational data from sites similar to those of interest for the modeler.

Archer *et al.* [23] used commonly recommended values for the wake decay coefficient of 0.04 for offshore and 0.075 for land-based farms. Peña and Rathman [51] found a more complex relationship for offshore wind farms related to roughness, stability, and turbulence and compared that to a simpler constant relationship of $\alpha \approx 0.4Ti$. Understanding the atmospheric stability is the key for more accurate wake loss predictions, as convective conditions have been shown to increase mixing and vertical momentum flux and reduce wake impacts, an effect modeled by increasing the wake decay coefficient in this model.

Further extensions to the Jensen model have been added through the years, such as adding ground effects and more realistic velocity deficit profiles. Bastankhah and Porté-Agel [35] replaced the top-hat velocity profile shape with a more realistic Gaussian shape and improved tuning of the wake decay constant using LES [52]. Xie and Archer [53] separated the rate of expansion in the vertical and horizontal directions and tuned their asymmetric model to LES of a single turbine in neutral stability conditions.

An alternate model developed around the same time as Jensen is the Larsen model. The model assumes axisymmetry and self-similar velocity profiles and is used in some engineering modeling tools. Details can be found in [54], as well as a paper on an extension [55] to include multiple turbine wake interactions and tuning to observational data.

Another set of models known as eddy-viscosity models are based on thin layer approximations of the Navier–Stokes equations. Ainslie [56] created such a model using an eddy-viscosity approach for turbulence closure that is also used in engineering tools today and continues to be refined [57]. In this model, the wake velocity profiles are assumed Gaussian in shape as in self-similar wake solutions.

$$\frac{U_{wake}(r)}{U_\infty} = 1 - D_M \exp\left(-3.56\left(\frac{r}{b}\right)^2\right) \tag{6.12}$$

where r is the radial distance from the wake centerline, b is the wake width, and D_M is defined as the velocity deficit at the wake centerline

$$D_M = 1 - \frac{U_c}{U_\infty} \tag{6.13}$$

where U_c is the wake centerline velocity; i.e., $U_{wake}(r = 0)$. Substituting this approximation into a reduced set of momentum and continuity equations, Ainslie derived a first-order differential equation for the centerline velocity that can be solved numerically.

$$\frac{dU_c}{d\hat{x}} = \frac{16\varepsilon(U_c^3 - U_c^2 - U_c + 1)}{U_c C_T} \tag{6.14}$$

where $\hat{x}$ is the downstream distance normalized by the rotor diameter and C_T is the rotor thrust coefficient. The eddy viscosity, ε, in this equation is empirically based and found using the following functions:

$$\epsilon = F(K_1 b(U_\infty - U_c) + \kappa^2 Ti) \tag{6.15}$$

$$F = \begin{cases} 1.0 & \hat{x} \geq 5.5 \\ 0.65 + \left(\dfrac{\hat{x} - 4.5}{23.32}\right)^{1/3} & \hat{x} < 5.5 \end{cases} \tag{6.16}$$

where κ is the von Karman constant ≈ 0.41 and $K_1 = 0.015$.

The initial centerline velocity deficit required to solve the equations in the far wake is determined 2D downstream and based on empirical wind tunnel observations.

$$D_M(x = 2D) = C_T - 0.05 - (16C_T - 0.5)\frac{Ti}{10} \tag{6.17}$$

where Ti is the turbulence intensity or the standard deviation of mean wind speed divided by the mean. Lastly, using conservation of momentum, the wake width at downstream locations is a function of the thrust coefficient and centerline wake deficit.

$$b = \sqrt{\frac{3.56C_T}{8D_M(1 - 0.5D_M)}} \tag{6.18}$$

Applied to the same wake, the Ainslie model provides a more rounded realistic wake contour compared to the Jensen model top-hat profile. Ainslie also includes a correction for wind-direction variability and its influence on wake meandering that can be added using formulas described in [56].

Although more advanced than self-similarity solutions, these axisymmetric models have limitations. For example, they are insufficient in the presence of shear and cannot predict the downward motion of wake centerline, for which more complex models are needed.

Many of these simpler models also do not include any influence of atmospheric stability, which is known to have a significant impact on wake behavior and dissipation [58]. To account for this lack of physicality, users modify coefficients that are functions of turbulence intensity and are closely correlated to stability. Highly stable models using LES [59] as well as observations [60] have shown that large veer can influence the wake shape into more of an oval that has been incorporated into some recent lower order axisymmetric models [61].

A more sophisticated steady wake model that does take into account atmospheric stability is FUGA, developed at the DTU by Ott *et al.* [62,63]. It solves the linearized RANS equations (see Section 6.3.2.2 for more explanation of RANS) cast in a mixed spectral form, along with a simple turbulence model. It relies on a table-look-up solution method that is extremely fast. Multiple wakes are linearly summed together to model wind farms (combining single wakes is discussed in the wake model combination subsection of this chapter). The FUGA model is included as an option within the WAsP wind-farm-planning tool.

Unsteady individual turbine wakes

Although the previously mentioned wake models approximate the mean nature of wakes and can be tuned to accurately predict power losses within wind farms, real wake behavior is unsteady. Wakes meander horizontally and vertically in the ABL and generate turbulence. The unsteadiness of wakes can increase structural loads and impact power production in a wind-farm environment. The half wake situation is shown to be the condition under which fatigue loads are increased most substantially although the likelihood of half wake situations decreases with downstream distance. The loading response is critical when examining wind-farm site suitability for individual turbines.

The unsteadiness of wakes is a result of forcing from large atmospheric turbulent structures and also self-induced turbulence generation. The length scale of the turbulence created by the wake is on the order of the rotor diameter, whereas unsteady length scales from atmospheric forcing are larger than the rotor diameter. Thus, these two methods of turbulence generation can and are modeled separately.

The simplest methods for modeling unsteadiness involve adding effective turbulence in areas impacted by wakes. The added turbulence is superimposed on the mean wake profiles of the models described earlier and existing atmospheric turbulence. A common model for this approach is the one recommended in the IEC design standard [14] based on the work of Frandsen [64]. The predicted added turbulence is a function of wind speed, ambient turbulence, wind-farm layout, and turbine thrust. In these models, the total turbulence intensity is assumed to be a sum of turbulent energy between the wake and atmospheric turbulence. Other added turbulence models include those by Quarton and Larsen; see also Duckworth *et al.* [65] for a comparison and validation of these methods.

More complex models for including wake unsteadiness assume the largest source of unsteadiness results from wake meandering as a result of atmospheric forcing. This meandering is included in the added turbulence models mentioned earlier and represented in steady wake models by the time-averaged shape of a meandered far wake. The difference between this and the added turbulence models is that in addition to increased turbulence levels, the structure of the turbulence evolves in time and space. The most popular model in this category is the DWM model developed first by Larsen *et al.* [39,66], which has been further augmented and refined in NREL's FAST.Farm tool [38]. The model consists of three components: the wake velocity deficit, wake meandering, and wake-added turbulence. The wake velocity deficit is calculated in the moving meandering frame of reference and allowed to grow with time. The velocity deficit is identical to those of the steady wake models, often using the Ainslie eddy-viscosity model [56] that is matched to blade element momentum solutions of the operating rotor.

Meandering is thought to be driven by large-scale turbulent structures in the atmosphere independent of turbine operation. The DWM wake is assumed to be a passive tracer convected by the overall flow field, assuming Taylor's frozen turbulence hypothesis, where the small-scale turbulence does not impact the wake shape or motion. Wake profiles are released and convected downstream at each simulation time step. The cutoff frequency to determine large versus small-scale is proportional to length scale of twice the rotor diameter.

$$f_c = \frac{U_\infty}{2D} \tag{6.19}$$

where f_c is the low-pass filter frequency, U_∞ is the free stream mean wind speed, and D is the rotor diameter.

In the model, the inflow is not modified by the presence of the turbine, but the meandering wake is superimposed upon it. The wake advection speed is based on a mean average wind speed over the rotor plane. The model is also sophisticated enough to reflect off of the ground, which may occur in certain atmospheric conditions. The atmospheric flow is calculated separately and can be derived from simple atmospheric inflow models or CFD solutions.

The added wake turbulence [67] is calculated in the moving meandering frame of reference and includes the effects from the wake shear layer and evolving blade shed vortices. The small-scale turbulent eddy size in this portion of the model is limited to about one rotor diameter. The strength of the turbulence is dependent on the wake velocity depth and radial gradient at a given downstream location.

The wake deficit, meandering behavior, and added turbulence contributions have been tuned to both lidar observations of unsteady wake and LES of operating turbines.

Turbine wakes in yaw

Another recent extension to individual turbine models has been additional capabilities for modeling turbine wakes in yawed conditions, which changes the wake behavior significantly. This was an issue first examined in the helicopter community for forward flight [68], and in the wind industry, it has been a concern for optimizing power production. Until recently, wind-farm operators have sought to minimize yaw error relative to unsteady incoming wind direction. The power loss of a yawed turbine is proportional to an exponential cosine function of the yaw angle.

$$P_{yawed} = P_{unyawed} \cos^n \gamma \tag{6.20}$$

where the observed exponent, n is between 1.4 and 2.0, and in practice less than the theoretically predicted value of 3.0 [16] for yaw angle, γ. To better inform the prediction of this flow condition, models have been adopted for yawed wind turbines [36,69–72] and validated by observations in wind tunnel studies [73].

There is new interest in yawed behavior of wind turbine wakes in the burgeoning area of wind-farm control [16,74]. A promising new control strategy modifies typical wake behavior by steering the wakes away from downstream turbines. The goal is to minimize wake loss downstream and increase the overall wind plant energy production to compensate for the upstream yawed turbine energy loss dictated by (6.20). A useful parameter for developing these control studies is the resulting wake yaw angle behind the turbine to determine in which direction and how far laterally the wake will propagate. Because of the conservation of lateral momentum, this angle is the opposite direction of the turbine yaw angle. A useful approximation developed by Bastankhah and Forté-Agel for the wake skew angle is

$$\chi \approx \gamma + \frac{0.3\gamma}{\cos\gamma}\left(1 - \sqrt{1 - C_T \cos\gamma}\right) \tag{6.21}$$

where χ is the wake angle relative to the rotor axis, γ is the rotor yaw angle relative to the wind direction, and C_T is the rotor thrust coefficient. This wake angle approximation is technically valid only at the rotor plane but has been shown to be accurate for multiple rotor diameters downstream, after which the velocity deficit is directed toward the free-stream velocity vector. As the yaw angle grows, so does the lateral flow, and an interesting area of current research is the interaction of multiple wakes in yaw and how they modify a larger portion of the flow field, creating induced vortices that entrain momentum from higher speed areas of the ABL [15] above the turbines.

Vortex methods

Vortex methods [75] are commonly used in the helicopter industry and provide a non-dissipative way to propagate blade tip vortices downstream of a turbine. These methods assume the vorticity created from the lift of the blade is released at the tip and hub and are propagated through incompressible, inviscid flow fields. The vorticity is then advected downstream in line filaments or particles to produce a coherent wake structure, and the equations of import are governed by the Biot–Savart law. The vorticity distribution along the blade is calculated using blade element momentum theory [50] including tip-loss models and is dependent on lift and drag coefficients of individual airfoils located along the blade. The shape of the wake is either prescribed [76] and fixed throughout the simulation or free [77], with the wake shape calculated every time step. Free vortex methods are computationally more expensive but are also applicable to a wider range of operating conditions than the more empirical prescribed wake shapes. Most models also include vortex breakdown models that approximate the dissipation of the convected vortices. These models tend to run faster than full CFD models of wakes, particularly when parallelized using fast multipole methods. The drawbacks of these methods are the added cost for unsteady calculations and the fact that the inclusion of turbulence can add exponentially more degrees of freedom, which is why these methods are most often used for steady inflow.

Wake model combination

The previous sections focused on individual wake models, but predicting wind-farm power output and turbine response requires aggregated wake impacts. Typically, wakes are combined through simple superposition [78]. The simplest approximation for wake model combination assumes the downstream turbine is only impacted by its closest upstream neighbor, which has been shown to agree reasonably well with observations from smaller land-based wind farms. From a loads perspective, data show that there is no substantial difference between single and multiple wake loading conditions [18], although this is still an active area of research. The methods for wake combination tend to be more sensitive in the offshore environment given the longer propagation of wakes in the low turbulence environments [21]. Given that wind farms are generally composed of more than two rows of turbines, different methods have been developed for modeling the interaction and combination of wakes as they propagate through the wind farm. And although many combination algorithms exist, the most accurate have been found to be linear and quadratic superposition methods.

Linear sum

$$\left(1-\frac{u_{n+1}}{U_\infty}\right)=\sum_{i=1}^{n}\left(1-\frac{u_{i+1}}{u_i}\right) \tag{6.22}$$

Quadratic sum

$$\left(1-\frac{u_{n+1}}{U_\infty}\right)^2=\sum_{i=1}^{n}\left(1-\frac{u_{i+1}}{u_i}\right)^2 \tag{6.23}$$

where n is the number of turbines upstream of turbine $n+1$, and u_i is the velocity at a turbine upstream. Initial attempts at superposition focused on linear methods [24] but were found to overestimate wake velocity deficits. Katic *et al.* [12] then proposed superposition of the squares of the velocity deficit, which resulted in improved agreement with observational data. Niayifar and Porté-Agel [52] have more recently suggested a slight modification of the linear sum model to improve agreement with observations. For wind directions where there is partial overlap between wakes, geometric approximations of overlapping areas are made to create composite wake combinations [9].

When flow is down a row of turbines within a wind farm, the largest loss of power is between the first and second row, with more losses occurring after the second row, depending on how the data are binned as shown in Figure 6.5. These losses can reach an asymptotic state once a balance between the momentum flux of the ABL and the momentum extraction of subsequent turbines is achieved. When the inflow is not aligned with a wind-farm row, the models are more difficult to calibrate; wake combination is an area of active research. Also, studies have shown that in large wind farms, the inclusion of wake model combination continued to underpredict overall wake losses, which led to the creation of deep array models, which is the subject of the next section.

Wind-farm models

As wind farms grew in size, researchers developed models of wind-farm interactions with the atmosphere where the wind farm was no longer decoupled from the atmosphere and could influence the inflow. These models began as distributed roughness elements to simulate turbine impacts on overall wind speed that decreases [79,80] using a simple boundary layer theory. These models are based on a single vertical

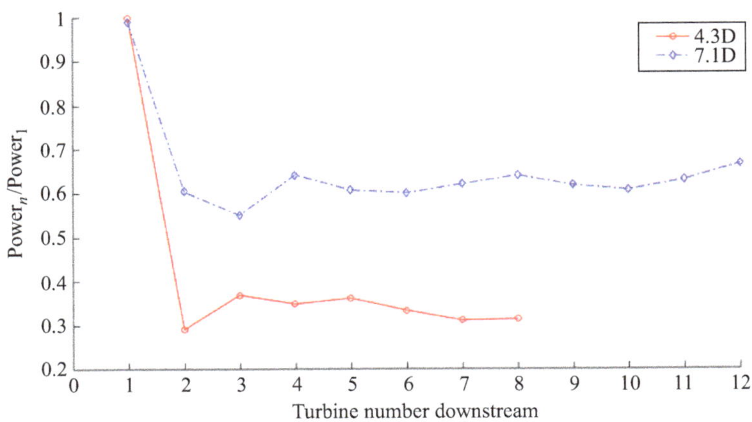

Figure 6.5 Power losses between turbines in single rows of two offshore wind farms with different turbine spacings as the wind blows down the row. One wind farm (red solid line) has 8 turbines spaced at 4.3 rotor diameters; the other (blue dash-dot line) has 12 turbines spaced at 7.1 rotor diameters

column slice through the wind farm where the mean velocities are assumed to be horizontally averaged. A commonly used example of the next generation of wind-farm models is the infinite wind-farm approximation from Frandsen [48], which is expressed as a modification to the ABL. A log layer develops both underneath and over the wind farm similar to a canopy model used to model forests. Friction velocities are calculated based on turbine thrust loading and wind-farm spacing for above and below the wind farm. This model has since been augmented by Peña and Rathman to include the impact of atmospheric stability [51]. These roughness approximations have recently been verified with higher fidelity LES [81] and validated against observations of operational offshore wind farms [82]. Ultimately, these models are combined with single turbine wake models [51,82] to couple the effects in the individual turbine wakes to the impact of the overall wind farm.

Models that capture the overall wind-farm wake production are of growing interest, particularly in areas with large growth in wind development, such as the North Sea. These large-scale interactions now interact with larger mesoscale driven phenomena across multiple wind farms. A commonly used model for this impact is that of Fitch *et al.* [83] where each turbine in a wind farm is modeled as an area of momentum extraction and turbulent kinetic energy augmentation within a mesoscale model. Other large-scale wake model considerations are the impact of Coriolis force [84] for modifying wind-farm wake direction, and the introduction of atmospheric gravity waves by the presence of the wind farm [6] can also impact unsteady performance. In addition to wind-farm interactions, wind-farm-scale models have been used to predict local and regional environmental impacts, although the accuracy of such estimates is still being evaluated.

The wind-farm simulation methods described in this section have significant issues as the flow fields become more complex. For example, most of the superposition methods have been well validated for offshore wind farms, but the accuracy in complex terrain is more suspect. And only recently have measurements provided some glimpses of how wakes behave in more complex environments [85]. Other complexities include land-sea interactions, forested sites, and Coriolis effects [84]. To capture these more unique environments, CFD models that better capture coupled nonlinear physical processes are required, and these are described in the next sections.

6.3.2 Computational-fluid-dynamics-based approaches

If two main ingredients of wind plant flow modeling are the inflow winds and wind turbine wakes, CFD enables the simultaneous simulation of both processes and all of their nonlinear interactions. To achieve such a solution, CFD-based approaches numerically solve the governing fluid-flow equations. These equations are highly nonlinear, coupled, and dictate not only the mean flow but also the turbulent flow. Therefore, many choices need to be made before even performing a simulation or choosing a CFD code about which form of the equations to use and how to treat turbulence. Also, to numerically solve the equations of fluid motion, one needs to discretize the equations using an appropriate numerical method on an appropriate computational mesh made up of points, cells, or elements. In this section, we outline some of the possibilities, and then we move on to how one generates realistic inflow winds and wind turbine wakes. Finally, we discuss how one puts those flow ingredients together to simulate the full wind plant.

6.3.2.1 Choice of equation set

The fluid-flow governing equations come in many different forms, and excellent texts covering this topic include those of Currie [86] and White [42]. In their most basic compressible form, these governing equations are

$$\frac{\partial \rho}{\partial t} + \frac{\partial}{\partial x_j}\left(\rho u_j\right) = 0 \tag{6.24}$$

$$\frac{\partial}{\partial t}\left(\rho u_i\right) + \frac{\partial}{\partial x_j}\left(\rho u_j u_i\right) + \rho \Omega_j u_k \varepsilon_{ijk} = -\frac{\partial p}{\partial x_i} + \frac{\partial}{\partial x_j}\left(\tau_{ij}\right) + \rho g_i + \rho f_i \tag{6.25}$$

$$\frac{\partial}{\partial t}\left(\rho h\right) + \frac{\partial}{\partial x_j}\left(\rho u_j h\right) = \frac{\partial p}{\partial t} + u_j \frac{\partial p}{\partial x_j} - \frac{\partial q_j}{\partial x_j} + \tau_{ij}\frac{\partial u_i}{\partial u_j} \tag{6.26}$$

which enforce the conservation of mass, momentum, and energy, respectively. The mass-conservation equation is often referred to as the continuity equation, and the momentum conservation equations are the Navier–Stokes equations. Equations (6.24)–(6.26) are given in the most general compressible form with reference frame rotation. In these equations, ρ is the density, u_i is the instantaneous velocity vector, Ω is the reference frame rotation vector, ε_{ijk} is the alternating tensor denoting a cross product, p is the pressure, τ_{ij} is the stress tensor, g_i is the gravity vector, f_i is a general body force vector, h is the enthalpy, and q_j is the heat flux vector. In engineering flows, often the frame rotation and gravity terms of the momentum equation are negligible, but in wind plant flows in which the ABL is inherently stratified and exists on the rotating Earth, they are very important and should almost always be retained.

Additional constitutive equations are necessary to close this set of governing equations. Commonly, the working fluid, air in this case, is assumed to follow the ideal gas law, which dictates that $p = \rho RT$, where R is the universal gas constant, and T is temperature. We can also relate enthalpy and temperature with $h = c_p T$, where c_p is specific heat at constant pressure. Further relations can be used to link viscosity and thermal diffusivity to temperature, but for wind-energy applications, these can generally be constant.

Depending upon the situation to be simulated, one may choose to simplify the form of governing fluid flow equations. For example, for low-Mach number applications, which is usually the case in wind-energy applications, one can assume that density is constant, yielding the incompressible set of fluid-flow equations. The assumption that viscosity and thermal conductivity do not vary with temperature is also valid. The result is that density and the time derivative vanish from (6.24); (6.25) simplifies slightly; and the pressure and stress work terms become negligible in (6.26), decoupling this equation from the equation set. Also, (6.26) may be cast in terms of temperature using $h = c_p T$. The incompressible equations are

$$\frac{\partial u_j}{\partial x_j} = 0 \tag{6.27}$$

$$\frac{\partial u_i}{\partial t} + \frac{\partial}{\partial x_j}\left(u_j u_i\right) + \Omega_j u_k \varepsilon_{ijk} = -\frac{1}{\rho}\frac{\partial p}{\partial x_i} + \frac{\partial}{\partial x_j}\left(\nu \frac{\partial u_i}{\partial x_j}\right) + g_i + f_i \tag{6.28}$$

$$\frac{\partial T}{\partial t} + \frac{\partial}{\partial x_j}\left(u_j T\right) = -\frac{\partial}{\partial x_j}\left(\kappa \frac{T}{\partial x_j}\right) \tag{6.29}$$

where ν is viscosity and κ is thermal conductivity.

The continuity condition is usually not enforced directly through the application of (6.27). More commonly, the momentum equation, (6.28), is numerically discretized, and then the divergence of it is taken. We are left with an equation in which there is divergence of velocity at the new time step level, which is set to zero to enforce continuity. The result is a Poisson equation for pressure. Essentially, we are solving for a pressure field that will enforce continuity.

However, if density stratification is important, which is almost always the case in the ABL, the incompressibility simplifications do not capture those buoyancy effects. Density stratification effects can be included through the use of the Boussinesq approximation for buoyancy in which the gravity term of (6.28) becomes

$$\frac{\rho + \rho'}{\rho} g_i = \left(1 - \frac{T - T_0}{T_0}\right) g_i \tag{6.30}$$

where T_0 is a reference temperature. This term states that fluctuations in density are due only to temperature fluctuations away from a reference state, and density has no dependence on pressure changes. This term also recouples the thermal transport equation with the momentum equation.

The Boussinesq approximation for buoyancy, which is well outlined by Stull [2] is commonly used for LES of the ABL, some examples of which include the works of Moeng [87], Deardorff [88], and Mason [89]. It is also used in wind plant LES, as can be seen in many works including those of Allaerts and Meyers [6,90], Churchfield *et al.* [37,91], Abkar *et al.* [92], and Lu and Porté-Agel [93]. However, it is a fairly gross approximation, and if large vertical motions are expected, as with flow over high terrain, or if the simulation domain is taller than a few kilometers, the approximation can break down. For example, if simulating a 10-km tall domain, the density at the top of the domain is roughly one-third of that at the surface because hydrostatic pressure variations dominate the density variation. In that case, one can resort to the fully compressible equation set or explore a low-Mach number compressible form of the equations, such as the anelastic form of the governing equations.

The anelastic equations allow for a variable density, but do not admit sound waves, which allows for larger simulation time steps. They assume a background density, temperature, and pressure profile and solve for perturbations in pressure and temperature from this base state. The anelastic equations are described in detail by Lipps and Hemler [94], Lipps [95], and Bannon [96].

It is also important to note that the governing equations described earlier in primitive and conserved variable form can be cast in alternate ways. One notable way is the velocity–vorticity formulation. It can be derived from the incompressible governing equations by taking the curl of (6.28), yielding a transport equation for vorticity and by taking the curl of the definition of vorticity, $\omega = \nabla \times u$, and applying the continuity relation of (6.27) yielding a Poisson equation relating velocity and vorticity. Velocity–vorticity formulations have the advantage that when solved numerically, vorticity-containing flow structures, like tip vortices, can be more easily preserved than with other formulations. A disadvantage is that appropriate boundary conditions are more difficult to define. Although this approach has been used more in the helicopter community than in the wind energy community, Fletcher and Brown [97] have applied the velocity–vorticity form of the governing equations to wind plant applications.

6.3.2.2 Treatment of turbulence

Beyond the form of the governing fluid-flow equations used, whether it is compressible, incompressible, anelastic, etc., one must consider the treatment of turbulence. Turbulence is ubiquitous in the ABL, on the turbine boundary layers, and in the turbine wakes. Because wind plant flows are of an extremely high Reynolds number, and hence contain a vast range of turbulent scales, it is currently, and for the foreseeable future, impossible to directly solve the Navier–Stokes equation, which in the field of CFD is called direct numerical simulations (DNS). Therefore, CFD-based methods applied to wind plant flows use some sort of averaging or filtering on the Navier–Stokes equations to reduce the resolution requirements.

The following is an example of why DNS is not used: if the smallest turbulent eddy within a wind plant is a millimeter in diameter, then a computational mesh with a resolution of less than a millimeter would be required. However, the wind plant may be ten or more kilometers on a side, and the ABL it lies within may extend a kilometer or two vertically. If we then say a DNS of the wind plant would require quarter millimeter resolution on a computational domain of 10 km in the horizontal directions and 2 km in the vertical, then a computational mesh of 1.28×10^{22} cells is required! With CFD, the time step size often scales with the grid resolution, so fine meshes mean small time steps. However, we are often interested in hours of wind plant simulation to obtain converged statistics. With the mesh described earlier, the time step size may be $\mathcal{O}(1 \times 10^{-5})$s, and we may desire to simulate a few thousand seconds of total time, meaning that of $\mathcal{O}(1 \times 10^{8})$ time steps may be needed! With typical CFD methods, this problem would require $\mathcal{O}(10^{17})$ compute cores; the largest high-performance computing system as of this writing, Summit, an IBM system at Oak Ridge National Laboratory in the United States, contains only 202,752 conventional cores (although it also contains numerous GPUs) [98], magnitudes less than required to use DNS on the wind plant flow problem.

To make the CFD problem tractable, the governing fluid-flow equations are filtered or averaged such that some or all of the turbulent scales are modeled, while the remaining larger scales are actually resolved. With LES, the equations are spatially filtered. The filter width can be chosen to provide a wave number above which turbulent scales are not well resolved, and it is often tied to the computational mesh. The resultant solution contains resolved turbulence down to some length scale, below which the effect of the smaller scales are all modeled.

One can also apply Reynolds averaging to the fluid-flow equations. This turbulence treatment technique is often referred to as solving the RANS equation. For wind plant flows, the most convenient form of Reynolds averaging is time averaging of the form

$$\bar{f} = \frac{1}{T}\int_{t}^{T} f(x,y,z,t)\mathrm{d}t \tag{6.31}$$

where f is the quantity to be averaged, $\bar{f}$ is the averaged quantity, and T is the length of time over which the average is performed. The interpretation of T is important. It is a timescale much larger than the turbulent timescales, but much smaller than nonturbulent, slowly varying motions, if they exist. With RANS, one solves for the Reynolds-averaged or "mean" component of the flow variables. If the slow-varying motions exist in the fluid system, the RANS equations retain an active time derivative, so they are often referred to as the unsteady RANS (URANS) equations.

In wind plant flows, the meaning of Reynolds averaging must be carefully considered. For example, in the background atmospheric flow driving the wind plant, there is often a spectral gap separating the ABL motions from the slower scale synoptic weather scales. Here, there is a clear interpretation of T that averages the boundary layer turbulence but leaves synoptic motions. There are conditions, such as highly convective daytime conditions, though, in which the ABL's largest scales become large enough and of sufficiently low frequency that there is no clear separation between them and synoptic scales. To further complicate matters, the rotor timescale is on par with the ABL timescales, yet with some rotor models, such as actuator lines, organized flow structures that cannot be considered turbulence, such as blade tip vortices, exist. To simultaneously capture rotor blade tip vortices and yet average the turbulence in the ABL seems impossible. However, simulations are indeed performed in this manner with unclear separation of turbulent scales. For example, URANS may be employed and the larger ABL scales (and unsteady near-wake features) are resolved. At that point, it is unclear if the simulation should be interpreted as URANS or what some researchers refer to as "very-large-eddy simulation."

In the case of LES and RANS, spatial filtering or Reynolds averaging yields the same form of the equations. For example, filtering or Reynolds averaging (6.27)–(6.29) for incompressible flow results in the equations,

$$\frac{\partial \bar{u}_j}{\partial x_j} = 0 \tag{6.32}$$

$$\frac{\partial \bar{u}_i}{\partial t} + \frac{\partial}{\partial x_j}\left(\bar{u}_j\bar{u}_i\right) + \Omega_j\bar{u}_k\varepsilon_{ijk} = -\frac{1}{\bar{\rho}}\frac{\partial \bar{p}}{\partial x_i} + \frac{\partial}{\partial x_j}\left(\nu\frac{\partial \bar{u}_i}{\partial x_j}\right) + \frac{\partial}{\partial x_j}R_{ij} + g_i + f_i \tag{6.33}$$

$$\frac{\partial \bar{T}}{\partial t} + \frac{\partial}{\partial x_j}\left(\bar{u}_j\bar{T}\right) = -\frac{\partial}{\partial x_j}\left(\kappa\frac{\bar{T}}{\partial x_j}\right) + \frac{\partial}{\partial x_j}Q_j \tag{6.34}$$

where the overbar denotes a filtered or Reynolds-averaged quantity. Applying the filtering or averaging operation to the nonlinear advection terms for momentum and temperature results in extra terms in (6.33) and (6.34), R_{ij} and Q_j. These terms represent the effect of the unresolved turbulent scales on the resolved or averaged scales. In the RANS equations, R_{ij} and Q_j are referred to as the Reynolds stresses and turbulent thermal flux, respectively. With LES, these quantities are referred to as the subgrid-scale (SGS) stresses and thermal fluxes, respectively. It is impossible to define these terms exactly without introducing new unknown terms of higher order, so a turbulence model is used to dictate the behavior of these terms.

The texts of Wilcox [99], Pope [100], and Sagaut [101] are excellent resources for LES and RANS turbulence treatments, but we will give some representative examples here.

The RANS equations (Equations (6.32)–(6.34)) are solved along with a turbulence model, such as the standard k–ε model, which is probably the most popular engineering RANS turbulence model. That model is given by

$$\frac{\partial k}{\partial t} + \frac{\partial}{\partial x_j}\left(\bar{u}_j k\right) = -R_{ij}\frac{\partial \bar{u}_i}{\partial x_j} - \varepsilon + \frac{\partial}{\partial x_j}\left[\left(\nu + \nu_t/\sigma_k\right)\frac{\partial k}{\partial x_j}\right] \tag{6.35}$$

$$\frac{\partial \varepsilon}{\partial t} + \frac{\partial}{\partial x_j}\left(\bar{u}_j \varepsilon\right) = -C_{\varepsilon 1}\frac{\varepsilon}{k}R_{ij}\frac{\partial \bar{u}_i}{\partial x_j} - C_{\varepsilon 2}\frac{\varepsilon^2}{k} + \frac{\partial}{\partial x_j}\left[\left(\nu + \nu_t/\sigma_\varepsilon\right)\frac{\partial \varepsilon}{\partial x_j}\right] \tag{6.36}$$

where k is the turbulent kinetic energy, ε is the turbulent dissipation rate, and $\nu_t = C_\mu k^2/\varepsilon$ is the turbulent viscosity. The model constants are C_μ, $C_{\varepsilon 1}$, $C_{\varepsilon 2}$, σ_k, and σ_ε.

The Reynolds stresses and turbulent thermal flux found in the RANS equations are evaluated using Boussinesq's hypothesis that states

$$R_{ij} = -\nu_t\left[\frac{\partial \bar{u}_i}{\partial x_j} + \frac{\partial \bar{u}_j}{\partial x_i}\right] + \frac{2}{3}k\delta_{ij} \tag{6.37}$$

$$Q_j = -\frac{\nu_t}{\mathrm{Pr}_t}\frac{\partial \bar{T}}{\partial x_j} \tag{6.38}$$

where δ_{ij} is the Kronecker delta and Pr_t is the turbulent Prandtl number.

It is interesting to note that within RANS turbulence modeling, there seems to be a bifurcation in modeling approaches between the atmospheric and engineering communities, but upon closer inspection, the typical models have the same basic roots. For example, a popular atmospheric RANS turbulence model used to model the effect of ABL turbulence within regional-scale weather models is that of Nakanishi and Niino [102] (often referred to as the Mellor–Yamada–Nakanishi–Niino [MYNN] model), which solves transport equations for four turbulence quantities, including turbulent kinetic energy (the other three turbulent quantities are related to turbulent temperature and moisture fluctuations). The parallels between atmospheric RANS models like the MYNN model and engineering variants like the k–ε model are apparent, but there are marked differences as well. Although the MYNN model is a RANS model used in the computation of the turbulent ABL, which is inflow to the wind plant, one would not use it within the wind plant itself because it contains simplifications that are not appropriate for wake flows. One of those simplifications is that it assumes the vertical variation of the flow is so much greater than the horizontal variation that only the vertical derivatives are retained. This assumption certainly does not hold for wakes.

Next, we discuss LES turbulence modeling, which is sometimes referred to as SGS modeling. Unlike with RANS turbulence models, atmospheric and engineering variants of LES turbulence models are not all that different. In fact, many LES turbulence models are used in both fields. A popular LES turbulence model is the one-equation model given by

$$\frac{\partial k_{sgs}}{\partial t} + \frac{\partial}{\partial x_j}\left(\bar{u}_j k_{sgs}\right) = -R_{ij}\frac{\partial \bar{u}_i}{\partial x_j} - C_\varepsilon \frac{k_{sgs}^{3/2}}{\Delta} + \frac{\partial}{\partial x_j}\left[\left(\nu + \nu_{sgs}\right)\frac{\partial k_{sgs}}{\partial x_j}\right] \tag{6.39}$$

where k_{sgs} is SGS turbulent kinetic energy, $\nu_{sgs} = C_k k^{1/2}\Delta$, and Δ is a filter length scale. The model constants are C_k and C_ε. The same relations for R_{ij} and Q_j shown in (6.37) and (6.38) are used here, except that ν_t is replaced with ν_{sgs}. The filter length scale is often tied to the computational grid resolution such that as the grid is refined, the SGS turbulent kinetic energy and viscosity are reduced, allowing the filtered equations (Equations (6.32)–(6.34)) to resolve more turbulent content.

6.3.2.3 Choice of numerical methods

Because CFD numerically solves some form of the governing fluid flow equations, we will briefly mention numerical methods used in CFD. There is a wide variety of numerical methods used in CFD, and each of them comes with advantages and disadvantages. We will mention some of the main methods here.

Pseudospectral methods are often used in atmospheric LES, such as in the work of Moeng [87] and Sullivan *et al.* [103], Now pseudospectral methods are being successfully used in wind plant LES, examples being the work of Allaerts [104], Allaerts and Meyers [6,90], Calaf *et al.* [81], Johnstone and Coleman [105], and Meyers and Meneveau [106]. Pseudospectral methods are used for solving the incompressible form of the governing equations. They assume the solution is in the form of a cosine/sine series. The advantages are that this method has exponential error convergence with grid refinement, and the pressure Poisson solve can employ efficient fast Fourier transform solvers. The disadvantages are that the method requires the use of a structured computational mesh (meshes with Cartesian grid lines, or ones that can map to Cartesian, with mesh volumes that are hexahedral) that is periodic in the spectral directions, which is limiting. Because the vertical direction in atmospheric simulation is not periodic, modelers often resort to finite-difference methods in that direction. This makes the method difficult or impossible for use in complex terrain. Wind plant flows are nonperiodic, but this is dealt with by placing a region near the outflow boundary that relaxes the flow toward some prescribed turbulent inflow.

Finite-volume methods are commonly used in engineering applications, but their use in the atmospheric community is less prevalent; however, they have gained significant traction in the wind community. Their computational meshes are seen as a collection of finite control volumes. They may be structured or unstructured (unstructured meshes may have shapes other than hexahedra). They have many advantages, including the fact that they can conserve mass and momentum to numerical precision, they are conceptually intuitive, they can handle a variety of boundary conditions, and they can handle very complex geometry or localized mesh refinement using unstructured meshes. DTU's Ellipsys3D CFD code [107,108] and the OpenFOAM CFD toolbox, upon which the NREL's SOWFA [37] is built, are examples of finite-volume based approaches.

Finite-difference methods are still used as well. For example, the code PALM [109], which has been used for wind-energy applications, uses a staggered grid finite-difference method. As mentioned earlier, many of the pseudospectral codes are only pseudospectral in the horizontal directions and use finite differences in the vertical.

More exotic is the lattice Boltzmann method, which solves the governing equations of fluid flow based on the lattice Boltzmann equation. Rather than looking at the fluid as a continuum, it is viewed as ensembles of molecules, where the propagation and collision between the molecules is modeled. An example of the lattice-Boltzmann method applied to wind energy is the work of Wood [110].

The study of numerical methods for CFD is a vast area of research. For the interested reader, we recommend consulting the many available texts, such as that of Ferziger and Perić [111] and Pletcher, Tannehill, and Anderson [112].

6.3.2.4 Inflow wind generation

In generating inflow winds for CFD-based wind plant flow simulation, some approaches rely on an auxiliary "precursor" CFD simulation to generate the wind field directly from the governing equations. Other approaches, though, use the non-CFD-generated inflow winds described in Section 6.3.1.1 as boundary conditions or forcings that then get advected downstream and adjust to the governing equations that act upon them. It is also important to consider that CFD methods used for generating inflow winds with RANS versus LES turbulence treatment are quite different from each other because RANS treatment resolves only the very largest turbulent scales or none at all, whereas LES treatment resolves a significant range of turbulent scales. In the end, though, no matter how the inflow is generated, an important objective is that this inflow satisfies the governing equations as well as possible so that it advects through the wind plant domain as seamlessly as possible.

Precursor approaches

One of the most natural methods for generating inflow winds with CFD is referred to as the precursor method. This method generates both the mean and turbulent part of the flow directly from the governing flow equations, so as to feed seamlessly into the wind plant domain. We illustrate this method from the LES point of view and show an overview of a precursor LES in Figure 6.6.

If one desires to simulate a wind plant with inflow winds with certain characteristics, a "precursor" or auxiliary simulation can be performed with no turbines. It is simply an atmospheric LES run before the wind plant LES used to generate inflow boundary data.

In this simulation, the domain is simply a box large enough to capture the full vertical extent of the ABL and some of the free atmosphere along with the largest horizontal turbulent scales. This domain is periodic on the sides to represent an infinite boundary layer.

The simulation is initialized with appropriate vertical profiles of temperature and velocity (other scalar quantities could also be included, such as moisture and CO_2)

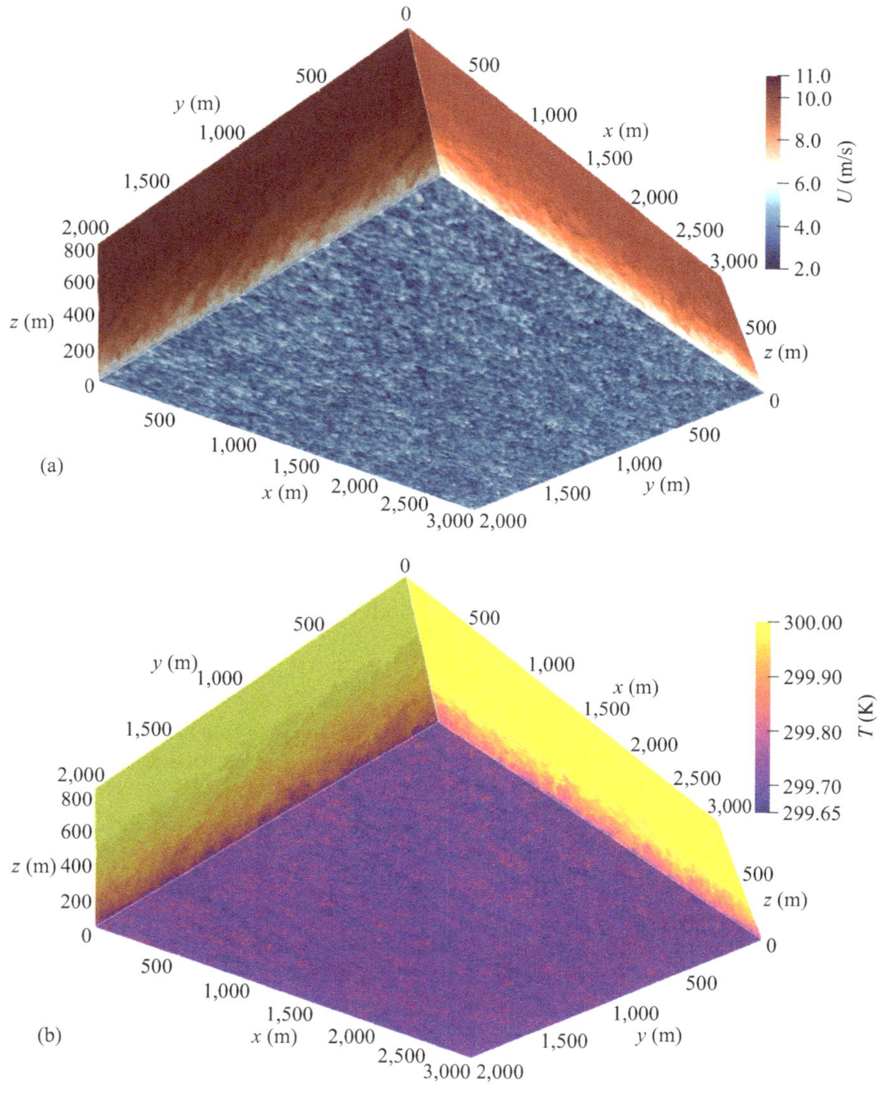

Figure 6.6 *Overview of a precursor large-eddy simulation during lightly stable atmospheric conditions: part (a) shows contours of instantaneous velocity and part (b) shows contours of potential temperature. The flow is viewed from below looking upward toward the surface that contacts the Earth*

along with some sort of perturbations near the surface. Surface conditions, such as roughness and heat flux are also specified. A background driving pressure gradient (or a geostrophic wind vector) is specified to drive the winds. Latitude information is also specified, which enters the Coriolis term.

The simulation is run until the turbulent ABL is fully developed. At that point, the precursor can be run longer while planes of data on the inflow boundary or boundaries are extracted and saved to be used later as boundary conditions to the nonperiodic wind plant simulation. We often call these precursors "canonical" if they are driven to some atmospheric stability state, such as neutral, daytime convective, or nighttime stable. We can also run "noncanonical" precursors and extract inflow data through an entire diurnal cycle. This is particularly interesting in that it allows one to then run a diurnal wind plant simulation using these data.

Alternately, one can run a single LES that contains both a precursor and a wind plant region of the domain. The precursor region is placed ahead of the wind plant region, and at the interface between the two regions is a recycling plane that is used to make the precursor region periodic. This approach is used by Stevens *et al.* [113]

One could also run a precursor using RANS turbulence treatment, but because precursor atmospheric simulations represent a horizontally infinite ABL and RANS does not resolve any turbulence, the domain need only be a single column and only vertical derivatives remain finite. Once reaching a fully developed state, such a RANS calculation would yield a vertical profile of velocity, temperature, and any other relevant scalars. This profile may even be time varying, for example, if steady surface cooling or a ramping geostrophic wind is applied. The one-dimensional profile could be then projected onto a two-dimensional inflow plane for a wind plant RANS calculation.

Precursors come with some computational cost because they are either an additional simulation performed before the wind plant flow simulation or they occupy an additional mesh region, creating a larger computational mesh. The cost is often a fraction of the cost of the wind plant flow simulation, and the turbulence generated automatically satisfies the governing equations. The cost can also be justified by the fact that any combination of different wind and surface conditions can be simulated, producing a mean and turbulent field compatible with those conditions. The precursor data can also be saved to disk and reused many times.

Nonprecursor approaches

If one does not want the additional cost of a precursor simulation, other methods exist for creating inflow for CFD-based wind plant flow simulations. These other methods are all variants on specifying mean inflow profiles that are applied to the inflow boundary conditions. Profiles of velocity and temperature (if atmospheric stability is taken into account) must be supplied at the inflow boundary. Often logarithmic or power law profiles as outlined in Section 6.3.1.1 are used.

More complicated mean profiles can be created as well. One way to do this is using a mesoscale numerical weather prediction code, a popular example being the Weather Research and Forecasting (WRF) code [114]. The term "mesoscale" refers to the regional-scale weather that spans hundreds to around a thousand kilometers horizontally with timescales ranging from tens of minutes to days. Mesoscale weather models do not resolve the smaller scale turbulence of the ABL, but rather they are URANS-like in which they use a turbulence model for this turbulence's effect on the mean flow. Therefore, one can extract time series of velocity, temperature, and turbulence statistics from the mesoscale model on planes corresponding to the inflow planes of the wind plant simulation. These data then are the mean inflow to the wind plant. This technique comes from a wider area of atmospheric flow simulation research called mesoscale–microscale coupling, where the term "microscale" refers to the ABL or wind plant-scale atmospheric flow. This subject is treated in detail in Chapter 4 of this book.

If RANS turbulence treatment is used in combination with prescribed mean inflow profiles, one also needs to provide profiles of turbulence quantities. For example, turbulent kinetic energy and turbulent dissipation rate must be supplied if the k–ε turbulence model is used. More details on inflow profiles for RANS cases are given by Parente *et al.* [115], Gorlé *et al.* [116], and Richards and Hoxey [117].

If LES turbulence treatment is used with prescribed mean inflow profiles, we have the problem that LES resolves turbulence, but the inflow information contains no resolved turbulence. Therefore, we must do something to either add resolved turbulence to the mean inflow data or perturb the flow such that resolved turbulence rapidly forms. Otherwise, it may take tens of kilometers before numerical errors accumulate to the point that resolved turbulence begins to form. There are many ways of creating turbulence in the LES given a mean inflow wind. Excellent overviews of these methods are given by Sagaut [101] in Section 10.3 of his text and by Wu [118] in his review article, and we will give a few examples here.

A novel approach to creating resolved inflow turbulence within a LES given a mean background wind profile is that employed by researchers at DTU [119]. They prescribe a plane that is upstream of the wind turbines over which space and time-varying body forces are applied to recover the synthetic turbulence of Mann [31] that is discussed in Section 6.3.1.1. Over the distance between this body-force plane and the turbine model, the turbulence adapts to the fact that the filtered Navier–Stokes equations are evolving it in space and time.

Another interesting and contrasting approach to generating resolved inflow turbulence is developed by Muñoz-Esparza *et al.* [120]. The testing of this idea is all done within the WRF framework, in which the mesoscale and microscale LES are handled within a single solver, but the method could be easily applied to any microscale or wind plant LES model. Here, mean inflow data from the mesoscale weather model is fed to the microscale domain. The method then applies eight-by-eight grid-cell wide horizontal slabs of uniform temperature perturbation, extending three slabs inward from the inflow boundary and stacked vertically to a fraction of the boundary layer height. Each slab has a random temperature perturbation from a uniform ± 0.5K distribution. These temperature perturbations slabs are more effective than using cell-wise random perturbations. These perturbations activate the buoyancy term in the vertical momentum equation and generated motion. Although kilometers of distance are still required for the perturbation-induced motions to reach realistic equilibrium turbulence, it outperforms other methods that were tried for comparison.

6.3.2.5 Wake generation

Modeling of wakes using CFD-based approaches differs greatly from that using non-CFD-based approaches. In non-CFD-based approaches, a wake model explicitly describes the wake evolution and behavior. In CFD-based approaches, though, some sort of model for the wind turbine's blades or rotor aerodynamics is applied, then a wake is automatically generated and propagated downstream by the governing fluid flow equations. Although CFD-based methods are more computationally expensive than non-CFD-based approaches, and the numerical methods, mesh generation, and turbulence modeling can be complex, wake generation is arguably simpler with CFD. Wake behaviors including transition from near-wake to far-wake, meandering, diffusion, merging, atmospheric-stability effect, interaction with complex terrain, and interaction with other turbines is naturally handled by the governing fluid flow equations.

There are two main methods for generating the wake using CFD: resolving the turbine blade geometry and representing the aerodynamic forces created by the blades using body forces, also known as actuator methods. Both methods can be simulated as rigid rotating structures or coupled with an aeroelastic tool to calculate and model structural deformation in response to loads, which we will also discuss here.

Blade-geometry-resolving methods

Blade-geometry-resolving CFD methods for generating wakes and rotor local flows make a complex topic, so much so that Chapter 2 of this book addresses the topic. We briefly discuss this topic here.

The main idea with blade-geometry-resolving methods is that the geometry of the rotor blades (the nacelle and tower geometry can be included as well) is accounted for in the CFD simulation. Most commonly, this is done by using a computational mesh that conforms to the blade geometry. In other words, there is a void in the mesh occupying the volume of the actual blade solid structure. At the surface of the blade, appropriate boundary conditions are specified. The computational mesh must rotate, and in more sophisticated cases, deform as the blades bend in response to time-varying aerodynamic loads. If nacelle yaw is to be accounted for, another degree of mesh rotation must be accounted for. To add to this, if modeling a floating offshore wind turbine, the mesh must account for the translational and rotational motions of the entire turbine structure in response to wind and waves.

The computational mesh is generally highly refined within the boundary layer of the turbine blades, tower, and nacelle themselves. These boundary layers are very thin, especially at the leading edges of the blades, sometimes requiring blade-surface-normal spacing of a micron adjacent to the surface. The mesh should also be refined to

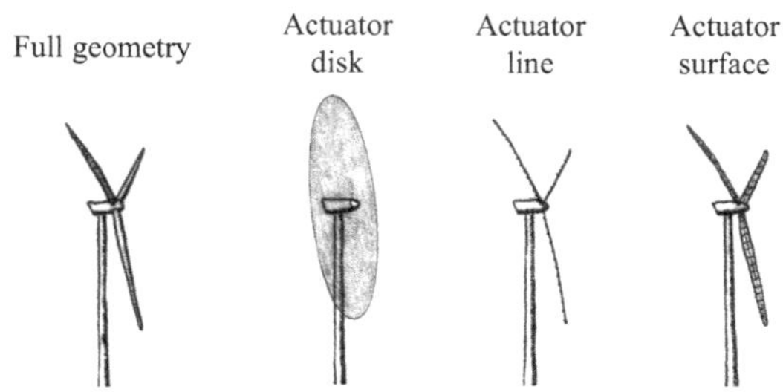

Figure 6.7 Schematic drawings of different actuator representations of wind turbine rotors for use with CFD-based flow modeling approaches

capture the individual blade wakes and tip and root vortices as they flow downstream. As one can see, creating the mesh alone is a daunting task.

In addition to meshing complexities, there are complexities in turbulence treatment. If one wishes to use LES in the wind plant simulation, one cannot reasonably perform LES within the blade boundary layers. The alternatives are wall-modeled LES and hybrid RANS-LES. With wall-modeled LES, a model for the shear stress exerted on the turbine surfaces by the flow is applied at the turbine surfaces, and the blade, nacelle, and tower boundary layers are coarsely resolved, which is still very computationally expensive. Less expensive, but more complex, is hybrid RANS-LES in which the turbulence is modeled using RANS treatment in the blade, nacelle, and tower boundary layers, and LES treatment away from these boundary layers. The RANS turbulence model smoothly blends to an LES turbulence model with distance away from the turbine structure. With any of these turbulence treatments, difficulties arise from how transition of the flow from a laminar to turbulent state is treated and from the fact that little work has been done in hybrid RANS-LES to blend an engineering/aerospace RANS turbulence models with atmospheric LES turbulence models.

An alternative to using geometry-conforming meshes in resolving the full turbine geometry is to use immersed boundaries, as is done by Yang *et al.* [121]. In the immersed boundary method, the mesh need not conform to the geometry. Instead, conditions are used that apply wall boundary conditions to the flow at the location of the geometry. For example, if performing wall-modeled LES of the blade geometry, this method will apply forces on the flow at the blade surface location such that no flow penetrates the flow surface and the proper shear stress is attained. Even though some of the grid complexity is relaxed, immersed boundary methods still require high resolution to be clustered in the region of the geometry boundary layers.

Because of the very high grid resolution used to resolve the turbine geometry boundary layers (and shapes like the thin trailing edges of blades), the time steps need to be very small to both smoothly resolve blade motion and maintain numerical stability. Time steps on the order of 10^{-4} s are not uncommon. Considering that with the full wind plant, we are often interested in simulation times of at least 30 min, that requires 18×10^6 time steps!

Actuator methods

With actuator methods, shown schematically in Figure 6.7, the blade geometry is not resolved, but instead, body forces are applied to the flow field in an attempt to replicate the force that the actual turbine blades apply to the flow. Not only can the blades be modeled with body forces, but also the tower and nacelle. There are varying degrees of fidelity of actuator methods, ranging from simple actuator disks to high-resolution advanced actuator lines and actuator surfaces, but all share the idea that body forces are geometrically distributed in some fashion within the flow field. These body forces appear as a source term in the Navier–Stokes equations. For example, in (6.33), the last term on the right-hand side, f_i, is the generalized density-normalized body force per unit volume.

A major advantage of actuator methods is that the body force location is independent of the computational mesh. The body force field can be applied anywhere within the mesh that one desires. Mesh regions need not move with the body force. Local grid refinement may be needed, though, to resolve these body-force distributions adequately. Actuator methods relieve the meshing complexity necessitated by geometry-resolving methods.

This advantage comes with some cost, though. Actuator methods do not capture thin boundary layers of the blades, nacelle, and tower of the turbine. Some actuator methods are able to capture features including the blade tip and root vortices and some representation of each individual blade's wake. But often we are not interested in the very blade-local flow feature—instead, we are interested in the far wake, and actuator methods capture this very well. Also, actuator methods rely on some sort of tabulated aerodynamic data. For example, actuator lines require airfoil coefficient of lift and drag versus angle of attack data. Simple actuator disks require a table of rotor thrust versus wind speed. Geometry-resolved methods naturally calculate their own surface pressures and stresses, so they are often seen as truly "predictive," but until RANS turbulence models, wall shear stress models, and turbulent transition models improve, the forces they predict have inherent error, and reliance on data tabulated from experiments is a viable alternative.

The simplest actuator method is the actuator disk. It uses a table of rotor global thrust versus wind speed, and perhaps rotor speed and blade pitch angle. At the simplest, a table of rotor thrust versus wind speed encapsulates all the information about the turbine's control system, such as tip-speed ratio and the way blade pitch changes with wind speed in above-rated wind conditions. The global thrust is then uniformly distributed over the rotor disk as a body force. The simple actuator disk is used in the simulations by Calaf, Meneveau, and Meyers [81].

The next level of fidelity is what is sometimes referred to as the advanced actuator disk. With the advanced actuator disk, a radially varying distribution of thrust and torque-generating forces are applied over the rotor disk. This requires information about how lift and drag along the blade vary with radius and other operating conditions, including tip-speed ratio and blade pitch angle. One way to provide such information is to know the twist, chord, and airfoil distribution along the blade, along with coefficient of lift and drag versus angle of attack tables for each airfoil type. Additionally, one would need to know the tip-speed-ratio and blade pitch angle at all wind speeds, or have a model of the turbine's torque and pitch controllers. The advanced actuator disk is shown to be a substantial improvement over the simple actuator disk by Wu and Porté-Agel [122].

Next comes the actuator line in which the three-dimensional blade is represented as a one-dimensional line. The main idea is that aerodynamic pressure and stress distributions exist over the blade surface. At each radius, those can be integrated into a force per unit length and applied along the radius of the actuator line. The actuator lines then rotate (and if coupled with a structural solver can flex) in the same way that the real blades do. Actuator lines require that the blade twist, chord, and airfoil type radial distributions are known, along with coefficient of lift and drag versus angle of attack tables for each airfoil type. Also, knowledge of the control system so that tip-speed variation with wind speed and blade-pitch response in above rated wind conditions is required. The wind-turbine actuator line was pioneered by Sørensen and Shen [123], but improvements have been suggested by Shives and Crawford [124], Churchfield *et al.* [125], Martínez-Tossas *et al.* [126], and Jha and Schmitz [127].

At the highest level of actuator method fidelity are actuator surface methods. They are similar to actuator lines, but rather than collapsing the three-dimensional blade to a line, they collapse it to a two-dimensional surface. Now, rather than integrating the surface pressure and stress to a force per unit length, a force per unit area can be applied over the surface. In this way, the chord-wise distribution of pressure and stress is captured. For example, airfoils have a suction peak near the leading edge that then gently decreases toward the trailing edge. That chordwise distribution can be captured with the actuator surface method. This means that the same information used for the actuator line must be provided to the actuator surface, but instead of airfoil coefficient of lift and drag tables, one must specify tables of coefficient of pressure and shear stress distribution along the chord of the airfoil at different angles of attack, which is harder to obtain.

An important detail concerning actuator methods is that earlier we describe taking the forces created by the rotor and collapsing them onto two-dimensional surfaces or one-dimensional lines, yet we must apply those onto a three-dimensional field. This is done using some sort of projection function. Sørensen and Shen [123] outlined a popular method for the actuator line. One can discretize the actuator line into a series of small line segments, each with its own force per unit length. Similarly, the actuator disk is discretized into a collection of small subsurfaces, each with its own force per unit area. The lumped force for that line segment or subsurface can be computed so that the actuator line or disk becomes a discrete series of point forces centered along the line segments or subsurfaces making up the actuator. Each of those point forces is then projected onto the three-dimensional flow field using a Gaussian-weighted function, where the Gaussian is a function of distance from the point force, becoming a cloud of force per unit volume. In aggregation, all of these clouds of force overlap to become a continuous smooth cloud of force per unit volume following the actuator line or disk, as shown in Figure 6.8. For more details on optimal ways to distribute this force for actuator lines, refer to the work of Martínez-Tossas *et al.* [126]

Actuators can also be used to model the nacelle and tower. At its simplest, the tower can be represented with a simple actuator line and the nacelle can be

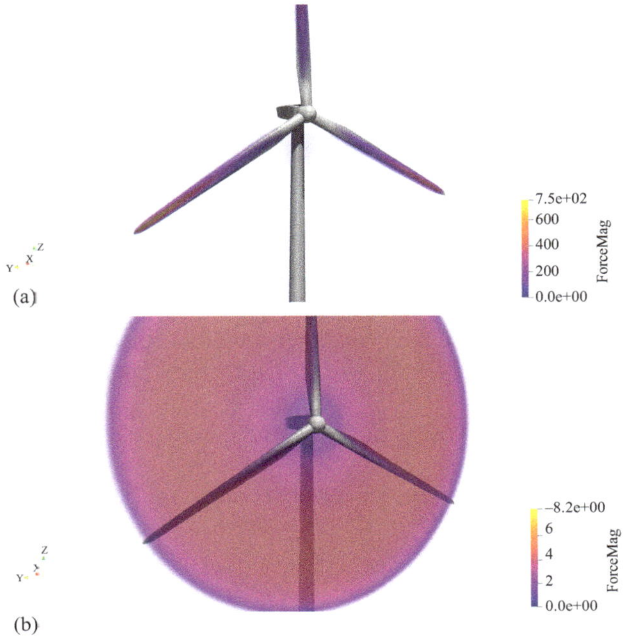

Figure 6.8 A depiction of the body forces projected onto the flow field using (a) the actuator line and (b) the actuator disk. The body forces are shown volume rendered in color, and the turbine geometry is only shown for reference

represented with a simple actuator disk, both of which only exert thrust force. Churchfield *et al.* [128] provide a comparison of different actuator approaches to modeling the tower and nacelle. Sarlak *et al.* [129] provide a novel actuator line-based tower model with an oscillating transverse force meant to mimic cylindrical-body shedding in which Strouhal frequency is an input.

The actuator disk methods are at the low-fidelity end of the spectrum because they only capture the gross rotor wake. The advanced actuator disk also captures gross wake rotation imposed by rotor torque. They do not capture blade-local flow effects. Actuator lines and surfaces, on the other hand, capture blade-local flow effects in addition to the gross wake. They capture effects including the tip and root vortices, the local circulation around the blade, and some representation of the blade wake. We cannot expect actuator methods to capture the fine details that blade-resolved methods can capture, such as the blade boundary layer, but they are quite capable. Troldborg *et al.* [108,130] showed that so long as the inflow is lightly turbulent, the wakes generated by blade-resolved versus actuator methods only slightly differ.

Coupling to structural dynamics

Aerodynamic models of the rotor, which then generate the turbine wake, can be coupled with wind turbine structural and system dynamics codes. Such codes treat the turbine structure with either finite-element analysis or modal analysis of the structure. Examples include the tools FAST, Bladed, and HAWC.

All of the previously mentioned rotor aerodynamics models can be coupled to structural and system dynamics tools. The coupling can be "loose" or "tight." A good example is the coupling of the actuator line model to a code, such as FAST, which is done in NREL's SOWFA code. This coupling is an example of "loose" coupling and works as follows:

- The flow velocity is sampled from the CFD field at discrete points along each actuator line and passed to FAST.
- FAST computes the aerodynamic loads as a response to the velocity.
- The aerodynamic loads are then combined with any other loads, such as gravity loads or torque from the generator, to calculate the structural response and motions of the turbine.
- Somewhere during the FAST call, the control system response is also computed as it impacts applied loads.
- The new blade positions and aerodynamic forces are passed back to the CFD solver.
- The forces are applied to the CFD field as body force.
- Finally, the CFD solution advances.

Tight coupling, on the other hand, would require that the CFD solution and the aeroelastic/system dynamic solutions iterate until they converge before proceeding to the next time step, which incurs more computational expense. Tight coupling may be necessary, though, for aeroelastic coupling with geometry-resolving simulations.

Coupling of CFD to aeroelastic tools is exciting in which a very complex aeroelastic system can be modeled. It is nearly a necessity for simulating floating offshore wind plants, which are becoming a reality, because of the added motion of the floating platform.

6.3.2.6 Putting the components together to use CFD to simulate the wind farm

With CFD-based wind plant modeling approaches, putting the components outlined previously together to perform a wind plant simulation may require complex computer code running on many cores with parallel communication and using state-of-the-art linear algebra packages and sophisticated turbulence models. However, CFD-based approaches are, perhaps, less conceptually complex than non-CFD-based methods. The conceptual simplicity comes from the fact that the governing fluid flow equations seamlessly and naturally handle the complex interactions of the different flow components. For example, with CFD-based methods, wakes naturally interact with one another, with the turbulent inflow wind, and with complex terrain. There is no need to decide if wake merging requires linear or quadratic sum superposition methods, for example. Figures 6.1 and 6.4 are examples of CFD simulations that have put all the previously described components together to represent the entire wind plant.

CFD-based simulations of the full wind plant are generally run as a single simulation in which the inflow is a simple profile with superimposed turbulence or as a

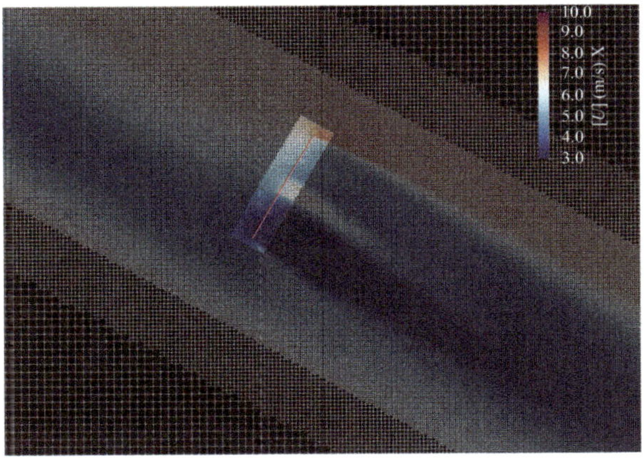

Figure 6.9 A horizontal cut-plane at turbine hub height through the mesh used for a wind-plant large-eddy simulation focusing on the local mesh refinement around a single turbine. The grid lines are colored by time-averaged velocity. The outer mesh resolution is 10 m, and the finest rotor-local resolution is 1.25 m. The rotor location is shown as a red line. The wake of an upstream turbine is advecting downstream into the turbine shown here

two-stage simulation with a precursor simulation first to generate a complex inflow turbulent wind field to be subsequently used as inflow to the wind plant simulation.

An important consideration in simulating the wind plant is the disparity in resolution needed to resolve wind turbine wakes and the background wind field. To well resolve turbulence in a neutral ABL using LES, one needs 5–10 m computational mesh resolution; to well resolve the turbulent structures within the wake of a turbine with a 100-m rotor diameter with LES, one needs 1–5 m resolution of the mesh (the resolution requirement is higher for smaller rotors). To carry 1–5 m resolution through the entire computational domain may be prohibitively expensive, so localized mesh refinement is often necessary. This localized grid refinement significantly adds to the mesh size and computational cost. Figure 6.9 shows a hub-height slice through the mesh of a wind plant LES.

A similar consideration is the range of timescales represented in a wind plant simulation. In simulating the background flow, if LES is being used, one typically does not want to exceed a Courant number (the ratio of length traveled by a fluid element in one time step to mesh cell length) of unity to well resolve the flow. For RANS turbulent treatment using an implicit-in-time solver, Courant numbers greater than one can be achieved. However, if actuator lines are used to represent the turbine, one desires to well resolve flow features, such as tip vortices, generated by the rotor. The rotor tip speed is somewhere between six and ten times the flow speed on modern turbines, and the grid cell lengths are usually refined around the turbines. Therefore, a rotor tip-speed-based Courant number constraint is often applied that means the time step must be much smaller than for the background flow. As an example, let us consider a simulation with a background mesh resolution of 10 m and a maximum wind speed of 10 m/s. A Courant number of one would allow for 1 s time steps. Around the turbine, though, the mesh resolution is 1 m, and the rotor tip speed is 80 m/s. A tip-speed-based Courant number would allow for a time step of 0.0125 s, which is eight times more restrictive than the time step restriction in the background flow. Most CFD codes do not allow one to use different time step sizes in different regions of the grid, unless it is a RANS-based code marching to steady state. To alleviate this tip-speed-based time step restriction, one may use actuator disks that do not resolve blade tip motion.

6.4 Validation efforts

Validation remains a critical aspect of wake model development, particularly as models become more complex and are applied to an ever growing variety of wind-farm applications. Validation studies have been ongoing since the 1970s and can be separated into two classes: wind tunnel testing and field testing in the real atmosphere. Wind tunnel tests provide a controlled environment where initial and boundary conditions are well quantified but suffer from scalability and physical mismatch issues that can only be addressed through full atmospheric-scale testing. Atmospheric observations are limited by coarse fidelity and an inability to easily control the wake-driving physical processes. Although imperfect, both approaches are helpful and can be used in combination to improve wake modeling accuracy.

Wind tunnel tests of single turbines have been performed using actuator disk models [131,132] and small-scale turbines with real rotating blades [133,134], some of which include the impacts of atmospheric stability [135–137]. There have also been studies of full wind farms within wind tunnels [138,139]. Such studies have been useful for calibration of wind-farm models and studies of parametric variations of wind turbine and wind-farm layout.

At full scale, measurements of smaller turbines have produced valuable data sets. For example, a series of observations at Sexbierum [140,141] have been used in the International Energy Agency Task 31 as an international wake model validation benchmark [142]. This is a wind farm of two 310 kW turbines and three meteorological masts. Wind speed measurements and impacts on the meteorological masts were taken over a period of 6 months. Wake observations for both single and multiple wake cases were gathered and wake impacts were estimated using changing wind direction over the time of the experiment. The data consist of wake velocity measurement profiles over a range of wind directions at multiple downstream points from the closest rotor. Comparisons between the Sexberium observations and simulations are shown in Figure 6.10.

In this validation example, it is difficult to tell the difference between velocity profiles, and some simulation tools perform better in different parts of the profile, meaning a definitive conclusion as to which model is most accurate remains elusive.

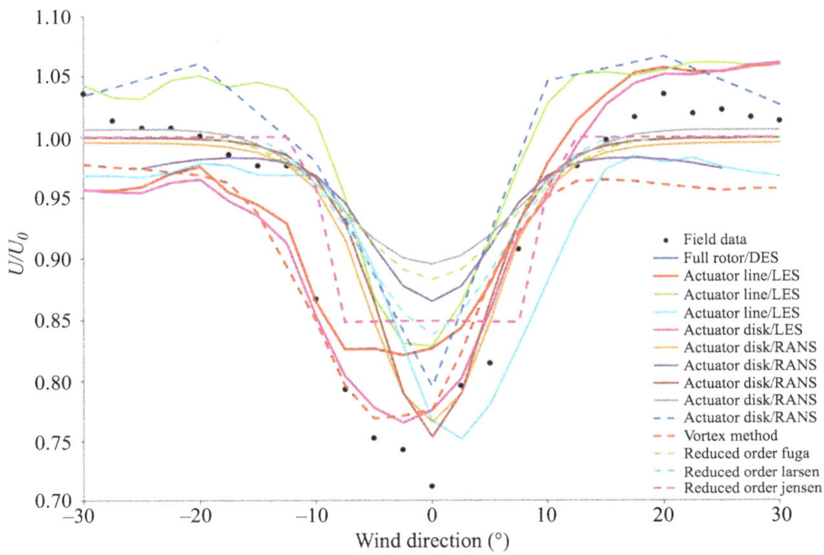

Figure 6.10 Comparison of velocity profiles for a wide range of single turbine wake models relative to observations at the Sexberium wind farm

As such, a different metric is used to show which models are performing the best for these validation profiles, the normalized mean average error (NMAE), defined as

$$NMAE(U_{wake}) = \frac{1}{U_{wake}^{obs}} \sum_{i=1}^{n} \left(U_{wake,i}^{obs} - U_{wake,i}^{sim} \right) \quad (6.40)$$

The calculated NMAE values for the Sexberium validation study are shown in Figure 6.11. In this figure, the error values for the most complex CFD models are on the left and the simplest engineering models are on the right. Most models perform poorly for the near wake validation case at 2.5D, which is expected. And usually the more complex models have lower error values than the simpler models. However, this is not always the case. Some LES models perform poorly even relative to the simplest of engineering tools; a discrepancy also seen by Göçmen *et al.* [21] in their validation studies. This suggests that the model setup and the user of the simulation tool are also critical variables for any validation study, and the best use practices for different fidelity models are also important.

In other validation studies, operational data from full-scale offshore wind farms have been used for validation; beginning with the ENDOW project [143,144] and the UpWind project [145]. Horns Rev, a wind farm in the Danish waters of the North Sea, has been particularly popular and serves as a validation case for many industry and research wind-farm wake models [58,142,146–148]. The significant focus on offshore model validation studies is because lower turbulence levels result

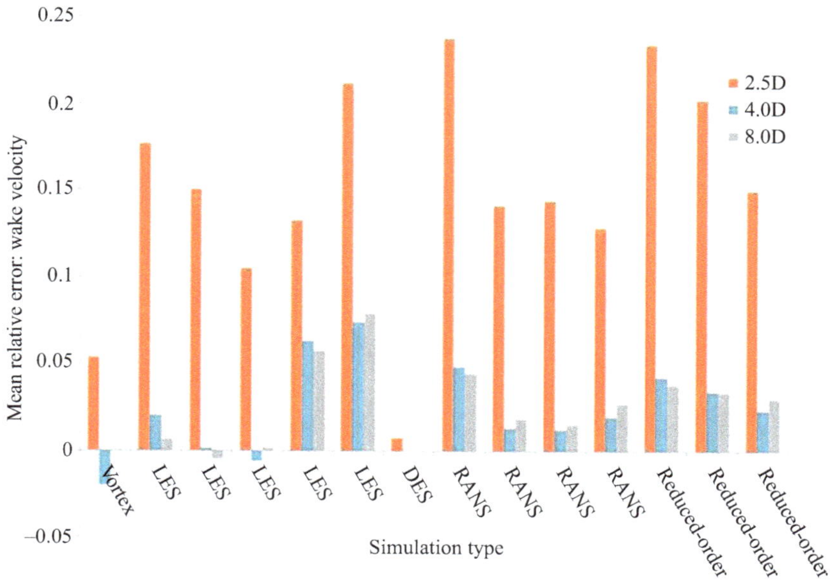

Figure 6.11 Comparison of mean average error of wake models to predict a single wake velocity profile at the Sexberium wind farm. Model complexity decreases from left to right

in longer propagation of wakes, which are thought to have greater impact offshore compared to wakes from land-based wind farms. These validation cases are based on meteorological observations coupled with turbine supervisory control and data-acquisition (SCADA) data, binned according to wind direction, atmospheric stability, and turbulence. For full-scale offshore wind farms, both Walker *et al.* [149] and Archer *et al.* [23] have found that engineering models tend to underpredict the wake loss. Both of these studies also found that predicted wake losses are more accurate for outer "near front" turbines in a wind farm than inner turbines. They also found models are less accurate with closer spacing, although these conclusions may be greatly influenced by the choice of wake superposition model.

SCADA data have also been useful for large-scale interfarm wake interactions [150,151], which are well predicted using simple wake models. In these studies, the largest wake model errors are again inside the wind farm. These studies also highlighted the importance of accounting for spatial variations in inflow conditions that must be observed and modeled as wind farm size increases.

The disadvantages of these offshore data sets are that the atmospheric conditions, detailed turbine characteristics, and operating behavior are not all well understood. Wind direction and stability are not always well established or modeled. Thus, modelers must be careful to match the atmospheric forcing functions within their models when doing validation. For example, Göçmen [21] uses the method of Gaumond [147] to include the impact of wind direction variability when comparing to Horns Rev and Lillgrund wind-farm data sets to improve the agreement during validation. Also turbine design information, particularly control system behavior, is often proprietary, meaning that modelers must make assumptions about turbine performance characteristics [152] that may impact the validation study results.

Validation data must also match the fidelity of models to ensure more rigorous validation. More high-fidelity observations with direct measurement of wakes using remote sensing are now being used to augment data taken from wind-farm SCADA systems. Remote sensing measurements made in offshore wind farms have provided direct observation of wakes [153,154], and similar validation studies were also done for land-based scale turbines [155]. Validation of engineering models using wake observation data from these data sets [65] showed reasonable agreement between observation and simulation.

New remote sensing measurements are becoming available that will serve as validation data sets for the near future. New observations with scanning lidars [156,157] are proving valuable; for example, Sandia National Laboratories has the Scaled Wind Farm Technology facility and researchers there have performed detailed observations of a single turbine wake using a high resolution scanning lidar [158]. Remote sensing techniques will continue to be important as turbines grow in size [159,160] and more traditional meteorological towers will become cost prohibitive at larger heights. New technologies, such as Ka- and X-band radars, are making real-time measurements of entire wind-farm interactions and wind-farm wakes possible [161,162]. And offshore synthetic aperture radar is a promising method for investigating wind-farm wakes [163] and wind-farm wake model validation [164]. Combining satellite with aircraft observations has also been shown to be useful [165].

In complex terrain, recent observations [85,166,167] and concurrent validation studies [168] of single-turbine wakes have begun. Terrain-specific wake models are also being developed [169] that will be needed as the number of wind farms grow in more complex environments.

Lastly, rigorous methods for validation must be developed and applied to wind-farm modeling tools [170]. Uncertainty quantification [171,172], direction uncertainty [147], and other sources of uncertainty in both observations and simulations will be critical for quantifying the accuracy of and improving wake models going forward. Bottom line, it remains difficult to validate models to a sufficient level of confidence and more research is required, including a new series of higher fidelity validation studies and affiliated observations campaigns over a range of wind-farm relevant environments.

6.5 Future development

Although the wind-farm models of today have impressive capabilities, these models and simulation tools will continue to evolve and capture additional physical processes not included in the current generation of tools. Physical processes studied using high-fidelity models in combination with new observations will flow into engineering models that are used in real-time control and optimization applications for the design and operation of the next generation of wind farms. For example, as described earlier, the interaction of multiple wakes in engineering tools could benefit from additional study at all levels of fidelity.

Larger wind farms necessitate models of larger scale phenomena, and future simulation tools will need to incorporate mesoscale physics through some degree of mesoscale–microscale coupling. This is a significant challenge in which mesoscale and microscale models have been developed independently for decades and combining them at the interface of what is known as the *terra incognita* is still not well studied. Model improvements in this area of study will require additional fundamental observations in addition to model development and validation.

Offshore wind energy will continue to grow worldwide, and special models will be required to interact with new turbine configurations, such as floating turbines with less yaw stability that may oscillate during times of wake interactions. New models will also be required for the land–sea interface [173]. And model improvements are needed for wind–wave and wave–wake interactions, which have been shown to be closely coupled [103], and under nonequilibrium conditions, velocity profiles deviate from Monin–Obukhov similarity theory.

Surface boundary conditions in complex terrain are also known to violate Monin–Obukhov theory as a result of nonhomogeneous velocity profiles caused by terrain. Similarly, there is much room for improvement in surface roughness, heat and moisture flux, and vegetation models. In general, surface boundary condition models will need to be improved for more accurate prediction of how turbulent winds interact with wakes. And initial and inflow conditions for flows in complex environments are needed as well.

Unsteady models will continue to grow in importance for predicting structural loading impacts and power losses within wind farms. Following the approach of the IEC design standard, the estimation of extreme winds (e.g., hurricanes) and how those are modified by the wind plant environment remains an unanswered question.

Each of these model developments will require new observational data sets for proper validation. In turn, this will require new instrumentation of higher fidelity and larger scale to capture all of the important physical processes that drive wind-farm performance and reliability. And within validation, new methods of uncertainty quantification and error metrics will be required to produce improved knowledge of model accuracy and acceptable uncertainty limits.

Wind-plant modeling methods themselves will also change, relying more on data assimilation and a closer integration of observations and simulation. Data science will see a growing role for analysis and assimilation of real-time data, statistical methods, and machine-learning techniques [4] filling in where physics are less well understood or model speed is critical. Together, these models and observations will help drive the design, development, and operation of the smart wind plant of the future [174], where the wind is no longer a passive input into the energy system but managed for maximal energy production, turbine reliability, and grid interaction [175].

Acknowledgment

This work was authored, in part, by the National Renewable Energy Laboratory, operated by Alliance for Sustainable Energy, LLC, for the U.S. Department of Energy (DOE) under Contract No. DE-AC36-08GO28308. Funding provided by the U.S. Department of Energy Office of Energy Efficiency and Renewable Energy Wind Energy Technologies Office. The views expressed in the article do not necessarily represent the views of the DOE or the U.S. Government. The U.S. Government retains and the publisher, by accepting the article for publication, acknowledges that the U.S. Government retains a nonexclusive, paid-up, irrevocable, worldwide license to publish or reproduce the published form of this work, or allow others to do so, for U.S. Government purposes.

References

[1] Clifton A, Smith A, Fields M. Wind Plant Preconstruction Energy Estimates. Current Practice and Opportunities. National Renewable Energy Laboratory (NREL). Golden, Colorado, United States; 2016.

[2] Stull RB. An Introduction to Boundary Layer Meteorology. Dordrecht: Kluwer Academic Publishers; 1988.

[3] Debnath M, Bhaganagar K. The Effects of Mean Atmospheric Forcings of the Stable Atmospheric Boundary Layer on Wind Turbine Wake. Journal of Renewable and Sustainable Energy. 2015;7(013124):1–21.

[4] King RN, Adcock C, Annoni J, *et al.* Data-Driven Machine Learning for Wind Plant Flow Modeling. Journal of Physics: Conference Series. 2018;1037:072004.

[5] Moon JS, Manuel L, Churchfield MJ, *et al.* Toward Development of a Stochastic Wake Model: Validation Using LES and Turbine Loads. Energies. 2017;11(1):53.

[6] Allaerts D, Meyers J. Gravity Waves and Wind-Farm Efficiency in Neutral and Stable Conditions. Boundary Layer Meteorology. 2018;166(2):269–299.

[7] Risø DTU. WAsP—The Wind Atlas Analysis and Application Program. Risø DTU; 2008.

[8] Garrad Hassan and Partners Ltd. WindFarmer, Wind Farm Design Software: Theory Manual. Garrad Hassan and Partners Ltd.; 2010.

[9] AWS Truepower. OpenWind Theoretical Basis and Validation, Version 1.3. AWS Truepower, Albany, New York, United States; 2010.

[10] Quick J, Dykes K, Graf P, *et al.* Optimization Under Uncertainty of Site-Specific Turbine Configurations. Journal of Physics: Conference Series. 2016;753:062012.

[11] Jensen NO. A Note on Wind Generator Interaction; 1983.

[12] Katic I, Højstrup J, Jensen NO. A simple model for cluster efficiency. In: European Wind Energy Association Conference and Exhibition; 1986. p. 407–410.

[13] Barthelmie RJ, Hansen KS, Pryor SC. Meteorological Controls on Wind Turbine Wakes. Proceedings of the IEEE. 2013;101(4):1010–1019.

[14] International Electrotechnical Commission. IEC 61400-1: Wind Turbines Part 1: Design Requirements. International Electrotechnical Commission; 2005.

[15] Fleming P, Annoni J, Churchfield M, *et al.* A Simulation Study Demonstrating the Importance of Large-Scale Trailing Vortices in Wake Steering. Wind Energy Science. 2018;3(1):243–255.

[16] Fleming P, Annoni J, Shah JJ, *et al.* Field Test of Wake Steering at an Offshore Wind Farm. Wind Energy Science. 2017;2(1):229–239.

[17] Churchfield M, Wang Q, Scholbrock A, *et al.* Using High-Fidelity Computational Fluid Dynamics to Help Design a Wind Turbine Wake Measurement Experiment. Journal of Physics: Conference Series. 2016;753:032009.

[18] Crespo A, Hernandez J, Frandsen S. Survey of Modelling Methods for Wind Turbine Wakes and Wind Farms. Wind Energy: An International Journal for Progress and Applications in Wind Power Conversion Technology. 1999; 2(1):1–24.

[19] Vermeer LJ, Sorensen JN, Crespo A. Wind Turbine Wake Aerodynamics. Progress in Aerospace Sciences. 2003;39:467–510.

[20] Sanderse B, van der Pijl SP, Koren B. Review of Computational Fluid Dynamics for Wind Turbine Wake Aerodynamics. Wind Energy. 2011;14:799–819.

[21] Göçmen T, Van der Laan P, Réthoré PE, *et al.* Wind Turbine Wake Models Developed at the Technical University of Denmark: A Review. Renewable and Sustainable Energy Reviews. 2016;60:752–769.

[22] Stevens RJ, Meneveau C. Flow Structure and Turbulence in Wind Farms. Annual Review of Fluid Mechanics. 2017;49:311–339.

[23] Archer CL, Vasel-Be-Hagh A, Yan C, *et al.* Review and Evaluation of Wake Loss Models for Wind Energy Applications. Applied Energy. 2018;226: 1187–1207.

[24] Lissaman PS. Energy Effectiveness of Arbitrary Arrays of Wind Turbines. Journal of Energy. 1979;3(6):323–328.

[25] Businger JA, Wyngaard JC, Izumi Y, *et al.* Flux-Profile Relationships in the Atmospheric Surface Layer. Journal of the Atmospheric Sciences. 1971;28(2):181–189.

[26] Dyer A. A Review of Flux-Profile Relationships. Boundary-Layer Meteorology. 1974;7(3):363–372.

[27] Paulson CA. The Mathematical Representation of Wind Speed and Temperature Profiles in the Unstable Atmospheric Surface Layer. Journal of Applied Meteorology. 1970;9(6):857–861.

[28] Walmsley JL, Troen I, Lalas DP, *et al.* Surface-Layer Flow in Complex Terrain: Comparison of Models and Full-Scale Observations. Boundary-Layer Meteorology. 1990;52(3):259–281.

[29] Troen I, De Baas A. A spectral diagnostic model for wind flow simulation in complex terrain. In: European Wind Energy Association Conference and Exhibition. A. Raguzzi; 1987. p. 243–249.

[30] Veers PS. Three-Dimensional Wind Simulation. Sandia National Laboratories, Albuquerque, New Mexico and Livermore, California, United States; 1988.

[31] Mann J. The Spatial Structure of Neutral Atmospheric Surface-Layer Turbulence. Journal of Fluid Mechanics. 1994;273:141–168.

[32] Ghate AS, Lele SK. Subfilter-Scale Enrichment of Planetary Boundary Layer Large Eddy Simulation Using Discrete Fourier–Gabor Modes. Journal of Fluid Mechanics. 2017;819:494–539.

[33] Sim C, Basu S, Manuel L. On Space-Time Resolution of Inflow Representations for Wind Turbine Loads Analysis. Energies. 2012;5(7):2071–2092.

[34] Annoni J, Fleming P, Scholbrock A, *et al.* Analysis of Control-Oriented Wake Modeling Tools Using Lidar Field Results. National Renewable Energy Lab. (NREL), Golden, Colorado, United States; 2018.

[35] Bastankhah M, Porté-Agel F. A New Analytical Model for Wind-Turbine Wakes. Renewable Energy. 2014;70:116–123.

[36] Bastankhah M, Porté-Agel F. Experimental and Theoretical Study of Wind Turbine Wakes in Yawed Conditions. Journal of Fluid Mechanics. 2016;806:506–541.

[37] Churchfield MJ, Lee S, Michalakes J, *et al.* A Numerical Study of the Effects of Atmospheric and Wake Turbulence on Wind Turbine Dynamics. Journal of Turbulence. 2012;13(14):1–32.

[38] Jonkman JM, Annoni J, Hayman G, *et al.* Development of FAST. Farm: a new multi-physics engineering tool for wind-farm design and analysis. In: 35th Wind Energy Symposium; 2017. p. 0454.

[39] Larsen GC, Madsen HA, Thomsen K, *et al.* Wake Meandering: A Pragmatic Approach. Wind Energy. 2008;11(4):377–395.

[40] Sørensen JN, Mikkelsen R, Sarmast S, *et al.* Determination of Wind Turbine Near-Wake Length Based on Stability Analysis. Journal of Physics: Conference Series. 2014;524:012155.

[41] Howard KB, Singh A, Sotiropoulos F, *et al.* On the Statistics of Wind Turbine Wake Meandering: An Experimental Investigation. Physics of Fluids. 2015;27(7):075103.

[42] White FM. Viscous Fluid Flow. McGraw-Hill Series in Mechanical Engineering. New York: McGraw-Hill; 1991.

[43] Johansson PBV. The Axisymmetric Turbulent Wake [dissertation]. Chalmers University of Technology. Gothenburg, Sweden; 2002.

[44] Johansson PBV, George WK, Gourlay MJ. Equilibrium Similarity, Effects of Initial Conditions and Local Reynolds Number on the Axisymmetric Wake. Physics of Fluids. 2003;15(3):603–617.

[45] Crespo A, Hernandez J. Turbulence Characteristics in Wind-Turbine Wakes. Journal of Wind Engineering and Industrial Aerodynamics. 1996;61(1):71–85.

[46] Wosnik M. Similarity theory for an axisymmetric turbulent wake with rotation. In: APS Meeting Abstracts; 2011. p. H12.009.

[47] Dufresne NP, Wosnik M. Velocity Deficit and Swirl in the Turbulent Wake of a Wind Turbine. Marine Technology Society Journal. 2013;47(4):193–205.

[48] Frandsen S, Barthelmie R, Pryor S, *et al.* Analytical Modelling of Wind Speed Deficit in Large Offshore Wind Farms. Wind Energy. 2006;9(1–2):39–53.

[49] Van Kuik GA. The Lanchester–Betz–Joukowsky Limit. Wind Energy: An International Journal for Progress and Applications in Wind Power Conversion Technology. 2007;10(3):289–291.

[50] Manwell JF, McGowan JG, Rogers AL. Wind Energy Explained: Theory, Design and Application. Chichester: John Wiley & Sons; 2010.

[51] Peña A, Rathmann O. Atmospheric Stability-Dependent Infinite Wind-Farm Models and the Wake-Decay Coefficient. Wind Energy. 2014;17(8): 1269–1285.

[52] Niayifar A, Porté-Agel F. A New Analytical Model for Wind Farm Power Prediction. Journal of Physics: Conference Series. 2015;625:012039.

[53] Xie S, Archer C. Self-Similarity and Turbulence Characteristics of Wind Turbine Wakes via Large-Eddy Simulation. Wind Energy. 2015;18(10): 1815–1838.

[54] Larsen GC. A Simple Wake Calculation Procedure. Risø National Laboratory Risø-M-2760; 1988.

[55] Larsen GC, Réthoré P. A Simple Stationary Semi-analytical Wake Model. Risø National Laboratory for Sustainable Energy, Technical University of Denmark: Roskilde, Denmark; 2009.

[56] Ainslie JF. Calculating the Flowfield in the Wake of Wind Turbines. Journal of Wind Engineering and Industrial Aerodynamics. 1988;27(1–3):213–224.

[57] Kim H, Kim K, Bottasso C, *et al.* Wind Turbine Wake Characterization for Improvement of the Ainslie Eddy Viscosity Wake Model. Energies. 2018;11(10):2823.

[58] Hansen KS, Barthelmie RJ, Jensen LE, *et al.* The Impact of Turbulence Intensity and Atmospheric Stability on Power Deficits Due to Wind Turbine Wakes at Horns Rev Wind Farm. Wind Energy. 2012;15(1):183–196.

[59] Vollmer L, Steinfeld G, Heinemann D, *et al.* Estimating the Wake Deflection Downstream of a Wind Turbine in Different Atmospheric Stabilities; an LES Study. Wind Energy Science. 2016;1(2):129–141.

[60] Bodini N, Zardi D, Lundquist JK. Three-Dimensional Structure of Wind Turbine Wakes as Measured by Scanning Lidar. Atmospheric Measurement Techniques. 2017;10(8):2881–2896.

[61] Gebraad PM, Churchfield MJ, Fleming PA. Incorporating Atmospheric Stability Effects Into the FLORIS Engineering Model of Wakes in Wind Farms. Journal of Physics: Conference Series. 2016;753:052004.

[62] Ott S, Berg J, Nielsen M. Linearised CFD Models for Wakes. Risø National Laboratory, Roskilde, Denmark; 2011.

[63] Ott S, Nielsen M. Developments of the Offshore Wind Turbine Wake Model Fuga. E-0046 Report; 2014.

[64] Frandsen ST. Turbulence and Turbulence-Generated Structural Loading in Wind Turbine Clusters. Risø National Laboratory; 2007.

[65] Duckworth A, Barthelmie R. Investigation and Validation of Wind Turbine Wake Models. Wind Engineering. 2008;32(5):459–475.

[66] Larsen GC, Madsen HA, Bingöl F, *et al.* Dynamic Wake Meandering Modeling. Risø National Laboratory, Technical University of Denmark, Roskilde, Denmark, Risø; 2007.

[67] Madsen HA, Larsen GC, Larsen TJ, *et al.* Calibration and Validation of the Dynamic Wake Meandering Model for Implementation in an Aeroelastic Code. Journal of Solar Energy Engineering. 2010;132(4):041014.

[68] Coleman RP, Feingold AM, Stempin CW. Evaluation of the Induced-Velocity Field of an Idealized Helicoptor Rotor. NASA Langley; 1945.

[69] Chaney K, Eggers AJ, Moriarty PJ, *et al.* Skewed Wake Induction Effects on Thrust Distribution on Small Wind Turbine Rotors. Journal of Solar Energy Engineering. 2001;123(4):290–295.

[70] Troldborg N, Gaunaa M, Guntur S. Modelling the influence of yaw using a simple vortex rotor model. In: EWEA 2012-European Wind Energy Conference & Exhibition. European Wind Energy Association (EWEA); 2012.

[71] Branlard E, Gaunaa M. Cylindrical Vortex Wake Model: Skewed Cylinder, Application to Yawed or Tilted Rotors. Wind Energy. 2016;19(2):345–358.

[72] Jiménez Á, Crespo A, Migoya E. Application of a LES Technique to Characterize the Wake Deflection of a Wind Turbine in Yaw. Wind Energy. 2010;13(6):559–572.

[73] Medici D, Alfredsson P. Measurements on a Wind Turbine Wake: 3D Effects and Bluff Body Vortex Shedding. Wind Energy: An International Journal for Progress and Applications in Wind Power Conversion Technology. 2006;9(3):219–236.

[74] Guntur S, Troldborg N, Gaunaa M. Application of engineering models to predict wake deflection due to a tilted wind turbine. In: EWEA 2012-European Wind Energy Conference & Exhibition. European Wind Energy Association (EWEA); 2012.

[75] Leishman JG. Challenges in Modelling the Unsteady Aerodynamics of Wind Turbines. Wind Energy. 2002;5(2–3):85–132.

[76] Currin HD. A Dynamic Prescribed Vortex Wake Model for the FAST/AeroDyn Wind Energy Conversion Simulation Code. University of Nevada, Reno; 2007.

[77] Gupta S, Leishman J. Stability of methods in the free-vortex wake analysis of wind turbines. In: 42nd AIAA Aerospace Sciences Meeting and Exhibit; 2004. p. 827.

[78] Machefaux E, Larsen GC, Leon JM. Engineering Models for Merging Wakes in Wind Farm Optimization Applications. Journal of Physics: Conference Series. 2015;625:012037.

[79] Newman B. The Spacing of Wind Turbines in Large Arrays. Energy Conversion. 1977;16(4):169–171.

[80] Frandsen S. On the Wind Speed Reduction in the Center of Large Clusters of Wind Turbines. Journal of Wind Engineering and Industrial Aerodynamics. 1992;39(1–3):251–265.

[81] Calaf M, Meneveau C, Meyers J. Large Eddy Simulation Study of Fully Developed Wind-Turbine Array Boundary Layers. Physics of Fluids. 2010;22(1):015110.

[82] Stevens RJ, Gayme DF, Meneveau C. Coupled Wake Boundary Layer Model of Wind-Farms. Journal of Renewable and Sustainable Energy. 2015;7(2):023115.

[83] Fitch AC, Olson JB, Lundquist JK, *et al.* Local and Mesoscale Impacts of Wind Farms as Parameterized in a Mesoscale NWP Model. Monthly Weather Review. 2012;140(9):3017–3038.

[84] Van der Laan MP, Hansen KS, Sørensen NN, *et al.* Predicting Wind Farm Wake Interaction with RANS: An Investigation of the Coriolis Force. Journal of Physics: Conference Series. 2015;625:012026.

[85] Wildmann N, Kigle S, Gerz T. Coplanar Lidar Measurement of a Single Wind Energy Converter Wake in Distinct Atmospheric Stability Regimes at the Perdigão 2017 Experiment. Journal of Physics: Conference Series. 2018; 1037:052006.

[86] Currie IG. Fundamentals of Fluid Mechanics. 2nd ed. New York: Marcel Dekker, Inc.; 1993.

[87] Moeng CH. A Large-Eddy-Simulation Model for the Study of Planetary Boundary-Layer Turbulence. Journal of the Atmospheric Sciences. 1984;41(13):2052–2062.

[88] Deardorff JW. Numerical Investigation of Neutral and Unstable Planetary Boundary Layers. Journal of the Atmospheric Sciences. 1972;29: 91–115.

[89] Mason PJ. Large-Eddy Simulation of the Convective Atmospheric Boundary Layer. Journal of the Atmospheric Sciences. 1989;46(11):1492–1516.

[90] Allaerts D, Meyers J. Boundary-Layer Development and Gravity Waves in Conventionally Neutral Wind Farms. Journal of Fluid Mechanics. 2017;814:95–130.

[91] Churchfield MJ, Lee S, Moriarty P, *et al.* A large-eddy simulation of wind-plant aerodynamics. In: American Institute of Aeronautics and Astronautics (AIAA) Aerospace Sciences Meeting, Nashville, Tennessee; 2012. AIAA Paper 2012-0537.

[92] Abkar M, Ahmad S, Porté-Agel F. Wake Flow in a Wind Farm During a Diurnal Cycle. Journal of Turbulence. 2016;17(4):420–441.

[93] Lu H, Porté-Agel F. Large-Eddy Simulation of a Very Large Wind Farm in a Stable Atmospheric Boundary Layer. Physics of Fluids. 2011;23(065101): 1–19.

[94] Lipps FB, Hemler RS. A Scale Analysis of Deep Moist Convection and Some Related Numerical Calculations. Journal of the Atmospheric Sciences. 1982;39:2192–2210.

[95] Lipps FB. On the Anelastic Approximation for Deep Convection. Journal of the Atmospheric Sciences. 1990;47(14):1794–1798.

[96] Bannon PR. On the Anelastic Approximation for a Compressible Atmosphere. Journal of the Atmospheric Sciences. 1996;53(23):3618–3628.

[97] Fletcher TM, Brown RE. Simulation of Wind Turbine Wake Interaction Using the Vorticity Transport Model. Wind Energy. 2010;13(7):587–602.

[98] Summit—Oak Ridge National Laboratory's Next High Performance Supercomputer. Available from: https://www.olcf.ornl.gov/olcf-resources/compute-systems/summit/ (accessed on 29 August 2018).

[99] Wilcox DC. Turbulence Modeling for CFD. La Cañada, CA: DCW Industries, Inc.; 1994.

[100] Pope SB. Turbulent Flows. Cambridge: Cambridge University Press; 2000.

[101] Sagaut P. Large Eddy Simulation for Incompressible Flows—An Introduction. 3rd ed. Berlin: Springer; 2006.

[102] Nakanishi M, Niino H. An Improved Mellor–Yamada Level-3 Model with Condensation Physics: Its Design and Verification. Boundary-Layer Meteorology. 2004;112(1):1–31.

[103] Sullivan PP, McWilliams JC, Patton EG. Large-Eddy Simulation of Marine Atmospheric Boundary Layers Above a Spectrum of Moving Waves. Journal of the Atmospheric Sciences. 2014;71(11):4001–4027.

[104] Allaerts D. Large-Eddy Simulation of Wind Farms in Conventionally Neutral and Stable Atmospheric Boundary Layers. PhD thesis, KU Leuven, Leuven, Belgium; 2016.

[105] Johnstone R, Coleman G. The Turbulent Ekman Boundary Layer Over an Infinite Wind-Turbine Array. Journal of Wind Engineering and Industrial Aerodynamics. 2012;100:46–57.

[106] Meyers J, Meneveau C. Optimal Turbine Spacing in Fully Developed Wind Farm Boundary Layers. Wind Energy. 2012;15:205–317.

[107] Sørensen NN. Numerical Modeling of Wind Turbine Wakes. Journal of Fluids Engineering. 2002;124:393–399.

[108] Troldborg N, Zahle F, Réthoré PE, *et al.* Comparison of Wind Turbine Wake Properties in Non-Sheared Inflow Predicted by Different Computational Fluid Dynamics Rotor Models. Wind Energy. 2015;18(7):1239–1250.

[109] Maronga B, Gryschka M, Heinze R, *et al.* The Parallelized Large-Eddy Simulation Model (PALM) Version 4.0 for Atmospheric and Oceanic Flows: Model Formulation, Recent Developments, and Future Perspectives. Geoscientific Model Development. 2015;8:1539–1637.

[110] Wood SL. Lattice Boltzmann Methods for Wind Energy Analysis. PhD thesis, University of Tennessee, Knoxville, Tennessee, United States; 2016.

[111] Ferziger JH, Perić M. Computational Methods for Fluid Dynamics. 3rd ed. Berlin: Springer-Verlag; 2002.

[112] Pletcher RH, Tannehill JC, Anderson DA. Computational Fluid Mechanics and Heat Transfer. 3rd ed. Boca Raton, FL: CRC Press; 2013.

[113] Stevens RJAM, Martínez-Tossas LA, Meneveau C. Comparison of Wind Farm Large Eddy Simulations Using Actuator Disk and Actuator Line Models with Wind Tunnel Experiments. Renewable Energy. 2018;116: 470–478.

[114] Skamarock WC, Klemp JB, Dudhia J, *et al.* A Description of the Advanced Research WRF Version 3. National Center for Atmospheric Research; 2008. NCAR Tech. Note NCAR/TN-475+STR.

[115] Parente A, Gorlé C, van Beeck J, *et al.* A Comprehensive Modelling Approach for the Neutral Atmospheric Boundary Layer: Consistent Inflow Conditions, Wall Function and Turbulence Model. Boundary-Layer Meteorology. 2011;140(3):411.

[116] Gorlé C, van Beeck J, Rambaud P, *et al.* CFD Modelling of Small Particle Dispersion: The Influence of the Turbulence Kinetic Energy in the Atmospheric Boundary Layer. Atmospheric Environment. 2009;43:673–681.

[117] Richards PJ, Hoxey PJ. Appropriate Boundary Conditions for Computational Wind Engineering Models Using the k–ε Turbulence Model. Journal of Wind Engineering and Industrial Aerodynamics. 1993;46–47:145–153.

[118] Wu X. Inflow Turbulence Generation Methods. Annual Review of Fluid Mechanics. 2016;49:23–49.

[119] Keck RE, Mikkelsen R, Troldborg N, *et al.* Synthetic Atmospheric Turbulence and Wind Shear in Large Eddy Simulations of Wind Turbine Wakes. Wind Energy. 2014;17(8):1247–1267.

[120] Muñoz Esparza D, Kosović B, Mirocha J, *et al.* Bridging the Transition from Mesoscale to Microscale Turbulence in Numerical Weather Prediction Models. Boundary-Layer Meteorology. 2014;153:409–440.

[121] Yang X, Sotiropoulos F, Conzemius R, *et al.* Large-Eddy Simulation of Turbulent Flow Past Wind Turbines/Farms: The Virtual Wind Simulator (VWiS). Wind Energy. 2015;18:2025–2045.

[122] Wu YT, Porté-Agel F. Large-Eddy Simulation of Wind-Turbine Wakes: Evaluation of Turbine Parametrisations. Boundary Layer Meteorology. 2011;138:345–366.

[123] Sørensen JN, Shen WZ. General Purpose Flow Solver Applied to Flow over Hills. Risø National Laboratory, Roskilde, Denmark; 1995 June. Risø-R-827-(EN).

[124] Shives M, Crawford C. Mesh and Load Distribution Requirements for Actuator Line CFD Simulations. Wind Energy. 2013;16: 1183–1196.

[125] Churchfield MJ, Schreck SJ, Martínez-Tossas LA, *et al.* An advanced actuator line method for wind energy applications and beyond. In: American Institute of Aeronautics and Astronautics (AIAA) SciTech Forum, Grapevine, Texas; 2017. AIAA Paper 2017-1998.

[126] Martínez-Tossas LA, Churchfield MJ, Meneveau C. Optimal Smoothing Length Scale for Actuator Line Models of Wind Turbine Blades. Wind Energy. 2017;20(6):1083–1096.

[127] Jha PK, Schmitz S. Actuator Curve Embedding an Advanced Actuator Line Model. Journal of Fluid Mechanics. 2018;834:1–11.

[128] Churchfield MJ, Wang Z, Schmitz S. Modeling wind turbine tower and nacelle effects within an actuator line model. In: American Institute of Aeronautics and Astronautics (AIAA) SciTech Forum, Kissimmee, Florida; 2015. AIAA Paper 2015-0214.

[129] Sarlak H, Meneveau C, Sørensen JN. Role of Subgrid-Scale Modeling in Large Eddy Simulation of Wind Turbine Wake Interactions. Renewable Energy. 2015;77:386–399.

[130] Troldborg N, Zahle F, Réthoré PE, *et al.* Comparison of the Wake of Different Types of Wind Turbine CFD Models. In: 50th AIAA Aerospace Sciences Meeting Including the New Horizons Forum and Aerospace Exposition, Nashville, Tennessee; 2012. AIAA Paper 2012-0237.

[131] Aubrun S, Loyer S, Hancock P, *et al.* Wind Turbine Wake Properties: Comparison Between a Non-rotating Simplified Wind Turbine Model and a Rotating Model. Journal of Wind Engineering and Industrial Aerodynamics. 2013;120:1–8.

[132] Espana G, Aubrun S, Loyer S, *et al.* Wind Tunnel Study of the Wake Meandering Downstream of a Modelled Wind Turbine as an Effect of Large Scale Turbulent Eddies. Journal of Wind Engineering and Industrial Aerodynamics. 2012;101:24–33.

[133] Krogstad PÅ, Adaramola MS. Performance and Near Wake Measurements of a Model Horizontal Axis Wind Turbine. Wind Energy. 2012;15(5):743–756.

[134] Pierella F. Experimental Investigation of Wind Turbine Wakes and Their Interaction. PhD thesis, Norwegian University of Science and Technology (NTNU), Trondheim, Norway; 2014.

[135] Chamorro LP, Porté-Agel F. Effects of Thermal Stability and Incoming Boundary-Layer Flow Characteristics on Wind-Turbine Wakes: A Wind-tunnel Study. Boundary-layer meteorology. 2010;136(3):515–533.

[136] Hancock P, Zhang S, Pascheke F, *et al.* Wind Tunnel Simulation of a Wind Turbine Wake in Neutral, Stable and Unstable Wind Flow. Journal of Physics: Conference Series. 2014;555:012047.

[137] Zhang W, Markfort CD, Porté-Agel F. Wind-Turbine Wakes in a Convective Boundary Layer: A Wind-Tunnel Study. Boundary-Layer Meteorology. 2013;146(2):161–179.

[138] Cal RB, Lebrón J, Castillo L, *et al.* Experimental Study of the Horizontally Averaged Flow Structure in a Model Wind-Turbine Array Boundary Layer. Journal of Renewable and Sustainable Energy. 2010;2(1):013106.

[139] Hamilton N, Cal RB. Anisotropy of the Reynolds Stress Tensor in the Wakes of Wind Turbine Arrays in Cartesian Arrangements With Counter-Rotating Rotors. Physics of Fluids. 2015;27(1):015102.

[140] Cleijne J. Results of Sexbierum Wind Farm; Double Measurements. TNO, The Netherlands; 1992. TNO Report 92-388.

[141] Cleijne J. Results of Sexbierum Wind Farm; Single Wake Measurements. TNO, The Netherlands; 1993. TNO Report 93-082.

[142] Moriarty P, Rodrigo JS, Gancarski P, *et al.* IEA-task 31 Wakebench: Towards a Protocol for Wind Farm Flow Model Evaluation. Part 2: Wind Farm Wake Models. Journal of Physics: Conference Series. 2014;524:012185.

[143] Rados K, Larsen G, Barthelmie R, *et al.* Comparison of Wake Models With Data for Offshore Windfarms. Wind Engineering. 2001;25(5):271–280.

[144] Schlez W, Umaña A, Barthelmie R, *et al.* ENDOW: Improvement of Wake Models Within Offshore Wind Farms. Wind Engineering. 2001;25(5): 281–287.

[145] Barthelmie RJ, Frandsen ST, Rathmann O, *et al.* Flow and Wakes in Large Wind Farms: Final Report for UpWind WP8;2011.

[146] Barthelmie R, Frandsen S, Hansen K, *et al.* Modelling the impact of wakes on power output at Nysted and Horns Rev. In: European Wind Energy Conference; 2009.

[147] Gaumond M, Réthoré PE, Ott S, *et al.* Evaluation of the Wind Direction Uncertainty and Its Impact on Wake Modeling at the Horns Rev Offshore Wind Farm. Wind Energy. 2014;17(8):1169–1178.

[148] Wu YT, Porté-Agel F. Modeling Turbine Wakes and Power Losses Within a Wind Farm Using LES: An Application to the Horns Rev Offshore Wind Farm. Renewable Energy. 2015;75:945–955.

[149] Walker K, Adams N, Gribben B, *et al.* An Evaluation of the Predictive Accuracy of Wake Effects Models for Offshore Wind Farms. Wind Energy. 2016;19(5):979–996.

[150] Nygaard NG. Wakes in Very Large Wind Farms and the Effect of Neighbouring Wind Farms. Journal of Physics: Conference Series. 2014;524:012162.

[151] Nygaard NG, Hansen SD. Wake Effects Between Two Neighbouring Wind Farms. Journal of Physics: Conference Series. 2016;753:032020.

[152] Churchfield MJ, Laursen J, Loeven A, *et al.* A comparison between wind turbine aerodynamics model output when using generic versus actual turbine characterization as input. In: 51st AIAA Aerospace Sciences Meeting, no. AIAA-2013-1208, Grapevine, Texas; 2013.

[153] Barthelmie RJ, Larsen G, Frandsen S, *et al.* Comparison of Wake Model Simulations With Offshore Wind Turbine Wake Profiles Measured by Sodar. Journal of Atmospheric and Oceanic Technology. 2006;23(7):888–901.

[154] Krishnamurthy R, Reuder J, Svardal B, *et al.* Offshore Wind Turbine Wake Characteristics Using Scanning Doppler Lidar. Energy Procedia. 2017;137:428–442.

[155] Machefaux E, Larsen GC, Troldborg N, *et al.* Investigation of Wake Interaction Using Full-Scale Lidar Measurements and Large Eddy Simulation. Wind Energy. 2016;19(8):1535–1551.

[156] Iungo GV, Porté-Agel F. Volumetric Lidar Scanning of Wind Turbine Wakes Under Convective and Neutral Atmospheric Stability Regimes. Journal of Atmospheric and Oceanic Technology. 2014;31(10):2035–2048.

[157] Aitken ML, Lundquist JK. Utility-Scale Wind Turbine Wake Characterization Using Nacelle-Based Long-Range Scanning Lidar. Journal of Atmospheric and Oceanic Technology. 2014;31(7):1529–1539.

[158] Herges T, Maniaci DC, Naughton BT, *et al.* High Resolution Wind Turbine Wake Measurements With a Scanning Lidar. Journal of Physics: Conference Series. 2017;854:012021.

[159] Bingöl F, Mann J, Larsen GC. Light Detection and Ranging Measurements of Wake Dynamics Part I: One-Dimensional Scanning. Wind Energy: An International Journal for Progress and Applications in Wind Power Conversion Technology. 2010;13(1):51–61.

[160] Trujillo JJ, Bingöl F, Larsen GC, *et al.* Light Detection and Ranging Measurements of Wake Dynamics. Part II: Two-Dimensional Scanning. Wind Energy. 2011;14(1):61–75.

[161] Hirth BD, Schroeder JL, Gunter WS, *et al.* Measuring a Utility-Scale Turbine Wake Using the TTUKa Mobile Research Radars. Journal of Atmospheric and Oceanic Technology. 2012;29(6):765–771.

[162] Nygaard NG, Newcombe AC. Wake Behind an Offshore Wind Farm Observed With Dual-Doppler Radars. Journal of Physics: Conference Series. 2018;1037:072008.

[163] Christiansen MB, Hasager CB. Wake Effects of Large Offshore Wind Farms Identified From Satellite SAR. Remote Sensing of Environment. 2005; 98(2–3):251–268.

[164] Hasager CB, Vincent P, Husson R, *et al.* Comparing Satellite SAR and Wind Farm Wake Models. Journal of Physics: Conference Series. 2015;625:012035.

[165] Emeis S, Siedersleben S, Lampert A, *et al.* Exploring the Wakes of Large Offshore Wind Farms. Journal of Physics: Conference Series. 2016;753: 092014

[166] Menke R, Vasiljević N, Hansen K, Hahmann AN, Mann J. Does the Wind Turbine Wake Follow the Topography? A Multi-lidar Study in Complex Terrain. Wind Energy Science. 2018;3:681–691.

[167] Barthelmie R, Pryor S, Wildmann N, *et al.* Wind Turbine Wake Characterization in Complex Terrain via Integrated Doppler Lidar Data From the Perdigão Experiment. Journal of Physics: Conference Series. 2018;1037:052022.

[168] Berg J, Troldborg N, Sørensen NN, *et al.* Large-Eddy Simulation of Turbine Wake in Complex Terrain. Journal of Physics: Conference Series. 2017;854:012003.

[169] Shamsoddin S, Porté-Agel F. Wind Turbine Wakes Over Hills. Journal of Fluid Mechanics. 2018;855:671–702.

[170] Oberkampf WL, Roy CJ. Verification and Validation in Scientific Computing. Cambridge: Cambridge University Press; 2010.

[171] Nygaard NG. Systematic quantification of wake model uncertainty. In: EWEA Offshore Conference; 2015. p. 10–12.

[172] Leon JPM, Réthoré PE, Natarajan A, *et al.* Uncertainty Quantification in Wind Farm flow Models. PhD thesis, Technical University of Denmark, Department of Wind Energy, Risø National Laboratory, Roskilde, Denmark; 2017.

[173] van der Laan M, Pena A, Volker P, *et al.* Challenges in Simulating Coastal Effects on an Offshore Wind Farm. Journal of Physics: Conference Series. 2017;854:012046.

[174] Dykes K, Hand M, Stehly T, *et al.* Enabling the SMART Wind Power Plant of the Future Through Science-Based Innovation. National Renewable Energy Laboratory (NREL), Golden, Colorado, United States; 2017.

[175] Shapiro CR, Bauweraerts P, Meyers J, *et al.* Model-Based Receding Horizon Control of Wind Farms for Secondary Frequency Regulation. Wind Energy. 2017;20(7):1261–1275.

Chapter 7

Wind-plant-controller design

Bart Doekemeijer[1], Sjoerd Boersma[1], Jennifer King[2], Paul Fleming[2], and Jan-Willem van Wingerden[1]

7.1 Introduction

According to the International Energy Agency's annual forecast on the topic of global energy production and consumption [1], wind energy is expected to continue its promising growth in the global energy production and is even projected to become the primary source of energy for Europe by 2030. Wind energy will become one of the key enablers for the European Union to achieve its goal of having at least 27% of Europe's energy consumption from renewable sources. By that time, approximately 80% of all new capacity added is projected to be from renewable energy sources.

However, in many places in the world, energy from wind remains more expensive than energy from nonrenewable sources such as coal and natural gas. One way to further reduce the levelized cost of wind energy is through wind farm control, in which turbines are operated in a coordinated fashion to increase their collective energy production and reduce the fatigue loads on the wind-turbine structures. Furthermore, as the penetration of wind energy in the electricity grid increases, there will be an increasing demand for wind farms to provide ancillary grid services, in order to prevent instability in the electricity grid (this may lead to, e.g., machine damage and power outages).

7.1.1 Structure of the chapter

The main purpose of this chapter is the introduction, explanation, and categorization of wind-farm-control algorithms and present their place in a detailed wind farm simulation framework. Specifically, this section will continue to give an introduction to the current practice in wind-farm operation, and the objectives and turbine control methods presented in the literature for coordinated wind farm control. Then, Section 7.2 is concerned with a general classification of past, current, and future

[1]Delft Center for Systems and Control, Delft University of Technology, Delft, The Netherlands
[2]National Wind Technology Center, National Renewable Energy Laboratory, Boulder, CO, USA

algorithms for coordinated wind farm control. The surrogate models necessary to synthesize such control algorithms will be discussed in Section 7.3. Note that a more detailed survey on the topics of Sections 7.2 and 7.3 can be found in [2] and [3]. Two examples of surrogate models and their application in control are outlined in Section 7.4. Furthermore, the place of a coordinated wind-farm-control algorithm in the general framework of a detailed wind farm simulation, including the inputs and outputs to the various submodels, will be described in Section 7.5. The chapter is concluded in Section 7.6.

7.1.2 Current practice in wind farm operation

In the existing commercial wind farms, turbines are operated by neglecting any coupling with the other turbines inside the farm. Specifically, in this so-called greedy control, the turbines are yawed in alignment with the mean wind direction, and the generator torque and blade pitch angles are changed to optimize power capture and minimize structural loads. However, as energy is extracted from the wind by the rotor, a slower, more turbulent flow develops behind the turbine, called the "wake." This wake often extends far behind the rotor (up to the equivalent of 15 rotor diameters downstream for the Nørrekær Enge II onshore wind farm [4], and even further for offshore wind farms). In a wind farm, these wakes will interact with turbines standing downstream. A photograph illustrating wake interaction is shown in Figure 7.1.

A turbine operating in the wake of an upstream turbine will capture less power (due to the decreased flow speed) and experience higher fatigue loads (due to the increased turbulence) than a turbine operating in freestream flow. Through high-fidelity simulations, wind tunnel experiments, and field tests, it has been shown that operating turbines at their locally nonoptimal settings can lead to a more efficient operation of the wind farm as a whole [2].

Figure 7.1 The Horns Rev offshore wind farm at the coast of Denmark under foggy conditions. Photograph by Christian Steiness

7.1.3 Degrees of freedom in the wind farm control problem

In the literature, two approaches have been followed to improve the operation of wind farms.* Both approaches take wake interactions into account by collectively operating the turbines for a joint objective.

1. *Axial induction control* or *derating control*. In this control methodology, the generator torque and blade pitch angles are operated differently from greedy control settings. A lot of research in the literature has focused on static axial induction control. The idea of power maximization using static axial induction control is to operate an upstream turbine at a fixed derated set point and compensate the loss in power capture of this turbine by the gain in power capture of downstream turbines. Further, initial work has been done toward dynamical axial-induction control for power maximization, in which the generator torque and blade pitch angles are changed on the seconds scale to, e.g., promote wake recovery downstream. Alternatively, these methods can be used to reduce fatigue loading or better distribute the fatigue loads over the various turbines in the farm. Furthermore, there is a trend toward the use of axial induction control for the integration of wind farms with the electricity grid, in which turbines are derated to follow a demanded power signal.

   Note that, depending on the controller hierarchy, one may not be able to change the local wind turbine control variables (e.g., generator torque and blade pitch angles) directly. Rather, wind turbines may be operated by a local wind turbine controller which is to receive set points from the supervisory farm control algorithm. These set points may be blade pitch angles, power set points, or even set points on the non-dimensionalized thrust coefficient, depending on the design of the local turbine controller.
2. *Yaw control* or *wake redirection control*. In this control methodology, the turbine is purposely misaligned with the mean wind direction, leading to a lateral displacement of the wake-structure downstream. The idea of yaw control is to sufficiently displace the wake such that it can be steered past a downstream turbine. Alternatively, yaw control may be used to avoid situations in which a wake partially overlaps a downstream rotor, as this situation typically leads to high fatigue loads. An example of the lateral wake deflection due to yaw misalignment is shown in Figure 7.2, with $\bar{u}$ the time-averaged wind speed non-dimensionalized by the time-averaged freestream wind speed at hub height $\bar{u}_h$.

   For wind turbines inside a farm, yaw control is highly dependent on the wind farm layout and the wind direction. Namely, improperly deflecting the wake inside a wind farm may lead to wake impingement on a different turbine, thereby displacing the problem rather than solving it.

*Improving wind-farm operation is not only limited to increasing the annual energy production but also covers the mitigations of structural loads and the integration with the electricity grid. The objectives of wind farm control will be discussed with more detail in Section 7.1.4.

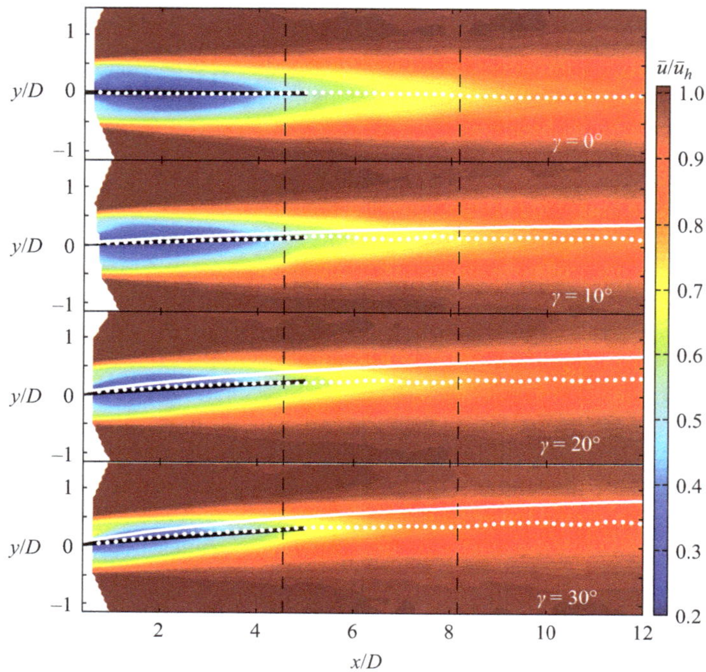

Figure 7.2 Experimental wind tunnel results for a turbine under different yaw angles $\gamma = 0$, 10°, 20°, 30°. Along the axes are the longitudinal x and lateral distance y, non-dimensionalized by the rotor diameter D, respectively. Reproduced, with permission, from [5]

7.1.4 Objectives of wind farm control

The goal of wind farm control is to reduce the levelized cost of wind energy by intelligently operating the turbines inside the wind farm. This high-level goal can be broken down into three subgoals: (1) maximization of the energy production, (2) minimization of the turbine fatigue loads, and (3) integration of wind energy with the electricity grid [2]. Each subgoal is described next.

7.1.4.1 Maximization of the farm's annual energy production

Turbines operating in the wake of an upstream turbine produce less power due to the lower mean inflow wind speed. For example, the Lillgrund offshore wind farm experienced an estimated total wind farm wake loss of 23% in energy production [6]. The literature on steady-state axial induction control (derating turbines to a constant operational point) for power maximization has been doubtful [2,7,8], but dynamical axial induction control in which turbines excite the flow has shown success in high-fidelity simulations [2,9–11]. Furthermore, yaw control has shown success in increasing the collective power production, both for steady-state operation [2,8,12,13] and more recently in dynamical operation as well [9].

7.1.4.2 Minimization of the turbines' structural degradation and fatigue

A second objective in wind farm control is the reduction of turbine fatigue loads to extend the turbine lifetimes or alternatively allow a lighter and cheaper turbine design. There have been initial simulation results on using yaw control to steer the wake symmetrically on top of a downstream rotor in order to avoid partial wake-rotor overlap situations [14], but the literature generally remains sparse on this topic [2], and it remains uncertain to what accuracy wakes can be steered. This is mainly due to the lack of accurate surrogate models that predict the effect of operational strategies on the loads of downstream turbines.[†] Further, the effects of yaw and dynamical axial induction control methods on the controlled turbine's structural loads have not been investigated sufficiently at the time of writing.

7.1.4.3 Provision of ancillary services for the electricity grid

As the share of energy production from renewable sources increases, challenges arise concerning the stability of the electricity grid. Namely, commercial wind turbines are currently disconnected from the electricity grid by their power electronics [15]. Furthermore, due to the variability in wind, the power capture of a wind turbine is by nature also varying. Thus, at its current pace, the increased penetration of electricity from wind will lead to a decreased stability in the electricity grid, which may lead to power outages and machine damage. Therefore, one may rather control the wind farm to meet a certain demanded (possibly time-varying) power, rather than solely optimizing for maximum power extraction. This so-called active power control may be challenging due to the changing atmospheric conditions and the large-scale interactions between turbines through their wakes. Active power control has been receiving an increasing amount of interest in the literature [16–18].

7.2 A classification of wind farm control algorithms

In this section, an overview of the various wind farm control methods is given according to the source of information (surrogate model and/or measurements) used in determining the next control policy. In this chapter, wind-farm control algorithms are classified as either greedy control, open-loop model-based control, closed-loop model-based control or closed-loop model-free control.

7.2.1 Current practice; greedy operation

The current practice in wind farm operation is "greedy control," in which turbines are controlled on a local level, as shown in Figure 7.3. The control policy of the

[†]A more exhaustive overview on surrogate models for wind farm control will be given in Section 7.3.

turbines, $\boldsymbol{q}$, aims to maximize each turbine's individual power capture and minimize loads on the mechanical and electrical systems [19], neglecting any interaction with neighboring turbines. A more detailed discussion on wind turbine control is out of the scope of this chapter, and the reader is referred to Chapter 6 of Volume 2. Note that this greedy turbine control policy is often suboptimal for the entire wind farm due to wake interaction.

In Figure 7.3, the measurements from the wind farm are denoted by z. These measurements may include the SCADA system measurements from each turbine, lidar system measurements, and met tower measurement data. In the greedy control framework, no measurements are used on a farm-level for the determination of a coordinated wind farm control policy.

7.2.2 Open-loop model-based controller synthesis

The first coordinated wind farm control algorithms in the literature were open loop, using a steady-state surrogate model of the wind farm (e.g., the 1983 Jensen model [20] or the 1988 Ainslie model [21]) which predicts the effect of a control policy on the flow and turbine dynamics, as shown in Figure 7.4. In the open-loop control

Figure 7.3 Greedy wind farm operation, in which turbines are controlled on an individual level, neglecting any coupling with other turbines

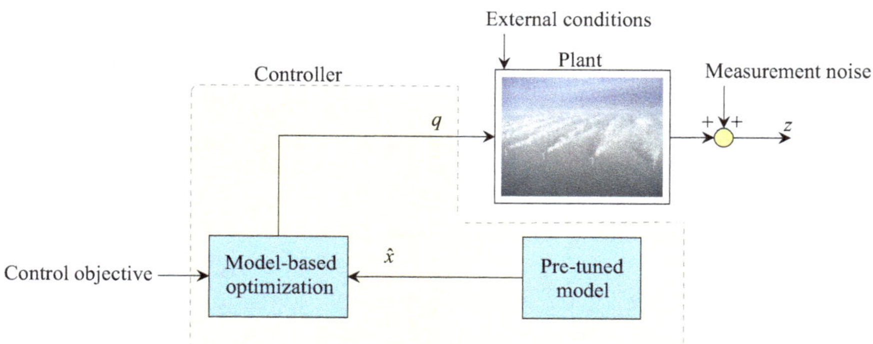

Figure 7.4 The open-loop framework in wind farm control, in which a surrogate model of the wind farm is tuned a priori and used to determine a coordinated control policy for the turbines

framework, the surrogate wind farm model used for optimization might be calibrated prior to operation. The model information $\hat{x}$ is then used for optimization. Control algorithms for steady-state surrogate models typically include gradient-based and nonlinear optimization algorithms (e.g., [7,12]), and predictive control methods for dynamical surrogate models (e.g., [9,10,22]). No measurements are used in the determination of the next control policy, and thus the performance highly depends on the accuracy of the model. Successful results for open-loop control have been shown situationally in the literature [12,13], yet it remains highly dependent on the model accuracy. Due to the complex dynamics on a range of temporal and spatial scales and the lack of knowledge about the environment, no surrogate model exists that is accurate for all the conditions that the farm experiences throughout annual operation [2]. To ensure robustness, a closed-loop control framework is necessary.

7.2.3 Closed-loop model-based controller synthesis

In the closed-loop model-based control framework, measurements from the wind farm z are used to adapt the surrogate wind farm model in real time to the current operational conditions in the wind farm. This real-time model calibration significantly adds to the accuracy of surrogate models and, therefore, adds to the accuracy and robustness of wind-farm-control algorithms. The framework is shown in Figure 7.5. As the wind-farm-control community is shifting toward closed-loop control, there is an increasing interest toward the real-time model calibration of steady-state surrogate models [23] and dynamical wind farm models [18,24,25] alike. Note that the same optimization algorithms as for open-loop model-based controller synthesis can be used.

7.2.4 Closed-loop model-free controller synthesis

Finally, a niche class is a closed-loop model-free wind-farm control. In this framework, measurements are used directly to decide the next control policy, without the

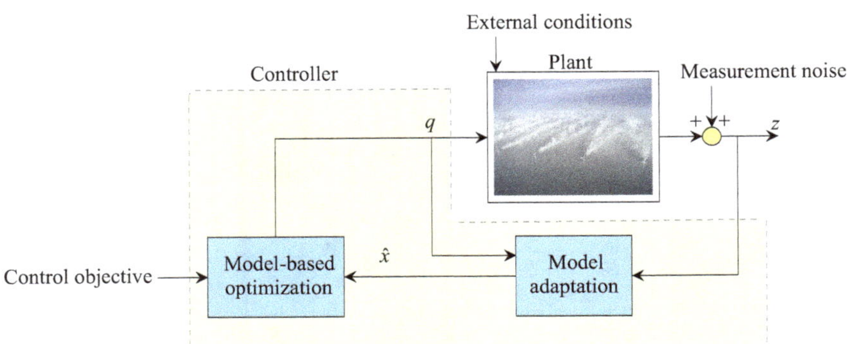

Figure 7.5 The closed-loop framework in wind farm control, in which a surrogate model is tuned in real-time using measurements from the farm, upon which a coordinated control policy for the turbines is determined

usage of a surrogate model, as shown in Figure 7.6. While the attention toward model-free optimization in the literature has decreased, it continues to be a topic of interest [26–31]. Popular model-free optimization algorithms are extremum-seeking control and dynamic programming. The convergence speeds of model-free optimization methods are questionable, but these algorithms avoid the challenge of requiring an accurate model of the flow and turbine dynamics.

7.3 Control-oriented modeling

The main focus in wind farm control has been on open-loop and closed-loop model-based controller synthesis (Sections 7.2.2 and 7.2.3, respectively). These methods rely on a simplified surrogate model of the flow and turbine dynamics inside the wind farm, predicting the effect of control actions on the wind farm output. There is a significant amount of literature on simplified wind-farm modeling, which can be separated in two categories: steady-state modeling and dynamical modeling. Steady-state surrogate models are static input–output mappings that attempt to predict the time-averaged wind-farm dynamics on a minutes scale, neglecting any temporal dynamics such as wake delays. Steady-state models are the topic of Section 7.3.1. On the other hand, dynamical wind farm models attempt to predict the flow and turbine dynamics of the wind farm as a function of time up to the seconds-scale. Dynamical models with a sufficiently low computational cost can be used for real-time control and are the topic of Section 7.3.2. Higher fidelity dynamical models such as large-eddy simulations are typically used for simulation and the validation of controller designs rather than for controller synthesis, and therefore out of the scope of this chapter. An example of each model type will be given in Section 7.4.

7.3.1 Steady-state surrogate models

The main focus in the literature of control-oriented modeling for wind farms has been on steady-state surrogate models [2]. These models are popular due to their

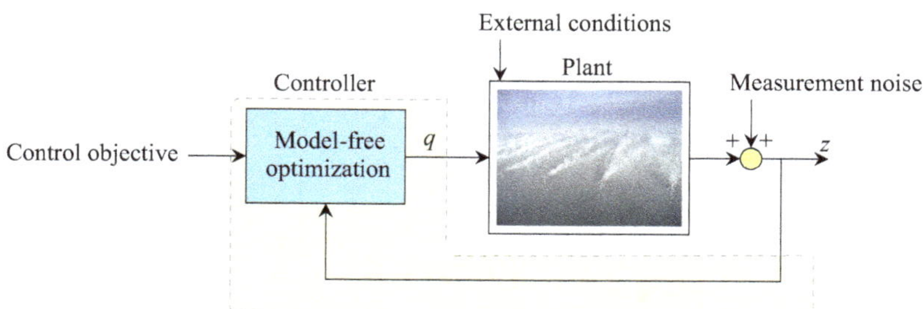

Figure 7.6 The model-free closed-loop framework in wind farm control, in which measurements from the farm are directly used to determine a coordinated control policy of all turbines

low computational cost and complexity. Fundamentally, these models neglect any instantaneous temporal dynamics (e.g., wake delays and wake meandering) and only model the time-averaged flow and turbine dynamics in the wind farm (in the orders of minutes). Mathematically, a steady-state model can be represented by

$$z = f(\boldsymbol{\phi}, \boldsymbol{q}), \tag{7.1}$$

where z is a vector containing the time-averaged outputs of the wind farm (e.g., flow fields and turbine power capture), $\boldsymbol{\phi}$ contains the model information (e.g., tuning parameters, turbine properties, and topology), $\boldsymbol{q}$ contains the control settings of all turbines in the farm, and $f(\bullet)$ is the static mapping from model parameters $\boldsymbol{\phi}$ and control settings $\boldsymbol{q}$ to the measurements z.

The most popular steady-state surrogate wind farm model is the 1983 Jensen's Park model [20], which employs momentum-deficit theory in which the wake is modeled as a linearly expanding region behind the turbine rotor, and includes a linear decay of the deficit. This surrogate model predicts the two-dimensional (2D) flow field at hub height under a uniform inflow, where the turbines are modeled using actuator disk theory [32]. The Jensen model has been used widely in the literature [2]. While the Jensen model can often be tuned to high-fidelity data accurately, the correct approach and frequency for tuning these parameters remains unclear, and therefore its applicability in a real wind farm remains questionable. Furthermore, the model does not include yaw capabilities. Another popular steady-state wind farm model is the Ainslie [21] model. Many modern steady-state surrogate models are derived from the Jensen or the Ainslie model. For example, the FLOw Redirection and Induction in Steady-state (FLORIS) model was developed by Gebraad *et al.* [12], extending on the Jensen's Park model by including yaw capabilities following Jiménez *et al.* [33] and implementing multiple discrete wake zones to increase fidelity [7,12]. More recently, the FLORIS model has been reformulated according to the work by Bastankhah and Porté-Agel [5] to improve the accuracy of derating control and to reduce the amount of tuning parameters. Specifically, they extend their previous work on a surrogate model for a single wake [34] by including the effects of atmospheric and turbine-induced turbulence following Crespo and Hernández [35], and extending the model to a wind farm following Niayifar and Porté-Agel [36].

7.3.2 Control-oriented dynamical surrogate models

While the main focus in the literature has been on steady-state surrogate models, there has been an increasing interest toward dynamical control-oriented wind farm models in recent years [2]. These models increase the fidelity by including temporal dynamics in their predictions. Since these models predict the instantaneous temporal and spatial dynamics of the flow and turbines inside a farm, they can be used for control at these timescales (on the order of seconds). Furthermore, control algorithms relying on a dynamical surrogate model have the potential to exploit dynamic effects inside the wind to promote wake recovery (e.g., [9]). However, dynamical surrogate models have an increased complexity originating from the fact that many dynamical models are nonlinear, often do not include analytical expressions for their functions and derivatives, and go paired with a relatively high computational cost. Control-oriented dynamical models attempt to trade-off these disadvantages with the increase in modeling accuracy. Mathematically, a dynamical model can be represented by

$$x_k = g(\boldsymbol{x}_{k-1}, \boldsymbol{\phi}_{k-1}, \boldsymbol{q}_{k-1}), \tag{7.2}$$

$$z_k = h(\boldsymbol{x}_k, \boldsymbol{\phi}_k, \boldsymbol{q}_k), \tag{7.3}$$

where k is a time index, $\boldsymbol{x}$ is the model state vector, $g(\bullet)$ is the forward-in-time state propagation equation, and $h(\bullet)$ is mapping from the state $\boldsymbol{x}$, model settings $\boldsymbol{\phi}$ and turbine settings $\boldsymbol{q}$ to the measurement vector $\boldsymbol{z}$.

Dynamical surrogate wind farm models vary from extensions of steady-state models to include temporal dynamics (e.g., [18], [37]), to models obtained through identification techniques leveraging higher fidelity simulation or experimental data (e.g., [38]), to time-efficient numerical solutions of the unsteady Navier–Stokes partial difference equations (e.g., [39,40]).

7.4 Examples

In this section, two control-oriented wind-farm models are described, including an application for each. Section 7.4.1 addresses a popular steady-state model from the literature, including a case study in which the yaw angles of the turbines inside a farm are optimized for maximum power extraction. Section 7.4.2 addresses a dynamical wind farm model derived from the physical Navier–Stokes equations, including the synthesis of a state estimator in pursuit of closed-loop wind farm control.

7.4.1 Steady-state wind farm model: FLORIS

7.4.1.1 Model description

This section showcases the steady-state FLORIS model, which is a popular surrogate model for steady-state wind farm control [41,42].

FLORIS is a modular framework that contains a collection of low-fidelity steady-state wind farm models developed by the Delft University of Technology and the US National Renewable Energy Laboratory, which capture the minute-to-minute dynamics in the wind farm. FLORIS is used for the purpose of real-time wind farm control, offline analysis, and layout optimization. Fundamentally, it includes several models from the literature for the wake deficit (e.g., Jensen [20]), and wake displacement as a function of the turbine yaw angle (e.g., Jiménez *et al.* [33]). In this article, the analytical wake model inspired by Bastankhah and Porté-Agel [5], also included in FLORIS, is employed for its strong physical basis, its validation with experimental data, and its relatively few tuning parameters compared to other models (e.g., the model by Gebraad *et al.* [12]).

For brevity, the focus in this section lies on the far-wake model. The reader is referred to [5] for the full derivation. Note that all equations described here are in the wind-aligned frame, with x aligned with the wind, y the lateral component, and z the vertical component, all centered at the turbine hub.

For a single wake, the near-wake region is modeled as a cone with its base at the rotor plane, and its tip at a distance x_0 downstream, where

$$x_0 = \frac{(1/2)\sqrt{2}D \cdot \cos(\gamma) \cdot (1 + \sqrt{1 - C_T})}{\alpha \cdot I + \beta\left(1 - \sqrt{1 - C_T}\right)}, \tag{7.4}$$

where C_T is the turbine thrust coefficient,‡ γ is the yaw angle, D is the rotor diameter, and α and β are tuning parameters. The turbulence intensity in front of the rotor, I, is calculated as a squared summation of the atmospheric turbulence intensity I_0 and the added turbulence intensities from upstream turbines I^+. Mathematically, this yields

$$I = \sqrt{\sum_{j=1}^{\aleph}\left(\frac{A_w^j}{(1/4)\pi D^2} I_j^+\right)^2 + I_0^2}, \quad \text{with} \quad I_j^+ = t_a \cdot a^{t_b} \cdot I_0^{t_c} \cdot \left(\frac{\Delta x_{\text{turb}}^j}{D}\right)^{t_d}, \tag{7.5}$$

where $\aleph$ is equal to the number of turbines, A_w^j is the relatively overlap area between the wake of the upstream turbine j and the downstream rotor, Δx_{turb}^j is the streamwise distance between the two turbines, and t_a, t_b, t_c, and t_d are tuning parameters.

For the far-wake region, $x \geq x_0$, the wake is modeled as a 2D Gaussian velocity deficit in y- and z-direction, symmetrical around a centerline. This centerline lies in the horizontal plane at hub height, displaced in y-direction from the turbine hub by δ_f, as

$$\begin{aligned}\delta_f &= \tan(\theta)\, x_0 + \frac{\theta}{5.2} \cdot \left(C_0^2 - 3e^{1/12} C_0 + 3e^{1/3}\right) \\ &\times \sqrt{\frac{\sigma_{y0}\sigma_{z0}}{k_y \cdot k_z \cdot C_T}} \cdot \ln\left[\frac{\left(1.6 + \sqrt{C_T}\right)\left(1.6 S_\sigma - \sqrt{C_T}\right)}{\left(1.6 - \sqrt{C_T}\right)\left(1.6 S_\sigma + \sqrt{C_T}\right)}\right] + \delta_r(x).\end{aligned} \tag{7.6}$$

In this equation, θ is the initial deflection angle, calculated as

$$\theta \approx \frac{0.3\gamma}{\cos\gamma}\left(1 - \sqrt{1 - C_T \cos\gamma}\right). \tag{7.7}$$

Furthermore, $C_0 = 1 - \sqrt{1 - C_T}$, k_y and k_z are linear wake expansion coefficients similar to that in Jensen [20], and S_σ is defined as $S_\sigma = \sqrt{(\sigma_y\sigma_z)/(\sigma_{y0}\sigma_{z0})}$, with σ_y and σ_z the standard deviations of the Gaussian in y- and z-direction, respectively, both a linear function of x. They are calculated as

$$\sigma_y = \sigma_{y0} + (x - x_0)k_y, \quad \text{with} \quad \sigma_{y0} = \frac{D}{2\sqrt{2}}\cos\gamma, \tag{7.8}$$

$$\sigma_z = \sigma_{z0} + (x - x_0)k_z, \quad \text{with} \quad \sigma_{z0} = \frac{D}{2\sqrt{2}}. \tag{7.9}$$

‡A mapping exists between C_T and physical quantities such as the generator torque and blade pitch angles.

Further, δ_r is the wake deflection induced by the rotation of the blades, approximated using a linear function following the idea of Gebraad *et al.* [12], by $\delta_r = a_d \cdot D + b_d \cdot x$, with a_d and b_d tuning parameters. The time-averaged wind speed in the far-wake region $\bar{u}$ at some location (x, y, z), with $x \geq x_0$, and with the origin at the turbine hub, is now defined as

$$\frac{\bar{u}(x,y,z)}{\bar{u}_h} = 1 - \left(1 - \sqrt{1 - \frac{\sigma_{y0}\sigma_{z0}}{\sigma_y \sigma_z} C_T}\right) \cdot \exp\left(\frac{(y - \delta_f)^2}{2\sigma_y^2} + \frac{z^2}{2\sigma_z^2}\right), \tag{7.10}$$

with $\bar{u}_h$ the time-averaged freestream wind speed, and σ_y, σ_z, and δ_f all functions of x. Finally, the time-averaged power capture of a turbine P_j is calculated as

$$P_j = \frac{1}{2} \cdot \rho \cdot \left(\frac{1}{4}\pi D^2\right) \cdot C_P \cdot \bar{u}_R^3 \cdot \eta \cdot (\cos \gamma)^{k_p}, \tag{7.11}$$

with η accounting for efficiency losses with a typical value of 0.90–0.95, ρ the air density, C_P the dimensionless power coefficient (which has a direct mapping from the control variable C_T), k_p a tuning variable as in Gebraad *et al.* [12], and $\bar{u}_R$ the rotor-averaged wind speed. The rotor-averaged wind speed is calculated by integrating the effect of multiple wakes over the turbine rotor, as

$$\bar{u}_R = \bar{u}_h \left(1 - \sqrt{\sum_{j=1}^{\aleph} \left(\frac{Q_j}{(1/4)\pi D^2}\right)^2}\right), \tag{7.12}$$

$$Q_j = \int_0^{D/2}\int_0^{2\pi} \left(r\left(1 - \frac{\bar{u}(r,\beta)}{\bar{u}_h}\right)\right) d\beta \, dr, \tag{7.13}$$

where (r,β) is the polar coordinate representation of (y, z) over the rotor.

7.4.1.2 Real-time wind farm optimization using FLORIS

The FLORIS model can be used to determine the minute-to-minute optimal control settings of the wind turbines inside a wind farm. In practice, the FLORIS model will be tuned a priori to high-fidelity simulation and/or experimental data to best capture the time-averaged dynamics inside the wind farm. Then, at every control loop, the atmospheric conditions in the FLORIS model are adjusted to match the present conditions inside the farm. These conditions include the atmospheric turbulence intensity I_0, the freestream wind direction, and the freestream wind speed $\bar{u}_h$. This information is derived from several-minute-averages of the flow field.

In the following case study, the FLORIS model is used for wind-farm-power maximization using wake deflection by turbine yaw. The following optimization problem is solved following an interior point method to address the nonlinearity and nonconvexity of the problem:

$$\overline{\boldsymbol{\gamma}}^{\text{opt}} = \arg\min_{\overline{\boldsymbol{\gamma}}} \left(-\sum_{j=1}^{\aleph} P_j(\gamma_j) \right), \tag{7.14a}$$

$$\text{with } -25^\circ \leq \gamma_j \leq 25^\circ, \tag{7.14b}$$

where $\overline{\boldsymbol{\gamma}} = \begin{bmatrix} \gamma_1 & \gamma_2 & \cdots & \gamma_\aleph \end{bmatrix}$. The yaw angle is constrained to suppress the increase in structural loading. The optimal yaw settings, $\overline{\boldsymbol{\gamma}}^{\text{opt}}$, are then distributed to the turbines and maintained for a 10-min period, upon which the cycle repeats.

In a three-by-three wind farm with DTU 10 MW wind turbines at 5D streamwise spacing and atmospheric conditions $\bar{u}_h = 11.5$ m/s, $I_0 = 0.05$, the difference between greedy control ($\gamma_j = 0$) and optimized control ($\gamma_j = \overline{\boldsymbol{\gamma}}_j^{\text{opt}}$) is investigated. Note that the parameter tuning procedure is not further outlined, and the case study in this section only serves as an example. The time-averaged flow fields are shown in Figure 7.7. From this figure, it can clearly be seen that the upstream turbines are strongly yawed to deflect their effect on the two downstream turbine rows. Turbines in the most downstream row are operating at $\gamma_7, \gamma_8, \gamma_9 = 0^\circ$, since their wakes do not affect any downstream turbines. More specifically, the optimal yaw angles are predicted to be

$$\overline{\boldsymbol{\gamma}}^{\text{opt}} = \begin{bmatrix} 24.8^\circ & 24.8^\circ & 24.7^\circ & 19.9^\circ & 19.8^\circ & 19.8^\circ & 0.0^\circ & 0.0^\circ & 0.0^\circ \end{bmatrix},$$

which yields a relative power increase of 11.4%: from 29.6 MW under greedy control to 32.9 MW under yaw-optimized control. Note that this significant gain in power

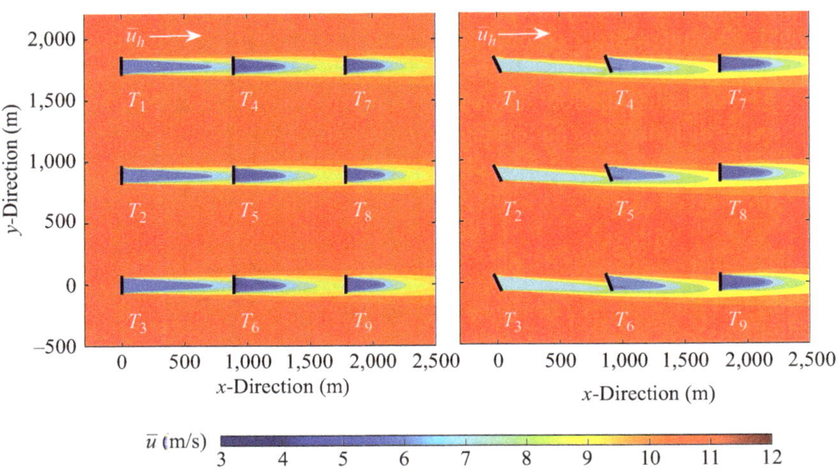

Figure 7.7 The time-averaged flow field predicted by FLORIS for the greedy case (left) and the yaw-optimized case (right)

production is due to the relatively close turbine spacing, the low atmospheric turbulence intensity and thus deep wakes, and the wind direction. Nonetheless, these situations are not uncommon in realistic wind farms [6].

Note that this increase in power production is predicted by FLORIS, and thus, it highly depends on the accurate tuning of the model parameters. Therefore, it is typically appropriate to test the optimized control policy in high-fidelity simulations or experiments. This is outside of the scope of this chapter.

7.4.2 Dynamical wind farm model: WFSim

7.4.2.1 Model description

WindFarmSimulator (WFSim) is a dynamical wind farm model developed at the Delft University of Technology for the purpose of real-time dynamical wind farm control. A brief summary of WFSim is given here, and the interested reader is referred to [40] for a more detailed description. The model is derived from the unsteady Navier–Stokes equations simplified to two dimensions (at hub height) under modifications of the continuity equation, to better match the flow dynamics found in higher fidelity, 3D models. The equations solved in the WFSim model are

$$\frac{\partial \boldsymbol{u}}{\partial t} + (\boldsymbol{u} \cdot \nabla_H)\boldsymbol{u} + \nabla_H \cdot \boldsymbol{\tau}_H + \nabla_H p - \boldsymbol{f} = 0, \tag{7.15}$$

$$\nabla_H \cdot \boldsymbol{u} = -\frac{\partial v}{\partial y}, \tag{7.16}$$

with $\nabla_H = \left[\partial/\partial x \;\; \partial/\partial y\right]^T$, $\boldsymbol{u} = \left[u \;\; v\right]^T$ the flow velocity in x- and y-direction, respectively, and p the pressure. It is emphasized that the WFSim model deviates from a traditional 2D Navier–Stokes model due to the difference in the continuity equation. Specifically, this modification allows for flow relaxation in the vertical dimension when, e.g., encountering slow down by a wind turbine. The turbines are modeled using actuator disk theory. The forcing term $\boldsymbol{f}$ is defined as

$$\boldsymbol{f} = \sum_{i=1}^{\aleph} \boldsymbol{f}_i, \quad \text{with} \tag{7.17}$$

$$\boldsymbol{f}_i = \frac{c_f}{2} C'_{T_i} [U_i \cos(\gamma_i)]^2 \begin{pmatrix} \cos(\gamma_i + \varphi) \\ \sin(\gamma_i + \varphi) \end{pmatrix} \mathrm{H}\left[\frac{D}{2} - ||\boldsymbol{s} - \boldsymbol{t}_i||_2\right] \delta\left[(\boldsymbol{s} - \boldsymbol{t}_i) \cdot \boldsymbol{e}_{\perp,i}\right],$$

with $\boldsymbol{s} = \left[x \;\; y\right]^T$, $\mathrm{H}[\,\cdot\,]$ the Heaviside function, $\delta[\,\cdot\,]$ the Dirac delta function, $\boldsymbol{e}_{\perp,i}$ the unit vector perpendicular to the ith rotor disk with position $\boldsymbol{t}_i$ and rotor diameter D, U_i the flow velocity at the rotor, and φ the mean wind direction in the farm. Further, C'_{T_i} is a variation on the traditional nondimensional thrust coefficient, as defined by [43]. In the WFSim model, C'_{T_i} and γ_i are considered as the control variables. Furthermore, the scalar c_f in (7.17) can be regarded as a tuning variable. The 2D tensor $\boldsymbol{\tau}_H$ contains the turbulence model and is defined as

$$\boldsymbol{\tau}_H = -\nu_t \boldsymbol{S}, \quad \text{with} \quad \nu_t = \ell_u(x,y)^2 \left| \frac{\partial u}{\partial y} \right|, \quad \boldsymbol{S} = \frac{1}{2}(\nabla_H \boldsymbol{u} + (\nabla_H \boldsymbol{u})^T), \tag{7.18}$$

where ν_t is the eddy viscosity, and $\ell_u(x,y)$ is the mixing length that has been made a linearly increasing function of the downstream distance, as suggested in [44].

The proposed mixing length parametrization is

$$\ell_u(x,y) = \begin{cases} G(x_i', y_i') * \ell_u^i(x_i', y_i') & \text{if } x \in \mathcal{X} \text{ and } y \in \mathcal{Y} \\ 0 & \text{otherwise} \end{cases} \tag{7.19}$$

with $G(x,y)$ a (smoothing) pillbox filter with radius 3, $*$ the 2D spatial convolution operator, $\mathcal{X} = \{x : x_i' \le x \le x_i' + \cos(\varphi)d\}$, $\mathcal{Y} = \{y : y_i' - (D/2) + \sin(\varphi)x_i' \le y \le y_i' + (D/2) + \sin(\varphi)x_i'\}$, and φ defined as the mean wind direction, which is bounded to $|\varphi| \le 45°$ in this work. In addition, d is constrained by $\cos(\varphi)d \le |x_q - x_i|$, with x_i a turbine's x-coordinate and x_q its downwind turbine's x-coordinate. The variable $\ell_u^i(x_i', y_i')$ can be seen as the local mixing length that belongs to turbine i, denoted as

$$\ell_u^i(x_i', y_i') = \begin{cases} (x_i' - d')\ell_s, & \text{if } x_i' \in \mathcal{X}_i' \text{ and } y_i' \in \mathcal{Y}_i'. \\ 0, & \text{otherwise,} \end{cases} \tag{7.20}$$

with $\mathcal{X}_i' = \{x_i' : d' \le x_i' \le d\}$, $\mathcal{Y}_i' = \{y_i' : |y_i'| \le D\}$, and ℓ_s being a tuning parameter that defines the slope of ℓ_u^i. This parameter may be related to turbulence intensity, i.e., the amount of wake recovery. It is assumed that the tuning parameters ℓ_s, d, and d' are equal for each turbine in the farm, which reduces the amount of tuning variables for the turbulence model to 3.

From the resolved flow field, the power generated by the farm is calculated as

$$P = \sum_{i=1}^{\aleph} \frac{1}{8} \rho \pi D^2 c_p C_{T_i}' [U_i \cos(\gamma_i)]^3. \tag{7.21}$$

In this equation, the scalar c_p can be seen as a tuning variable and will be set equal for all turbines in the farm. Adding the previously defined tuning variables c_f, ℓ_s, d, and d', results in a total of five tuning parameters in the WFSim model. Spatial and temporal discretization of (7.15) and (7.16) by employing the finite volume method and implicit method, respectively, results in similar equations as defined in (7.2) and (7.3). The measurement vector z_k may contain longitudinal and lateral flow velocities at hub height, and turbine power capture signals. The turbine control settings C_{T_i}' and γ_i for each turbine are collected in the input vector $\boldsymbol{q}_k$.

The WFSim model has been validated in various high-fidelity simulations, showing good agreement in terms of hub-height flow fields and power capture in two-turbine, six-turbine, and nine-turbine wind-farm case studies [40]. A snapshot of the flow field from a six-turbine simulation in WFSim is shown in Figure 7.8.

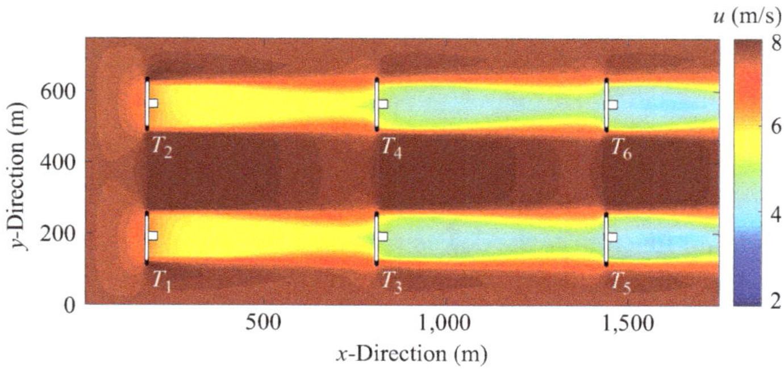

Figure 7.8 Example of an instantaneous flow field at hub height evaluated with the WFSim model for a six-turbine wind farm. The wind is flowing from left to right

7.4.2.2 Real-time model adaptation for the WFSim model

An important aspect of any model-based closed-loop controller is the model adaptation algorithm, which leverages measurements for the estimation of the model states $\boldsymbol{x}$ (e.g., local wind and turbine conditions) and/or a subset of the model parameters $\boldsymbol{\phi}$ (e.g., freestream wind conditions) in real time. Real-time model adaptation for the WFSim model has been the topic of research of Doekemeijer *et al.* [24,25]. In [24], the authors propose a two-step model adaptation algorithm:

1. First, the wind-farm-wide freestream wind speed $\bar{u}_h$ and wind direction $\bar{\theta}_h$ at hub height are estimated. The latter is derived by time- and turbine-averaging measurements from the wind vane on each turbine's hub, after removing outliers. Knowing $\bar{\theta}_h$, the turbines operating in a wake can be distinguished from the turbines operating in freestream flow through the use of a simplified wake model, following the example of [45]. Finally, the generated power measurements and control settings of each upstream turbine i can be used to provide an estimate of the freestream wind speed following an inversion rule of actuator disk theory, by

$$u_h^i = \sqrt[3]{\frac{2P^i}{C_P^i A_D^i \rho}}, \tag{7.22}$$

with u_h^i the instantaneous freestream wind speed according to upstream turbine i, P^i the generated power measurement, C_P^i the turbine's power coefficient (derived from the control settings), A_D the rotor swept surface area, and ρ the air density. Then, these estimates are averaged and low-pass filtered to provide a single wind-farm-wide estimate of $\bar{u}_h$, by

$$\tau \cdot \frac{\partial \bar{u}_h}{\partial t} = \left(\frac{1}{N_{\text{upstream}}} \cdot \sum_{i \in \text{upstr. turbines}}^{N_{\text{upstream}}} u_h^i \right) - \bar{u}_h, \tag{7.23}$$

with τ the time constant of the first-order low-pass filter, and N_{upstream} the number of upstream turbines.

An important remark is that this methodology for the estimation of $\bar{u}_h$ relies solely on power measurements and therefore only works for below-rated conditions. For estimation of $\bar{u}_h$ in above-rated conditions, one may, for example, require the implementation of a wind speed estimator on each individual turbine, from which the local wind speed in front of each turbine can be estimated, as demonstrated by [46].

2. Second, the system states $\boldsymbol{x}$ (in this case: the 2D flow and pressure fields) and a subset of model parameters $\boldsymbol{\phi}$ are estimated through the use of an ensemble Kalman filter (KF) (EnKF). KFs are the default choice for recursive state estimation in the area of closed-loop control. The EnKF deviates from the standard KF by approximating the covariance matrices by a finite set of so-called ensemble members, which are propagated through the nonlinear system dynamics as a Monte Carlo approach in each timestep. This has shown to better represent the true system dynamics compared to linearizing the system dynamics for a sufficiently large ensemble size [47,48]. Furthermore, the EnKF has shown good performance for large systems using very small ensemble sizes [49], yielding a very low computational cost.

Note that the estimation performance highly depends on the measurements available. For this estimation framework, the effect of the measurement source on the estimation performance is further investigated in [24]. In general, a balance has to be found between estimations which are accurate enough for control and sensor cost.

A case study

This model adaptation concept has been tested using high-fidelity simulation data from SOWFA (see Chapter 6) in [24,25], and the core results are summarized here. For a more detailed description, see [24].

The simulation is of a nine-turbine wind farm in which the flow is excited by rapid variations on the amount of energy extracted by each turbine, identical to the simulation in [16]. The wind farm is arranged in a three-by-three formation of NREL 5 MW baseline turbines with a streamwise and lateral spacing of $5D$ and $3D$, respectively, with D the rotor diameter. The simulation is of a neutral ABL with a freestream wind speed $\bar{u}_h$ of 12 m/s and a freestream turbulence intensity of 5%, in which the turbines are modeled using the actuator line model. In SOWFA, the simulated domain is spatially discretized at 3 m $\times$ 3 m $\times$ 3 m near the rotors, and 12 m $\times$ 12 m $\times$ 12 m at the outer regions of the farm.

This simulation setup has been implemented in WFSim [40] on a domain of 2.5 km $\times$ 1.6 km, spatially discretized at 25 m $\times$ 38 m, with a temporal discretization of $\Delta t = 1$ s. The control settings are mapped following a similar approach as [50].

In this experiment, the adaptation algorithm has to estimate the freestream wind speed $\bar{u}_h$, the model state vector $\boldsymbol{x}$ (local flow field), and the mixing length slope of the turbulence model, $\ell_s \subset \boldsymbol{\phi}$. The only measurements available are the instantaneous generated power signal of each turbine P^i, artificially disturbed by zero-mean white Gaussian noise with $\sigma_P = 10$ kW. The WFSim parameters $\bar{u}_h$ and ℓ_s are purposely initialized with poor values of 9 m/s and 0.5, respectively, in order to investigate convergence. The steady-state optimal value for ℓ_s is found to be 0.92, following from a multi-objective optimization considering the error in flow fields and the error in the turbine power signals of WFSim and SOWFA. Similarly, the initial state vector $\boldsymbol{x}_0$ is initialized as a uniform flow of 9 m/s along the x-direction. The results are shown in Figures 7.9 and 7.10.

Figure 7.9 shows the estimated model states $\boldsymbol{x}$, which are the 2D flow fields at turbine hub height. This figure is very insightful to see the balance between state and parameter estimation. It shows the estimated flow field at times $t = 10$ s, $t = 150$ s and $t = 400$ s. At time $t = 10$ s, the error is very significant, since the model is initialized with wind speeds of 9 m/s instead of the true 12 m/s. However, as time progresses, the error quickly reduces, both due to corrections in the inflow ($\bar{u}_h$) and due to corrections to the state vector (flow field) itself, as seen at $t = 150$ s. After approximately 100–200 s, ℓ_s also starts to converge (see Figure 7.10), and the error further reduces, as seen at $t = 400$ s. At $t = 400$ s, all parameters have more or less converged, and the error in flow fields between SOWFA and WFSim is negligible.

Figure 7.10 shows the successful convergence of both parameters. Specifically, $\bar{u}_h$ converges within 300–400 s, but could be made to converge faster or slower by changing the time constant τ in (7.23). Furthermore, ℓ_s converges within 250 s toward its optimal value of 0.92.

For these simulations, the computational cost of the EnKF was 1.2 s/iteration on an octacore CPU. This makes the EnKF at least one order of magnitude faster than other KF algorithms [24]. Furthermore, the low computational cost enables the use of WFSim for closed-loop controller synthesis. Finally, the algorithm does not require additional information—it relies solely on readily available SCADA data (power measurements) for the adaptation of the WFSim model to the present conditions in the atmosphere and inside the farm. The next step, closed-loop controller synthesis with the WFSim model, is ongoing at the time of writing.

7.5 Software architecture

In order to demonstrate the position of a wind-farm control module in the bigger picture of a high-fidelity wind-farm simulation, this section addresses the software architecture. First, a distinction will be made between centralized and distributed farm-control algorithms. Second, an overview of the various incoming and outgoing signals for such a control module will be provided.

7.5.1 Centralized vs. distributed control

In terms of software architecture, a distinction in wind farm control algorithms can be made based on their communication protocols. Currently, almost all coordinated wind-farm-control algorithms proposed in the literature are centralized (e.g., [7,9,

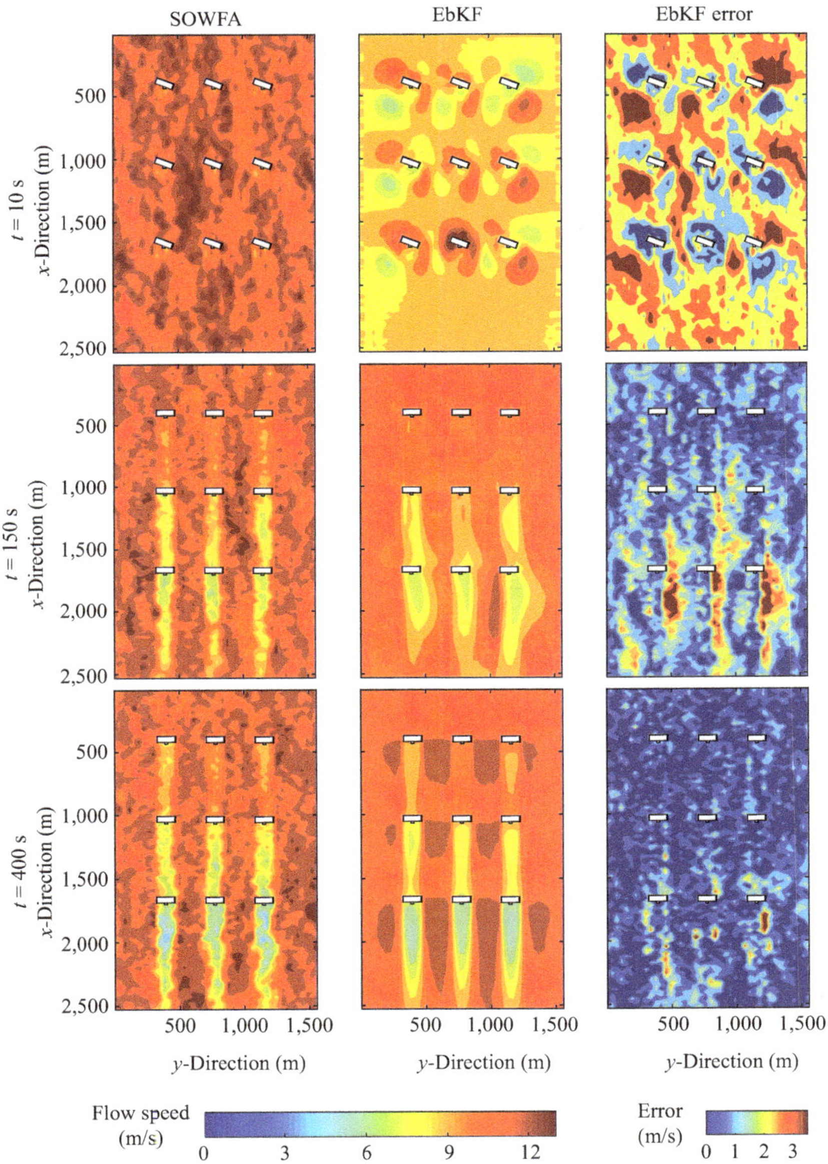

Figure 7.9 *The estimated model state at times $t = 10$ s, $t = 150$ s, and $t = 400$ s. Instantaneous power measurements of each turbine are leveraged for estimation of $\mathbf{x}$, ℓ_s, and $\bar{u}_h$. The initial error is large due to a purposely introduced error in the initial conditions in WFSim (0.5 and 9.0 m/s for ℓ_s and $\bar{u}_h$, respectively). Convergence of the estimates is reached within 400 s*

10,12,16,28,51,52]), demonstrated in Figure 7.11. Centralized controllers rely on a central node that collects all the information from the various nodes (turbines, but also, e.g., measurement towers or lidar systems), determines the next control action for each node, and then distributes these control actions back to the turbines. In terms of controller synthesis, this is the easiest and most straightforward way for wind farm control, especially since most control-oriented wind farm models are difficult to decentralize [2].

On the other hand, there is a small amount of literature available on distributed wind-farm-control algorithms (e.g., [22,53–55]), demonstrated in Figure 7.12. In

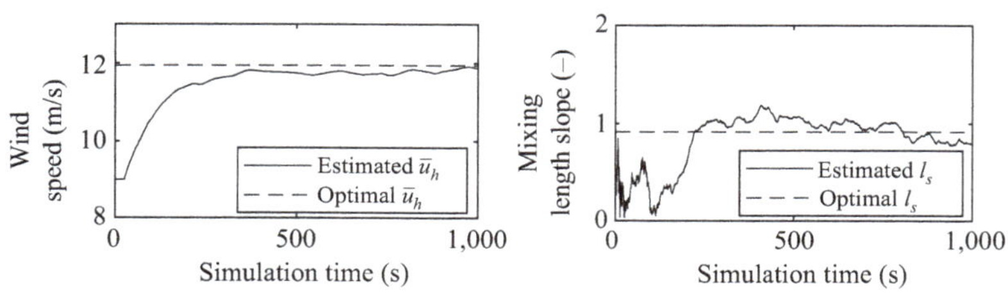

Figure 7.10 Parameter convergence of $\bar{u}_h$ and ℓ_s in the nine-turbine case study. The freestream wind speed converges within 400 s, while the turbulence model ℓ_s converges within approximately 250 s

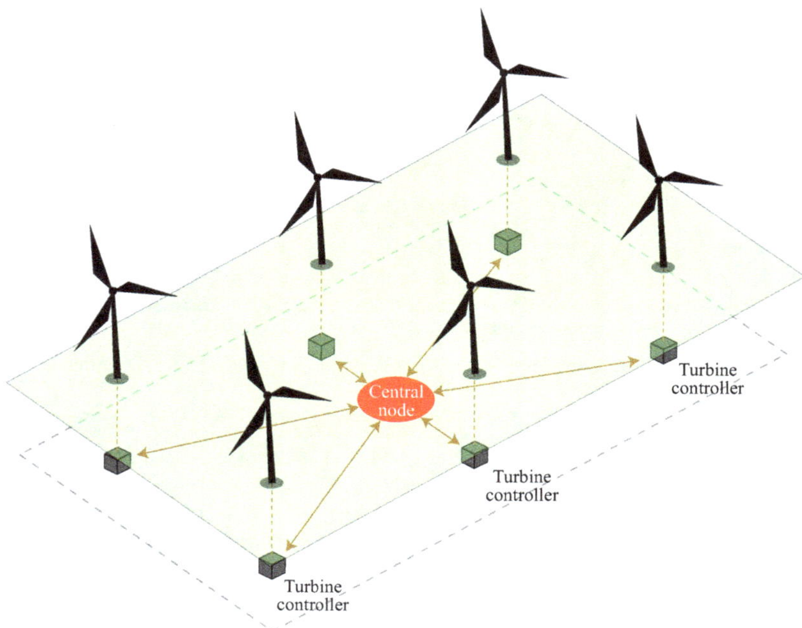

Figure 7.11 The communication infrastructure for centralized control

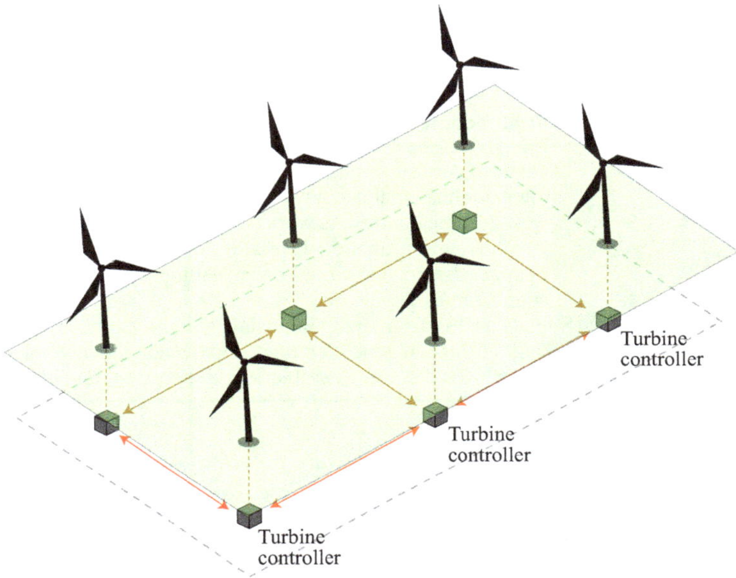

Figure 7.12 The communication infrastructure for distributed control

the distributed control architecture, each node (turbine) exchanges information with their neighboring nodes and locally determines its own next control policy. While this means that the amount of information available for each node is limited (and thereby the control policy may be more conservative), distributed control has a number of benefits. First, it provides a modular approach to the wind farm control problem, and it would be straightforward to add or remove wind turbines to the farm. Second, the computational resources are also distributed, and this avoids the need for a centralized node with a high computational power. Specifically, surrogate models for wind farms with many turbines usually grow high-dimensional and thereby often go paired with a high computational cost, which is challenging for real-time control. Because of these reasons, distributed farm control may be preferred over centralized farm control for large wind farms.

7.5.2 Communication with other simulation submodels

In a high-fidelity simulation environment, the supervisory wind farm control algorithm will communicate with several other submodels to exchange information. The signals sent from the control module, whether it is a single central module or several distributed modules, are displayed in Table 7.1. The signals received by the control module are displayed in Table 7.2. In both tables, the signals are ordered by their relevance from top (very common) to bottom (rare).

Table 7.1 Generalized information infrastructure for wind farm control: outputs. The signals are ordered by their relevance from top to bottom

Module	Sent to module
Local wind turbine controller	Demanded control settings (e.g., reference yaw angle, reference pitch angle, demanded power capture. Note that the current wind turbine controllers work with so-called swap-variables, which are multifunctional variables used for the communication of supervisory wind-farm-control settings to the local turbine controllers)
Trans. sys. operator (TSO)	Signals useful for maintaining balance in the electricity grid (e.g., the actual generated power by the collective wind farm, the predicted power available over a specified time horizon)

Table 7.2 Generalized information infrastructure for wind farm control: inputs. The signals are ordered by their relevance from top to bottom

Module	Received from module
Internal	Synchronization information (e.g., system time), time-invariant wind farm specifications (e.g., turbine specifications, turbine numbering, and wind farm topology)
Local wind turbine controller	SCADA data (e.g., power capture, measured yaw angle, wind vane measurement, anemometer measurement, and LSS rot. speed)
Wind farm flow model	Measurements used for model adaptation (e.g., flow fields from lidar systems and met mast measurements)
Aero-elastic turbine model	SCADA data (if not available from the turbine controller), measurements from turbine loads if available (e.g., deformations and accelerations)
Trans. sys. operator (TSO) or electricity grid model	Control objective related to APC (e.g., electricity grid frequency and demanded farm-wide power signal)

7.6 Conclusion

In order to further promote the growth of wind energy in the global energy production, the levelized cost of wind energy is to be reduced. One way to achieve this is by improving the annual energy capture and reducing the structural degradation of turbines inside a farm. Additionally, wind farms are to be regulated such that they can be integrated with the electricity grid in order to avoid stability issues and its resulting consequences, such as power outages and machine damage. Wind farm control tackles these issues by controlling wind turbines inside a farm in a coordinated fashion, taking into account the interactions between turbines through the formation of wakes.

In the literature, wind farm control has mainly been tackled by the use of steady-state surrogate models that predict the effect of a control policy on the time-averaged farm's power production. These models are popular for their low complexity and computational cost, and control using steady-state models is limited to the scale of minutes. Further, there has been a growing interest toward dynamical surrogate models, which predict the spatial and temporal dynamics of the wind and turbines inside a farm. These models go paired with a significant increase in complexity and computational cost, yet control up to the scale of seconds is allowed. Additionally, dynamical control has the potential to exploit dynamic behavior in the wind, such as, e.g., the use of periodic control signals to excite the flow and promote wake recovery. Both steady-state and dynamical surrogate models are active topics of research, largely because the degree of model dynamics that needs to be accounted for in wind farm controller synthesis is still an open question in the literature.

While a wide variety of wind farm models is available in the literature, no model is consistently accurate over the annual wind farm operation due to the lack of knowledge about the environment, and the tough-to-model flow dynamics on a range of spatial and temporal scales. Therefore, model-based wind-farm control demands a closed-loop approach, in which measurements from the wind farm are used in real time to calibrate the surrogate model, upon which this surrogate model is used to evaluate the optimal control policy. Synthesis of such calibration algorithms is becoming increasingly popular in the literature.

In terms of software architecture, most control algorithms are implemented in a centralized manner, in which all information is processed at a single central node. An alternative to centralized control is a distributed implementation, in which turbines communicate exclusively with neighboring turbines to determine their collectively optimal control policy. Distributed control allows a modular approach to wind-farm control and additionally reduces the local computational cost compared to one centralized processing node. However, more effort will have to be invested in the development of distributed wind farm models to further promote the development of distributed wind-farm control algorithms. The information received by both centralized and distributed farm control algorithms typically includes various measurement data (e.g., SCADA, met mast, and strain gages). The information transmitted by wind farm control algorithms include the control actions or reference signals sent to the local wind-turbine controllers.

Acknowledgment

This work was authored, in part, by the National Renewable Energy Laboratory, operated by Alliance for Sustainable Energy, LLC, for the U.S. Department of Energy (DOE) under Contract No. DE-AC36-08GO28308. Funding provided by the U.S. Department of Energy Office of Energy Efficiency and Renewable Energy Wind Energy Technologies Office. The views expressed in the article do not necessarily represent the views of the DOE or the U.S. Government. The U.S. Government retains and the publisher, by accepting the article for publication, acknowledges that the U.S. Government retains a nonexclusive, paid-up, irrevocable, worldwide license to publish or reproduce the published form of this work, or allow others to do so, for U.S. Government purposes.

References

[1] International Energy Agency. World Energy Outlook 2017. International Energy Agency; 2017.

[2] Boersma S, Doekemeijer BM, Gebraad PMO, *et al.* A tutorial on control-oriented modeling and control of wind farms. Proceedings of the American Control Conference (ACC). 2017; pp. 1–18.

[3] Knudsen T, Bak T, Svenstrup M. Survey of wind farm control – power and fatigue optimization. Wind Energy. 2015;18(8):1333–1351.

[4] Højstrup J. Spectral coherence in wind turbine wakes. Journal of Wind Engineering and Industrial Aerodynamics. 1999;80(1):137–146.

[5] Bastankhah M, Porté-Agel F. Experimental and theoretical study of wind turbine wakes in yawed conditions. Journal of Fluid Mechanics. 2016;806: 506–541.

[6] Barthelmie RJ, Pryor SC, Frandsen ST, *et al.* Quantifying the impact of wind turbine wakes on power output at offshore wind farms. Journal of Atmospheric and Oceanic Technology. 2010;27(8):1302–1317.

[7] Annoni J, Gebraad PMO, Scholbrock AK, *et al.* Analysis of axial-induction-based wind plant control using an engineering and a high-order wind plant model. Wind Energy. 2016;19(6):1135–1150.

[8] Campagnolo F, Petrović V, Bottasso CL, *et al.* Wind tunnel testing of wake control strategies. Proceedings of the American Control Conference (ACC). 2016; pp. 513–518.

[9] Munters W, Meyers J. Dynamic strategies for yaw and induction control of wind farms based on large-eddy simulation and optimization. Energies. 2018;11(177):1–32.

[10] Munters W, Meyers J. An optimal control framework for dynamic induction control of wind farms and their interaction with the atmospheric boundary layer. Philosophical Transactions of the Royal Society of London A: Mathematical, Physical and Engineering Sciences. 2017;375(2091):1–19.

[11] Frederik J, Kröger L, Gülker G, van Wingerden JW. Data-driven repetitive control: wind tunnel experiments under turbulent conditions. Control Engineering Practice. 2018;80:105–115.

[12] Gebraad PMO, Teeuwisse FW, van Wingerden JW, *et al.* Wind plant power optimization through yaw control using a parametric model for wake effects – a CFD simulation study. Wind Energy. 2016;19(1):95–114.

[13] Fleming PA, Annoni J, Shah JJ, *et al.* Field test of wake steering at an offshore wind farm. Wind Energy Science. 2017;2(1):229–239.

[14] van Dijk MT, van Wingerden JW, Ashuri T, *et al.* Wind farm multi-objective wake redirection for optimizing power production and loads. Energy. 2017;121:561–569.

[15] Aho J, Buckspan A, Laks J, *et al.* A tutorial of wind turbine control for supporting grid frequency through active power control. Proceedings of the American Control Conference (ACC). 2012; pp. 3120–3131.

[16] Van Wingerden JW, Pao LY, Aho J, *et al.* Active power control of waked wind farms. IFAC World Congress. 2017; pp. 4570–4577.

[17] Fleming PA, Aho J, Gebraad PMO, *et al.* Computational fluid dynamics simulation study of active power control in wind plants. Proceedings of the American Control Conference (ACC). 2016; pp. 1413–1420.

[18] Shapiro CR, Bauweraerts P, Meyers J, *et al.* Model-based receding horizon control of wind farms for secondary frequency regulation. Wind Energy. 2017;20(7):1261–1275.

[19] Pao LY, Johnson KE. Control of wind turbines. IEEE Control Systems. 2011;31(2):44–62.

[20] Jensen NO. A Note on Wind Generator Interaction; 1983. RISØ-M-2411.

[21] Ainslie JF. Calculating the flowfield in the wake of wind turbines. Journal of Wind Engineering and Industrial Aerodynamics. 1988;27(1):213–224.

[22] Bay CJ, Annoni J, Taylor T, *et al.* Active power control for wind farms used distributed model predictive control and nearest neighbor communication. Proceedings of the American Control Conference (ACC). 2018.

[23] Bottasso CL, Schreiber J. Online model updating by a wake observer for wind farm control. Proceedings of the American Control Conference (ACC). 2018.

[24] Doekemeijer BM, Boersma S, Pao LY, Knudsen T, van Wingerden JW. Online model calibration for a simplified LES model in pursuit of real-time closed-loop wind farm control. Wind Energy Science. 2018;3(2):749–765.

[25] Doekemeijer BM, Boersma S, Pao LY, *et al.* Ensemble Kalman filtering for wind field estimation in wind farms. Proceedings of the American Control Conference (ACC). 2017; pp. 19–24.

[26] Marden JR, Ruben SD, Pao LY. A model-free approach to wind farm control using game theoretic methods. IEEE Transactions on Control Systems Technology. 2013;21(4):1207–1214.

[27] Ciri U, Rotea MA, Santoni C, Leonardi S. Large-eddy simulations with extremum-seeking control for individual wind turbine power optimization. Wind Energy. 2017;20(9):1617–1634.

[28] Rotea MA. Logarithmic power feedback for extremum seeking control of wind turbines. IFAC-PapersOnLine. 2017;50(1):4504–4509. 20th IFAC World Congress.

[29] Ciri U, Rotea MA, Leonardi S. Model-free control of wind farms: a comparative study between individual and coordinated extremum seeking. Renewable Energy. 2017;113:1033–1045.

[30] Gebraad PMO, van Wingerden JW. Maximum power-point tracking control for wind farms. Wind Energy. 2015;18(3):429–447.

[31] Campagnolo F, Petrović V, Schreiber J, *et al.* Wind tunnel testing of a closed-loop wake deflection controller for wind farm power maximization. Journal of Physics: Conference Series. 2016;753(3):032006.

[32] Bianchi FD, de Battista H, Mantz RJ. Wind Turbine Control Systems: Principles, Modeling and Gain Scheduling Design. 1st ed. London: Springer; 2007.

[33] Jiménez A, Crespo A, Migoya E. Application of a LES technique to characterize the wake deflection of a wind turbine in yaw. Wind Energy. 2010;13(6): 559–572.

[34] Bastankhah M, Porté-Agel F. A new analytical model for wind-turbine wakes. Renewable Energy. 2014;70:116–123.

[35] Crespo A, Hernández J. Turbulence characteristics in wind-turbine wakes. Journal of Wind Engineering and Industrial Aerodynamics. 1996;61(1): 71–85.

[36] Niayifar A, Porté-Agel F. Analytical modeling of wind farms: a new approach for power prediction. Energies. 2016;9(741):1–13.

[37] Gebraad PMO, Fleming PA, van Wingerden JW. Wind turbine wake estimation and control using FLORIDyn, a control-oriented dynamic wind plant model. Proceedings of the American Control Conference (ACC). 2015; pp. 1702–1708.

[38] Annoni J, Seiler P. A method to construct reduced-order parameter-varying models. International Journal of Robust and Nonlinear Control. 2016; 27(4):582–597.

[39] Soleimanzadeh M, Wisniewski R, Brand A. State-space representation of the wind flow model in wind farms. Wind Energy. 2013;17(4):627–639.

[40] Boersma S, Doekemeijer B, Vali M, *et al.* A control-oriented dynamic wind farm model: WFSim. Wind Energy Science. 2018;3(1):75–95.

[41] Boersma S, Doekemeijer BM, Siniscalchi-Minna S, van Wingerden JW. A constrained wind farm controller providing secondary frequency regulation: an LES study. Renewable Energy. 2018;134:639–652.

[42] Annoni J, Fleming P, Scholbrock A, *et al.* Analysis of control-oriented wake modeling tools using lidar field results. Wind Energy Science Discussions. 2018;2018:1–17.

[43] Meyers J, Meneveau C. Large eddy simulations of large wind-turbine arrays in the atmospheric boundary layer. Aerospace Sciences Meeting. 2010.

[44] Iungo GV, Santoni-Ortiz C, Abkar M, *et al.* Data-driven reduced order model for prediction of wind turbine wakes. Journal of Physics: Conference Series. 2015;625(1):012009.

[45] Mittelmeier N, Blodau T, Kühn M. Monitoring offshore wind farm power performance with SCADA data and an advanced wake model. Wind Energy Science. 2017;2(1):175–187.

[46] Simley E, Pao LY. Evaluation of a wind speed estimator for effective hub-height and shear components. Wind Energy. 2016;19(1):167–184.

[47] Wan EA, Van Der Merwe R. The unscented Kalman filter for nonlinear estimation. Proceedings of the IEEE 2000 Adaptive Systems for Signal Processing, Communications, and Control Symposium. 2000; pp. 153–158.

[48] Houtekamer PL, Mitchell HL. Ensemble Kalman filtering. Quarterly Journal of the Royal Meteorological Society. 2005;131(613):3269–3289.

[49] Evensen G. The ensemble Kalman filter: theoretical formulation and practical implementation. Ocean Dynamics. 2003;53(4):343–367.

[50] Doekemeijer BM, Van Wingerden JW, Boersma S, *et al.* Enhanced Kalman filtering for a 2D CFD NS wind farm flow model. Journal of Physics: Conference Series. 2016;753(5):052015.

[51] Goit JP, Meyers J. Optimal control of energy extraction in wind-farm boundary layers. Journal of Fluid Mechanics. 2015;768:5–50.

[52] Siniscalchi-Minna S, Bianchi FD, Ocampo-Martinez C. Predictive control of wind farms based on lexicographic minimizers for power reserve maximization. Proceedings of the American Control Conference (ACC). 2018.

[53] Biegel B, Madjidian D, Spudić V, *et al.* Distributed low-complexity controller for wind power plant in derated operation. 2013 IEEE International Conference on Control Applications (CCA). 2013; pp. 146–151.

[54] Spudić V, Conte C, Baotić M, *et al.* Cooperative distributed model predictive control for wind farms. Optimal Control Applications and Methods. 2014;36(3):333–352.

[55] Annoni J, Bay C, Taylor T, *et al.* Efficient optimization of large wind farms for real-time control. Proceedings of the American Control Conference (ACC). 2018.

Chapter 8

Forecasting wind power production for grid operations

John W. Zack[1,2]

8.1 The role of wind-power forecasting

The non-dispatchable variability of wind power production presents a substantial challenge to electric system operators who are assigned the task of balancing the demand and generation at each moment at the lowest possible cost while maintaining ultra-high system reliability. The economic factor is the key component of this problem since one can always adequately manage variability and maintain reliability if cost is not considered.

The challenge becomes increasingly difficult as the percentage of demand served by variable generation assets increases. Although the focus in this discussion is on value and impact of forecasting wind-power variations, the increasing variability of other components of the grid system increases the importance of forecasting wind power generation. A number of tools or approaches are potentially available to system operators to assist in meeting this challenge. However, many require substantial long-term planning and significant implementation costs. These include a shift to more flexible (i.e., quicker response) generation assets, demand–response programs, implementation of one or more forms of energy storage, more accommodative market structures and the planning of the geographic diversity of variable generation assets. However, one of the most cost-effective and easily implemented tools to assist in the management of the non-dispatchable variability of wind power generation is the short-term forecasting of the production. This can provide system operators with the lead-time and "large-change event" visibility to make more economical decisions while maintaining the required high levels of reliability.

There are four key components to a forecasting solution that should each be optimized in order to provide maximum value to the end user: (1) high-quality and representative measurement data for input into the forecasting procedure, (2) skillful forecasting models, (3) effective communication of the critical forecast information to automated or manual decision-makers and (4) meaningful assessment of forecast

[1]MESO, Inc., Troy, NY, USA
[2]AWS Truepower, a UL Company, Albany, NY, USA

performance to provide users with confidence in using the forecast information for decision-making. This four-component approach can be concisely summarized into four key words: sense, model, communicate and assess. A compromise in any of these components can result in less than the maximum value being obtained from a forecast for the application. This chapter will address each of these four components of an optimal wind-power forecasting solution.

This chapter is organized into five sections. This introductory overview of the nature of the wind forecasting problem is the first section. The second section discusses the input data requirements and opportunities to improve forecast performance by acquiring additional input data. The third section presents an overview of the methods that are widely used in producing wind power forecasts and how they are typically applied. The fourth section provides a summary of the various types of forecast products that are frequently employed. The fifth section discusses the issues associated with the evaluation of forecast performance and the metrics that are commonly used for this purpose.

8.2 Sense: gathering and ingestion of predictive information

There are two fundamental types of data used by a forecast system: (1) quasi-real-time data used as input in each prediction cycle, and (2) historical data used to train the statistical components of the forecast system. The historical data is also used to evaluate the performance of the forecasts generated by the integrated forecast system as well as by its individual components, which can also serve as a basis for tuning the system.

Ultimately, all of the predictive information for each forecast cycle is provided to a forecast system via the quasi-real-time input data. From the broadest perspective, there are an enormous number of data types and data elements that are ingested by a state-of-the-art forecast system. The input data includes the vast array of global atmospheric sensor data that is employed by the various types of atmospheric prediction systems as well as the meteorological and generation data from the forecast target facility and nearby locations. However, different segments of the input data pool are used by each component of the forecast system. For example, the physics-based atmospheric models utilize a multitude of data types from regional and global domains to specify their three-dimensional initial states. On the other hand, statistical time-series models typically only use recent data from the forecast target site and perhaps nearby off-site locations.

Forecast providers and users typically only have control (if they have any at all) over the measurement data gathered at the forecast target site (usually the generation facility) and perhaps in the vicinity of the target site. Beyond that, they generally must work with the vast (but not optimally sited for their application) array of data that is available from a multitude of public and private sources. This section provides an overview of the potential for users and providers to deploy a network of sensors that are targeted to improve forecasts for their target variable and look-ahead time range.

8.2.1 Area of influence

An important concept is the space-time envelope of data influence that determines which data locations and variables have impact on a forecast for a particular or a typical scenario. This has important implications for how to gather additional data to improve forecast accuracy. This concept is illustrated with data for the Columbia River Valley in the Pacific Northwest region of the United States in Zack *et al.* [1] in Figure 8.1. The images depict the fraction of the variance of numerical weather prediction (NWP) forecasts of the 80-m wind speed (i.e., the predictand) that is explained (the R^2 parameter) by variations in a "measurement" (i.e., a predictor) variable (also the 80-m wind speed in this example) at regional locations around the forecast target site (the white box) for a 1-h (left panel) and 3-h (right panel) forecast for a statistically significant sample of cases. The purple and blue colors represent very low values of R^2, which indicate that there is essentially no relationship between changes in the predictor variable (simulated sensor measurement) and changes in the forecast value at the target location. That is, the forecast for the specified look-ahead time scale is not sensitive to measurements of the specified variable at these locations. In contrast, the yellow and red shading depict relatively high values of R^2, which indicate a stronger statistical relationship between variations in the simulated measurement data and the resulting forecast. Thus, the area covered by the green to red colors can be considered to be the forecast-sensitivity region. Measurements in this region have a significant impact on the forecast, but measurements in the purple and blue regions have negligible impact.

A comparison of the left and right panels of Figure 8.1 indicates that the area of forecast sensitivity increases by a large amount from a 1-h to a 3-h forecast. The implication is that measurements must be made over a much larger area to improve a 3-h forecast than to improve a 1-h forecast. Of course, this expansion in the area of forecast sensitivity continues beyond 3 h and becomes very large as the look-ahead period expands to one day and multiple days. Ultimately, the area of forecast sensitivity for a long-range forecast (e.g., 10 days or more) is the entire global atmosphere.

There are several points to note about the concept of a spatial domain of forecast sensitivity. The first point is that the shapes of the areas of sensitivity are generally quite complex and not symmetric around the forecast target location. They are typically skewed to the prevailing regional upstream direction, which is typically to the west for the location depicted in Figure 8.1. The patterns are also strongly influenced by the terrain as shown by channeling of the high sensitivity along the Columbia River Valley and the appearance of high sensitivity along the Cascade Mountains. Second, the charts in Figure 8.1 represent an average (or climatology) of the sensitivity over a sample of many cases. The area and magnitude of the sensitivity is flow dependent, and therefore, varies within a set of forecast cases. In some situations (weak and slow-moving atmospheric features), the area of sensitivity expands slowly with increasing forecast look-ahead time, while in others (rapidly moving atmospheric features), it grows very rapidly. Third, the charts in Figure 8.1 only depict the sensitivity of the forecast to a single predictor variable, which in this case is the same type of

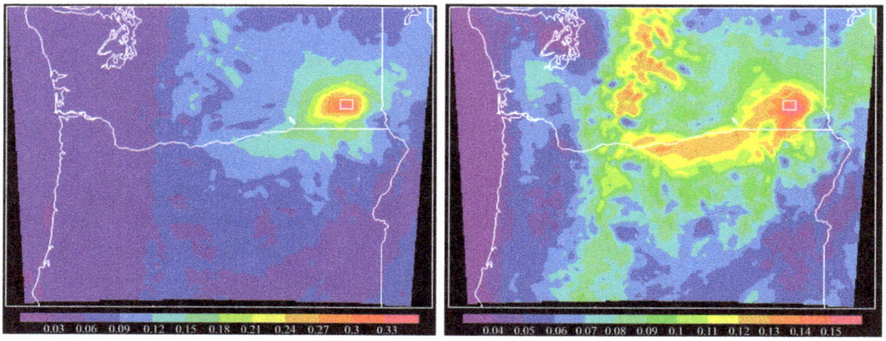

Figure 8.1 An example from [1] of the increase in the area of forecast sensitivity as the look-ahead time for an 80-m wind-speed forecast increases for a forecast site in Southeastern Washington based on an ensemble of simulations from an NWP model. The white box represents the forecast target area. The color shading represents the R^2 values of the relationship between the 80-m wind-speed values at each location (the predictors) at the forecast issue time and the forecasted value in the target box (the predictand) for 1 h (left) and 3 h (right) later

variable as the predictand. Of course, the forecast of 80-m wind speed over the white box is, in general, sensitive to variations in a wide range of atmospheric variables (e.g., pressure and temperature) and at locations in the vertical as well as the horizontal. Thus, there are many possible measurements to which the forecast can be sensitive. A fourth point is that the variables in their respective sensitivity regions are generally correlated to each other to some degree. Thus, it is usually not beneficial to measure each variable at each sensitive location.

8.2.2 Observation targeting

The concept of forecast sensitivity can be used to provide guidance for the optimum placement of sensors to improve the performance of forecasts. The basic objective is to identify the combination of variables and locations that provide the greatest positive impact on forecast performance at a specified cost level. Intuitively, one might think that this would be a very challenging task and it is.

A key challenge, as previously noted, is that the area of forecast sensitivity typically expands rapidly with look-ahead time. This means that measurements must be made over a bigger region (i.e., more measurement locations) to get the same impact on forecast performance. This is visually depicted in the comparison of the left and right panels of Figure 8.1. Thus, the most value from a specified number of sensors (and therefore at a specified cost level) is obtained for shorter look-ahead periods.

A second significant challenge is the flow dependence of the sensitivity. The patterns in Figure 8.1 were constructed by combining the sensitivity data from a modestly large sample of cases. However, the forecast-sensitivity patterns vary substantially among forecast scenarios depending on factors such as the large-scale wind direction and speed. Therefore, the optimal location of sensing devices also varies substantially. This leads to the not unexpected but rather useless conclusion that it is necessary to make measurements of all of the primary weather variables at all surrounding locations if one is to have optimal measurement data for all possible forecast scenarios. However, many scenarios occur very infrequently or do not have much value for the application (e.g., low-wind-speed cases for wind-power-production forecasting). Thus, a more practical objective is to try to identify the forecast-sensitivity patterns and the implied optimal measurement locations for the most frequently occurring scenarios that have significant value for the application.

A number of approaches have been formulated and employed to address this objective. The simplest approach is a subjective method in which one looks at the prevailing upstream direction and the average speed and estimates the direction and distance (for a specific look-ahead time) at which a measurement should be made to provide predictive information about features propagating toward the target site. A second approach is to infer the spatial time-lagged correlation pattern between the target site and surrounding locations from the available measurements. This can be useful if there are a sufficient number of sensor sites available to define spatial patterns from which the value of potential additional sites can be inferred. Unfortunately, such data is often not available and that is typically the motivation for considering the deployment of a targeted sensor network.

A third approach does not rely on existing measurement data but instead employs an ensemble of physics-based NWP simulations to diagnose the spatial patterns of forecast sensitivity. This method is called ensemble sensitivity analysis (ESA). It has been used by [1–3] to analyze the wind-forecast-sensitivity patterns in the Tehachapi Pass of California and the Columbia River Basin of the Pacific Northwest. In the ESA approach, an ensemble of NWP simulations is generated for each case in a sample of cases over a target region. The size of the ensemble must be large enough to provide a statistically significant sample of forecasts (e.g., >30) for each case. The forecast sensitivity is diagnosed for each target site for each case. The sensitivity is the strength of the relationship between changes in the forecast variable at the target site and time-lagged (i.e., prior) changes of variables at other locations (i.e., each NWP model grid point not at the target site). Parameters that measure the sensitivity (such as the slope of a regression line or the R^2 value) at each point can be mapped (as in Figure 8.1) for each case, and ultimately these can be compiled into composites for scenarios of interest (e.g., west wind cases, high wind cases, etc.). This type of analysis provides guidance for the deployment of sensors for targeted forecast scenarios.

The results of the Tehachapi Pass analysis by Zack *et al.* [3] were later used to guide the deployment of a targeted sensor network in this region. The impact of that sensor network on short-term forecast performance is presented in subsequent sections of this chapter.

8.3 Model: translating predictive information into a forecast

There are generally three broad classes of prediction techniques that are employed in a state-of-the-art wind forecast system: (1) physics-based meteorological models, (2) statistical methods applied to meteorological variables and (3) power-output models, which are statistical or physics-based relationships between meteorological variables and electrical power generation. The following sections discuss each of these classes of prediction techniques.

8.3.1 Physics-based techniques

Physics-based forecasting techniques are based on broadly applicable physical principles. A key attribute is that they do not require a training sample to generate the prediction equations. Therefore, skillful predictions can be made even in situations in which there is no historical data available from the forecast target entity or nearby locations. This also means that their range of predictions is not limited to what has been observed in a historical sample. Thus, they are theoretically capable of predicting events with little or no historical precedent.

There are several types of physics-based models that have been used in the forecasting of wind power production, but the previously noted NWP models are the dominant type. The NWP approach [4] is based upon the application of the fundamental principles of conservation of mass, momentum and energy and the equation of state for moist air to the atmosphere. These principles are formulated as a set of differential equations, which are then "solved" by numerical methods such as finite-difference approximations or spectral techniques. The most basic set of NWP model equations accounts for basic processes in an atmosphere during which the energy content of a parcel of air does not change in time (adiabatic) on the scale of the NWP grid. The form of these equations is well known from basic physics, and there is very little uncertainty in their formulation.

Additional terms are then added to the NWP model equations to account for processes that change the energy content of an air parcel. These are processes such as long- and short-wave radiative transfer, the physics of water phase changes, motions (i.e., turbulence and moist convection) that occur on scales smaller than the NWP grid and the fluxes of heat, moisture and momentum from the underlying surface of the earth. The mathematical formulation (often referred to as "parametrizations") of these processes is heavily based on physical principles, but they also have a significant empirical component. The empirical component is mostly related to processes (such as the formation of a raindrop or the impact of a turbulent eddy) that occur on too small of a scale to explicitly model on the NWP grid. Therefore, the bulk effects of these processes are modeled through statistical relationships with grid-scale variables. While an effort is made to develop sets of relationships that are universally applicable, they often have some location or atmospheric regime dependence. As a result, these relationships often introduce biases (systematic errors) into the NWP forecasts.

NWP prediction systems are actually composed of two major components: (1) an NWP model and (2) a data assimilation system. The NWP model contains the physics-based prediction equations and operates on a specified initial atmospheric state for a model's spatial domain to generate time-dependent predictions.

The function of the data assimilation system is to create an accurate representation of the initial state (i.e., the starting point for the NWP forecast process) from a large and diverse set of atmospheric sensor data as well as previous NWP forecasts. The primary issues faced by an NWP data assimilation system are: (1) measured data is not available for every model variable at every model grid point, (2) the atmospheric state at any moment in time must be inferred from data from a very diverse set of atmospheric sensors that measure different subsets of atmospheric variables and have a broad range of spatial and temporal resolution and coverage and large variations in characteristic measurement errors, (3) atmospheric variables are tightly coupled via the physics-based equations; therefore, a modification made to the value of one variable must be accompanied by appropriate modifications to other variables in order to maintain a physically realistic state. The most typical approach used by data assimilation systems is to start with the three-dimensional state provided by a forecast from the previous cycle of the same model (referred to as a warm start). This atmospheric state is sometimes referred to as the "first guess" or "background state." The available sensor data is then used to update this first guess. These updates to the first guess essentially represent corrections of the model forecast errors.

The key issue is how to spread the influence of a measurement from the point of measurement to nearby model grid points. The spreading of the influence typically uses some estimate of the spatial pattern of the model's error covariance. These provide the information about how much connection there is between the model error diagnosed at the measurement point and the error at nearby grid points where there is no measurement data (hence, errors cannot be explicitly computed). The spatial error covariance is typically flow dependent. It is difficult to estimate these although techniques based on ensembles of NWP simulations have been developed to estimate the flow-dependent variations in the spatial error covariance. However, in practice, a climatological spatial error covariance dataset is typically compiled from a historical set of NWP forecasts, and this is used in the process of spreading the influence of measurement data when updating the first-guess state. The improvement in the specification of the flow-dependent variations of the spatial error covariance is an area in which NWP forecasts are likely to be improved in the future.

The forecast errors and the associated uncertainty of NWP model predictions are associated with three primary factors: (1) resolution of the grid and the numerical methods used to solve the equations, (2) uncertainty in statistical relationships and approximations of physics-based principles in the formulation of the NWP model physics and (3) uncertainty in the specification of the initial state due to spatially sparse or unrepresentative data, sensor error and unrepresentative spatial error covariance models used in the initialization process. These issues produce a combination of systematic and random errors. As will be discussed later, the presence of systematic errors provides an opportunity to improve NWP forecasts via statistical techniques that can diagnose and partially correct these errors.

In addition to the correction of systematic errors via statistical post-processing, the uncertainty associated with NWP predictions is also frequently addressed through the use of the ensemble concept. The basic concept of a forecast ensemble is that a set of forecasts is created for a given forecast period by varying the input data or mathematical formulation of the model physics within their respective ranges of uncertainty. This yields a set of forecasts. A given ensemble may contain members of only one type or a combination of both types. It seems reasonable to expect that a combination of both types would more accurately represent the forecast uncertainty. However, in practice, some forecast scenarios are more sensitive to the initial state, and others are more sensitive to the model physics.

If all the significant sources of input uncertainty are appropriately represented in the formulation of an ensemble, then the forecast ensemble should accurately represent the uncertainty in the forecast. However, there are so many individual sources of input uncertainty in an NWP system that it is virtually impossible to have a comprehensive representation of all of the significant sources. Hence, the variations of the forecast values from a typical NWP ensemble will underestimate the forecast uncertainty. This is typically addressed by statistically calibrating the forecast distribution produced by the ensemble through the use of historical data from a forecast ensemble and the associated outcomes, so that it more accurately represents the true probability distribution of the forecast.

One of the most significant attributes of NWP models is that they supply physically consistent predictions of virtually the entire set of meteorological variables. Thus, a single NWP run can be used for virtually all types of atmospheric forecast applications. An example of a high-resolution NWP forecast of 80-m wind speed over the island of Hawaii is shown in Figure 8.2.

NWP prediction systems can be applied to a forecast problem on many different space and time scales with a wide variety of model configurations. United States National Weather Service (USNWS) systems are typically configured to operate in one of three modes: (1) global, (2) regional or (3) rapid update. Global NWP systems have a forecast domain that encompasses the entire world, and they are typically used to generate forecast with look-ahead periods of 1–2 weeks. However, the global NWP forecasts are now also periodically produced for periods of 30 days or longer to provide guidance for monthly or seasonal forecasts. Global NWP models are typically run on a 6-h cycle although some NWP centers use a 12-h cycle. Current examples of global NWP models that are widely used for wind-forecast applications are the Global Forecast System (GFS) operated by the USNWS, the Global Deterministic Prediction System run by Environment Canada (EC), the global system run by the European Centre for Medium-Range Weather Forecasts (ECMWF) and the global prediction system of the United Kingdom's Meteorological Office.

Regional NWP systems operate on a limited-area domain such as a continent or part of a continent. Current examples of regional NWP models are the North American Mesoscale (NAM) model operated by the USNWS and EC's Regional Deterministic Prediction System.

Rapid-update NWP models are similar to regional NWP systems, but they are operated on shorter update cycles. These are typically run on 1- or 2-h cycles and frequently cover a smaller domain with a higher resolution grid than the standard regional models. Their primary objective is to frequently assimilate the latest sensor

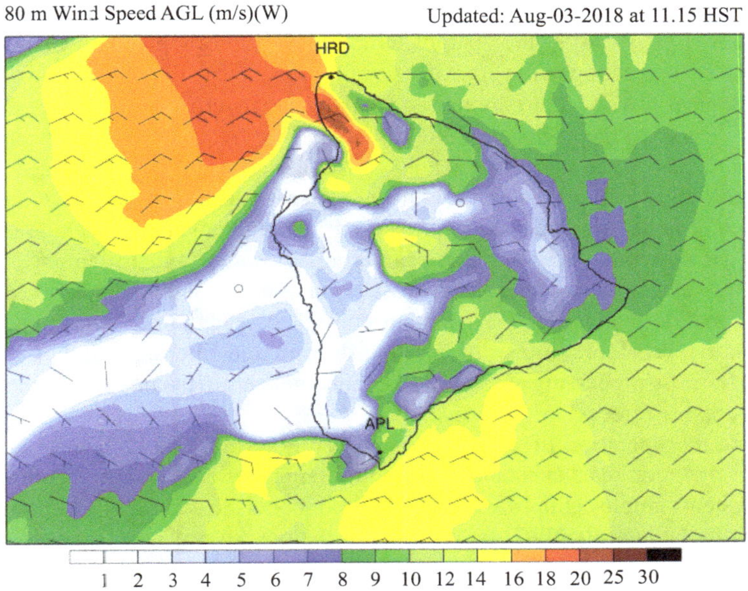

Figure 8.2 An example of a high-resolution (horizontal grid spacing of 3 km) NWP model forecast of 80-m wind speed over the island of Hawaii (also known as the Big Island). Color shading depicts the wind speed in m/s

data and make a very short-term forecast. The forecast duration is typically less than 24 h. Current examples of rapid-update models are the Rapid Refresh (RAP) and High-Resolution Rapid Refresh (HRRR) models run by the USNWS. The RAP is run hourly on a 13 km grid while the HRRR employs a 3 km grid for its hourly update cycle. These forecasts are usually available within an hour after the initialization time.

The NWP data that are most widely used for wind power forecasting applications are from NWP systems that are run at major government-run forecast centers (e.g., USNWS and ECMWF). However, it should be noted that none of these systems are specifically optimized for wind (or solar) power forecasting applications. They are primarily designed for general weather forecasting and extreme weather warning applications with an emphasis on temperature and precipitation prediction. In order to address this limitation, some wind power forecast systems also employ customized NWP systems that are run "in-house" by commercial forecast providers and in a few cases by forecast users themselves.

A large fraction of the customized systems is based on the weather-community-supported open-source NWP system known as the Weather Research and Forecasting (WRF) model [5]. However, other NWP modeling systems are also used for this purpose. There are several potential benefits of the using a customized NWP system to supplement the NWP data from government-run models: (1) higher horizontal and/or vertical resolution (i.e., smaller grid cells) can be used to better represent small scale features in the vicinity of the forecast target site, (2) the configuration of the model physics (i.e. the "parameterizations") can be optimized for wind (or solar) power forecasting in the area of interest, (3) local data (e.g., from wind facility meteorological towers) that may not be available to government-center models can be assimilated into the in-house NWP system, (4) the frequency of the update cycle (e.g. 1 h, 2 h, 6 h, etc.) and the length of the forecast can be customized for the application and (5) large datasets with the same model configuration (since it is controlled in-house) can be produced for use in training advanced machine learning models. However, the value of these benefits varies substantially among forecast applications and scenarios, and it is necessary to have sufficient computational power and information technology (IT) infrastructure to gather the necessary input data, execute the NWP forecast over the desired domain and forecast time horizon, and perform the post-processing of the NWP output within a sufficiently short time period to yield useable forecast information.

Since the operation of a full real-time NWP system is a somewhat complex endeavor that requires a substantial amount of computational resources, some wind power systems employ simplified physics-based models to add some additional local detail to the large-scale NWP forecasts from the government forecast centers. The most common approach for wind power forecasting systems is to employ a mass-conservation wind flow model or a computational fluid dynamics (CFD) model along with a very high-resolution dataset of terrain elevation and possibly other surface parameters (e.g., roughness height). These are much simpler and require less computational resources than a full NWP system, and therefore they can typically be run with much higher spatial resolution (100s or even 10s of meters). However, their simplified physics impose limitations on the types of atmospheric processes that can be simulated (i.e., forecasted). A mass-conservation-type model that has been widely used in the wind power community is WAsP [6]. An example of a CFD model that has been used for wind power forecasting applications is msMicro [7].

8.3.2 Statistical approaches

Statistical prediction techniques are applied to a very wide range of forecasting problems, and the forecasting of atmospheric variables is one of them. Many textbooks have been published, which supply an overview of statistical prediction techniques (e.g., [8]), the details of specific methods (e.g., [9]) and the application of statistical methods to atmospheric prediction problems (e.g., [10]). A number of methods have been employed for short-term wind power prediction. This section provides a high-level overview of the most widely used methods in wind power prediction, and how they are typically applied.

8.3.2.1 Methods

This section provides a high-level survey of the methods that have been frequently employed in the statistical components of wind power forecast systems. There are two fundamental types of statistical tools used in prediction applications: (1) data preprocessing methods and (2) prediction-model generation (training) and application techniques.

In prediction applications, data preprocessing tools serve several functions. These include the identification and elimination of erroneous or inconsistent data (i.e., quality control), handling of missing data elements, normalization of the data and the possible grouping of data into categories or classes to facilitate the construction of better prediction models. The grouping methods include approaches such as principal components analysis (PCA) and clustering techniques. Some of the most advanced prediction-model generation techniques implicitly incorporate data-grouping concepts into the prediction-model training process; therefore, the value of performing explicit grouping as a preprocessing step is somewhat dependent on the method subsequently used to train a prediction model.

An enormous number of prediction-model generation techniques have been formulated, tested and widely used in operational forecasting applications. Many of these have been and are being used in wind-generation forecasting activities. The performance results from a wide range of wind-forecasting applications indicate that no single method consistently exhibits the best performance although a small set of advanced methods has produced models that are consistently among the best performers.

One of the oldest and most basic statistical prediction methods is linear regression. Typical applications employ multiple predictor variables. The use of linear regression with two or more predictor variables is widely referred to as multiple linear regression (MLR). MLR is extensively documented in most textbooks on basic statistical methods (e.g., [10]). Many variations of MLR have been formulated. Although MLR generally does not perform as well as many of the recently formulated advanced prediction methods when large high-quality training datasets are available, it is still a valuable tool in many situations when only limited amounts of data are available. However, the coupling of MLR with the application of sophisticated data preprocessing tools can sometimes achieve prediction performance levels that are similar to the most advanced prediction methods.

Another set of basic empirical prediction methods that have been employed for many years is based on the analog concept. The analog approach has implicitly been widely used by human forecasters throughout history by noting that a particular situation is "similar" to a set of previous scenarios; therefore, the outcome of the previous scenarios can be used as the basis for a forecast for the current situation. Prior to the economic availability of high levels of computing power, this approach was generally subjectively employed with graphical displays to facilitate subjective pattern matching. However, the recent availability of high-performance computing has enabled this prediction concept to be used in a more rigorous and quantitative manner. For example, it has been applied to wind power forecasting under the name of "analog ensemble" (AE) by Delle Monache *et al.* [11]. This approach performs a search through the available historical records to find the cases that most closely match the current forecast situation based on a set of matching variables. It is typically (but not exclusively) applied to the output of NWP models and thus represents the outcomes of past NWP forecasts that are similar to the current NWP forecast. The historical cases that most closely match the current situation based on a specified threshold value for a closeness parameter are identified, and the outcomes of those cases are used to construct an ensemble of outcomes (i.e., the "AE"). A deterministic forecast can be created by constructing a composite of this ensemble, and a probabilistic forecast can be constructed by generating a probability density function from the distribution of the ensemble members. The fundamental strength of the AE approach is that it essentially creates a custom grouping of historical cases based on each forecast situation. One of the key issues with this approach is that it does not have a basis for the prediction of cases for which there is no satisfactory match in the historical sample. Other issues are the specification of the matching variables and the definition of closeness.

A relatively recent advancement in the regression concept that has been used for wind-power forecasting applications is support vector regression (SVR). The SVR method [12] is rooted in the support-vector-machines (SVMs) concept [13], which originated as a tool for classification problems. SVMs are supervised learning models with associated learning algorithms that analyze data and recognize patterns. An SVM model is a representation of the examples as points in space, mapped so that the examples of the separate categories are divided by a clear gap that is as wide as possible. New examples are then mapped into that same space and predicted to belong to a category based on which side of the gap they fall on.

SVR was proposed by Drucker *et al.* [14]. SVR is the SVM tool that is most commonly used for wind-power forecasting applications. The key attribute of SVR is that its model training and optimization process utilizes a prediction error margin (a band) to reduce the importance of data points with small prediction errors in the training process. In other words, it places more weight on the minimization of the errors that are beyond the margin. An example of the use of SVR in a time-series-based wind-power-prediction scheme is provided in Kramer and Gieseke [15].

One of the advanced prediction methods that has a long history of use in prediction applications is artificial neural networks (ANNs) [9,16]. ANNs are a family of statistical learning models inspired by biological neural networks (e.g., the central nervous systems of animals). ANNs are generally presented as systems of interconnected "neurons," which send messages to each other. Mathematically, neurons accept inputs and have an activation function that determines their output. The connections have numeric weights that can be tuned based on experience, making ANNs adaptive to inputs and capable of learning. A common formulation of ANNs, which is employed in wind-power forecasting applications, is the multilayer perceptron (MLP) [9,17]. The MLP is a class of feed-forward ANNs that consist of at least three layers of nodes. ANNs can be very powerful tools for building complex prediction models. However, their exceptional ability to model complex nonlinear functions also makes them prone to overfitting relationships in training datasets and thus producing prediction models that do not generalize well. This is especially an issue in noisy and small datasets (relative to the number of predictor variables and neurons in the ANN). An example of the use of ANNs in wind power prediction is presented in Castellani *et al.* [18].

In recent years, a considerable amount of research and development effort has been focused on the application of the decision-tree concept [19] for the training of prediction models. Tree models in which the target variable can take a discrete set of values are called classification trees. In these tree structures, leaves represent class labels, and branches represent conjunctions of features (predictors) that lead to those class labels. Decision trees in which the target variable can take continuous values are called regression trees. The term classification and regression tree is an umbrella term first used by Breiman *et al.* [19], to refer to both of the previous procedures. Trees used for regression and trees used for classification have some similarities as well as some differences.

Many of the best-performing tree-based techniques construct more than one decision tree and are often referred to as ensemble methods. One class of tree-based ensemble techniques is called "boosted trees," and these methods iteratively build an ensemble by training each new instance to emphasize the training instances previously mis-modeled. A second class of tree-based ensemble techniques is referred to as bootstrap aggregated (or bagged) trees. This approach builds multiple decision trees by repeatedly resampling (with replacement) the training data and then constructing a composite of the results of the ensemble of trees to produce a prediction.

One method based on the decision-tree-ensemble approach that has seen considerable use in wind power forecasting is called "random forests" (RFs). RF falls into the previously noted class of methods called bootstrap aggregated trees. It operates by constructing a multitude of decision trees at training time, and for the regression application, the prediction is the mean of the predictions of the individual trees. The algorithm for inducing an RF was developed by Leo Breiman and Adele Cutler [20], and they created the name "RFs." The subsampled ensemble approach used in RF corrects for the tendency of decision trees to overfit their training set. An example of the application of RF to wind power forecasting is provided in Lahouar and Ben Hadj Slama [21].

Another concept that been incorporated into recently developed advanced statistical prediction techniques is "gradient boosting" [22,23]. In general, the boosting concept combines a set of weak prediction (learning) models into a single strong prediction model in an iterative fashion. The concept can be applied in a number of ways with a wide range of prediction models. The term "gradient boosting" refers to the use of gradient descent algorithms to iteratively develop the set of weak prediction models.

The use of decision-tree constructs as the basis for the prediction models within the gradient boosting approach has yielded some of the best-performing prediction-model generation algorithms currently available. One of the most popular implementations is known as gradient boosted machine (GBM). Although GBM is based on the same prediction model structure (decision trees) as RF, the two approaches have significant differences when considering which to employ for a specific application. In the RF approach, an ensemble of decision trees is constructed from subsamples of the training dataset. The final prediction is the mean of the result of each tree. However, GBM employs an iterative approach that produces the final results in a sequentially additive manner (rather than the conceptually parallel approach of random subsamples in RF). The GBM algorithm tries to fit the residual (i.e., the error) of the previous set of trees. The result is based on the combination of a sequential set of trees with each tree in the sequence attempting to model the residuals (errors) from the previous set of trees. GBM typically produces shorter trees.

The GBM approach has been enhanced by a number of developers. The most widely used enhanced version is a python-based implementation called extreme gradient boosting, which is commonly referred to as XGBoost [24]. In contrast to GBM, XGBoost uses regularized gradient boosting formulation to control overfitting, which gives it better performance. In addition, XGBoost is formulated to achieve much higher computational performance through parallel processing.

A number of recent machine-learning-oriented prediction competitions have been won by competitors using models generated by XGBoost [25]. It has also produced models that have consistently been among the best performers in a range of wind-power forecasting applications although models based on this method have not always been the best performer. As a result, XGBoost is considered by many renewable energy forecasters to be the preferred method for the statistical components of the state-of-the-art wind-power forecast systems when large high-quality datasets are available for training. Some examples of its performance are presented in the following sections.

8.3.2.2 Applications

Statistical techniques are applied to the short-term wind power prediction problem in several ways. There are four prominent types of applications: (1) time-series prediction using local area data, (2) refining the forecast output from NWP models (such as reducing the systematic errors for specific target variables or making forecasts of related but not explicitly predicted variables) via a procedure widely known as Model Output Statistics (MOS), (3) construction of a composite of an ensemble of forecasts produced by a set of individual methods and (4) power-output models that convert predictions of meteorological variables to forecasts of power output. These are described in the following sections.

Time-series models

The objective of time-series prediction techniques is to use the predictive information contained in the values, trends or patterns of the recent history of the forecast variable or related variables. This is typically the best-performing approach for wind forecasts with look-ahead times from a few minutes ahead to approximately 2–3 h ahead because it has a number of advantages over NWP-based approaches for this look-ahead time frame.

There are a wide variety of statistical time-series models that have been used for short-term wind power prediction. They are differentiated by (1) the amount and types (sources) of input data used to train the models and generate predictions, (2) the statistical methods used to construct the prediction model and (3) the specific configuration (e.g., values of the method's internal parameters) of the statistical methods or the algorithm used to switch among methods or method configurations (e.g., selecting different methods or predictors by scenario such as weather regime or time of day).

The results from a recent experiment [26,27] conducted in the Tehachapi Wind Resource Area (TWRA) of California illustrate the variations in the skill level of 0–3 h-ahead predictions of wind power production using only time-series data (i.e., without NWP inputs) associated with changes in input data and statistical methods. The fundamental objective of this experiment was to assess the impact on forecast performance of data from a targeted atmospheric sensor network deployed in the vicinity of the TWRA based on a forecast-sensitivity analysis. The forecast-sensitivity study was based on the use of the ESA method with an ensemble of high-resolution NWP forecasts over the TWRA [2,3]. The results of this analysis provided guidance for the selection of the sensor locations and the variables to be measured to improve the 0–3 h-ahead wind power forecasts in the TWRA. The actual site selection also had to consider real-world issues such as site suitability, site access, availability of power, cost to deploy at a site and other factors.

The location of the venue (the TWRA) for this experiment is depicted in Figure 8.3. The turbine symbols denote the location of six aggregates of wind generation facilities that were the target for the time-series forecast experiments. The regional generation capacity considered in this project was 2,319 MW (i.e., the sum of all six aggregates) although the total generation capacity for the TWRA at the time of the experiment was over 3,000 MW. However, some of the facilities in the region were excluded from the experimental dataset because of their limited availability of data or poor data quality. The locations of the eight sensors deployed in this project are also shown in Figure 8.3. They consisted of three microwave radiometers, two SODARs (600-m vertical range), one mini-SODAR (200-m vertical range), a radar wind profiler and a radio acoustic sounding system.

The TWRA experiments analyzed the impact of several factors on forecast performance including (1) the type (source) of predictor data, (2) the definition of the predictand variable and (3) the type of statistical method used to build the prediction model. All of the experiments were based on a 25-month dataset. This was the full deployment period of the project's targeted sensors. The forecast evaluation was

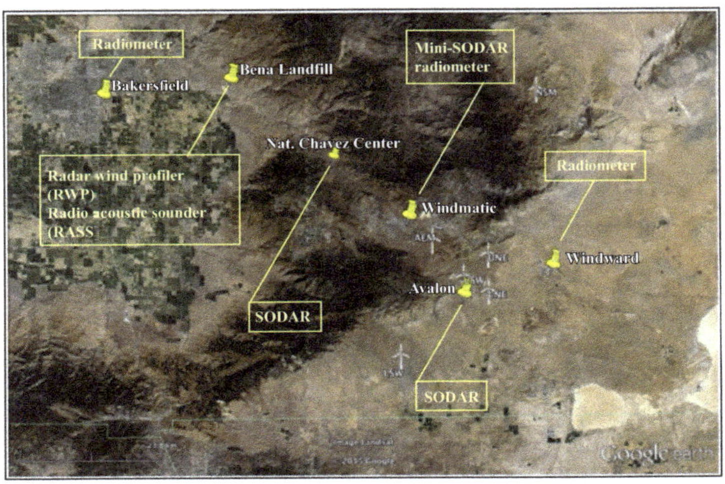

Figure 8.3 The locations and types of targeted atmospheric sensors (yellow markers) deployed in the TWRA short-term forecasting project and the centroid of six aggregates of wind generation facilities (wind-turbine icons) that served as the target of the experimental forecasts produced in the project

performed on a 12-month subset of that period that extended from October 2015 to September 2016. In order to maximize the size of the training sample used to build the statistical models, a rolling 24-month training procedure was employed. This procedure excluded the month for which forecasts were to be produced and used data from the other 24 months to train the statistical models. Thus, 12 different statistical models were trained for the 1-year forecast evaluation period.

The impact of different sources (types) of predictor data on the performance of 0–3 h-ahead time-series forecasts for the regional aggregate (composite of the six individual aggregates) 15-min average TWRA wind power production (capacity of 2,319 MW) is illustrated in Figures 8.4–8.6. All of the predictor source experiments employed the XGBoost method to train the prediction models. The chart in Figure 8.4 depicts the mean absolute error (MAE) of five sets of predictions that use different sets of predictors for look-ahead periods of 15, 30, 60, 90, 120 and 180 min. Figure 8.5 illustrates the reduction in MAE for each non-persistence prediction relative to a persistence forecast. Figure 8.6 illustrates the reduction in MAE for predictor sets that employ external data relative to the predictors from only the wind-facility data.

The five sets of predictors are cumulative. That is, each set in the sequence employs all of the previous subsets of predictors plus a new subset. The first predictor subset is labeled "persistence" and uses the measured 15-min average power production for the period ending at forecast time zero as the forecast for all periods in the forecast look-ahead time window (0–3 h). This is a forecast of no change in production.

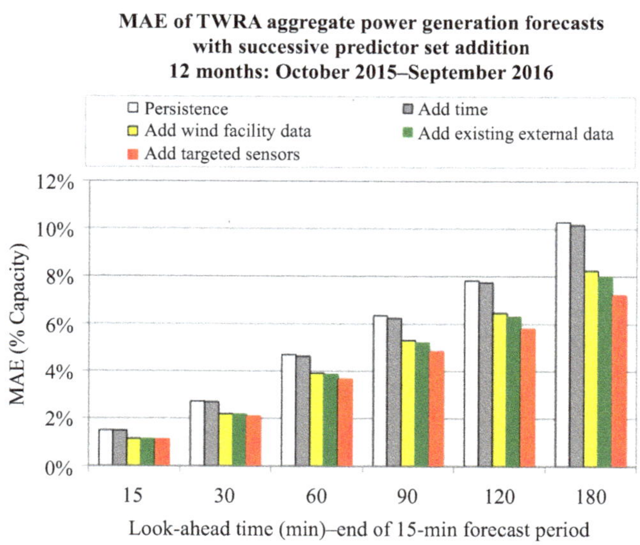

Figure 8.4 MAE (% of capacity) versus look-ahead time for 0–3 h forecasts of the 15-min average wind power production from the TWRA aggregate over the 1-year period from October 2015 to September 2016 for each of five source-dependent sets of predictors

The second subset is labeled "add time" and is based on a predictor pool of the time of the day and day of the year in addition to the persistence value. The addition of the date and time information to the predictor pool results in a slight reduction in MAE, especially for the longer look-ahead periods. These predictors essentially represent a type of production change climatology. Thus, it can be viewed as persistence plus an approximation of a climatological change.

The third predictor subset adds predictors from the recent time series of TWRA power production to the persistence plus climatology model. This is a type of autoregressive model and accounts for recent trends in production but uses no information from external sources (i.e., outside of the wind generation facilities). This model produces a substantial reduction in MAE relative to the persistence plus climatology model for all look-ahead times. However, the impact (i.e., the MAE reduction) is greatest for the 15-min look-ahead period for which a reduction of 23.2% is achieved. The impact decreases with increasing look-ahead time to a minimum of 16.5% for a 90-min-ahead forecast. The MAE reduction is slightly higher for 2-h and 3-h-ahead forecasts. These results indicate that there is considerable predictive information in the recent trends in power production, but the greatest value of this information is for very short look-ahead periods (0–30 min). This is well known, and this type of information is widely used for operational very-short-term forecasts.

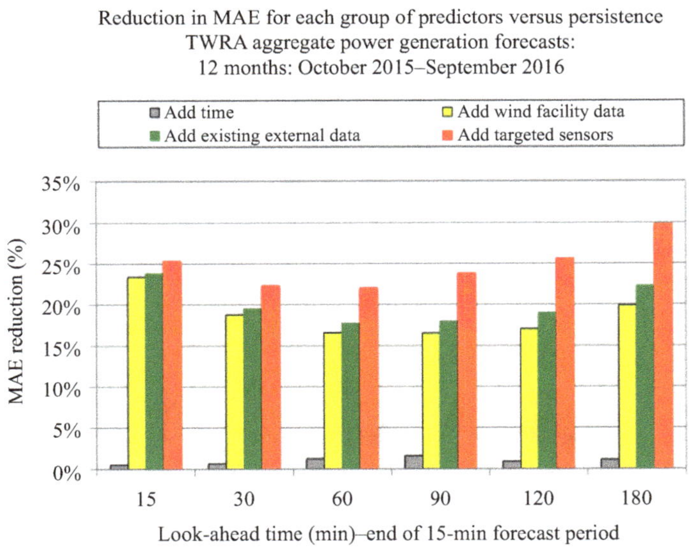

Figure 8.5 Percentage of MAE reduction versus a persistence forecast by look-ahead time achieved by each source-dependent set of predictors for 0–3 h forecasts of the 15-min average TWRA aggregate (capacity of 2,319 MW) wind power production over the 1-year period from October 2015 to September 2016

The fourth subset is labeled "add existing external data" and utilizes predictors derived from the external time-series data from existing (i.e., not deployed by the project) meteorological monitoring stations in the TWRA area. These included airport and so-called mesonet sites. All of these provided only in situ near-surface measurements (i.e., no remotely sensed data) such as the 10-m wind speed and direction, 2-m temperature and surface pressure. The MAE reduction relative to predictor set #3 (add wind facility data) achieved by using this data is depicted by the green columns in Figure 8.6. The addition of this data to the predictor pool produced an average MAE reduction relative to the "add wind facility data" pool of 2.1% for the 0–3 h period. It ranged from a minimum reduction of 0.5% for a 15-min forecast to a maximum of a 2.9% for a 180-min forecast.

The fifth subset is denoted as "add targeted sensors" and incorporates a set of predictors derived from the array of remote-sensing devices depicted in Figure 8.3. The addition of this data produces a substantial reduction in MAE (Figure 8.4) and increase in skill over the persistence forecast (Figure 8.5) relative to predictor sets #3 and #4 for look-ahead times of 60 min and longer. The percentage of MAE reduction relative to predictor set #3 (add wind facility data) is shown in Figure 8.6. The average MAE reduction relative to the use of only the wind facility data (set #3) is 9.3%. It ranges from approximately 2.5% for a 15-min forecast to about 12.4% for

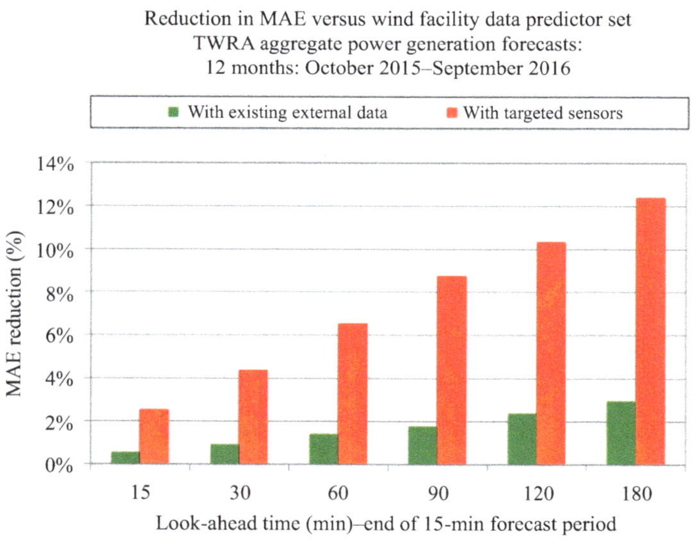

Figure 8.6 Percentage of MAE reduction by look-ahead time achieved by adding external predictors from (green) existing TWRA-area sensors and (red) the project's targeted sensor network to the pool of all on-site predictors for 0–3 h forecasts of the 15-min average TWRA aggregate (capacity of 2,319 MW) wind power production over the 1-year period from October 2015 to September 2016

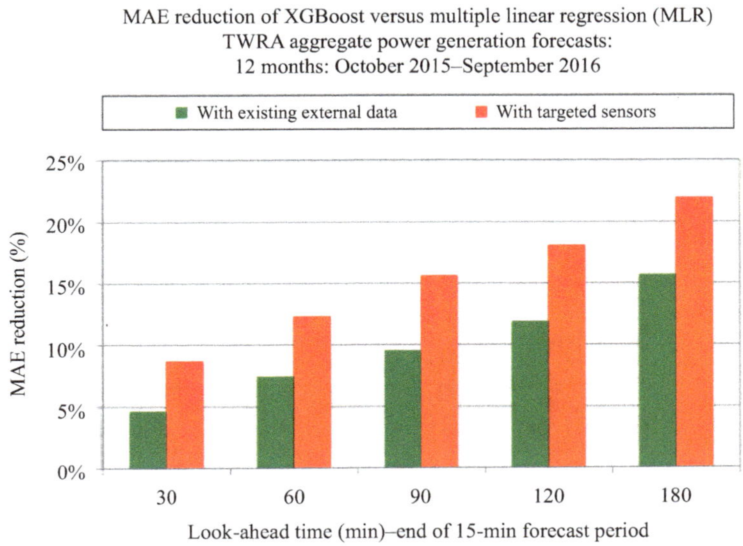

Figure 8.7 Percentage of MAE reduction by look-ahead time achieved by building prediction models with the XGBoost method versus the MLR approach for the "add existing external" (set #4) and "add targeted sensors" (set #5) predictor sets for 0–3 h forecasts of the 15-min average TWRA aggregate (capacity of 2,319 MW) wind power production over the 1-year period from October 2015 to September 2016

a 180-min forecast. There is also a considerable reduction in the MAE relative to the use of predictor set #4 (the data from the wind facilities and existing external sensors) as indicated by the relative size of the red and green columns in Figure 8.6. This demonstrates the potential forecast benefit of data from remote-sensing devices deployed at targeted locations in the vicinity of wind generation facilities.

As noted previously, the prediction performance results shown in Figures 8.4–8.6 were produced with an advanced machine-learning method called XGBoost. The benefit obtained from using a state-of-the-art machine-learning method was investigated by training a traditional MLR model with the same predictors and training sample as the XGBoost model for predictor sets #4 (add existing external) and #5 (add targeted sensors). The percentage reduction in MAE achieved by XGBoost relative to MLR for predictor sets #4 and #5 by look-ahead time is shown in Figure 8.7. XGBoost provides a substantial MAE reduction relative to MLR for both predictor sets and for all look-ahead times. However, it is interesting to note that the benefit of XGBoost versus MLR is significantly greater for predictor set #5. This suggests that advanced machine-learning algorithms provide more benefit over simpler traditional statistical prediction methods when larger and more complex datasets are available.

A third issue that was investigated in the TWRA project was the impact of the formulation of the predictand on forecast performance. While there is no doubt that many

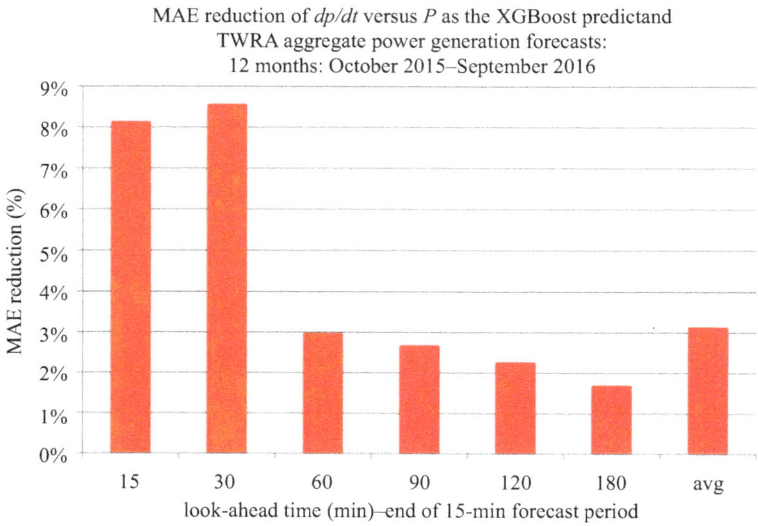

Figure 8.8 Percentage of MAE reduction by look-ahead time achieved by using the rate of change (indirect prediction) versus the 15-min average power generation (direct prediction) as the target predictand for the XGBoost model for 0–3 h forecasts of the 15-min average TWRA aggregate wind power production over the 1-year period from October 2015 to September 2016

complex predictand formulations could be concocted, two straightforward approaches are (1) direct statistical prediction of the power production for the future time interval and (2) statistical prediction of the change in production from time zero to the future interval and then the calculation of the future power production by adding the predicted change to the measured value at time zero. An experiment was conducted to assess the relative performance of these two approaches. The same 25-month dataset and rolling-sample training procedure were used to train two XGBoost models. The first model employed the power production as the predictand and the second employed the change in power production as the predictand. A comparison of the MAE for these two sets of forecasts indicated that use of the change in production as the predictand yielded better performance. The percentage reduction in MAE by look-ahead time using this approach relative to directly predicting the power production is shown in Figure 8.8. The impact is most significant for the very short look-ahead periods (15 and 30 min) for which the MAE reduction is over 8%. The impact decreases to about 3% for a 60-min forecast and continues to monotonically decrease with increasing look-ahead time.

A significant practical issue with the operational use of advanced statistical prediction models that utilize many predictors from different sources for short-term wind power prediction is that the real-time availability of data from different sources varies considerably and that it is likely that some of the data required for a complex

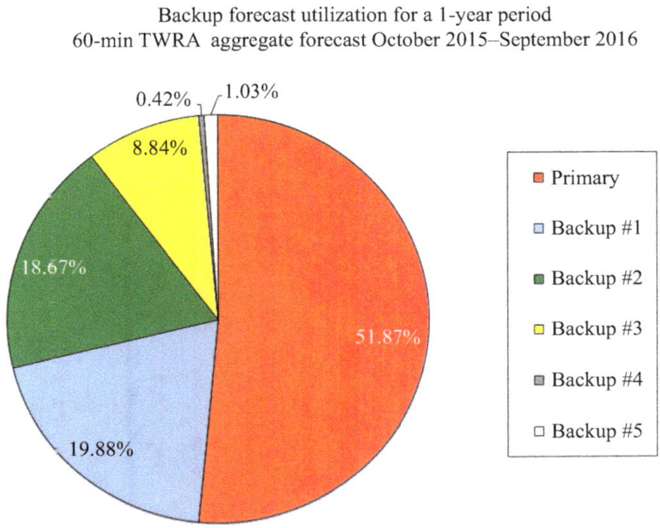

Figure 8.9 Percentage of forecast cycles for which the primary (best) model configuration and each backup model configuration were used during a 1-year forecast evaluation period based on the availability of input data for each model

model with many predictors will not be available for many operational forecast cycles. In order for an operational system to always produce a short-term forecast, it is necessary to have a hierarchy of statistical prediction models that function with a range of input datasets from most frequently missing to least frequently missing and also a method that can yield a forecast with no real-time inputs. In order to examine the impact of this issue on forecast performance, a hierarchy of forecast procedures was formulated. It consisted of a primary model (the best performing model with all selected predictors) and five backup models. The backup models were structured in a hierarchical manner based on data availability. Thus, backup #1 excluded the use of the most unreliable data employed in the primary model. Backup #2 excluded the most unreliable data contained in backup #1 and so on. Backup model #5 was structured to not depend on real-time data flow and thus always be capable of producing a forecast. The operational procedure for each 15-min forecast cycle is to use the predictions from the highest available model in the hierarchy. Since each look-ahead time employs a different model with different predictors, a separate hierarchy is used for each look-ahead time. As an example, the forecast model utilization for 1 year of TWRA 60-min forecasts is shown in Figure 8.9. In this case, the primary model (all predictors available) was used about 52% of the time. The first backup model was used in approximately 20% of the cases, and the second backup was used in 19% of the cycles. Thus, the top three models in the hierarchy were employed for about 91% of the forecast cycles. The last resort backup (i.e., #5) had to be used for only about 1% of the cycles.

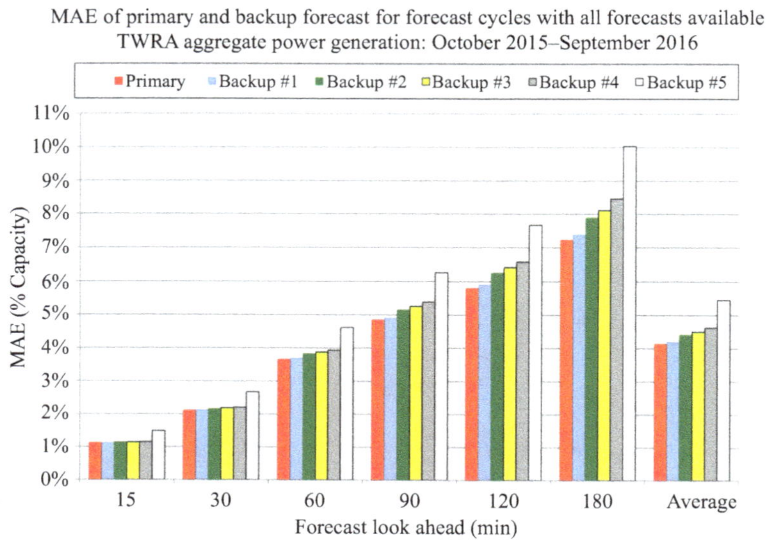

Figure 8.10 MAE of each configuration in the hierarchy of forecast contingency methods for the times when forecasts were available from the primary and all backup prediction models (i.e., the 51.87% of the sample indicated in Figure 8.9)

The obvious issue in the use of backup models with fewer predictors is the degree of degradation in forecast performance. The MAE by look-ahead time for each model in the hierarchy for forecast cycles for which all models were available (i.e., the 52% of the cycles noted in Figure 8.9) is shown in Figure 8.10. This depiction indicates that the degradation is generally greater for longer look-ahead times. This is associated with the fact that the forecasts for the shortest look-ahead times are dominated by the persistence and the trends in power production from the wind generation facilities. This data generally has the highest availability. The lowest data availability rates are from the off-site remote sensing devices and as shown in Figures 8.4 and 8.5, their greatest value is for the longer look-ahead times. The largest degradation in performance is when the last resort backup (backup #5) must be used.

Model Output Statistics (MOS)

A second major application of statistical techniques in wind power forecasting is MOS. The fundamental objectives of MOS are to reduce the magnitude of systematic errors (biases) in the forecasts from an underlying (often physics-based) prediction model and/or generate predictions of variables that are not explicitly produced by the underlying model. The term "MOS" originated in a paper published in the early 1970s by Glahn and Lowry [28] that described their method to predict a set of meteorological variables at airport measurement sites using output variables from early NWP models as predictors in a screening MLR model. A more recent example of an operational MOS procedure is provided in [29].

As a result of its origins, the term MOS is often associated with the application of MLR methods to the output of NWP models. This approach is still widely used in general meteorological forecasting and in the prediction of wind power production. However, the concept of MOS should be viewed in a much broader perspective. First, a broad set of advanced statistical techniques, such as those described in previous sections, can be employed for MOS applications. Second, the approach can be applied to many types of underlying non-statistical prediction models, not only NWP systems.

There are a number of factors that should be considered in the application of MOS in the prediction of wind power production. An initial issue is whether to use the MOS approach to directly predict the power production from a facility or to employ MOS to improve upon the NWP prediction of the relevant meteorological variables (such as the wind speed and direction) and subsequently employ an explicit power-output model to create the power forecasts.

A second fundamental issue is the selection of the type of statistical prediction method that is to be employed and the configuration of that method. MLR is the most basic method that is typically employed. However, in recent years, many of the advanced machines learning methods such as GBM and XGBoost have been used in MOS-type components of wind-power prediction systems. It may seem logical to simply use one of the most advanced methods. However, they do not always yield better performance than MLR and typically are much more computationally intensive. Experience has indicated that the benefits of the advanced methods are more frequently realized when large samples of high-quality training data are available.

The impact of the choice of the statistical method used in a MOS application for wind power prediction is illustrated in Figure 8.11. The chart depicts the reduction in MAE for MOS forecasts from 26 different statistical methods applied to the output of two NWP models relative to an MLR technique applied to the output of the same two models. The two NWP models are the National Weather Service's HRRR model and a custom-configured high-resolution (1 km) version of the WRF model [5] run as part of a commercial wind forecasting service. The performance results are for 0–15 h wind power forecasts for the aggregate of all generation resources in the TWRA of California. The 26 methods are based on statistical prediction modules available in the Python-based Scikit-learn package [30]. All of the MOS forecasts are based on the use of the same set of nine predictor variables from each model and an identical training sample with a size of 11 months. The predictand (the target variable) was the 15-min average aggregated TWRA (capacity of 2,319 MW) wind power production. Forecasts were evaluated over the 1-year period extending from October 2015 to September 2016.

The evaluation results indicate that most of the statistical prediction-model generation methods available in the Scikit package do not perform significantly better than linear regression (i.e., MLR) for either NWP model. The methods that substantially outperform MLR are highlighted with red and blue shading. The red shading denotes methods based on the decision-tree concept, and the blue shading denotes an ANN model (i.e., MLP). The color scheme is the same for both NWP models although the method that achieves the greatest MAE reduction is not the same for the two models. It is interesting to note that a number of the methods perform substantially

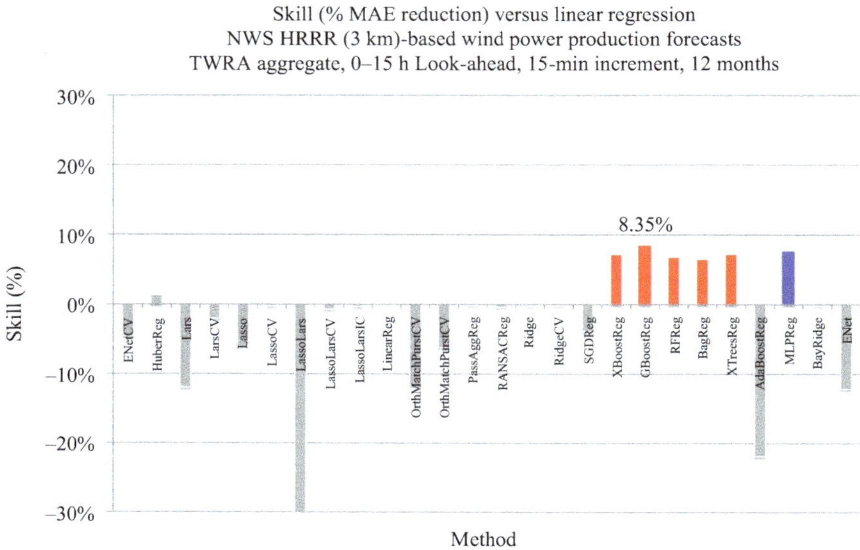

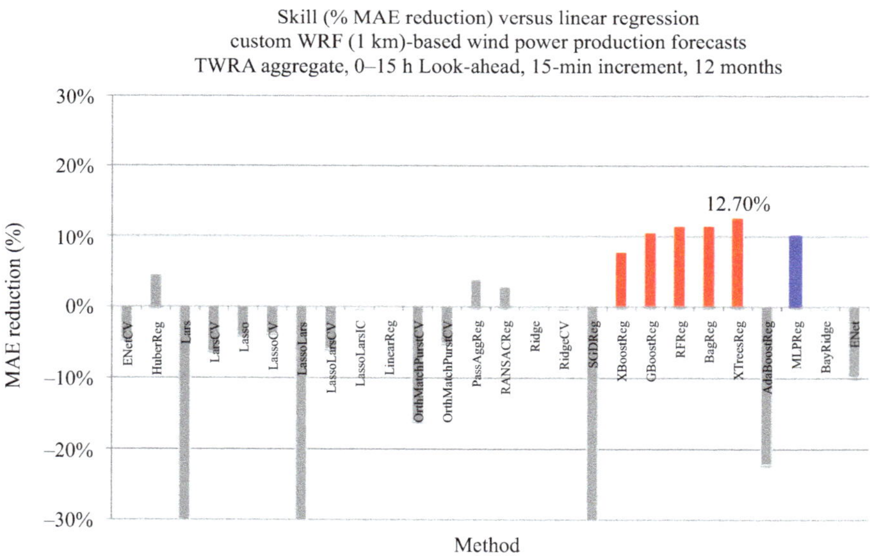

Figure 8.11 Percentage reduction in the MAE relative to the use of multiple linear regression of NWP-based 0–15 h wind power forecasts for the TWRA aggregate over a 1-year period resulting from the application of 26 statistical prediction methods to the output from the USNWS's (top) High-Resolution Rapid Refresh (HRRR) model that utilizes a 3 km horizontal grid and (bottom) a custom-configured version of the WRF model with a 1 km horizontal grid

worse than MLR. In general, the improvement over MLR is greater for the customized high-resolution WRF model than for the HRRR model.

A third issue is the representativeness of the data in the training sample. The most common version of this issue is that the attributes of the underlying model (e.g., the NWP model) that supplies the predictors to the MOS procedure are not constant during the period for which data is available for use in a training sample. This often is the result of upgrades such as an increase in NWP model resolution or changes to the formulation of the NWP model physics. These types of changes can significantly alter the patterns of systematic errors in the model forecasts. Therefore, the use of data from before the change in the underlying model may not represent the error patterns that occur after the change. If one chooses to use only data from after the model change, the training sample size will be more limited. Since upgrades to government-run NWP systems routinely occur, this effectively eliminates the possibility of using a very long training sample.

There are two approaches that are widely used to mitigate this limitation. First, many of the changes to the underlying NWP model are minor and have little impact on the systematic error patterns of particular variables. Therefore, data from before and after the model change can be used with little or no negative impact on MOS performance. However, a considerable amount of caution must be used when applying this approach. The impact of a NWP model change typically varies substantially among model variables and even between geographic areas for a specific variable. A second approach is to employ predictors from an in-house model that is controlled by the user. This enables the user to generate a historical training dataset that is produced by a model with an unchanging configuration although even with this approach the characteristics of the input data, which are typically not under the control of the user, may have changed.

A fourth issue is the training sample strategy. There are an almost infinite number of possible training sample configurations. The training sample strategies are frequently classified into three broad categories: (1) static, (2) dynamic and (3) regime-based.

In the static approach, a single training sample is used to generate a fixed set of MOS equations that are infrequently updated. The training sample typically covers a long period of a year or more. The same set of equations is used for all subsequent forecast cycles.

In contrast, the dynamic approach typically uses a much smaller training sample that is frequently updated. A typical dynamic MOS configuration is a rolling 60-day sample that ends on the forecast cycle before the current forecast cycle. In this approach, the oldest data element is deleted on each cycle, and the most recent one is added. In this configuration, the composition of the training sample changes for each forecast cycle, and a new set of MOS prediction equations is used for each cycle. There are several advantages to this approach. First, the training sample is typically more representative of the current error patterns in the model forecasts since it is drawn from the current season and often from the current weather regime. Second, the use of a short training sample avoids many of the issues associated with impact of changes in the underlying model except immediately around the time of the model change. The disadvantages of this approach are that the MOS equations are less stable because they are derived from a short sample and it is unlikely that maximum benefit will be obtained from advanced machining learning methods because they generally exhibit the greatest improvement over simpler methods when a very large training sample is employed.

The regime-based MOS approach is based on the concept of defining groups of cases for which the underlying model has similar error patterns. The use of the word "regime" in this context is often misinterpreted to mean "weather regime," but the clustering should be based on "underlying model (e.g., NWP) error patterns" to have value. Certainly, in many cases, the "error regimes" are correlated with "weather regimes," and this information may be useful in designing a regime-based MOS approach, but ultimately it is the error regimes that are critical to identify. There are also many possible implementation strategies for the regime-based concept itself. A commonly employed strategy is to define a fixed set of error regimes that are defined subjectively or via an objective approach such as a clustering algorithm or PCA. The training sample is then divided into "N" regimes (clusters) by one of these approaches, and a separate set of MOS equations is derived for each regime by employing a statistical prediction method. The production of a forecast is then accomplished by assigning the current forecast scenario to the most appropriate regime and using the MOS equations for that regime to generate the prediction. The primary benefit of this approach is that more case-specific error regimes can be identified, which can yield more effective error correction. However, a substantial disadvantage is that the division of the training sample into several subgroups results in the training sample being smaller for each regime. It is often the case that the number of data points exhibits considerable variation among the regimes (i.e., some regimes occur much more frequently than others) and, therefore, some of the regimes may not have a statistically meaningful sample size. Another approach is to formulate a dynamic regime-based strategy in which there is no predefined set of regimes. Instead, a custom regime is created for each forecast scenario by selecting historical cases that are the best matches for the current case. This is essentially the approach employed by the AE method. In general, it is difficult to get the maximum benefit from the advanced machine-learning methods with the regime-based approach because of the small sample sizes that typically occur in the subdivided samples. However, a substantial portion of the advantage of the advanced machine learning methods for MOS is often related to their ability to implicitly find and exploit error regimes during the training process and the regime-based approach with a simpler statistical method is often a more explicit and transparent way to extract the same predictive information.

The overall impact of MOS on the performance of unadjusted NWP forecasts can be quite large and highly dependent on the characteristics of the underlying NWP forecasts. An example of the impact of a MOS procedure on wind power forecasts is shown in Figure 8.12. This chart depicts the percentage reduction in MAE for MOS 0–15 h forecasts for the aggregated wind power production from the TWRA from each of two statistical methods (MLR and the best machine-learning method for each model) relative to a baseline raw NWP forecast for each of three NWP models. The baseline raw NWP forecast was generated by interpolating the hub-height (about 80-m in the TWRA) wind speed and direction from the three-dimensional NWP grid-point

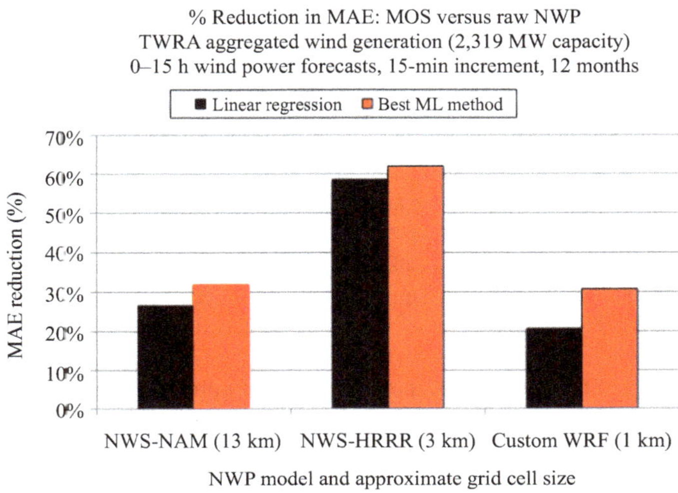

Figure 8.12 Percentage reduction in the MAE of wind power forecasts relative to a baseline of a raw NWP forecast for three NWP models when a MOS procedure based on multiple linear regression (black) and the best machine learning method for each model (red) are applied to the NWP output. The data is for 0–15 h wind power forecasts for the aggregated wind generation for the TWRA region over a 1-year period

dataset. The wind data is then used as input into a facility-scale power-output model statistically derived from measured wind speed and power generation data. The three models are the USNWS's NAM and HRRR models and a custom configuration of the WRF model run with a grid cell size of 1 km. The percentage of MAE reduction is 25%–30% for the NAM model, 20%–30% for the custom WRF model and about 60% for the HRRR model. As might be expected, NWP models with large biases will obtain more benefit from the application of MOS. In this case, the HRRR model had a large positive hub-height wind speed bias of about 3 m/s for the TWRA. The bias for the other two models was less than 1 m/s. Linear regression is quite effective at removing much of the bias. The benefit of the machine-learning methods is largest for the custom WRF model.

Ensemble Composite Models (ECM)

The objective of the ECM is to construct the best-performing composite forecast from an ensemble of individual forecasts. Although the term "ensemble" is frequently used to describe a set of NWP forecasts, an ensemble for a typical state of the art wind forecast system is often composed of forecasts produced by several types of methods including raw NWP forecasts, MOS-adjusted NWP forecasts, time-series models and feature detection and tracking models. The ECM is actually a form of MOS with the predictors coming from many prediction models rather than a single model. Therefore, most of the same techniques that are employed for the single-model MOS are applicable to the ECM. But there are some issues that distinguish the ECM application from the typical single-model MOS application.

One issue in the ECM application is which members of a forecast ensemble are distinguishable and which are indistinguishable. The members are indistinguishable if the members are generated by using the identical forecast model and randomly perturbing some aspect of the modeling system in each forecast cycle. This is typically applied to the initialization datasets for an NWP run in order to simulate initialization uncertainty. In this case, there is no basis to distinguish one member of the ensemble from another, and it is not useful to attempt to assign differential weighting to each of these members. However, the ensemble mean of an indistinguishable ensemble will typically produce a lower forecast error than an individual member. Therefore, a typical approach is to create ensemble-mean variables and apply a MOS procedure to the ensemble-mean variables. In contrast, the distinguishable members are produced through the use of characteristically different prediction methods (e.g., different models or different configurations of the same model) or input datasets (e.g., systematically omitting or adding specific input datasets). In this case, there is a basis for differential weighting of the members since it is possible that some methods or input datasets may yield different error patterns under specific circumstances (e.g., regimes). The means of the indistinguishable ensembles and the individual members of the distinguishable ensembles can then be used as input into a statistical multi-method ensemble composite, some members of which are ensembles themselves. This is sometimes referred to as a "super ensemble."

A second issue in the formulation of an ECM is whether to use all of the raw predictors from each member method as input into the ECM (a "super-MOS" approach) or to first create a forecast for the ultimate target variable (e.g., hub-height wind speed) from each method and then use only the target variable forecasts as input into the ECM. In general, it is better to apply MOS to each individual model and then combine the resulting forecasts since each model has its own unique error patterns, which may be difficult to distinguish in a training sample with forecasts from many different models.

A third consideration is the selection or filtering of member inputs into the ECM. It is tempting to use as many methods as could be available to the ECM. However, an indiscriminate application of that philosophy can be detrimental. This issue is based on two factors: (1) the correlation of the errors among the individual forecasts and (2) the size of the historical sample available to the ECM. The first factor is a reflection of the intuitive fact that the construction of a composite will have no benefit if the errors of all the methods are the same for each forecast period. In that case, any composite of the methods will of course yield the same error as any individual method since all the errors are the same and there is no basis for distinguishing the performance of the methods. A less extreme and more typical occurrence is that the errors of individual methods are highly correlated. In this case, there will be minimal benefit in the construction of an ensemble. The point is that ensemble members with high error correlations to other members do not provide much value in the construction of the composite. However, the result can be worse than no impact. Indiscriminate use of highly correlated members in an ensemble composite can have detrimental effects. For example, in the case when some members have highly correlated errors and some do not, a simple equally weighted ensemble average will have the beneficial impact of less intercorrelated members diluted by the highly intercorrelated members. In effect, one is implicitly placing heavier weight on the forecast represented by the highly correlated members since it essentially represents a multiple occurrence of the same forecast in the ensemble. On the other hand, the use of poor-performing uncorrelated forecasts will not be beneficial either.

The use of an appropriate statistical technique will serve to minimize the weighting of the highly correlated members and achieve an optimal blending of the highly correlated and less correlated methods. This is where the training sample size can become an issue. The use of a large number of input methods along with an advanced statistical technique (with many adjustable parameters) can result in an overfitting issue if only a small training sample is available. The availability of a long representative training sample can minimize this issue, but this can often be difficult to assemble because many of the input models will change periodically.

8.3.3 Power output models

Once a high-quality prediction of the key meteorological variables is generated via a composite of the methods discussed in the previous sections, the meteorological data must be transformed into a prediction of wind power production. This is accomplished with the use of a power-output model. This type of model can be constructed on different scales (single turbines, clusters of turbines, a generation facility (i.e., "wind farm"), regional aggregate, etc.). However, the most common approach is to construct a facility-scale output model. The facility-scale power-output model not only represents the relationship between the meteorological variables and the facility-scale power generation but also implicitly or explicitly accounts for non-meteorological effects.

Although the basic concepts and objectives are the same, a number of different modeling strategies can be employed, and there are notable differences between them. The most fundamental option is whether to use an explicit or implicit power-output model. The implicit approach predicts the power output at the MOS or ECM step by training the statistical model to go directly from predictions of meteorological variables to power output in the MOS or ECM process. In this case, the equivalent of a power-output model is implicit within the MOS and/or ECM equations. This simplifies the prediction process and also removes some of the need for high-quality meteorological data from the generation facility. The disadvantage of this approach is that it makes it more difficult to separate the components of the forecast error that are associated with the meteorological predictions from those that are associated with the power-production model. This makes it harder to analyze the performance and refine the prediction system. Results suggest that the explicit approach generally yields better forecast performance for facilities that supply high-quality meteorological data, while the implicit approach may be as good or better when high-quality meteorological data is not available from the facility.

If the explicit approach is pursued, then the modeling method and the granularity of the model must be selected. There are two fundamental types of power-output models: (1) process (physics)-based and (2) bulk statistical. The process-based models attempt to simulate the behavior of the facility based on the physical layout and hardware specifications of the facility. These models use meteorological and operating data and the hardware and layout specifications as input into quasi-physics-based model equations to calculate the power output. These models have considerable detail, and the engineering processes are generally modeled quite well. However, the detail of input data required for these models to perform well is generally more than is typically available in operational applications. This generally limits their performance in an operational setting.

In most operational applications, statistical power-output models are employed because they provide better performance. These models are statistical relationships between measured meteorological data and actual power output. Any of the statistical methods previously described can be employed for this purpose. However, the data may be noisy, and in many cases, simple models will perform as well or better than more sophisticated methods. The facility-scale models can be constructed at different levels of granularity. For example, in the case of a wind generating facility, statistical relationships could be constructed for the output of each turbine or for the aggregated output of the facility. The aggregated approach is more typically employed because the data is often not available at higher granularity, and even in cases where such data is available, the impact of modeling with additional granularity on forecast performance is often minimal, but the level effort is somewhat greater.

8.3.4 Integrated forecast system

A typical state-of-the-art wind power (or more generically a variable renewable energy generation) forecast is based on a multi-method-ensemble approach that incorporates prediction system components based on the methods discussed in previous sections. Although the focus here is on wind power forecasting, many of the same components are used for solar power production forecasts. The components and data flow in a generic version of such a forecast system are depicted in Figure 8.13.

The vast majority of external or internal providers of wind-power forecast information employ a system that is a variant of this general structure. The major variations are: (1) the number and source of datasets from government-run NWP models that are injected into the system; (2) the number, type and configuration of customized in-house NWP systems or other physics-based models such as CFD models included in the forecast process; (3) the number, type and configuration of post-processing (i.e., MOS) algorithms applied to the NWP output as well as the size and composition of the training samples; (4) the number, type, configuration and input data types of the time-series prediction models; (5) the number (if any) and type of feature detection and tracking models such as the cloud motion vector model heavily used in solar power forecasting; (6) the type and configuration of the algorithm(s) that constructs a probabilistic or deterministic composite forecast from the ensemble of forecasts produced by the individual methods and (7) the methods and data used to construct explicit power-output models if they are used by the system as an alternative to the implicit approach.

There are numerous factors to consider in the design of a wind-power forecast system for a specific application such as serving as the basis for a commercial forecast

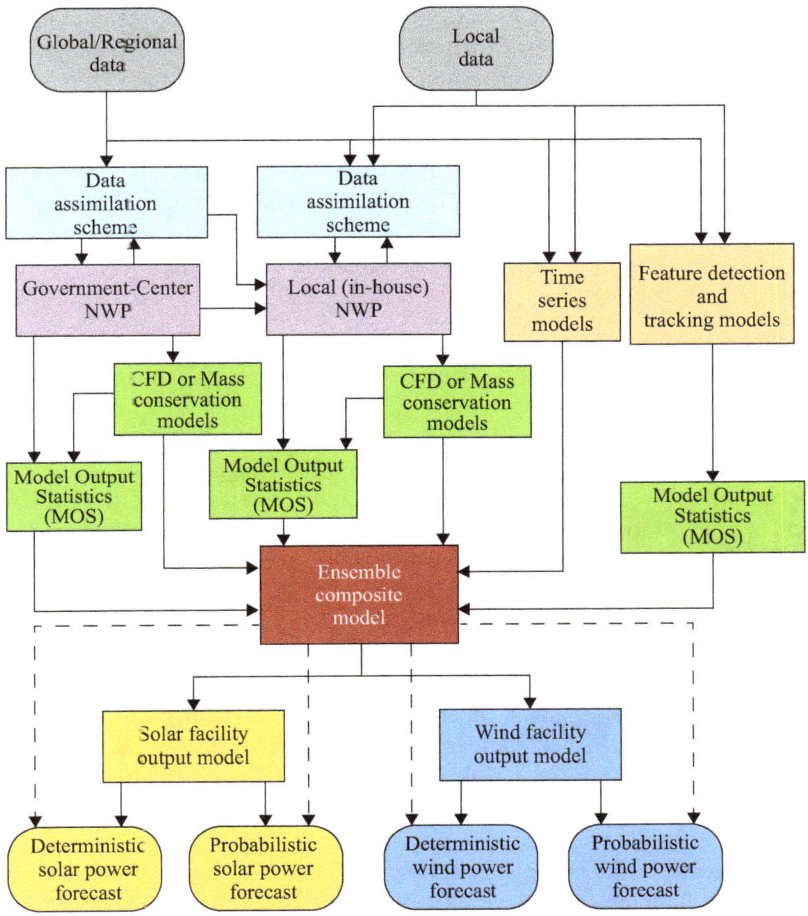

Figure 8.13 A schematic depiction of the components and data flow of a typical multi-method state-of-the-art wind and solar forecast system

service or the generation of in-house predictions for an end user. The factors include: (1) the types of government-run NWP data that are available to the provider due to access cost, bandwidth limitations or access permission issues, (2) the availability of computational resources and expertise to run in-house NWP models and the assessment of whether the target forecast application (e.g., wind facilities in complex terrain) would benefit from the use of such models, (3) the range of experience with advanced statistical prediction model algorithms and supporting tools, (4) knowledge of forecast-performance patterns of specific physics-based or statistical tools, (5) type and amount of data available for the forecast target site and its vicinity and (6) the needs (e.g., emphasis on hours ahead or day ahead) and characteristics (e.g., complex versus simple terrain, mid-latitude versus tropical, and coastal versus in-land) of a specific forecast application. It is evident from this partial list of factors that some are related to the knowledge and experience of the system implementer and operator, and others are related to the requirements and characteristics of a specific application. However, experience has indicated that there is no single system design that is optimal for all applications, and furthermore, different configurations may be optimal for the same application at different times due to changes in weather regimes, availability of specific data types and other factors. This provides support for the common practice of sophisticated forecast users to use predictions from more than one forecast system (e.g., forecasts from multiple providers) as input into critical subjective or objective decision-making processes.

8.4 Communicate: inform the user for decision-making

The third key component of the wind forecast value chain is the communication of the wind forecast information to the forecast user for use in the targeted application. The manner in which forecast information is communicated to the user will significantly impact (1) what information a user assimilates into their decision-making processes, (2) how quickly the user incorporates it into their processes and (3) the degree of confidence a user has in the forecast information.

There is a substantial range in the types of users and applications; therefore, there is no single optimal way to communicate forecast information to users. The primary users of wind power forecasts are system operators, distribution utilities, transmission capacity providers, power traders, generation facility owners, operators or schedulers and power aggregators. Each of these users may have one or more applications for the wind-power forecast information. In fact, some of the users may actually be algorithms or models that ingest the forecast information and provide application information to the end user. While the manner in which forecast information is communicated may be somewhat different for human-based and machine-based users, the format and content of the forecast information will impact the value that either type of user extracts from a set of forecast information.

8.4.1 Deterministic versus probabilistic

Perhaps, the most fundamental issue in communicating forecast information to a user is the choice between deterministic and probabilistic presentations. Deterministic forecasts supply one type of information: the predicted future values of the target variable. On the other hand, probabilistic forecasts provide two types of information: the future values of the target variable and an estimate of the uncertainty in the predicted future values. Thus, a probabilistic format is a more comprehensive representation of forecast information. However, it is generally more complex and harder to use in applications.

An example of the alternative perspectives provided by deterministic and probabilistic forecasts is provided in Figure 8.14. The top panel provides a deterministic view of the anticipated wind power production for the next 7 days. No uncertainty information is provided. The bottom panel depicts a probabilistic forecast for the same period. This consists of ten probability of exceedance (POE) lines that range

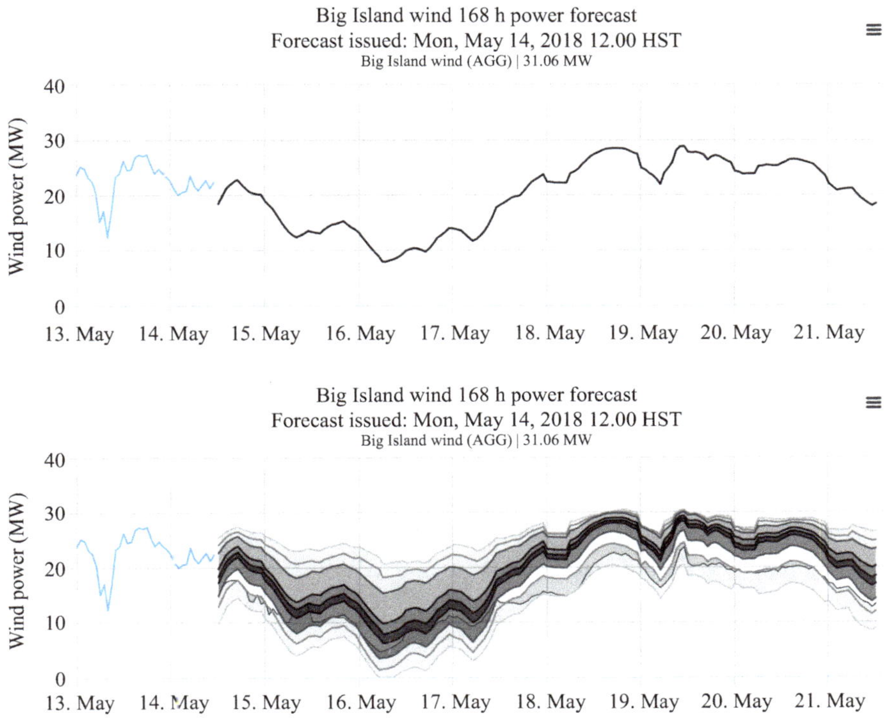

Figure 8.14 An example of a deterministic (top) and a probabilistic (bottom) presentation of a 7-day (168-h) forecast of the system-wide wind power production on an island grid system in Hawaii. The multiple forecast lines on the bottom chart depict ten probability of exceedance (POE) levels, which from bottom to top represent the 96%, 90%, 80%, 60%, 50%, 40%, 20%, 10% and 4% POE values

from 96% to 4% (i.e., the actual production should exceed a specific POE line for a percentage of the cases equal to the specified probability). The spread between these lines provides a measure of the uncertainty in the forecast and can be demonstrated to be related to the forecast error (i.e., larger spreads on average yield larger forecast errors).

8.4.2 Time series versus event-based

A second key attribute is whether the forecast information is presented in a time-series or an event-based format. As implied by its name, the time-series format presents information as a sequence of numbers that represent the values of the target variable at specified intervals during the forecast look-ahead period. The time-series format is the most widely used format, but event-based forecasts are often more aligned with how many human users utilize forecast information.

A comparison of the time-series and event-based forecast formats is presented in an example from the system operated by the Electric Reliability Council of Texas (ERCOT) that is shown in Figures 8.15 and 8.16. The charts depict a forecast issued by the ERCOT Large Ramp Alert System at 16.00 Central Time (CT) on a day in May 2018. It provides a 6-h look-ahead period that extends to 22.00 CT. Both depict predictions of large changes in wind power production that are known as "wind ramps." A time-series format is displayed in Figure 8.15. This depicts the probability of exceeding a specified ramp rate for each of three different time scales (15, 60 and 180 min) beginning at the time shown on the chart. In this example, the probability of large upward ramp rates becomes very large during the middle and latter part of the forecast period.

The chart in Figure 8.16 depicts the same forecast scenario from an event perspective. In this view, the forecast is for the occurrence or nonoccurrence of the event, and if an event is anticipated than values for the event parameters are predicted. For wind ramps, the parameters are amplitude, duration, maximum ramp rate and start time. In the depicted example, an upward ramp event is predicted. The red circles provide a deterministic prediction of the attribute parameters. In this case, the amplitude is expected to be about 5,200 MW, a duration of about 255 min is expected, the maximum rate is predicted to be about 600 MW in 15 min, and the start time is forecasted to be 17.45. The red bars adjacent to the circles provide an 80% confidence interval for each parameter.

8.5 Assess: evaluation of forecast performance

The evaluation of forecast performance is the fourth component of the wind forecast value chain. While the evaluation of forecast performance is typically viewed as a measurement of forecast accuracy, it should be assessed within the broader context of measuring the value to the application. A well-designed forecast assessment program should serve to (1) estimate the value of the forecasts for the user's specific applications to guide decisions on the level of resources that should be allocated to acquire, produce or enhance forecast information, (2) develop and guide the confidence of the forecast user, (3) monitor the progress of efforts to increase forecast value and (4) provide guidance for the formulation of future efforts to increase forecast value. The first three objectives should be of interest to both the forecast user and provider. The fourth is primarily of value to the forecast provider but may be of interest to sophisticated forecast users.

Forecast performance can be evaluated in many ways. However, the most useful forecast evaluation protocols will address all of the four objectives noted in the previous paragraph. The core issue in the design of a forecast evaluation procedure that achieves these objectives is the determination of the key attributes of the forecast that are to be evaluated and then the formulation of metrics that effectively evaluate them. This will depend on two core factors: (1) the content and format of the forecast dataset and (2) the way in which a user's application is sensitive to forecast error.

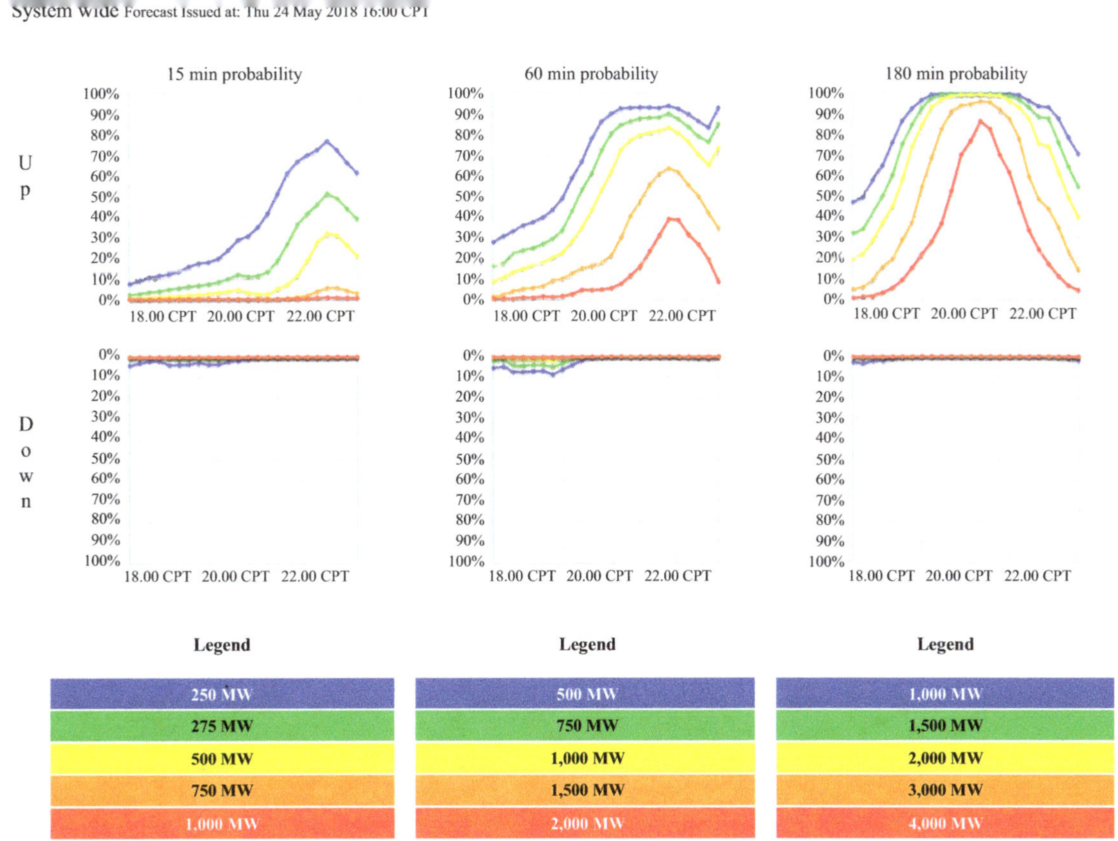

Figure 8.15 A time-series forecast of the probability of specified system-wide wind-power ramp rates on the system operated by ERCOT issued at 16.00 Central Time on May 24, 2018. The forecast time horizon is 0–6 h. Predictions are presented for three ramp rate time scales: (from left to right) 15, 60 and 180 min. The lines indicated the probability that the ramp rate for the specified time scale beginning at the time on the horizontal axis will exceed the MW threshold denoted by the color. Upward ramp probabilities are shown in the top row and downward ramp probabilities in the bottom row

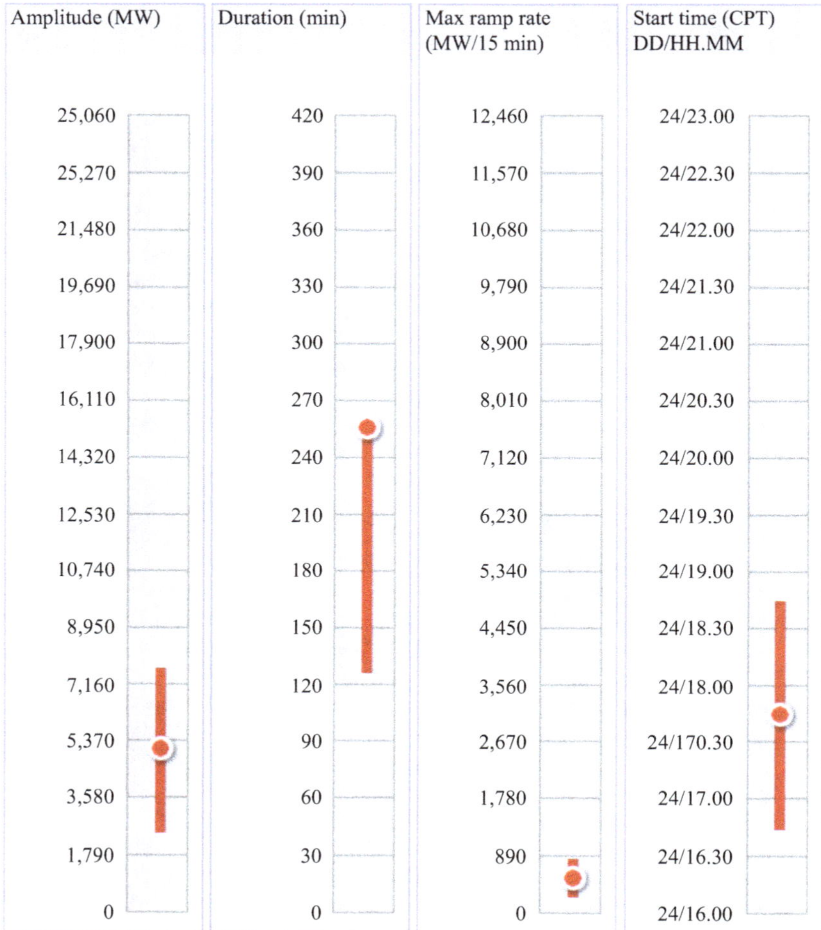

Figure 8.16 An event-based forecast of system-wide wind ramp events on the system operated by ERCOT issued at 16.00 LT on May 24, 2018. Four parameters are forecasted for each predicted event: (1) amplitude (MW), (2) duration (min), (3) maximum ramp rate (MW/15 min) and (4) start time (local time). The red circle denotes the deterministic prediction and the red bars depict the 80% confidence interval for each parameter

Economic value can be derived from the forecasting of wind power production in a number of ways over a wide range of look-ahead time scales by a diverse group of users. Ahlstrom *et al.* [31] note that different types of wind forecast users have very different motivations and different sensitivities to forecast errors. They provide an overview of the motivations of the key classes of users. System operators (which in some cases are utilities) are primarily concerned with the reliable and economic operation of their power system, and better forecasts have the potential to lower operating costs while maintaining the required levels of reliability. Wind plant owners look to maximize the revenue from their investment in the wind plants but may prefer long-term hedge positions to being subject to price risk in the energy market. Utilities purchasing wind power through long-term power purchase contracts may also have a low tolerance for price risk or be averse to active trading in the market. Financial traders are on the other side of all these issues. Since financial traders are willing to deal with higher levels of risk on individual transactions—winning some and losing some—they can exploit information that is better but not perfect. They are also willing to pay for marginally better information, such as a more accurate wind power forecast, if it provides them with even a slight advantage in the market.

The alternate perspectives that can be obtained by using different approaches to forecast evaluation will be briefly illustrated by an application example from a small isolated grid system with a high percentage of variable renewable (wind, solar and hydro) generation on one of the Hawaiian Islands. The system's generation assets include 31 MW of wind generation provided by two facilities, 16.2 MW of run-of-river hydro-based generation from three facilities and approximately 90 MW of distributed (i.e., behind-the-meter (BTM)) solar-based generation.

The average daily net load profile for the system for each quarter of 2017 is depicted in Figure 8.17. The net load profile is what is actually served by the utility's generation assets and consists of the actual (gross) demand minus the BTM generation. The net load profile is characterized by two daily minima and maxima. There is a morning peak of about 135 MW at approximately 08.00 Hawaiian Standard Time (HST) and an evening peak near 175 MW around 18.00 HST. There is a night-time minimum typically near 100 MW at about 03.00 HST and a mid-day minimum typically near 120 MW that is associated with BTM rooftop solar generation.

Several key decision-making time frames and scenarios are associated with this net load profile. One of these is just before sunrise at about 05.00 HST. At this time, the system operators must determine the mix of generation that will be used to meet the morning peak and the following mid-day minimum. There are two key questions:

- Will mid-day net loads be low enough to shut down a conventional generation unit after the morning peak?
- Will a late morning or mid-day excess energy situation occur due to high "as available" variable renewable generation?

The expected amount of wind and BTM solar generation for the period between the morning net load peak and the mid-day minimum plays a key role in the pre-sunrise-planning process. Updated wind and solar forecasts for the next 6 h are available to

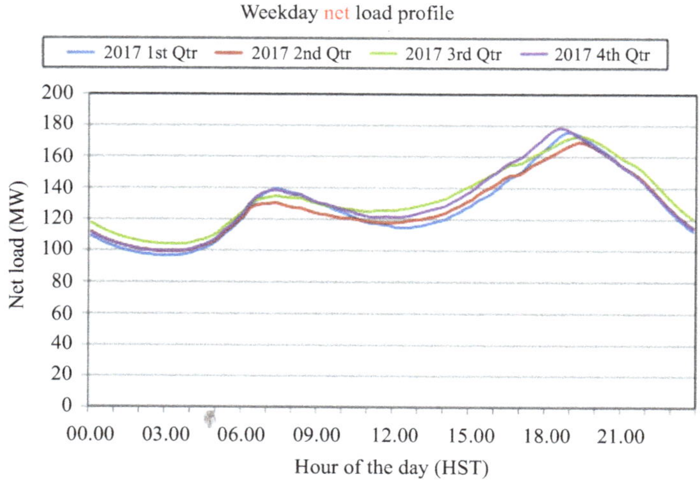

Figure 8.17 Net load (actual load minus distributed (also known as 'behind the meter' or BTM) generation)) profile for each quarter of 2017 for an island grid system in Hawaii

the system operator every 15 min. The forecast issued at 05.00 HST is generally a key input into the morning decision-making process.

One common way to assess the performance of the wind generation forecasts is to measure the typical magnitude of the error over all of the 15-min intervals in the 0–6 h look-ahead period, which at 05.00 HST corresponds to the period from 05.00 to 11.00 HST. This is usually done with metrics such as the MAE or the root-mean-square error (RMSE). The MAE of the operational multi-method-ensemble forecasts and a reference persistence (i.e., no change) forecast of the island-wide wind generation for this 6-h period for all 12 months of 2017 is depicted in Figure 8.18. The MAE of the multi-method ensemble is just over 8% of capacity for this period and is about 17% lower than a persistence forecast for the same period. This is in line with the state-of-the-art performance levels for the look-ahead period and the time resolution of the forecasts as well as the types of wind regimes that are the target of this application. However, this metric provides little insight on the performance of the attributes of the forecast that are important to the operational decision-making process.

In the case of the renewable generation forecasts used in the pre-sunrise decision-making on this island in Hawaii, the key issues are (1) whether the wind generation will significantly higher or lower than its level at 05.00 HST during the late morning period (08.00–11.00 HST) and (2) whether the morning rise (also 08.00–11.00 HST) in the BTM photovoltaic (PV) generation will be significantly greater or less than the typical values. In order to define "significant" for each of these two decision factors, their respective distributions for all the days of 2017 were examined. These are shown in Figure 8.19.

The upper panel depicts the distribution of the change in the island-wide wind production (MW) from 05.00 HST to the 3-h average for 08.00–11.00 HST.

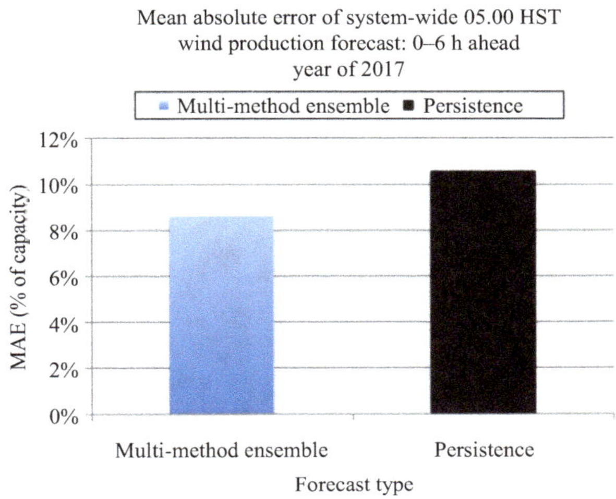

Figure 8.18 MAE of 0–6 h forecasts of the 15-min average system-wide wind generation (% of the installed capacity of 31 MW) issued each day at 05.00 HST for an island grid system in Hawaii for 2017

The majority of the days have a slight increase of 1–3 MW, but the distribution has a long tail on both ends, which indicates a small number of cases with larger increases or decreases. The highest and lowest 10% of the days were defined as significant from an operational perspective. This corresponded to a wind generation change of −3.3 MW on the low end and an increase of +8.0 MW on the high end.

The lower panel of Figure 8.19 depicts the change in BTM solar generation between 08.00 and 11.00 HST (the morning rise period). The peak in the distribution (i.e., the typical increase in PV production during this 3 h period) is near 27 MW. Analogous to the approach used to define "significant" for the wind changes, the 10% on the low and high ends of the distribution was used to classify the days as significantly below or above the typical PV increase. The threshold values were 21.9 MW on the low end and 33.4 MW on the high end.

The key forecast-performance question for this particular operational decision-making scenario is how well does the forecast system anticipate the occurrence of events on the tails of the distributions (e.g., the top and bottom 10%). This is not well measured by metrics such as the MAE or root mean square error (RMSE), which are dominated by the cases in the middle of the distribution. Some insight into this question can be obtained by examining scatter plots of the actual and forecasted values for each of the two key parameters. The scatter plot in Figure 8.20 depicts the predicted (from the 05.00 HST forecast) and actual morning change in wind power generation for each day of 2017 for which both forecasted and actual data were available. The diagonal black line represents perfect forecasts (i.e., forecasted=observed). The further a point is from this line, the larger the forecast error.

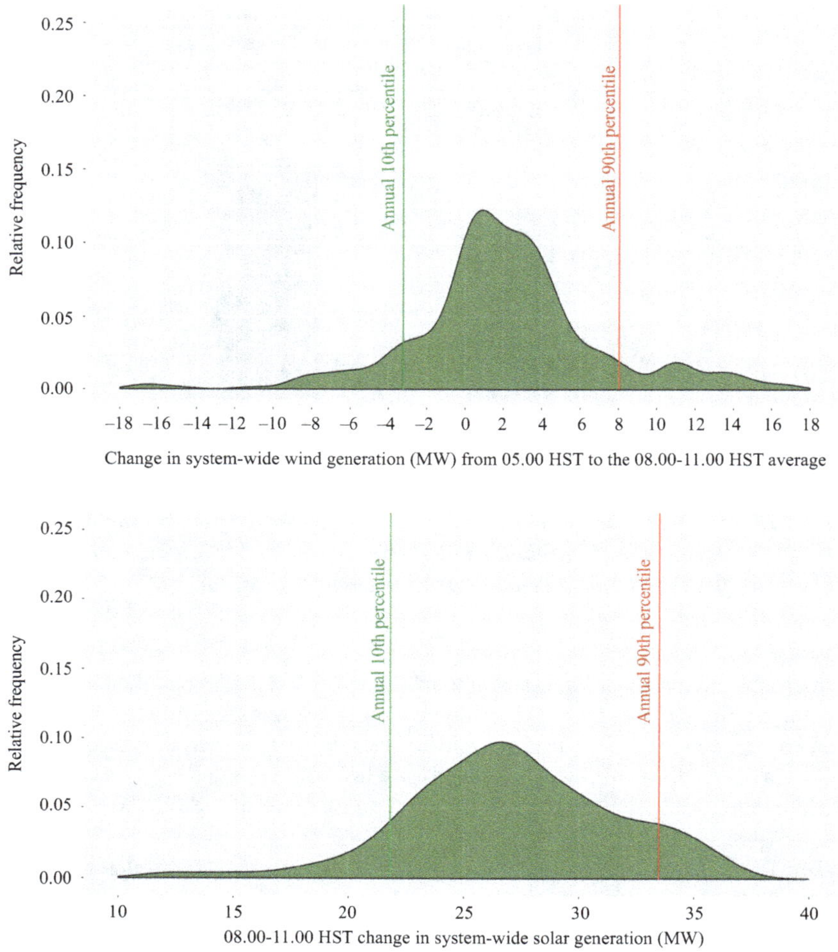

Figure 8.19 Distribution of changes in system-wide wind-based (top) and solar-based (bottom) generation that are relevant to operator decision-making at 05.00 HST for the year 2017 on an island grid system in Hawaii. The change in wind-based generation (MW) is the average generation in the 3-h period from 08.00 to 11.00 HST minus the generation at 05.00 HST. The change in solar-based generation is the increase in the estimated behind-the-meter photovoltaic (PV) generation from 08.00 to 11.00 HST

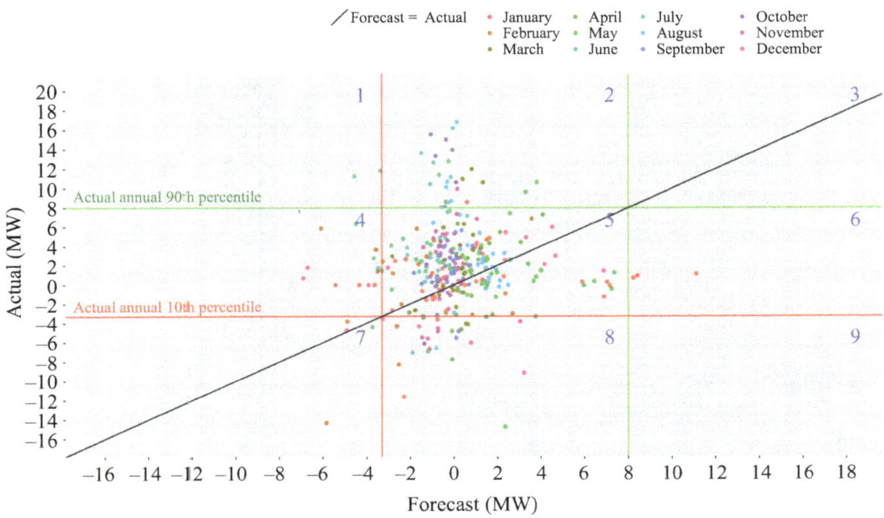

Figure 8.20 Daily forecasted versus actual system-wide morning wind power change (08.00–11.00 HST average minus 05.00 HST value) for 2017 for an island grid system in Hawaii. The red lines represent the 10th percentile and the green lines represent the 90th percentile of the distribution of actual changes. The diagonal black line represents the line of perfect forecasts (i.e., forecasted change = actual change). Each data point represents 1 day in 2017. The bold blue numbers denote the contingency table boxes used to calculate the event-based forecast performance metrics

The red and green lines denote the boundaries of the lower and upper 10% of the actual distribution. These lines define nine boxes (numerically labeled in blue font) on the chart, which implicitly define a contingency table of forecasted values versus observed outcomes. The center box (#5) represents cases for which the forecasts correctly predict the observed outcomes that are within the middle 80% of the distribution (i.e., the typical cases). As expected, the majority of the points fall in this box. However, these cases are not very significant for operational decision-making. As noted, the most critical boxes for that application are the lower left (#7) and upper right (#3). These represent correct forecasts of the lower and upper 10% of the changes in wind power generation event distribution. There are very few points in these boxes indicating despite their respectable performance in the MAE metric, the forecasts do not provide accurate guidance for the large change events that are important to operational decision-making. It can be noted that a set of the points extend upward and downward from the central cluster. The points that are above the green line (box #2) or below the red line (box #8) represent cases for which a tail event was observed, but the forecast indicated that conditions would be in the typical range. The cases for which the forecast indicated a tail event are to the left and right of the vertical red

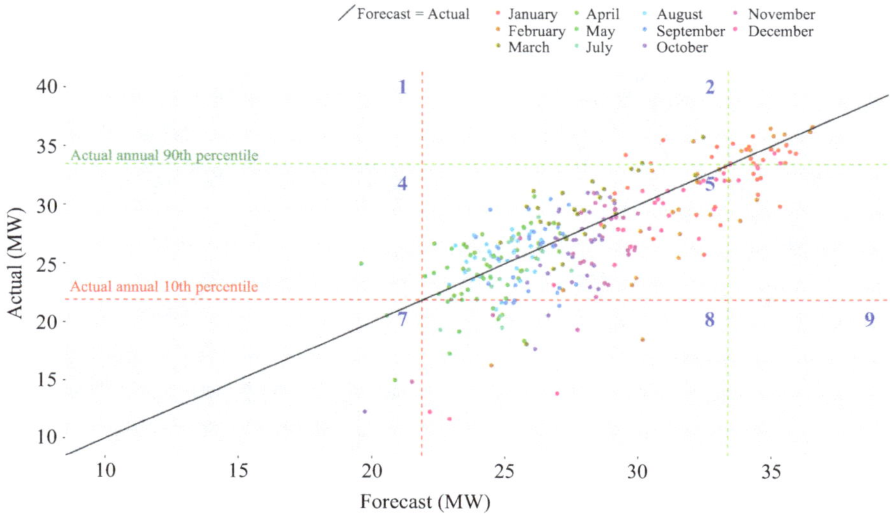

Figure 8.21 Daily forecasted versus actual system-wide morning solar power generation change (11.00–08.00 HST) for 2017 for an island grid system in Hawaii. The red lines represent the 10th percentile and the green lines represent the 90th percentile of the distribution of actual changes. The diagonal black line represents the line of perfect forecasts (i.e., forecasted change=actual change). Each data point represents 1 day in 2017. The bold blue numbers denote the contingency table boxes used to calculate the event-based forecast performance metrics

and green lines. It is obvious that there are many fewer points to the left and right of the vertical green and red lines than there are above and below the horizontal red and green lines. This indicates that the forecasts are biased toward the prediction of typical conditions.

The analogous scatter plot for the forecasts of the 08.00–11.00 HST change in solar generation is shown in Figure 8.21. As expected, the majority of the points are in the central box, which depicts the days for which the forecasted and observed morning rise in BTM solar generation were in the middle 80% of the actual distribution. As with the wind forecast plot, correct forecasts of the upper and lower 10% of the distribution are in the lower left (#7) and upper right (#3) boxes. There are many more points in the upper right box than in the lower left box. This indicates that the forecast system exhibited more skill at identifying days with a much above average morning increase in solar generation (i.e., much less than normal cloudiness—the very sunny days) than days with a much below average increase in solar generation (i.e., much cloudier than normal days).

The forecast performance information contained in the scatter plots can be transformed into a performance metric in a number of ways. One straightforward possibility

Wind	Forecasted				
Observed	*Category*	Down	No change	Up	Obs freq
	Up	2	27	0	10.3%
	No change	13	243	3	80.7%
	Down	4	29	0	9.0%
	Fore freq	**5.9%**	**93.1%**	**0.9%**	**100.0%**

Solar	Forecasted				
Observed	*Category*	Below	Typical	Above	Obs freq
	Above	0	8	24	10.1%
	Typical	1	232	11	79.5%
	Below	4	27	0	10.4%
	Fore freq	**1.6%**	**87.0%**	**11.4%**	**100.0%**

Figure 8.22 Forecasted versus observed outcome contingency table for system-wide 05.00 HST (top) wind generation and (bottom) solar generation forecasts for all days in 2017 with available forecasted and observed data

is to define an "event" as the occurrence of a tail case (i.e., outcome within the upper or lower 10% of the observed outcomes) and then employ event-based metrics to quantify the forecast performance. Event-based forecast performance metrics can be computed from a contingency table, which essentially represents a count of the points in each box of Figures 8.20 and 8.21. The corresponding contingency tables are shown in Figure 8.22. The following metrics expressed in terms of the box numbers employed in Figures 8.20–8.22 can be used to quantify the performance:

Hit rate (HR) = (forecasted observed events)/(# of observed events)

$$= (\#3 + \#7)/(\#1 + \#2 + \#3 + \#7 + \#8 + \#9)$$

Miss rate (MR) = (unforecasted observed events)/(# of observed events)

$$= (\#1 + \#2 + \#8 + \#9)/(\#1 + \#2 + \#3 + \#7 + \#8 + \#9)$$

False alarm rate = (forecasted events with no occurrence)/(# of forecasted events)

$$= (\#1 + \#4 + \#6 + \#9)/(\#1 + \#4 + \#7 + \#3 + \#6 + \#9)$$

Critical success index (CSI) = (number of correctly forecasted events)/(sum of observed and forecasted events)

$$= (\#3 + \#7)/(\#1 + \#2 + \#3 + \#4 + \#6 + \#7 + \#8 + \#9)$$

The values of these metrics for the 05.00 HST wind and solar generation forecasts for 2017 are shown in Figure 8.23. As suggested by the scatter plots, the solar forecasts exhibited more skill in forecasting the tail events than the wind generation forecasts with an overall CSI of 37.3% for the predictions of the solar events versus a CSI of only 5.1% for the wind event forecasts.

Metrics	Wind	Solar
Hit rate	6.5%	44.4%
Miss rate	93.5%	55.6%
FA rate	81.8%	30.0%
CSI	5.1%	37.3%

Figure 8.23 Values of event-based metrics for system-wide 05.00 HST wind and solar generation forecasts for 2017 for an island grid system in Hawaii

The key point is that much of the value to operational decision-making in a reliability-focused environment is based on the ability of the forecast to anticipate significant changes in generation associated with atypical conditions (i.e., tail events), but typical forecast evaluation approaches tend to be heavily weighted to the measurement of forecast performance under typical conditions and therefore most forecasts are optimized (i.e., tuned) to maximize performance under these conditions as measured by the MAE or RMSE metrics. However, it is likely that forecasts for the tail events would exhibit more skill if a forecast system was optimized for this purpose.

Of course, the attributes of the forecast that are most critical to a user will depend upon the application. The attributes that are important to system operators will likely be different from those that are critical to other users such as energy traders. The key point is that the optimization objective for the forecasts and the method of forecast evaluation should be consistent and focused on the way in which the targeted user's application is sensitive to forecast error. This will maximize the likelihood that optimal value will be obtained from the forecasts for each application.

References

[1] Zack, J., E. J. Natenberg, S. Young, *et al.*, 2010: "Application of ensemble sensitivity analysis to observation targeting for short term wind speed forecasting in the Washington-Oregon region," LLNL Technical Report LLNL-TR-458086.

[2] Zack, J., E. J. Natenberg, S. Young, J. Manobianco, and C. Kamath, 2010: "Application of ensemble sensitivity analysis to observational targeting for short term wind speed forecasting," LLNL Technical Report LLNL-TR-424442.

[3] Zack, J., E. J. Natenberg, S. Young, *et al.*, 2010: "Application of ensemble sensitivity analysis to observation targeting for short term wind speed forecasting in the Tehachapi region winter season," LLNL Technical Report LLNL-TR-460956.

[4] Coffier, J., 2012: Fundamentals of Numerical Weather Prediction (Reprint ed.), Cambridge, Cambridge University Press, 368 pp. ISBN-10: 110700103X, ISBN-13: 978-1107001039.

[5] Skamarock, W., B. K. Joseph, J. Dudhia, *et al.*, 2008: "A Description of the Advanced Research WRF Version 3," NCAR Tech. Note NCAR/TN-475+STR, 113 pp.

[6] WAsP Manual: Wind Analysis and Application Program (WAsP), 1993: Vol 2: Users Guide. Roskilde, Denmark: Risø National Laboratory, ISBN 87-550-178.

[7] Walmsley, J. L., P. A. Taylor, and T. A. Keith, 1986: Simple model of a neutrally stratified boundary-layer flow over complex terrain with surface roughness modulations (MS3DJH/3R). Boundary-Layer Meteorology, 36, 157–186.

[8] Geisser, S., 2017: Predictive Inference: An Introduction. New York: Chapman & Hall. 240 pp. ISBN 0-412-03471-9.

[9] Haykin, S., 1998: Neural Networks: A Comprehensive Foundation (2nd ed.). Upper Saddle River, NJ, Prentice Hall. ISBN 0-13-273350-1.

[10] Wilks, D. S., 2011: Statistical Methods in the Atmospheric Sciences, Volume 100 (International Geophysics Series), 3rd ed. Cambridge, MA, Academic Press, 704 pp, ISBN 0123850223.

[11] Delle Monache, L., T. Eckel, D. Rife, B. Nagarajan, and K. Searight, 2013: Probabilistic weather prediction with an analog ensemble. Monthly Weather Review, 141, 3498–3516.

[12] Smola, A. and B. Scholkopf, 2004: A tutorial on support vector regression. Statistics and Computing, 14, 199–222.

[13] Cortes, C. and V. Vapnik, 1995: Support-vector networks. Machine Learning, 20 (3), 273.

[14] Drucker, H., J. C. Burges, L. Kaufman, A. J. Smola, and V. N. Vapnik, 1997: Support Vector Regression Machines. In: Advances in Neural Information Processing Systems 9, NIPS, 1996, pp. 155–161, MIT Press.

[15] Kramer, O. and F. Gieseke, 2011: Short-Term Wind Energy Forecasting Using Support Vector Regression. In: Corchado E., Snášel V., Sedano J., Hassanien A. E., Calvo J. L., Ślęzak D. (eds) Soft Computing Models in Industrial and Environmental Applications, 6th International Conference SOCO 2011. Advances in Intelligent and Soft Computing, vol 87. Springer, Berlin, Heidelberg.

[16] Wasserman, P. D. and T. Schwartz, 1988: Neural networks. II. What are they and why is everybody so interested in them now? IEEE Expert, 3 (1), 10–15.

[17] Hastie, T., R.Tibshirani, J. Friedman 2009:The Elements of Statistical Learning. Springer-Verlag New York, ISBN 978-0-387-84857-0, doi 10.1007/978-0-387-84858-7, 745 pp.

[18] Castellani, F., D. Astolfi, M. Mana, M. Burlando, C. Meiner, and E. Piccioni, 2016: Wind Power Forecasting techniques in complex terrain: ANN vs. ANN-CFD hybrid approach. Journal of Physics: Conference Series 753, 082002. https://doi.org/10.1088/1742-6596/753/8/082002.

[19] Breiman, L., J. H. Friedman, R. A. Olshen, and C. J. Stone, 1984: Classification and Regression Trees. Monterey, CA: Wadsworth & Brooks/Cole Advanced Books & Software. ISBN 978-0-412-04841-8.

[20] Breiman, L. 2001: Random forests. Machine Learning, 45 (1), 5–32. doi:10.1023/A:1010933404324.

[21] Lahouar, A. and J. Ben Hadj Slama, 2017: Hour-ahead wind power forecast based on random forests. Renewable Energy, 109, 529–541, ISSN 0960-1481.

[22] Friedman, J. H., 2001: Greedy function approximation: a gradient boosting machine. The Annals of Statistics, 29 (5), 1189–1232.

[23] Friedman, J. H., 2002: Stochastic gradient boosting. Computational Statistics & Data Analysis, 38 (4), 367–378.

[24] Chen, T. and C. Guestrin, 2016: Xgboost: a scalable tree boosting system. In: Proceedings of the 22nd ACM SIGKDD International Conference on Knowledge Discovery and Data Mining, KDD '16, pp. 785–794, New York, NY, USA. ACM.

[25] Jin, Y. and J. Chuang, 2017: Tree Boosting With XGBoost?—?Why Does XGBoost Win "Every" Machine Learning Competition? AI Technology & Industry Review – www.syncedreview.com. Available at https://medium.com/syncedreview/tree-boosting-with-xgboost-why-does-xgboost-win-every-machine-learning-competition-ca8034c0b283.

[26] Cooperman, A., C. P. van Dam, J. Zack, S-H. Chen, and C. MacDonald, 2018: "Improving short-term wind power forecasting through measurements and modeling of the Tehachapi wind resource area," California Energy Commission. Publication Number: CEC-500-2018-002.

[27] Zack, J. and S. Young, 2017: Impact of Targeted Measurements and Advanced Machine Learning Techniques on 0–3 hr Ahead Rapid Update Wind Generation Forecasts in the Tehachapi Wind Resource Area. In: Proceedings of the 16th International Workshop on Large-Scale Integration of Wind Power into Power Systems, Berlin, Germany, 25–27 October 2017, ISBN: 978-3-9816549-6-7.

[28] Glahn, H. R. and D. A. Lowry, 1972: The use of Model Output Statistics (MOS) in objective weather forecasting. Journal of Applied Meteorology, 11, 1203–1211.

[29] Vannitsem, S., 2008: Dynamical properties of MOS forecasts: analysis of the ECMWF operational forecasting system. Weather and Forecasting, 23, 1032–1043.

[30] Pedregosa, F., G. Varoquaux, A. Gramfort, *et al.*, 2011: Scikit-learn: machine learning in Python. Journal of Machine Learning Research, 12, 2825–2830.

[31] Ahlstrom, M., D. Bartlett, C. Collier, *et al.*, 2013: Knowledge is power: efficiently integrating wind energy and wind forecasts. IEEE Power and Energy Magazine, 11 (6), 45–52. doi: 10.1109/MPE.2013.2277999.

Chapter 9
Cost of wind energy modeling

M. Maureen Hand[1,2], Volker Berkhout[3], Paul Schwabe[1], David Weir[4], and Ryan Wiser[5]

9.1 Introduction

Cost of wind energy is a critical aspect of wind-power-plant modeling and simulation. This metric is useful when evaluating the potential impact of technology design innovation. When combined with consideration of the value of wind energy, it enables the analysis of investment decisions, informs policy design, and facilitates comparison of electricity-generation technologies.

A wind power plant is a complex, interconnected, system of systems. The incremental cost of technology innovation must be balanced by incremental energy capture in order to provide an overall system benefit. For example, innovative designs that allow longer wind-turbine blades to be produced with reduced material inputs enable a wind turbine to increase energy yield at a lower cost. Investment in equipment that monitors the health of individual components to signal maintenance or scheduled replacement can avoid unplanned turbine downtime and lost energy production. The cost of wind energy, estimated over the expected lifetime of the wind power plant, captures this fundamental relationship and indicates the possible value of new design or operation concepts.

Investment decisions require transparency around initial and future costs as well as projected revenue. The timing of expenditures and the corresponding time value of money are important in both the construction and operational phases of the life of a wind power plant. A full understanding of the required investment to bring a wind power plant to commercial operation, along with all possible revenue streams, provides insight into the commercial viability of wind-power-plant deployment. This applies to new wind power plants as well as to evaluating options of refurbishing, or repowering, existing wind power plants. Differences between cost and revenue are often addressed through policy mechanisms such as taxes or direct payments like

[1]National Renewable Energy Laboratory, Golden, CO, USA
[2]California Air Resources Board, Sacramento, CA, USA
[3]Fraunhofer Institute for Energy Economy and Energy System Technology, Kassel, Germany
[4]Norwegian Water Resources and Energy Directorate, Oslo, Norway
[5]Lawrence Berkeley National Laboratory, Berkeley, CA, USA

feed-in premiums. Cost of wind energy is a necessary component of investment decisions and incentive design.

Wind power plants operate in the context of the electricity system among a number of different electricity-generation technology options. The lifetime cost of any electricity-generating technology must be weighed against the value that technology provides to the electricity system, in terms of energy as well as ancillary services required to maintain electricity system operation at all times.

In each of the applications mentioned earlier, cost of wind energy is a critical input. There are four primary aspects required to estimate the cost of energy over the lifetime of a wind power plant. These include initial investment costs to bring a plant to commercial operation, operating costs throughout the lifetime of the wind power plant, financing costs for investors to recover capital investments with an appropriate return, and the primary product—energy produced over the lifetime of the wind power plant.

This chapter focuses on considerations made when modeling cost of energy to be used in any of the abovementioned applications. The level of fidelity required to model each of the four, primary cost of energy elements depends on the characteristics of the wind power plant under consideration. Some "top-down" models utilize empirical data to reflect currently installed wind power plants, while physics-based, or "bottom-up," models may be implemented to estimate aspects of future wind-power-plant technology. Regardless of the ultimate application of cost of energy modeling, it is important to carefully assess the wind-power-plant characteristics to determine the type of model and the level of fidelity required to represent each of the four cost of energy elements.

For example, evaluating the impact of an innovative offshore wind-plant-maintenance strategy may include the following cost of energy input considerations:

- Empirical data to represent initial investment costs based on a sample of recently installed wind power plants;
- bottom-up modeling of wind-power-plant operating costs that include wind-power-plant downtime associated with a combination of service vessel characteristics and weather conditions expected over the life of the wind power plant;
- physics-based energy production modeling that reflects representative wind-speed characteristics and incorporates the operation and maintenance downtime associated with service vessel access and weather conditions; and
- empirical data to represent financing costs based on an assessment of interest rates and investor interviews.

A levelized cost of energy (LCOE) estimate for the innovative offshore wind-plant-maintenance strategy compared with the LCOE for the status quo technology may provide relevant information.

In contrast, when modeling an investment decision for a wind power plant in a specific geographic location, the cost of energy input considerations may include

- empirical data, such as vendor quotes, to represent initial investment costs and/or projected operating costs over the life of the wind power plant;

- physics-based energy production models that capture site terrain effects and daily or annual wind speed and direction variations to obtain a distribution of estimated energy production over the life of the wind power plant; and
- return on investment requirements associated with each individual investor.

Discounted cash-flow models that explicitly represent cash flow to each individual investor, and expected revenue over the life of the wind power plant may be applied to solve for power-purchase-agreement prices.

Estimating wind-power-plant cost of energy elements for applications such as policy incentive design or comparison with other electricity-generation technologies over the next decade may include considerations such as

- projections of future initial investment costs and operating costs based on empirical data trends and/or assessments of expected technology advancements conducted using physics-based models;
- physics-based energy production models that reflect high-level, annual wind-speed characteristics over a range of relevant geographical locations; and
- empirical data to represent financing costs based on cost of capital expectations.

LCOE estimates may provide an initial indicator. Policy design studies will require the analysis of expected revenue over the period. Comparison of wind power plants with other electricity-generation options will require an assessment of the wind power plant in the broader electricity system.

To isolate the cost of energy from anticipated revenue or subsidy schemes, a commonly used metric that captures these four elements is the LCOE.

9.2 Levelized cost of energy (LCOE)

LCOE represents all investment and operating costs over the lifetime of the wind power plant as well as the lifetime energy produced by the plant as shown in the following equation:

$$\mathrm{LCOE} = \frac{\sum_{n=1}^{N} (C_n + M_n)(1+r)^{-n}}{\sum_{n=1}^{N} E_n(1+r)^{-n}} \tag{9.1}$$

where C_n is the capital expenditures in year n, M_n is the operating expenditures (OPEX) in year n, E_n is the energy produced in year n, N is the expected lifetime of wind power plant, and r is the discount rate.

Investment costs, or capital expenditures (CAPEX), include all expenses required to bring a wind power plant to commercial operation. Development costs incurred before construction begins as well as costs incurred throughout the construction phase are included. Decommissioning costs set aside at project initiation are also included.

OPEX that are expected throughout the life of the wind power plant are estimated. These costs include property taxes and land leases, regularly scheduled maintenance, and estimated costs for unplanned activities that may include component replacement.

Maintenance strategies affect the amount of time a wind turbine or wind power plant is able to operate, and thereby the estimated annual energy production (AEP).

Energy production estimates over the life of the wind power plant reflect expected annual averages. Consideration of interannual variability in wind resource as well as planned and unplanned downtime is included, typically in terms of probability of occurrence. Wind-turbine availability is directly related to the operation and maintenance strategy.

The cost of capital, or cost to finance a wind power plant, reflects the debt and equity contributions from all wind-power-plant investors. The cost of capital reflects the risk profile of various investors and is often reflected in the discount rate.

The expected lifetime of the wind power plant is inherently represented in each of the four primary input parameters. Capital investments may be made at project initiation to accommodate a longer lifetime. Operational strategies may be implemented based on expected component-replacement schedules. AEP is influenced by downtime for component replacement as components age. The cost of capital reflects the investor time horizon for capital recovery.

There are many variations of (9.1), which capture a wide range of detail. Comparing LCOE values requires care to assure that similar aspects are represented. Typically, LCOE excludes revenue and the impacts of subsidies; LCOE represents only the expenditures associated with installing and operating a wind power plant.

As noted earlier, the four, primary cost of energy elements required to compute LCOE or to model cost of energy are also relevant when considering applications that include assessment of the value of wind energy.

9.3 Overview of cost of energy modeling

Modeling the cost of wind energy requires translating physical wind-power-plant design characteristics, such as wind-turbine geometry, wind-power-plant layout, and wind-turbine component design into lifetime cost and energy production estimates as illustrated in Figure 9.1.

Evaluation of cost of energy, particularly with respect to the introduction of innovative technologies and operation strategies, requires tradeoffs between investment cost (including capital costs and balance-of-station (BOS) costs), operational costs, and energy yield (expressed as AEP in Figure 9.1).

Wind-power-plant design to increase energy capture must ultimately have lower incremental cost than incremental energy gained. Wind-power-plant energy production may be substantially increased by siting a plant further from shore in stronger wind resources. However, the additional cost to design, manufacture, and install innovative substructures to accommodate deeper water must be offset by the additional energy capture.

During the design phase, a large number of decisions are made that affect the maintainability of the turbine, the durability of components, and the reliability of the system. While an automatic lubrication system or additional sensor equipment increases initial investment cost, the investment may pay off over the operational

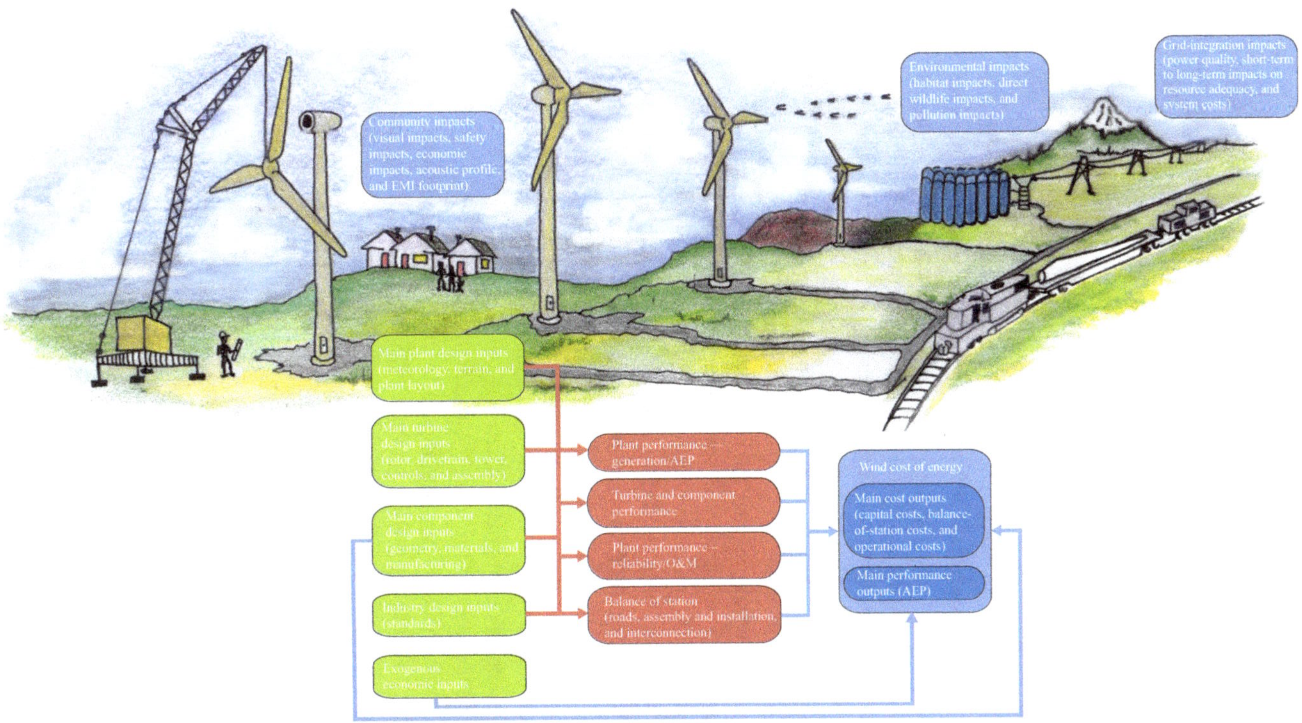

Figure 9.1 Wind-power-plant systems, including cost of energy (artist: Rick Hinrichs) [1]

lifetime of the wind power plant through reduced operational expenditures. Components may be designed to exactly meet the given load case criteria or to exceed the criteria with larger safety margins or redundancies, thereby reducing failure risks and hence, operational expenditures. In addition, these design considerations may increase the probability of extending the wind-power-plant lifetime, energy yield, and/or revenue, again at the cost of higher initial investment costs.

The introduction of new technology designs and wind-power-plant operation strategies requires demonstration and experience in order to be widely deployed. The financial community may perceive increased risk associated with emerging technologies, which results in higher cost of capital to finance these wind power plants. As technologies are proven, the cost of capital may be reduced.

Investment cost, operating cost, cost of capital, and energy production are dependent on the wind-power-plant characteristics and, as such, highly interrelated. The interdependency among these cost of energy parameters poses a significant modeling challenge. Models often focus on one parameter, but the combined effect of all parameters ultimately affects business decisions related to the cost of wind energy. Because LCOE represents the life cycle costs and energy production of a wind power plant, it is particularly a useful metric for comparing wind-power-plant characteristics. The level of fidelity in modeling each of the four parameters required to estimate LCOE depends upon the wind-power-plant characteristics.

Understanding the impact of a specific technology innovation applied to an existing wind power plant requires a different modeling approach than quantifying the impact of that same innovation on a conceptual wind power plant to be installed at a future date. Determining the cost of energy of a wind power plant to be installed in a specific geographic location differs from understanding how the cost of energy varies across a range of geographical sites. For each of these applications of cost of energy modeling, the level of fidelity required to estimate each of the four, primary cost of energy components may differ.

As described later in this chapter, there are two basic characteristics of models used to estimate the four primary inputs to cost of energy. Some models rely on empirical data representing wind power plants that are currently installed and operating or were at one time. These models typically reflect annual average values and may be considered "top-down." Other models rely on physics-based or design-based relationships that provide a predictive capability based on fundamental wind-power-plant characteristics that may or may not represent existing wind power plants. These models can incorporate time-varying or stochastic inputs and may be considered "bottom-up." Most cost of energy estimates include some combination of empirical data and physics-based predictive capability.

Empirical data is useful for the high-level assessment of cost of wind energy within a given country or region that may be compared to other countries or regions. Empirical data reflects current wind-power-plant technology and is often reported through market reports or other financial disclosures. The use of plant-level statistics representing currently installed and operating wind power plants may provide sufficient information for comparing cost of energy among countries and identifying trends over time [2–5]. In the case of using average values, it is important to recognize the inherent variation among individual wind power plants associated with wind resource conditions and project ownership structures, among other things. However, empirical data can provide valuable insights.

Estimating the cost of future wind-power-plant technology typically requires physics-based models. Empirical data does not accommodate innovation that changes fundamental scaling relationships. The level of detail required depends on the innovation considered For example, estimating the impact of scaling to higher rated capacity wind turbines in offshore wind applications (e.g., from 5 MW wind turbines to 10 MW wind turbines) requires detailed investment cost models to capture the relationships between the turbine substructure size, material content, fabrication time, and installation time—as well as the wind turbine itself. While detailed investment cost models may be required, lower fidelity OPEX, energy production, and financing cost estimates may suffice for this example.

The level of fidelity of physics-based models used to estimate each of the four cost of energy elements will vary based on the wind-power-plant characteristics. Representation of wind power plants that may be installed globally or in a large geographic region will have lower fidelity investment cost, operating cost, energy production, and cost of capital estimates than representation of the cost of energy elements of a wind power plant intended for a specific wind-power-plant technology, wind resource climate, or geographic location.

9.4 Modeling investment costs

Investment costs, or CAPEX, represent all costs required to bring a wind power plant to commercial operation. This includes the development phase, the installation phase, and the commissioning phase. To the extent that funds are set aside at project initiation to cover costs of decommissioning the wind power plant at the end of the project life, these costs are also represented as investment costs. Investment costs are typically expressed in terms of currency per unit of wind-power-plant capacity, \$/kW or \$/MW.

Figure 9.2 provides examples of investment cost categories for an offshore wind power plant. In addition to the wind turbine and all subsidiary components, BOS and financial costs contribute to CAPEX. The BOS costs include expenditures associated with the substructure that supports the wind turbine, assembly and installation of the wind turbines, the electric infrastructure that connects the turbines to each other and the wind power plant to the electric grid, wind-power-plant infrastructure to support operations and maintenance of the wind power plant, and costs associated with developing and designing the wind power plant. Financial costs include insurance, carrying charges during construction, and funds reserved for contingencies or future expenditures. The system cost breakdown structure shown in Figure 9.2 identifies six levels of capital and operating cost elements, more than 300 distinct components [6] for both land-based and offshore wind power plants.

Decommissioning costs for wind power plants are not typically set aside in reserve when the project is initiated. These costs are often excluded in cost of energy analysis because there are options as a wind power plant ages. In some cases, wind power plants

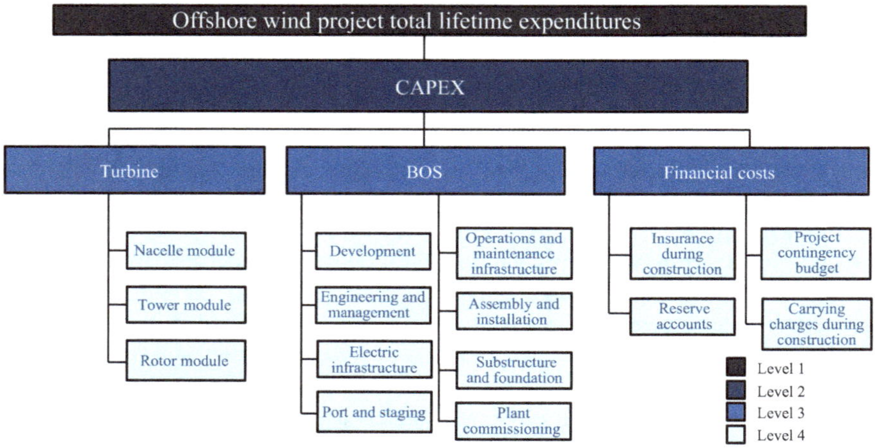

Figure 9.2 Wind-power-plant system cost breakdown structure: CAPEX levels 1–3 [6]

are fully decommissioned including removal of turbines and supporting infrastructure. In other cases, wind turbines may be refitted with new blades, towers, and/or drive trains; this option is often called repowering and essentially gives a wind power plant an extended life. Another example of repowering includes complete removal of several small turbines and installation of fewer, larger turbines in a nearby location. These options are typically evaluated as a "new" wind power plant, and costs are estimated independent of the original wind power plant.

Investment costs for wind power plants that are currently operating or are in planning stages may be available from a variety of sources. This empirical data provides one valuable method for modeling cost of energy, primarily related to historical or existing wind power plants. Empirical data reflecting the investment costs of wind power plants installed to date is collected and presented by a number of organizations [4,7–9]. In some countries, wind-power-plant owners are required to provide this data to governments; and in other countries, this data is not readily available. It is important to ascertain that investment cost data represents the same scope of development, installation, and construction costs when comparing data from different sources.

When modeling cost of energy of plants to be installed in the future, new technologies that differ from existing wind power plants may be incorporated. A bottom-up or component-level modeling approach is needed. Empirical data related to existing wind power plants may be used to validate component-level models.

Capital investment costs for wind power plants that incorporate new or different wind-power-plant technologies may be estimated using models that estimate individual component costs based on wind-power-plant characteristics (see also Volume 2, Chapters 3, 4, 5, and 7). Empirical cost data related to weight, size, water depth, and distance from shore or other technical characteristics can be used to develop curve fits [10–13]. Component sizing tools incorporate physical design characteristics, such as design-load conditions to estimate the dimensions and mass of components and thereby the cost of individual components [14,15]. For example, a drive-train sizing tool may be used to estimate the component size and mass to compare a range of drive-train topologies such as a standard gearbox and generator or a direct drive, permanent magnet topology or a novel concept not currently deployed in wind power plants [16]. Process-based tools reflect the type of equipment and time required to estimate the cost of completing tasks associated with wind-power-plant construction. For example, installation of a wind turbine in an offshore wind power plant may require different vessels to transport, lift, and install the substructure, tower, nacelle, and blades. Manufacturing models associate the quantity of material, fabrication method, and time to estimate the cost of producing wind-power-plant components [17]. In some cases, the cost estimates include macroeconomic changes to prices of commodities such as steel or concrete [12,18].

9.5 Modeling energy production

Wind power production is inherently variable on both short and long time scales. The main challenge in modeling energy production for use in calculating LCOE is to achieve the best estimate of the average AEP, which can be sold, for the lifetime of the wind power plant, generally expressed in MW h/year. When modeling AEP, it is important to account for production losses and uncertainties.

AEP can be modeled both pre-construction and post-construction. High-fidelity pre-construction estimates of AEP are central to financial investment decisions for wind power plants. Lower fidelity pre-construction estimates based on generalized wind-resource conditions are frequently used in other applications. Post-construction AEP estimates can serve to assess the value of existing windfarm assets based upon their performance, or as a means to identify shortcomings of pre-construction AEP estimates. Both pre and post-construction AEP estimates are relevant in cost of energy modeling.

While different input data are available for pre- and post-construction modeling of AEP, the general modeling approach lends itself to characterization as a measure-correlate-predict problem. In situ *measured* wind or production data from a limited time period must be *correlated*/referenced to long-term data in order to correct for seasonal or interannual variability. After correcting for short-term effects in the data, further modeling and assumptions about uncertainties allow *prediction* of the future AEP of a wind power plant.

The AEP of a single wind turbine will depend on wind conditions at the location of the turbine and the energy capture, which depends on the type of turbine deployed. This, in itself, presents an enormous modeling challenge, traversing scales from global weather patterns down to microscale, boundary-layer interaction between air molecules and wind-turbine blade surfaces. Lacking any single, computationally feasible model for solving this problem, energy production for wind turbines, in practice, is estimated using a combination of empirical and model data.

Wind-turbine manufacturers measure power curves for different wind-turbine models, which summarize effects happening across many of the smaller scales by showing the measured gross electrical output of the wind turbine (in kW or MW) vs. the hub-height wind speed (in m/s). A key aspect of estimating energy production for a wind turbine is thus to accurately model the wind conditions at the site. The output of a wind turbine will also vary based upon other factors than wind speed, such as air density, wind shear, or turbulence. It is important to understand how these conditions will affect the production of the wind turbine over its lifetime, but presently they tend to be estimated linearly or through modified power curves. Going from a single wind turbine to a wind farm, wake effects can significantly impact the AEP. Layouts for wind farms should take into account prevailing wind directions and give adequate distance between turbines to reduce wake losses for the wind farm as a whole. Modeling of wake effects from wind turbines is a field of its own (see Volume 1, Chapter 6), but despite significant research efforts on advanced wake modeling, the AEP for most wind farms is calculated using simple, linear models for wind velocity deficits such as the Jensen model [19].

In pre-construction AEP modeling, the general approach will be to accurately measure the wind speed at several points within the planned wind farm to assess terrain effects on the wind, and at several heights above ground to study the wind shear conditions. Wind measurements must be made over a long enough time period that they are representative of the range of different seasonal wind conditions that can arise at the site, generally at least 1 year [20]. Even with all seasons represented in a measurement series, wind conditions alone can lead to variations in the production of ±30% from 1 year to another. In using shorter term wind measurements to estimate the average annual production of a planned wind farm over a 20-year lifetime, the seasonal and interannual variability of wind must be taken into account. This can be done by correlating the measurements to another, overlapping but longer term wind-speed data source such as a long-term weather station or output for a meteorological model hindcast. The optimum combination and length of measured and reference data series, as well as the approaches used, are site specific since different wind regimes can present different levels of complexity.

Once wind measurements are long-term corrected, they need to be extrapolated to the position of all of the wind turbines in the planned wind farm to make estimates of the production for each turbine. Average annual wind speeds can vary considerably over short distances due to the effects of, e.g., terrain or forest canopy. This creates the challenge of correctly modeling the microscale flow in the wind farm area and its implications for local average wind conditions. Traditionally, this flow modeling has been done with models based upon linearized versions of the Navier–Stokes equations, but increasingly, in later years, the more advanced nonlinear methods of computational fluid dynamics, such as Reynolds averaged Navier–Stokes and large eddy simulation, are being applied in the wind energy industry.

Each wind site presents challenges in terms of accurately modeling the wind resources. At some sites, most of the modeling efforts need to be focused on understanding local thermal phenomena such as low-level jets or sea breezes, whereas other sites may present challenges with turbulence arising from complex terrain or forest canopy. An ongoing challenge for wind resource modeling is to find ways to efficiently identify which local meteorological or terrain effects will have the highest impact on the wind power production for the site, and how to measure and model the wind resources in the most cost-effective but accurate manner possible.

After sufficiently understanding the wind resources at a site, the gross AEP can be estimated using wind speeds and power curves (standard or modified) for the individual wind turbines. The gross AEP represents the best guess potential production before losses, and predicted losses must be applied to this value to arrive at the net AEP, which is the production relevant to cost of energy modeling. Figure 9.3 provides a conceptual illustration of gross and net energy estimates. Typical losses include wake losses, as mentioned before, and losses arising from downtime/system unavailability, electrical losses, high wind shear, turbulence, high-wind hysteresis, icing, and blade degradation. The individual sources of energy losses must be summarized to a total percent loss in a way that accounts for their simultaneity or lack thereof. Predicting percentage energy losses from different sources presents its own set of modeling challenges and may overlap with the modeling of operations expenditures. In cases where losses can be avoided or reduced by maintenance, cost-benefit modeling may be a method for deciding or negotiating maintenance strategies.

In addition to losses, pre-construction AEPs often account for uncertainty. Uncertainties arise from the various measurements and models used in calculating the AEP, such as the wind measurements, modeling vertical and horizontal extrapolation to wind-turbine positions, and power curves. The sum of the uncertainties is often assumed to be a probability distribution for the AEP with different AEPs having different probabilities of exceedance. The gross AEP minus the total losses will usually constitute the central estimate (P50) AEP, meaning that the uncertainties suggest a 50% chance that the actual production will exceed this value, and a 50% chance that it will not exceed this value. This is the most relevant estimate for cost of energy modeling, although banks and other financiers will typically be interested in additional levels, such as the P90 AEP level, where there is a 90% chance that the actual production will be higher, so that they can understand the risk presented by the sum of the uncertainties. Understanding the uncertainties in this way can allow financiers to factor in the AEP uncertainties in their risk modeling. Figure 9.3 illustrates the use of predicted losses in going from a gross production estimate to a net production estimate. The figure also illustrates a range of uncertainties assumed around a central estimate of AEP.

Post-construction estimates of AEP can be made with varying levels of complexity in cost of energy modeling. In reassessing the value of a wind farm, an owner may conduct a detailed analysis of production data compared to wind conditions and known information about maintenance downtime or other factors to reevaluate the AEP for the wind power plant. On a higher level, a government or policy analyst may analyze the AEP for several wind farms based on empirical operational data in order to track the reduction of wind-power LCOE over time. Despite different levels of complexity, each problem presents the challenge of using measured production data for one or several years to predict the average annual production over the lifetime of the wind power plant. In each case, the interannual variability of the wind will be

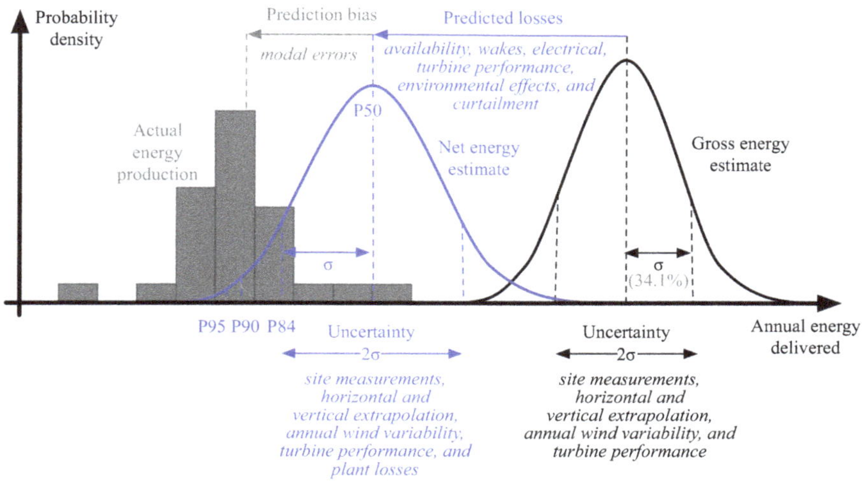

Figure 9.3 The net energy estimate is a key part of the pre-construction energy estimate. This figure illustrates a case where the actual energy production is less than the net energy estimate [21]

important to take into account, and this is often done by dividing a year's production by a calculated wind anomaly (in percent of long-term average) for that year. Examples of other factors that should be considered where known are instances of extraordinary downtime due to maintenance or curtailment due to grid restrictions. When accounting for these factors, an assessment should be made of whether these problems are expected to persist over the lifetime of the wind power plant.

9.6 Modeling operational expenditures

Expenditures for the operation of a wind farm (OPEX) represent all costs required to maintain and operate a wind power plant throughout its life cycle. This includes costs such as insurance, property taxes, and technical management, which may be roughly the same from one year to the next. It also includes maintenance and repair costs, which vary annually as the project ages. The present value of operational costs estimated over the lifetime of the wind power plant is typically expressed in terms of currency per unit of wind-power-plant capacity per year, \$/kW-year, or in terms of currency per unit of energy produced per year, \$/kW h-year or \$/MW h-year.

Operational expenditures include all payments, which occur after the installation and commissioning of a wind farm. There is no standardized break down of cost categories defined. A common cost split includes [22]

- maintenance and repair;
- land (or sea) lease;
- management, administration, and technical operations;

- insurance;
- reserves; and
- other.

Maintenance and repair is the largest cost category and reflects the greatest variability over the life of the wind power plant. Other operational expenditures tend to be less variable over the project life but may be significantly different based on the geographic location of the wind power plant and the operator's choice of contractual structure. Additional detailed categorization and more detailed insights into OPEX are available from [4,6,23] and with a specific offshore focus in [24,25].

A key decision made by the wind-power-plant owner is whether the operation and maintenance activities should be performed by the owner as an in-house service or contracted to a service provider. The service provider could either be the service branch of the original equipment manufacturer or an independent service provider. An original equipment manufacturer has access to unique information regarding wind-turbine component characteristics, and therefore is the most common type of operation and maintenance service provider. The service provider guarantees a negotiated availability of the wind power plant to the operator; this guaranteed wind-power-plant availability is often a critical risk assessment criterion for financial institutions when determining whether or not to finance a project. Wind-power-plant availability can be defined in different ways, either time-based or energy-based. For cost of energy modeling, energy-based availability is more relevant as it accounts for the timing of system unavailability and whether it occurs during wind periods when the wind power plant could be producing power. At the same time, the service provider is incentivized to optimize maintenance activities for cost reduction as long as the guaranteed availability can be maintained. Service contracts differ in the range of services and risks they cover as well as the contract duration.

Lost energy production and the corresponding lost revenue during unplanned downtimes due to component failures are significant risks that are mitigated through service contracts with guaranteed availability. The lost revenue, particularly for offshore wind power plants with limited accessibility, will quickly outweigh the cost for additional detection and prevention measures.

As more risks are covered in the service contract, the maintenance and repair payments are regarded as a type of insurance rather than as costs that can be associated with specific operation and maintenance activities. This effect makes it difficult to estimate and categorize specific expenditures. Table 9.1 illustrates the share of OPEX for the primary categories for the first decade of wind-power-plant operation and the second decade. The split between fixed and variable costs depends on the contractual structure. Over time, the proportion of annual costs associated with maintenance and repair increases while the proportion of costs that are largely fixed decreases.

Little information is publicly available that allows for a direct, bottom-up estimation of the use of resources as most of the service is contracted and not explicitly disclosed. Available public data often references industry insights, expert interviews or just states assumptions. There is little information available on a regular time scale, and there is a lack of transparency in the quality of the data.

Empirical OPEX data may come from the financial statements of wind power plants which are owner-operated. Data may be reported on a per plant or even per turbine basis. Statistics based on financial reporting to small investors in wind power plants in Germany is shown in Figure 9.4 [27]. In the UK, data was made public under the capital grant schemes [28]. Some organizations operate internal databases and are able to provide regular findings and updates on that basis for some markets [4,9].

Similarly to modeling investment cost, empirical OPEX data is most useful when representing wind power plants currently in operation. In contrast to CAPEX, OPEX is a dynamic cost that varies from year to year. Empirical data may show cost trends such as the end of the guarantee period or the rise of failures and downtimes due to fatigue and wear effects as the project ages. Newer wind power plants do not yet have a long record of empirical OPEX data, and new technologies are often introduced to reduce future OPEX. The use of empirical data to model OPEX must be done with

Table 9.1 Share of annual OPEX by cost category and decade of operation [26]

Cost category	Average share of OPEX	
	Year 1–10	**Year 11–20**
Maintenance and repair (%)	38	46
Land lease (%)	26	21
Management, administration, and technical operations (%)	15	13
Insurance (%)	3	3
Reserves (%)	7	8
Other (%)	11	9

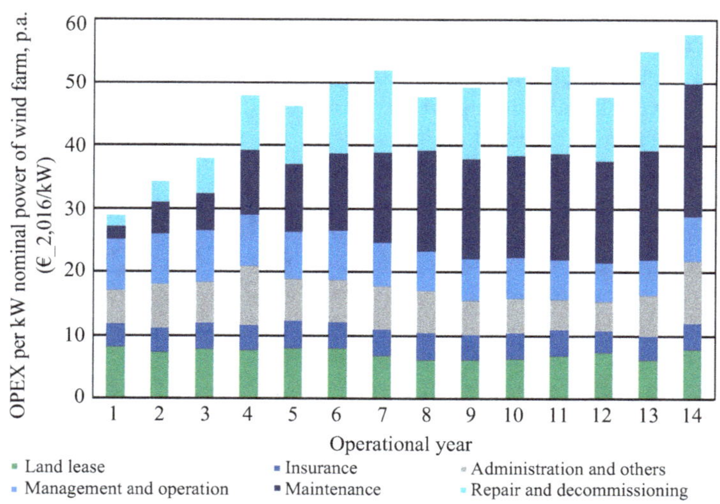

Figure 9.4 OPEX by year of operation for wind power plants in Germany [27]

care to assure representation of life cycle costs for wind power plants with similar technologies, characteristics, and/or ages.

When modeling OPEX for wind power plants to be installed in the future that may incorporate technologies that differ from existing wind power plants, a bottom-up modeling approach is needed. These models focus on required maintenance activities as well as assessing the failure rates of components and the associated repair activities. The resulting actions in terms of resources, such as time required for employees, cranes, vessels, and tools, as well as spare parts, can be estimated in cost terms.

For offshore wind power plants, the logistics cost are a major consideration. The main drivers are the cost for crane vessels, which are needed for major repair and replacement work. Again, these costs depend on the structure of the project and the contractual setup with offshore logistics providers. The price for vessel rentals may strongly vary with the advance time, the vessel availability, and the weather conditions. The accessibility of turbines based on the wind and wave conditions is an important consideration. In this kind of simulation, the assumed failure rates have a large impact on the result as they determine the number of necessary service actions on the turbine, so this choice has to be made with great care.

As shown in Figure 9.5, the wind-power-plant operator must weigh a great deal of information to determine an operation and maintenance strategy. Models have been developed to simulate and evaluate a variety of operation and maintenance strategies [29,30] and compared with each other in reference cases [31]. These models represent the stochastic behavior of component failures, weather-related impacts on operation and maintenance strategies, and availability of staff and spare parts. The failure behavior over time of turbine components may be broken down by component, failure root cause, and failure severity categories [32–35]. Wind and wave conditions follow seasonal patterns and may be correlated. Distributions, such as Weibull, are frequently used to reflect this stochastic behavior in the absence of detailed empirical data.

Figure 9.5 Wind-power-plant-operator data inputs and operation choices. Source: Fraunhofer IEE

Simulations can optimize different objectives, such as reduced LCOE over the life of the wind power plant or increased annual profits. These objectives do not necessarily result in the same OPEX. Optimization schemes balance options depending on the objectives that may include constraints associated with shifting downtime to periods with less energy output and corresponding revenue loss, the ability of vessels to operate safely in certain wave conditions, and the forecast quality of metocean conditions. The use of technicians, vessels, and spare parts has to be prioritized according to different strategies or objectives. The scheduling and routing of vessels and teams to offshore turbines is an optimization problem of its own [36].

The operational phase of a wind power plant is subject to a wide range of stochastic inputs. Stochastic inputs may result in values that vary by day, month, and year throughout the lifetime of the wind power plant. However, modeling wind-power-plant operations is frequently done using statistics that exclude stochastic variation or incorporate some aspects of stochastic variation. Statistics often reflect the levelized, annualized, net present value of all expected operational expenditures.

Considering the tradeoffs between OPEX, wind-power-plant availability, and wind-power-plant lifetime is even more challenging. As the complexity of models increase, the number and type of input parameters grow. The feasibility of obtaining empirical data representing this wide range of inputs to validate models diminishes. The level of detail and complexity required to estimate operational expenditures for a given simulation is a critical modeling decision.

9.7 Modeling cost of capital

Modeling the cost of capital is an important factor in estimating the cost of wind energy. Cost of capital generally refers to the expenses associated with funding a wind power plant and can be composed of a number of distinct capital types including grants, various forms of debt, equity, and other sources. The term "financing costs" is used herein to represent the cost of capital and to distinguish it from similar sounding "capital costs" or "investment costs" intended to represent the CAPEX of a wind power plant. The cost of capital is often selected as the discount rate when computing LCOE.

Determining the appropriate financing costs to use in cost of energy modeling is in exact. Financing costs are influenced by numerous factors, some of which are measurable (e.g., global economic indicators such as benchmark interest rates and lending conditions) while others are more subjective or relationship-based (such as the deal history and experience between an investor and developer that may impact rates). Financing costs are also fundamentally influenced by the supply and demand of investment. In periods of relative capital scarcity, financing costs can rise because of the limited availability of dollars rather than the fundamentals of the wind power plant. Conversely, during periods of high capital availability, steep competition among investors to fund a project can cause investors to accept greater risks or lower rates of return compared to normal practices. Because of these reasons and more, modeling of financing costs tends to be project and time dependent.

Despite the relative subjectivity of determining an appropriate financing cost assumption, it is critically important to include market-based and technology-specific financing costs in cost of energy modeling analysis. The exclusion of financing costs generally represents a noncommercial development such as a demonstration-scale wind power plant funded by grant dollars. Modeling financing costs based on commercial financing rates can result in an LCOE that is more likely to attract private investors and is repeatable and scalable.

It is also important to use financing costs that are current and within the range of accepted market conditions. Modeling wind power plants with overly high financing costs will result in an LCOE that does not appear to be commercially viable compared to other competing energy generation sources. Wind power plants with unrealistically low representation of the financing costs may not reflect the actual risks of a project and would be unlikely to generate sufficient investor interest.

Previous research has attempted to quantify the impact of financing costs on a wind power plant's LCOE using different approaches. Schwabe *et al.* [37] calculate how differences between a low- and high-cost financing environment could impact LCOE and find an impact of approximately \$9/MW h (or a 20% LCOE change relative to the base case). Wiser *et al.* use an extensive expert elicitation to assess the relative cost-reduction opportunities from financing costs from 2014 to 2030 and find up to 12% reduction opportunity for onshore wind and up to 25% for offshore wind owing to its less mature technology state [38].

There are multiple reasons for variations in financing costs for wind power plants. For one, financing rates are based on benchmark lending rates. As these rates vary over time so do the rates for financing a wind project. Therefore, a wind project's financing costs will ultimately depend on when the project is seeking investment. Second, there are several wind-industry-wide risk factors such as regulatory risks (i.e., a changing remuneration scheme nationally) that influence the perception of how wind energy project risks compare to other industry competing for investment (i.e., other electricity generation sources). Lastly, there are also wind-project-specific risks that reflect unique elements of a project, such as experience of the developer, local energy market risks, and perception of the quality of the turbine technology. In aggregate, these factors contribute to the variation in financing costs for wind power plants. Importantly, some of these risk factors may be addressed by actions like regulatory support and R&D activities, while others are more systemic such as the long-term direction of benchmark interest rates and thus not influenced by wind-industry-cost-reduction strategies.

There are several sources of capital that collectively comprise the financing rates of a wind power plant and provide a basis for forming the input assumptions around financing costs. Each of these sources of capital has a different financing cost associated with it based on the respective acceptance of risks compared to other sources of capital. These sources of capital can include the following:

- Equity capital represents the owner of a wind power plant and is typically repaid only after all other expense obligations have been met. Equity terms are variable

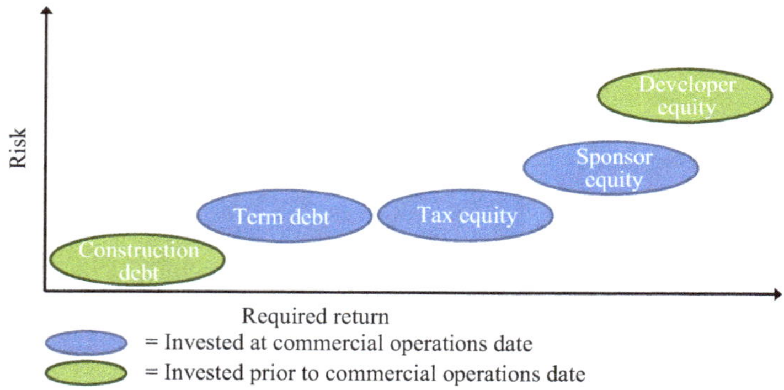

Figure 9.6 Relative risk and return of capital sources investing in wind power plants [37]

based on the performance of the project and generally have the largest risk and return of any investors.

- Debt capital is contractual loan with set repayment terms including interest rate and schedule. The lenders provide the principal to the project and receive principal and interest payments. Lenders typically have risk-protection mechanisms and high priority in repayment so are a lower risk capital source, but also limited upside.
- Tax-equity capital is a US-specific capital source that acts as a hybrid between traditional equity and debt capital sources. Tax-equity investors provide capital in exchange for tax benefits such as tax-based credits and deductions.

The metrics used to describe the relative combination of these capital sources for a wind power plant include the following:

- Capital ratio or debt-to-income rate is the proportion of each of the capital sources that make up the financing of the project as a whole.
- Wind-power-plant lifetime refers to the period in which a project is able to operate and generate cash flow. This lifetime can be longer than the contracted cash-flow period, which means plant owners may seek to extend power purchase contracts or sell power on the open market.
- Weighted average cost of capital (WACC) incorporates the different allocation of capital sources and costs into a singular financing metric. The proportion of each capital source is multiplied by the return requirements for each source to provide a weighted average metric. In most cases, WACC is considered on an after-tax basis.

Figure 9.6 illustrates the relative risks and returns of different capital sources common to wind power plants. Capital sources with higher risks require higher rates of returns to compensate investors for the added risk.

There are several different sources for empirical data for financing costs for wind power plants. These sources can have subtle variations in reporting terms, so care is warranted when comparing financing metrics. Moreover, the fidelity of the data presented varies across sources; some will provide a single metric like WACC or discount rate, while others explicitly break out WACC or similar metrics into its respective components. Empirical data sources for financing include

- public company financial filings from wind-energy developers;
- industry-trade publications from outside consultancies involved in wind-energy finance transactions including legal, accountancy, independent engineer, and owner representatives;
- regulatory filings to the state and regional energy commissions particularly where a local electric utility is seeking approval to either purchase the output or own some or all of the wind power plant;
- benchmark financial metrics such as government-reported debt interest rates; and interviews with industry experts.

9.8 Calculating cost of energy

Calculating LCOE based on (9.1) is typically done using one of two methods: a simplified recovery factor methodology or using a discounted cash-flow model. Assuming that CAPEX occurs in year 0 (C) and that AEP does not vary over time ($E_n = E$), (9.1) can be simplified. Analysis of annual variations in expenditures and complex project finance structures is best accomplished using discounted cash-flow (DCF) models.

A recovery-factor approach is a simplified approach. In this method, a single value or multiplier "recovers" the project's CAPEX and financing costs by converting the project's CAPEX into a series of recurring, equally sized payments over the assumed project lifetime. To estimate LCOE, additional periodic expenditures such as annual OPEX costs can be then added to the recurring payment stream and then divided by estimated AEP. The result shows the annual costs on a per unit of energy produced basis which would be made for the life of the wind power plant. The multiplier or recovery factor is typically estimated using the project's overall financing costs such as the WACC as the discount rate and the period over which the capital investment is repaid as the expected wind-power-plant lifetime. It is important that WACC is set at a level that ensures sufficient recovery of financing costs so that the LCOE is sufficient to represent a scenario when all investors are repaid their specific return requirements. A familiar example of the recovery-factor approach is a basic fixed mortgage payment that provides fixed regular payments over the life of an asset such as a home mortgage. The monthly payment amount is calculated to recover the initial cost of the home as well as the total interest payments made. The benefit of the recovery-factor approach is the reduced financial inputs required and simplified modeling approach. The recovery-factor approach is often used to show highlevel impacts of financial modeling without getting into year-by-year cash-flow analyses [39,40].

The second approach used for modeling cost of energy involves the use of a discounted cash-flow model. These models are ubiquitous analysis tools used by investors, developers, policymakers, and other industry stakeholders to forecast a wind power plant's periodic financial performance over the life of the asset. Like the recovery-factor approach, DCF models incorporate both upfront and ongoing cost

Table 9.2 Illustrative example of discounted cash-flow expenditures (red) and revenue (green) [39]

Period (Year)	**0**	**1**	**5…**	**10…**	**20**
Investment	CAPEX				
Energy sales		Revenue	Revenue	Revenue	Revenue
Less					
		OPEX	OPEX	OPEX	OPEX
		Debt	Debt	Debt	
		Taxes	Taxes	Taxes	Taxes
				Gearbox	
Plus					
	Incentive 1				
		Incentive 2	Incentive 2		
Equals					
After-tax cash flow to equity	Cash flow	Cash flow	Cash flow	Cash flow	Cash flow

steams (i.e., CAPEX and OPEX) as well as energy production to calculate LCOE or other metrics. Discounted cash-flow models, however, explicitly show how cost and revenue streams change over the life of the wind power plant rather than simplifying with a single representative multiplier. Discounted cash-flow models are often referred to as free cash flow to equity models as they forecast the cash flow to equity investors after all other expenses have been paid. The cash flow going to the equity investor is also typically discounted to present value terms with a calculated net present value metric. The project's internal rate of return is also shown and can be calculated on a before- or after-tax basis. Because discounted cash-flow models account for various cash-flow streams over time they can be used for modeling irregular maintenance activities (e.g. gearbox refurbishment or replacement), varying incentive types (upfront or production based), tax environments and depreciation schedules, changing investment ratios, and investor constraints among other factors. An illustrative example of a DCF is shown in Table 9.2; expenditures are illustrated in red while revenue is presented in green text [39].

9.9 Estimating future cost of wind energy

As described earlier, wind power plants are complex systems. Although individual cost of energy parameters may be estimated independently, ultimately the tradeoffs between these parameters provide the valuable insights needed to understand how wind technology may change in the future. Approaches that seek to accommodate design tradeoffs in a cost of energy framework, reflecting all relevant input parameters, provide the greatest insight in the future cost of wind energy.

Empirical data necessarily reflects the historic or current state of wind technology, and learning curves rely upon empirical data, typically correlating historical cost developments with historical wind-power deployment amounts. Although the resulting historical relationships have been used to estimate future wind energy costs

[41,42], learning curves provide little detailed insight into the causal mechanisms that result in cost reduction [43–46]. Moreover, most learning curve analyses have focused exclusively on understanding trends in capital cost [46–48]. Only a few more-recent studies have emphasized the more relevant design metric of the cost of wind energy [4,42,49]. The use of learning curves based on historical data is particularly challenging for estimating the cost of energy of technologies with little or no commercial deployment such as fixed-bottom and floating offshore wind power plants.

Expert knowledge, obtained through interviews, workshops, or formal elicitation, is another method of obtaining information from which to estimate the future cost of wind energy. Expert knowledge has proven valuable for offshore wind technologies that lack a long historic record but can also be usefully employed for land-based wind and for broader wind sector assessments [49–51].

Physics-based models typically reflect both cost and energy production to account for design tradeoffs that lead to future cost of energy reductions. A detailed representation of specific possible technology advancements can be evaluated [18,52]. The complex interactions in both the design and operation phase of wind power plants can be modeled using system engineering techniques and multidisciplinary analysis and optimization schemes [53,54] as discussed in Volume 2, Chapter 7.

A combination of empirical data, physics-based modeling and expert knowledge is often utilized when estimating the future cost of wind energy [55,56]. The approach of combining empirical data and modeled estimates with expert insights yields insight into the mechanisms of future cost reduction in terms of technology advances; it also improves the reliability and provides a sense of the uncertainty of a given estimate. Figure 9.7 provides one example of an assessment of the potential sources of cost reduction associated with a wind power plant anticipated to begin operation in 2020 compared to a base case reflecting typical wind-power-plant technology in 2010. This example highlights the impact on LCOE associated with individual cost component changes (blue, green, and pink bars) as well as the total cost reduction (red bar) that incorporates interaction among the individual cost components.

9.10 Considering the value of wind energy

The cost of wind energy is a primary decision variable for the wind-power industry, and reducing LCOE has historically been a core goal of the sector. However, the LCOE does not reflect commercial viability of a particular wind-power-plant concept, nor is LCOE sufficient on its own to assess whether investors may be attracted to a wind power plant or when to repower or decommission a plant. Questions about whether policy incentives or subsidies for wind energy are needed or are set at appropriate levels also cannot be answered with LCOE alone; additional considerations must be included. LCOE is an imperfect measure of economic competitiveness and is not used as the sole decision variable when comparing alternative sources of electricity supply.

For each of these decisions, the cost of wind energy is a critical input, including investment costs, operating costs, energy production, and financing costs. However, in each case, the cost of wind energy must be compared with the value of wind energy to make comparative judgments.

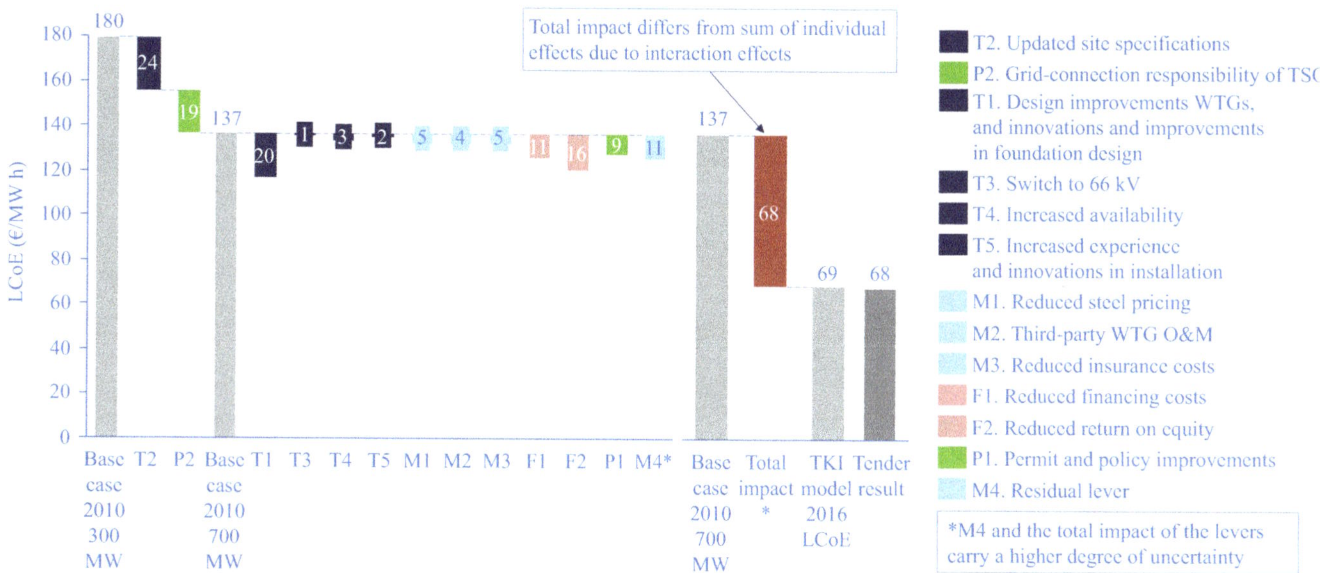

Figure 9.7 LCOE impact of individual and combined cost-reduction levers [56]

The timing and magnitude of costs along with the timing and magnitude of revenues becomes critical when assessing individual investor perspectives or policy schemes. Discounted cash-flow models that capture both costs and revenues can be usefully employed. For example, these models can solve for power purchase agreement prices that result in adequate value to all investors in a wind power plant to fully cover the costs of the project and return adequate profit [57]. They can be used to help inform repowering, refurbishment, and decommissioning decisions as wind power plants age. Anticipated future revenue must exceed anticipated future costs if investors are to continue to maintain plant operations [58–60]. Such models can also be used to evaluate various policy incentives to determine the amount of incentive required such that all revenue earned by the wind power plant (whether from the private market or through incentives) is adequate to cover necessary costs and profit [61].

Outside of the wind power sector, considerations of direct costs and system value are also important when assessing the relative attractiveness of different types of electricity generation. A primary function of electric planners and utilities is to minimize the total cost of reliably meeting demand for electricity, subject to other policy, societal, and legal constraints. A primary function of wholesale electricity markets, meanwhile, is to assist in meeting these goals by maximizing societal welfare in delivering electricity services, primarily through creating markets that enable efficient dispatch and investment decisions. Minimizing total system costs requires consideration of not just LCOE, but also the "system value" of various generation sources.

As described in [62], if all generation sources were homogenous, decision-making by planners, utilities, and power plant investors would be simple: purchase from or invest in the source with the lowest LCOE. However, power plants have widely varying technical and economic characteristics, and so deliver different services, e.g. a natural gas combustion turbine may operate only in the 5% of peak hours in a year, whereas a nuclear plant may operate on a 24×7 basis for the majority of the year. Understanding the LCOE of a technology is critical, and seeking ways to reduce technology LCOE is a proper function for industry and R&D managers. However, *comparing* the LCOE of *different* technologies that provide varying services is misleading. This is true when comparing a natural gas combustion turbine to a nuclear plant and is also true when comparing the LCOE of a wind power plant to the LCOE of a dispatchable thermal power plant [63–66]. The role of power sector planners and wholesale power markets is to balance the various characteristics of different generation sources and create an optimal least-cost portfolio mix of the available options.

Electricity planners, investors, and forecasters often use complex computer models (or combinations of models) to simulate dispatch and capacity expansion, and to assess what combination of resources will best minimize total costs, considering generation, transmission, and distribution. These models are able to capture not only the LCOE of each resource, but also the specific values (or costs) that each resource brings to the grid—energy, capacity, balancing, and transmission. Internally, the models compare the technology *cost* and marginal system *value* of each technology choice to find the portfolio of resources that minimize total system costs, subject to reliability and other constraints. These same models are also sometimes used to assess the "system value" of wind in providing electricity services, and to assess how that system value changes with increasing penetration and altered conditions. Furthermore, some utilities use these system value calculations to help compare different technology options in procurement processes where transparency needs limit the ability to solely rely on complex models. Specifically, utilities determine which option is most attractive by finding the technology with the lowest LCOE net of its system value (or conversely the technology with the highest system value net of its LCOE).

"System value" refers to the ability of an electricity generator to contribute to meeting demand such that it avoids the cost of instead using other bulk-power system assets. In general, costs that might be avoided include those related to energy, capacity, balancing, and transmission. Of these, *energy value* is often based on the ability of a resource to offset the production costs of other resources; in competitive markets, wholesale market locational-marginal prices are sometimes used to estimate this value. *Capacity value* measures a generator's contribution to overall long-term resource adequacy via an ability to offset the need (and cost) for capacity from other sources. *Balancing* (sometimes also called integration) needs originate from the required flexibility to maintain a continuous operational balance between supply and demand, with some resources increasing these needs and related costs and others offsetting those needs and costs. Finally, *transmission* refers to the differential impact of resources on the need for additional transmission infrastructure. *System value* ideally would comprehensively embed the marginal impacts of all four of the considerations noted earlier. *Market value*, meanwhile, refers to the subset of these factors that are actually priced in markets in the form of generator-specific revenue.

An idealized perfectly competitive wholesale electricity market would mimic the decisions described previously, with efficient investment decisions occurring when a generator is able to cover its total fixed and variable costs (in effect, total LCOE) through the revenue it earns in the market [67]. The prices for electricity services (energy, capacity, and ancillary services) and the quantity of services provided by the generator dictate the revenue earned, in effect its "market value." If an investor has a technology LCOE which is lower than the market value as revealed in wholesale markets, then an investment would be profitable for the investor and efficient in the sense of maximizing welfare. Similarly, the revenue earned by a wind generator in a perfectly designed competitive wholesale market reveals its direct economic value, and investment would be welfare maximizing if that revenue exceeds the LCOE of the wind power plant.

To be clear, modeled outcomes and wholesale electricity prices do not necessarily reflect the *full* "system value" of individual resources. Both models and wholesale markets are limited in their ability to account for all impacts, particularly for issues around the uncertainty, variability, and locational resource dependence of variable renewable generators such as wind. Additionally, wholesale electricity markets as well as the models used to simulate them may not be designed to reveal the complete set of values for all energy resources at all times, particularly broader societal values, such as differential environmental costs and benefits. As such, broader policy considerations also sometimes come into play, with revenue for wind generators affected by those decisions—whether through feed-in tariffs or premiums, tax incentives, markets for renewable energy certificates, or otherwise.

Though the wind sector has historically focused on minimizing LCOE, as wind energy supply increases there will be a growing need to also assess the impact of wind technology choices on system value. Research has shown that the system value of wind will tend to decline as wind penetrations increase (e.g., [66,68]). At the same time, various options can be used to help stem that decline and boost the system value of wind. Some of the options that can, in part, be controlled by the wind sector, include the use of larger rotors and taller towers to increase overall capacity factors, optimizing the siting of wind power plants to areas where the temporal output profile of wind power plants better matches system needs, pairing storage with wind power plants, and enabling wind power plants to participate in ancillary service markets (e.g., [69–71]).

9.11 Conclusion

Modeling the cost of wind energy requires an estimation of four primary parameters: investment costs, operating costs, cost of capital, and AEP. These parameters inherently reflect the expected lifetime over which the wind power plant will operate. Both empirical data (top-down) and physics-based (bottom-up) models are useful. Empirical data primarily is associated with existing wind power plants while exploration of future wind-power-plant technologies requires some level of physics-based modeling analysis. The level of fidelity required to model each of the four primary parameters depends upon the characteristics of the wind power plant and frequently varies for each of the four parameters. LCOE is a useful metric for evaluating wind-power-plant characteristics over time and geography or the potential impact of technology design innovation.

The cost of wind energy, in terms of the four primary inputs, may be combined with wind-power-plant revenue, policy interventions, and electricity system characteristics to contrast LCOE with the value of wind energy. This combination of cost of wind energy and value of wind energy enables the analysis of investment decisions, informs policy design, and facilitates comparison of electricity generation technologies.

Acknowledgment

This work was authored, in part, by the National Renewable Energy Laboratory, operated by Alliance for Sustainable Energy, LLC, for the U.S. Department of Energy (DOE) under Contract No. DE-AC36-08GO28308. Funding provided by the U.S. Department of Energy Office of Energy Efficiency and Renewable Energy Wind Energy Technologies Office. The views expressed in the article do not necessarily represent the views of the DOE or the U.S. Government. The U.S. Government retains and the publisher, by accepting the article for publication, acknowledges that the U.S. Government retains a nonexclusive, paid-up, irrevocable, worldwide license to publish or reproduce the published form of this work, or allow others to do so, for U.S. Government purposes.

References

[1] Dykes, K., Meadows, R., Felker, F., *et al.*, 2011. Applications of Systems Engineering to the Research, Design, and Development of Wind Energy Systems. NREL/TP-5000-52616. https://www.nrel.gov/docs/fy12osti/52616.pdf.

[2] Vitina, A., Luers, S., Wallasch, A.-K., *et al.*, 2015. IEA Wind Task 26: Wind Technology, Cost and Performance Trends in Denmark, Germany, Ireland, Norway, the European Union, and the United States: 2007–2012. NREL/TP-6A20-64332. https://www.nrel.gov/docs/fy15osti/64332.pdf.

[3] Hand, M.M., ed. 2018. IEA Wind TCP Task 26—Wind Technology, Cost, and Performance Trends in Denmark, Germany, Ireland, Norway, Sweden, the European Union, and the United States: 2008–2016. NREL/TP-6A20.71844. National Renewable Energy Laboratory, Golden, CO USA. https://www.nrel.gov/docs/fy19osti/71844.pdf.

[4] IRENA 2018. Renewable Power Generation Costs in 2017, International Renewable Energy Agency, Abu Dhabi. https://www.irena.org/-/media/Files/IRENA/Agency/Publication/2018/Jan/IRENA_2017_Power_Costs_2018.pdf.

[5] Lazard 2017. Lazard's Levelized Cost of Energy Analysis – Version 11.0. https://www.lazard.com/media/450337/lazard-levelized-cost-of-energy-version-110.pdf

[6] Mone, C., Stehly, T., Maples, B., Settle, E., 2015. 2014 Cost of Wind Energy Review. NREL/TP-6A20-64281. https://www.nrel.gov/docs/fy16osti/64281.pdf.

[7] Bolinger, M. and Wiser, R., 2018. 2017 Wind Technologies Market Report. https://emp.lbl.gov/wind-technologies-market-report.

[8] Musial, W., Beiter, P., Schwabe, P., *et al.*, 2017. 2016 Offshore Wind Technologies Market Report. DOE/GO-102017-5031. https://energy.gov/sites/prod/files/2017/08/f35/2016%20Offshore%20Wind%20Technologies%20Market%20Report.pdf

[9] IEA Wind https://community.ieawind.org/task26/dataviewer (accessed 12 August 2018).

[10] Noonan, M., Stehly, T., Mora, D., *et al.*, IEA Wind TCP Task 26 – Offshore Wind International Comparative Analysis NREL/TP-6A20.71558. https://www.nrel.gov/docs/fy19osti/71558.pdf.

[11] Maness, M., Maples, B., Smith, A., 2017. NREL Offshore Balance-of-System Model. NREL/TP-6A20-66874. https://www.nrel.gov/docs/fy17osti/66874.pdf.

[12] TKI Wind-op-Zee, 2016. FLOW-Competitive through Cooperation. TKI Wind-op-Zee. http://www.flow-offshore.nl/images/flow-openbaar/FLOW_Book-11_EN.pdf.

[13] Dicorato, M., Forte, G., Pisani, M., and Trovato, M., 2011. Guidelines for assessment of investment cost for offshore wind generation. Renewable Energy 36, 2043–2051.

[14] Dykes, K., Damiani, R., Roberts, O., and Lantz, E., 2018. "Analysis of Ideal Towers for Tall Wind Applications." Proceedings of the 2018 Wind Energy Symposium at the AIAA SciTech Forum, 8–12 January 2018, Kissimmee, FL: 11 pp. New York, NY: American Institute of Aeronautics and Astronautics, Inc. (AIAA). https://doi.org/10.2514/6.2018-0999.

[15] Guo, Y., Parsons, T., King, R., Dykes, K., Veers, P., 2015. Analytical Formulation for Sizing and Estimating the Dimensions and Weight of Wind Turbine Hub and Drivetrain Components. NREL/TP-5000-63008. https://www.nrel.gov/docs/fy15osti/63008.pdf.

[16] Naughton, B.T., Koplow, J.P., Vanness, J.W., Sethuraman, L., Maness, M., and Dykes, K., Fri. Twistact Techno-Economic Analysis for Wind Turbine Applications United States. doi:10.2172/1414654. https://www.osti.gov/servlets/purl/1414654.

[17] TPI Composites, Inc. 2003. Cost Study for Large Wind Turbine Blades: WindPACT Blade System Design Studies. SAND2003-1428. https://energy.sandia.gov/download/23611/

[18] Fingersh, L., Hand, M., and Laxson, A., 2006. Wind Turbine Design Cost and Scaling Model. NREL/TP-500-40566. https://www.nrel.gov/docs/fy07osti/40566.pdf

[19] Jensen, N.O., 1983. A Note on Wind Generator Interaction. Risø National Laboratory, Report M-2411.

[20] Liléo, S., Berge, E., Undheim, O., Klinkert, R., and Bredesen, R., 2013. Long-Term Correction of Wind Measurements Elforsk Report 13:18.

[21] Clifton, A., Smith, A., and Fields, M., 2016. Wind Plant Preconstruction Energy Estimates: Current Practice and Opportunities. NREL/TP-5000-64735. https://www.nrel.gov/docs/fy16osti/64735.pdf

[22] Wallasch, A.-K., Lüers, S., and Rehfeldt, Dr.-Ing.K., 2015. Kostensituation der Windenergie an Land in Deutschland: Update. (Cost Situation of Land-based Wind Energy in Germany: Update). BWE and VDMA Power Systems. https://www.windguard.de/veroeffentlichungen.html?file=files/cto_layout/img/unternehmen/veroeffentlichungen/2015/Kostensituation%20der%20Windenergie%20an%20Land%20in%20Deutschland%20-%20Update.pdf

[23] Shafiee, M., Brennan, F., and Espinosa, I.A., 2016. A parametric whole life cost model for offshore wind farms. International Journal of Life Cycle Assessment 21, 961. https://doi.org/10.1007/s11367-016-1075-z

[24] DNV GL (formerly GL Garrad Hassan) 2013. A Guide to UK Offshore Wind Operations and Maintenance. s.l.: The Crown Estate

[25] ORE Catapult, 2017. 'Portfolio Review 2016 – System Performance, Availability and Reliability Trend Analysis'. https://ore.catapult.org.uk/app/uploads/2018/02/SPARTAbrochure_20March-1.pdf

[26] Wallasch, A.-K., Lüers, S., Rehfeldt, Dr.-Ing.K., Vogelsang, K. (Deutsche WindGuard), and Jachmann, H. (Zentrum für Sonnenenergie- und Wasserstoff-Forschung Baden-Württemberg), 2018. Vorbereitung und Begleitung bei der Erstellung eines Erfahrungsberichts gemäß §97 Er-neuerbare-Energien-Gesetz, Teilvorhaben II e): Wind an Land, Report to the Federal Ministry of Economics and Energy Germany, https://www.erneuerbare-energien.de/EE/Redaktion/DE/Downloads/bmwi_de/bericht-eeg-6-wind-an-land.pdf

[27] Daldorf, Werner in Windenergie Report Deutschland 2016 (2017). Hg. v. Kurt Rohrig. Stuttgart: Fraunhofer Verlag. http://publica.fraunhofer.de/dokumente/N-445609.html

[28] UK Department for Business Innovation & Skills (Hg.), 2006. Offshore Wind Capital Grants Scheme. http://webarchive.nationalarchives.gov.uk/20090609012622/http://www.berr.gov.uk/energy/environment/etf/offshore-wind/page45496.html

[29] Hofmann, M., 2011. A review of decision support models for offshore wind farms with an emphasis on operation and maintenance strategies. Wind

Engineering, 35, 1–16. http://journals.sagepub.com/doi/abs/10.1260/0309-524X.35.1.1.

[30] Welte, T.M., Sperstad, I.B., Espeland Halvorsen-Weare, E., Netland, Ø., Nonås, L.M., and Stålhane, M., 2018. Operation and maintenance modelling, in: Offshore Wind Energy Technology, O. Anaya-Lara, J.O. Tande, K. Uhlen, and K. Merz, Eds., ed Chichester: Wiley. https://books.google.de/books?id=uIxODwAAQBAJ.

[31] Dinwoodie, I., Endrerud, O-EV., Hofmann, M., Martin, R., and Sperstad, I.B., 2015. Reference cases for verification of operation and maintenance simulation models for offshore wind farms. Wind Engineering 39 (1), S.1–S.14. DOI: 10.1260/0309-524X.39.

[32] Carroll, J., McDonald, A., and McMillan, D., 2016. Failure rate, repair time and unscheduled O&M cost analysis of offshore wind turbines. Wind Energy 19, 1107–1119. http://dx.doi.org/10.1002/we.1887

[33] Pfaffel, S., Faulstich, S., and Rohrig, K., 2017. Performance and reliability of wind turbines: a review. Energies 10 (11), 1–27. http://www.mdpi.com/1996-1073/10/11/1904

[34] Artigao, E., Martín-Martínez, S., Honrubia-Escribano, A., and Gómez-Lázaro, E., 2018. Wind turbine reliability: a comprehensive review towards effective condition monitoring development. Applied Energy 228, 1569–1583. https://doi.org/10.1016/j.apenergy.2018.07.037.

[35] Faulstich, S., Berkhout, V., Mayer, J., Siebenlist, D., 2016. Modelling the failure behaviour of wind turbines. Journal of Physics: Conference Series 749, S.12019. DOI: 10.1088/1742-6596/749/1/012019.

[36] Irawan, C.A., Ouelhadj, D., Jones, D., Stålhane, M., and Sperstad, I.B., 2017. Optimisation of maintenance routing and scheduling for offshore wind farms. European Journal of Operational Research, 256, 76–89. http://dx.doi.org/10.1016/j.ejor.2016.05.059.

[37] Schwabe, P., Feldman, D., Fields, J., and Settle, E., August 2017. Wind Energy Finance in the United States: Current Practice and Opportunities. National Renewable Energy Laboratory. https://www.nrel.gov/docs/fy17osti/68227.pdf

[38] Wiser, R., Jenni, K., Seel, J., *et al.*, June 2016. Forecasting Wind Energy Costs and Cost Drivers: The View of the World's Leading Experts Lawrence Berkeley National Laboratory and National Renewable Energy Laboratory. https://emp.lbl.gov/publications/forecasting-wind-energy-costs-and

[39] Gifford, J., Grace, R., and Rickerson, W., May 2011. Renewable Energy Cost Modeling: A Toolkit for Establishing Cost-Based Incentives in the United States National Renewable Energy Laboratory. https://www.nrel.gov/docs/fy11osti/51093.pdf

[40] Short, W., Packey, D.J., and Holt, T., 1995. Manual for the Economic Evaluation of Energy Efficiency and Renewable Energy Technologies. NREL/TP-462-5173. https://www.nrel.gov/docs/legosti/old/5173.pdf

[41] Luderer, G., Krey, V., Calvin, K., *et al.*, 2014. The role of renewable energy in climate stabilization: results from the EMF27 scenarios. Climatic Change, Springer Verlag, 123(3/4), 427–441.

[42] Williams, E., Hittinger, E., Carvalho, R., and Williams, R., 2017. Wind power costs expected to decrease due to technological progress. Energy Policy 106, 427–435.

[43] Ferioli, F., Schoots, K., and Van der Zwaan, B.C.C., 2009. Use and limitations of learning curves for energy technology policy: a component-learning hypothesis. Energy Policy 37, 2525–2535.

[44] Ek, K. and Söderholm, P., 2010. Technology learning in the presence of public R&D: the case of European wind power. Ecological Economics 69, 2356–2362.

[45] Yeh, S. and Rubin, E.S., 2012. A review of uncertainties in technology experience curves. Energy Economics 34, 762–771.

[46] Samadi, S. 2018. The experience curve theory and its application in the field of electricity generation technologies – a literature review. Renewable and Sustainable Energy Reviews 82, 2346–2364.

[47] Lindman, Å. and Söderholm, P., 2012. Wind power learning rates: a conceptual review and meta-analysis. Energy Economics 34, 754–761.

[48] Rubin, E.S., Azevedo, I., Jaramillo, P., and Yeh, S., 2015. A review of learning rates for electricity supply technologies. Energy Policy 86, 198–218.

[49] Wiser, R., Jenni, K., Seel, J., *et al.*, 2016. Expert elicitation survey on future wind energy costs. Nature Energy 1, 16135.

[50] Neij, L. 2008. Cost development of future technologies for power generation – a study based on experience curves and complementary bottom-up assessments. Energy Policy 36, 2200–2211.

[51] Wüstemeyer, C., Madlener, R., and Bunn, D.W., 2015. A stakeholder analysis of divergent supply-chain trends for the European onshore and offshore wind installations. Energy Policy 80, 36–44.

[52] Sieros G., Chaviaropoulos, P., Sørensen, J., Bulder, B., and Jamieson, P., 2010. Upscaling wind turbines: theoretical and practical aspects and the impact on the cost of energy. Wind Energy 15, 3–17.

[53] Dykes, K., Graf, P., Scott, G., *et al.*, 2015. Introducing WISDEM: An Integrated System Model of Wind Turbines and Plants (Presentation). NREL (National Renewable Energy Laboratory). NREL/PR-5000-63564. https://www.nrel.gov/docs/fy15osti/63564.pdf.

[54] Sanchez Perez-Moreno, S. *et al.*, 2016. Journal of Physics: Conference Series 753 062011 http://iopscience.iop.org/article/10.1088/1742-6596/753/6/062011/pdf

[55] Dykes, K., Hand, M., Stehly, T., *et al.*, 2017. Enabling the SMART Wind Power Plant of the Future Through Science-Based Innovation. NREL/TP-5000-68123. https://www.nrel.gov/docs/fy17osti/68123.pdf.

[56] Vree, B. and Verkaik, N., 2017. Offshore Wind Cost Reduction Progress Assessment. TKI Wind op Zee. https://www.topsectorenergie.nl/sites/default/files/uploads/Wind%20op%20Zee/20170220_Rap_TKI_Offshore_Wind_Cost_reduction.pdf

[57] Bruck, M., Sandborn, P., and Goudarzi, N., 2018. A levelized cost of energy (LCOE) model for wind farms that include power purchase agreements (PPAs). Renewable Energy 122, 131–139. https://doi.org/10.1016/j.renene.2017.12.100.

[58] Ziegler, L., Gonzalez, E., Rubert, T., Smolka, U., and Melero, J.J., 2018. Lifetime extension of onshore wind turbines: a review covering Germany, Spain, Denmark, and the UK. Renewable and Sustainable Energy Reviews, 82, 1261–1271. https://doi.org/10.1016/j.rser.2017.09.100.

[59] Rubert, T., McMillan, D., and Niewczas, P., 2018. A decision support tool to assist with lifetime extension of wind turbines. Renewable Energy 120, 423–433. https://doi.org/10.1016/j.renene.2017.12.064.

[60] Luengo, M.M. and Kolios, A., 2015. Failure mode identification and end of life scenarios of offshore wind turbines: a review. Energies 8. http://www.mdpi.com/1996-1073/8/8/8339.

[61] Bolinger, M., 2014. An Analysis of the Costs, Benefits, and Implications of Different Approaches to Capturing the Value of Renewable Energy Tax Incentives. Lawrence Berkeley National Laboratory, Berkeley, CA http://eta-publications.lbl.gov/sites/default/files/lbnl-6610e.pdf

[62] Wiser, R., Mills, A., Seel, J., Levin, T., and Botterud, A., 2017. Impacts of Variable Renewable Energy on Bulk Power System Assets, Pricing, and Costs. Lawrence Berkeley National Laboratory, Berkeley, CA.

[63] Joskow, P.L., 2011. Comparing the costs of intermittent and dispatchable electricity generating technologies. The American Economic Review 101 (3), 238–241. https://doi.org/10.1257/aer.100.3.238.

[64] Borenstein, S., 2012. The private and public economics of renewable electricity generation. Journal of Economic Perspectives 26 (1), 67–92. https://doi.org/ 10.1257/jep.26.1.67

[65] Edenhofer, O., Hirth, L., Knopf, B., *et al.*, 2013. On the economics of renewable energy sources. Energy Economics 40 (Supplement 1), S12–S23. https://doi.org/10.1016/j.eneco.2013.09.015.

[66] Hirth, L., 2013. The market value of variable renewables: the effect of solar wind power variability on their relative price. Energy Economics 38 (July), 218–236. https://doi.org/10.1016/j.eneco.2013.02.004

[67] Stoft, S., 2002. Power System Economics: Designing Markets for Electricity. Wiley/IEEE.

[68] Mills, A. and Wiser, R., 2014. Changes in the economic value of wind energy and flexible resources at increasing penetration levels in the rocky mountain power area. Wind Energy 17 (11), 1711–1726. https://doi.org/ 10.1002/we.1663

[69] Riva, A.D., Hethey, J., and Vitina, A., 2017. Impacts of Wind Turbine Technology on the System Value of Wind in Europe. Prepared for IEA Wind Task 26.

[70] Mills, A. and Wiser, R., 2015. Strategies to mitigate declines in the economic value of wind and solar at high penetration in California. Applied Energy 147 (June), 269–278. https://doi.org/10.1016/j.apenergy.2015.03.014.

[71] Hirth, L. and Müller, S., 2016. System-friendly wind power: how advanced wind turbine design can increase the economic value of electricity generated through wind power. Energy Economics 56 (May), 51–63. https://doi.org/10.1016/j.eneco.2016.02.016.

Index

www.ingramcontent.com/pod-product-compliance
Lightning Source LLC
LaVergne TN
LVHW011831200926
848484LV00002B/345

* 9 7 8 1 7 8 5 6 1 5 2 1 4 *